ANNALS OF THE NEW YORK ACADEMY OF SCIENCES

Volume 1012

EDITORIAL STAFF

Director, Publishing and New Media
SARAH GREENE

Managing Editor
JUSTINE CULLINAN

Associate Editors
STEFAN MALMOLI
TRUMBULL ROGERS

The New York Academy of Sciences
2 East 63rd Street
New York, New York 10021

THE NEW YORK ACADEMY OF SCIENCES
(Founded in 1817)

BOARD OF GOVERNORS, September 2003 – September 2004

TORSTEN N. WIESEL, *Chairman of the Board*
GERALD D. FISCHBACH, *Vice Chairman*
JOHN T. MORGAN, *Treasurer*
ELLIS RUBINSTEIN, *Chief Executive Officer* [ex officio]

Honorary Life Governors
WILLIAM T. GOLDEN   JOSHUA LEDERBERG

Governors

KAREN E. BURKE	PETER B. CORR	R. BRIAN FERGUSON
RONALD L. GRAHAM	MARNIE IMHOFF	WENDY EVANS JOSEPH
JACQUELINE LEO	RODERT W. LUCKY	PAUL MARKS
BRUCE McEWEN	RONAY MENSCHEL	JOHN F. NIBLACK
SANDRA PANEM	PETER RINGROSE	DAVID D. SABATINI
	DEBORAH WILEY	

VICTORIA BJORKLUND, *Counsel* [ex officio]   LARRY R. SMITH, *Secretary* [ex officio]

REDOX-ACTIVE METALS IN NEUROLOGICAL DISORDERS

ANNALS OF THE NEW YORK ACADEMY OF SCIENCES
Volume 1012

REDOX-ACTIVE METALS IN NEUROLOGICAL DISORDERS

*Edited by Steven M. LeVine, James R. Connor,
and Hyman M. Schipper*

The New York Academy of Sciences
New York, New York
2004

Copyright © 2004 by the New York Academy of Sciences. All rights reserved. Under the provisions of the United States Copyright Act of 1976, individual readers of the Annals are permitted to make fair use of the material in them for teaching or research. Permission is granted to quote from the Annals provided that the customary acknowledgment is made of the source. Material in the Annals may be republished only by permission of the Academy. Address inquiries to the Permissions Department (editorial@nyas.org) at the New York Academy of Sciences.

Copying fees: For each copy of an article made beyond the free copying permitted under Section 107 or 108 of the 1976 Copyright Act, a fee should be paid through the Copyright Clearance Center, Inc., 222 Rosewood Drive, Danvers, MA 01923 (www.copyright.com).

♾ The paper used in this publication meets the minimum requirements of the American National Standard for Information Sciences—Permanence of Paper for Printed Library Materials, ANSI Z39.48-1984.

Library of Congress Cataloging-in-Publication Data

Redox-active metals in neurological disorders / edited by Steven M. LeVine, James R. Connor, and Hyman M. Schipper.
   p.; cm. — (Annals of the New York Academy of Sciences; v. 1012)
   Includes index.
   ISBN 1-57331-502-8 (cloth: alk. paper) — ISBN 1-57331-503-6 (pbk.: alk. paper)
   1. Neurotoxicology—Congresses. 2. Iron—Toxicology—Congresses. 3. Manganese—Toxicology—Congresses. 4. Transition metals—Toxicology—Congresses. 5. Nervous system Degeneration—Etiology—Congresses. 6. Oxidative stress—Congresses.
   [DNLM: 1. Nervous System Diseases—etiology. 2. Iron Metabolism Disorders—complications. 3. Manganese Poisoning. WL 140 R319 2004] I. LeVine, Steven M. II. Connor, James R. III. Schipper, Hyman M., 1954– . IV. Series.
   Q11.N5 vol. 1012
   [RC347.5]
   500 s—dc22
   [616.8/04

2004001532

GYAT / PCP
Printed in the United States of America
ISBN 1-57331-502-8 (cloth)
ISBN 1-57331-503-6 (paper)
ISSN 0077-8923

ANNALS OF THE NEW YORK ACADEMY OF SCIENCES

Volume 1012
March 2004

REDOX-ACTIVE METALS IN NEUROLOGICAL DISORDERS

Editors
STEVEN M. LEVINE, JAMES R. CONNOR, AND HYMAN M. SCHIPPER

CONTENTS

Preface. *By* STEVEN M. LEVINE, JAMES R. CONNOR, AND
 HYMAN M. SCHIPPER ix

Part I. Redox-Active Metals in Neurobiology: Iron, Copper, and Zinc

Iron Metabolism and the IRE/IRP Regulatory System: An Update.
 By KOSTAS PANTOPOULOS 1

The Metabolism of Neuronal Iron and Its Pathogenic Role in Neurological
 Disease: Review. *By* TORBEN MOOS AND EVAN H. MORGAN 14

Fluorescence Detection of Redox-Sensitive Metals in Neuronal Culture:
 Focus on Iron and Zinc. *By* IAN J. REYNOLDS 27

Metal-Catalyzed Disruption of Membrane Protein and Lipid Signaling in the
 Pathogenesis of Neurodegenerative Disorders. *By* MARK P. MATTSON .. 37

Iron, Atherosclerosis, and Neurodegeneration: A Key Role for Cholesterol in
 Promoting Iron-Dependent Oxidative Damage? *By* WEI-YI ONG
 AND BARRY HALLIWELL 51

Severity of Neurodegeneration Correlates with Compromise of Iron
 Metabolism in Mice with Iron Regulatory Protein Deficiencies.
 By SOPHIA R. SMITH, SHARON COOPERMAN, TIM LAVAUTE, NANCY
 TRESSER, MANIK GHOSH, ESTHER MEYRON-HOLTZ, WILLIAM LAND,
 HAYDEN OLLIVIERRE, BERNARD JORTNER, ROBERT SWITZER III,
 ALBEE MESSING, AND TRACEY A. ROUAULT 65

Heme Oxygenase-1: Transducer of Pathological Brain Iron Sequestration
 under Oxidative Stress. *By* HYMAN M. SCHIPPER 84

Nutritional Iron Deprivation Attenuates Kainate-Induced Neurotoxicity in Rats: Implications for Involvement of Iron in Neurodegeneration. *By* S. SHOHAM AND M. B. H. YOUDIM 94

Part II. Redox-Active Metals in Neurobiology: Manganese and Cadmium

Manganese Neurotoxicity. *By* ALLISON W. DOBSON, KEITH M. ERIKSON, AND MICHAEL ASCHNER 115

Oxidative Basis of Manganese Neurotoxicity. *By* DIEM HAMAI AND STEPHEN C. BONDY 129

Divalent Metal Transporter 1 in Lead and Cadmium Transport. *By* JOSEPH P. BRESSLER, LUISA OLIVI, JAE HOON CHEONG, YONGBAE KIM, AND DESMOND BANNON 142

Part III. Redox-Active Metals and CNS Disorders: Alzheimer's Disease

Redox-Active Metals, Oxidative Stress, and Alzheimer's Disease Pathology. *By* XUDONG HUANG, ROBERT D. MOIR, RUDOLPH E. TANZI, ASHLEY I. BUSH, AND JACK T. ROGERS 153

Selective Cu^{2+}/Ascorbate-Dependent Oxidation of Alzheimer's Disease β-Amyloid Peptides. *By* CHRISTIAN SCHÖNEICH 164

Redox Metals in Alzheimer's Disease. *By* BOZHO M. TODORICH AND JAMES R. CONNOR 171

Oxidative Stress and Redox-Active Iron in Alzheimer's Disease. *By* KAZUHIRO HONDA, GEMMA CASADESUS, ROBERT B. PETERSEN, GEORGE PERRY, AND MARK A. SMITH 179

Magnetic Iron Compounds in Neurological Disorders. *By* JON DOBSON 183

Part IV. Redox-Active Metals and CNS Disorders: Parkinson's Disease

The Relevance of Iron in the Pathogenesis of Parkinson's Disease. *By* MARIO E. GÖTZ, KAY DOUBLE, MANFRED GERLACH, MOUSSA B. H. YOUDIM, AND PETER RIEDERER 193

Manganese-Induced Parkinsonism and Parkinson's Disease. *By* C. W. OLANOW 209

Brain Ferritin Iron as a Risk Factor for Age at Onset in Neurodegenerative Diseases. *By* GEORGE BARTZOKIS, TODD A. TISHLER, IL-SEON SHIN, PO H. LU, AND JEFFREY L. CUMMINGS 224

Part V. Redox-Active Metals and CNS Disorders: Other Disorders

Hematoma Removal, Heme, and Heme Oxygenase Following Hemorrhagic Stroke. *By* KENNETH R. WAGNER AND BARNEY E. DWYER 237

The Role of Iron in the Pathogenesis of Experimental Allergic
  Encephalomyelitis and Multiple Sclerosis. *By* STEVEN M. LEVINE
  AND ANURADHA CHAKRABARTY 252

Hereditary Causes of Disturbed Iron Homeostasis in the Central Nervous
  System. *By* PREM PONKA 267

Mitochondrial Localization of Human PANK2 and Hypotheses of Secondary
  Iron Accumulation in Pantothenate Kinase–Associated
  Neurodegeneration. *By* MONIQUE A. JOHNSON, YIEN MING KUO,
  SHAWN K. WESTAWAY, SUSAN M. PARKER, KATHERINE H. L. CHING,
  JANE GITSCHIER, AND SUSAN J. HAYFLICK 282

Aceruloplasminemia: An Inherited Neurodegenerative Disease with
  Impairment of Iron Homeostasis. *By* XUEYING XU, SOKHON PIN,
  MURAYA GATHINJI, RALPH FUCHS, AND Z. LEAH HARRIS 299

Part VI. Metal Chelation Therapy in Neurological Disorders

Ironing Iron Out in Parkinson's Disease and Other Neurodegenerative
  Diseases with Iron Chelators: A Lesson from 6-Hydroxydopamine and
  Iron Chelators, Desferal and VK-28. *By* MOUSSA B. H. YOUDIM,
  GALIA STEPHENSON, AND DORIT BEN SHACHAR 306

Novel Chelators for Central Nervous System Disorders That Involve
  Alterations in the Metabolism of Iron and Other Metal Ions.
  By DES R. RICHARDSON. 326

Redox Neurology: Visions of an Emerging Subspecialty.
  By HYMAN M. SCHIPPER 342

Index of Contributors 357

The New York Academy of Sciences believes it has a responsibility to provide an open
forum for discussion of scientific questions. The positions taken by the participants in
the reported conferences are their own and not necessarily those of the Academy. The
Academy has no intent to influence legislation by providing such forums.

Preface

A mounting body of evidence indicates that redox-active metals participate in the pathophysiological processes of numerous disorders of the central nervous system in humans. During disease states, redox-active metals generate toxic reactive oxygen species. Redox-active metals are present in high concentrations in the brain, making this organ particularly vulnerable to oxidative stress. Because the brain has a limited capacity for recovery after injury, even mild to moderate tissue damage can have lifelong consequences. Thus, an understanding of the role of metals is important in understanding the normal functioning of the human brain as well as the pathophysiological processes that underlie many brain disorders. Studies that illuminate the role of redox-active metals in neuropathologic processes have the potential to facilitate development of effective therapies for the management of diverse neurological conditions. While a pathogenic role for redox-active metals has been implicated in several rare neurological disorders for decades, only recently have scientists come to appreciate that redox-active metals also may be critical components in the promotion of pathology in common neurological disorders including Alzheimer's and Parkinson's disease.

Although there is a growing consensus on the role of redox-active metals in the pathogenesis of both common and rare neurological disorders, many challenges remain. The challenges include (1) understanding how redox-active metals progress from serving vital functions within brain cells to becoming key participants in disease processes; (2) identifying mechanisms by which redox-active metals cause tissue damage in various neurological diseases; (3) identifying protective mechanisms that go awry or become overwhelmed during disease states; (4) identifying neurological diseases in which redox metals act as primary pathogenic factors and those diseases in which they are involved in secondary pathogenic reactions; and (5) identifying plausible strategies for therapeutic interventions directed at the pathogenic processes that redox-active metals are involved in. It is the primary objective of this volume to disseminate some of the latest developments on research related to these above stated challenges.

The papers have been divided into three main sections: (1) Redox-Active Metals in Neurobiology [*Parts I and II listed in the Table of Contents*], (2) Redox-Active Metals and CNS Disorders [*Parts III, IV, and V*], and (3) Metal Chelation Therapy in Neurological Disorders [*Part VI*]. In the first section, the papers cover the management of iron and other redox-active metals in normal and disease states. Neuropathogenic mechanisms of oxidative tissue damage are also considered. In the second section, the role of redox-active metals in common neurodegenerative disorders such as Alzheimer's and Parkinson's disease are examined in several papers, while other conditions such as stroke, multiple sclerosis, and hereditary conditions like pantothenate kinase–associated neurodegeneration and aceruloplasminemia are addressed in individual papers. In the third section, the potential application of chelation therapy for the treatment of neurological disorders is discussed and a proposal for the development of Redox Neurology as an independent subspecialty of Neurology is put forth.

ACKNOWLEDGMENTS

Dr. LeVine: I would like to express my gratitude to many institutions and individuals. Included among the institutions are the University of Kansas Medical Center for providing a positive and supportive environment for the pursuit of scientific discovery and educational excellence; the Department of Molecular and Integrative Physiology for a congenial and stimulating working environment; the Mental Retardation Research Center for supplying infrastructure for our scientific endeavors; and several funding agencies including the National Institutes of Health, National Science Foundation, National Multiple Sclerosis Society, Hunter's Hope Foundation, United Leukodystrophy Foundation, and others. Among the many invaluable individuals are my wife, Lori; my son, Nicholas; members of the Department of Molecular and Integrative Physiology; and the students and postdoctoral fellows that have worked in my lab.

Dr. Connor: I express my gratitude to the postdoctoral fellows and students who have given their time and talents to the study of iron and oxidative stress, and especially Sharon Menzies, my research assistant for 16 years. Sharon passed away this summer at the age of 38. The research that we have been able to pursue would not have been possible without the support of the National Institutes of Health, the G. M. Leader Family, the Jane B. Barsumian Trust, and the Soter S. and Carolyn C. Harbolis Alzheimer's Research Endowment. Most of all, I am grateful for the love and support of my wife, Judy; my daughter, Jennifer; and my son, Jonathan.

Dr. Schipper: I am deeply indebted to numerous mentors, colleagues, and students who have nurtured my interest in the "redox neurosciences" and to the S. M. B. D. Jewish General Hospital and McGill University for providing a stimulating work environment. I also wish to express my heartfelt appreciation to my wife, Rachel, and the boys, Joshua and David, for their unflagging encouragement and support.

Finally, the editors would like to thank all the authors who have put forth thought-provoking manuscripts illustrating the relevance and importance of redox-active metals in the pathogenesis of neurological disorders.

—STEVEN M. LEVINE
—JAMES R. CONNOR
—HYMAN M. SCHIPPER

Iron Metabolism and the IRE/IRP Regulatory System

An Update

KOSTAS PANTOPOULOS

Lady Davis Institute for Medical Research, Sir Mortimer B. Davis Jewish General Hospital, and Department of Medicine, McGill University, Montreal, Quebec, Canada

ABSTRACT: Cellular iron homeostasis is accomplished by the coordinated regulated expression of the transferrin receptor and ferritin, which mediate iron uptake and storage, respectively. The mechanism is posttranscriptional and involves two cytoplasmic iron regulatory proteins, IRP1 and IRP2. Under conditions of iron starvation, IRPs stabilize the transferrin receptor and inhibit the translation of ferritin mRNAs by binding to "iron responsive elements" (IREs) within their untranslated regions. The IRE/IRP system also controls the expression of additional IRE-containing mRNAs, encoding proteins of iron and energy metabolism. The activities of IRP1 and IRP2 are regulated by distinct posttranslational mechanisms in response to cellular iron levels. Thus, in iron-replete cells, IRP1 assembles a cubane iron-sulfur cluster, which prevents IRE binding, while IRP2 undergoes proteasomal degradation. IRP1 and IRP2 also respond, albeit differentially, to iron-independent signals, such as hydrogen peroxide, hypoxia, or nitric oxide. Basic principles of the IRE/IRP system and recent advances in understanding the regulation and the function of IRP1 and IRP2 are discussed.

KEYWORDS: iron metabolism; ferritin; transferrin receptor; iron responsive elements; IRP1; IRP2; oxidative stress; aconitase; DMT1; ferroportin

IRON, OXIDATIVE STRESS, AND DISEASE

While virtually all cells and organisms utilize iron as a cofactor in a multitude of biochemical activities, excess iron promotes the generation of reactive radicals, which in turn damage cells and tissues. This condition of "oxidative stress" is encountered in diseases of hereditary and secondary iron overload,[1] such as hereditary hemochromatosis and secondary (transfusional) siderosis, which eventually result in liver damage and heart failure. Oxidative stress is also a hallmark of neurodegenerative disorders,[2] such as Parkinson's and Alzheimer's diseases, where iron accumulates pathologically in the substantia nigra[3] or in senile plaques,[4] respectively. Even

---

Address for correspondence: Kostas Pantopoulos, Lady Davis Institute for Medical Research, Sir Mortimer B. Davis Jewish General Hospital, 3755 Cote-Ste-Catherine Road, Montreal, Quebec H3T 1E2, Canada. Voice: 514-340-8260, ext. 5293; fax: 514-340-7502.
  kostas.pantopoulos@mcgill.ca

though it is still not clear whether this is a primary cause or a secondary effect, iron is increasingly being considered as a pathogenic cofactor.[5,6]

OVERVIEW OF IRON METABOLISM

The human body contains ~3–5 g iron, from which the vast majority (~70%) is utilized in erythroid cells for heme and hemoglobin synthesis.[7] Senescent erythrocytes are phagocytosed by macrophages and iron recycles in the circulation for a new round of utilization. There are no specific mechanisms for iron secretion from the body, and iron loss can only occur via bleeding or cell desquamation (e.g., in sloughed mucosal cells). This nonspecific iron loss is compensated by dietary iron absorption, which takes place in the duodenum. The underlying mechanism has begun to be understood in the last few years, following the cloning and characterization of molecules with critical function in this process, including iron transporters and oxidoreductases (reviewed in refs. 8 and 9).

According to a widely accepted model (depicted in FIG. 1A), the first step of the pathway involves reduction of ferric Fe(III) to ferrous Fe(II) in the intestinal lumen by the reductase Dcytb (duodenal cytochrome b), and transport of Fe(II) across the duodenal epithelium by the apical transporter DMT1 (divalent metal transporter 1). The export of Fe(II) to the portal circulation is mediated by the basolateral transporter ferroportin. This step is followed by reoxidation of Fe(II) by the plasma ferroxidase ceruloplasmin (or its membrane-associated homologue, hephaestin), and scavenging of Fe(III) by the plasma iron carrier transferrin, which delivers it to the erythroid bone marrow and tissues. Iron-loaded transferrin binds to the cell surface transferrin receptor (TfR) and undergoes endocytosis[10,11] (FIG. 1B). Iron release is accomplished by acidification of the endosome to pH ~5.5. Iron is then transported across the endosomal membrane by DMT1 and utilized for the synthesis of iron-containing proteins. Excess of intracellular iron is sequestered into ferritin, a shell-like protein composed of 24 subunits of H- and L-chains that can store up to 4500 Fe(III) atoms. A fraction of cytosolic iron, known as "labile iron pool" (LIP),[12] presumably remains bound to low-molecular-weight chelates and reflects the overall iron status of the cell.

Some important aspects of tissue-specific iron transport, which may also involve the cellular uptake of nontransferrin bound iron (NTBI), are far from being defined. For example, while it is clear that the transferrin circulating in the brain interstitial fluid does not derive from plasma, but is rather synthesized and released by oligodendrocytes and choroid plexus epithelial cells, the actual mechanism for iron transport across the blood-brain barrier has not yet been firmly established.[5,6] Nonetheless, the pathways for intracellular iron utilization and storage and the regulatory principles of cellular iron metabolism are conserved in virtually all cells of the body.

REGULATION OF CELLULAR IRON METABOLISM: THE IRE/IRP SYSTEM

The expression of TfR and ferritin is coordinately and reciprocally controlled in response to iron supply at the posttranscriptional level.[13,14] The mRNAs encoding

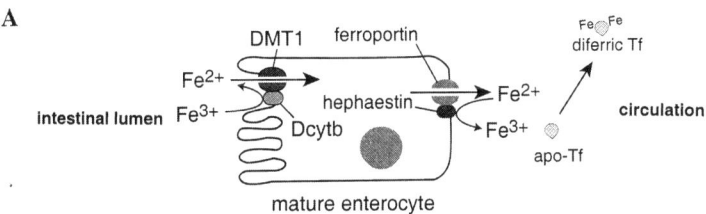

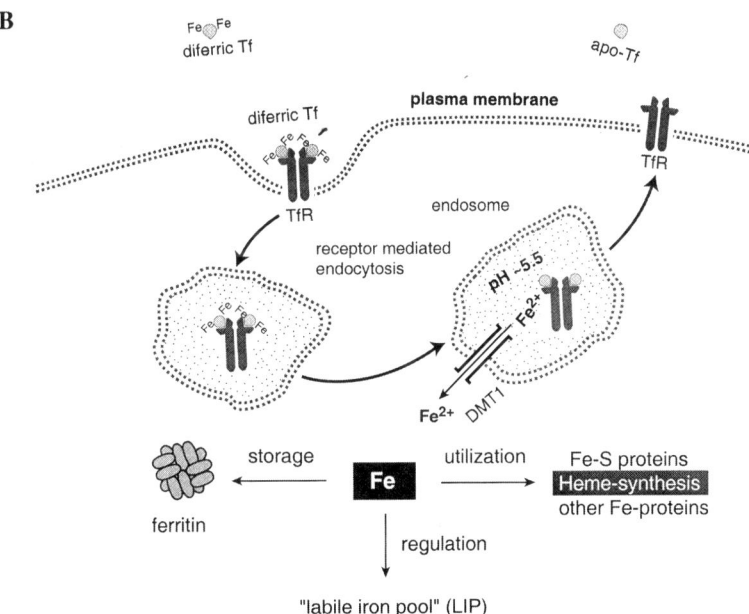

FIGURE 1. (A) A model for iron transport across duodenal epithelial cells. Ferric iron is reduced in the intestinal lumen by Dcytb. Ferrous iron is then transported across the apical membrane of the mature enterocyte by DMT1. The export of iron across the basolateral membrane of enterocytes to circulation is mediated by ferroportin. This step is coupled with reoxidation of ferrous to ferric iron by membrane-bound hephaestin. Plasma iron is immediately scavenged by transferrin (Tf). **(B)** Schematic representation of the Tf cycle. Plasma Tf binds to its cell surface receptor (TfR), and the complex is internalized by endocytosis. Acidification of the endosome results in the release of ferric iron from Tf and reduction and subsequent transport of ferrous iron across the endosomal membrane by DMT1. The pathway is completed by recycling of the apoTf-TfR complex to the cell surface and release of apoTf. Intracellular iron is utilized for the synthesis of iron-containing proteins and the excess is stored into ferritin. A fraction of chelatable iron defined as "labile iron pool" (LIP) determines the cellular iron status.

A

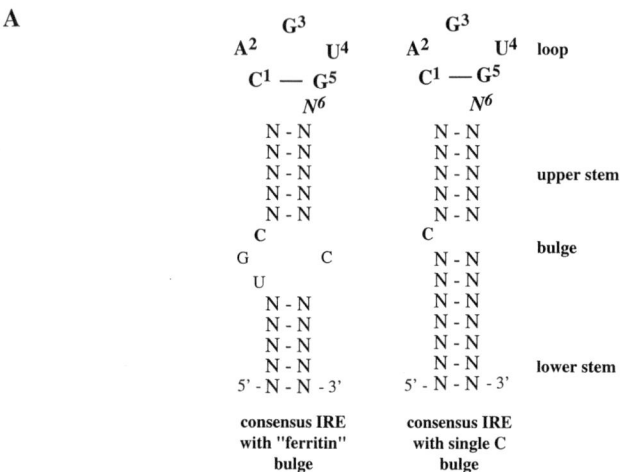

B

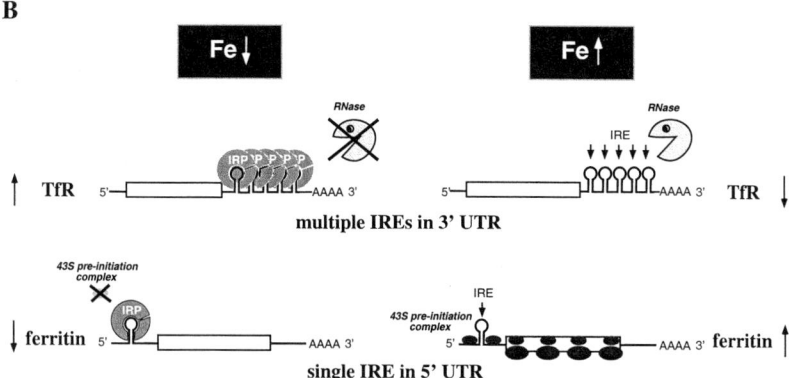

FIGURE 2. (**A**) The consensus IRE motif. It consists of a hexanucleotide loop (5'-CAGUGN-3') and a stem, interrupted by a bulge with an unpaired C residue. Base pairing between C^1 and G^5 is functionally important. N^6 could be any nucleotide, but not G, which would potentially disrupt C^1–G^5 interaction by C^1–G^6 pairing. The bulge may consist of an asymmetric tetranucleotide, as in many ferritin mRNAs (*left*), or a single C residue (*right*). (**B**) Regulation of TfR and ferritin expression by the IRE/IRP system. Decreased iron supply activates binding of IRPs to IREs, resulting in stabilization of TfR mRNA and translational inhibition of the mRNAs encoding H- and L-ferritin. These responses lead to increased iron uptake and reduced iron storage. Conversely, increased iron supply inactivates binding of IRPs to IREs, resulting in degradation of TfR mRNA and translation of the mRNAs encoding H- and L-ferritin. These responses lead to decreased iron uptake and elevated iron storage.

TfR and ferritin (both H- and L-chains) contain structural motifs known as "iron responsive elements" (IREs) in their untranslated regions (UTRs). These hairpin structures consist of ~30 nucleotides and are phylogenetically conserved in vertebrates and some insects and bacteria[15] (FIG. 2A). The hairpin in a typical IRE forms a 5'-CAGUGN-3' loop (the underlined C and G interact by hydrogen bonding) and a stem with moderate stability ($\Delta G \approx -7$ kcal/mol), interrupted by an unpaired C residue. The TfR mRNA contains five IRE copies in its 3' UTR and is, thus far, the only identified IRE-containing mRNA with multiple IREs. By contrast, the mRNAs encoding H- and L-ferritin contain a single IRE in their 5' UTR. A characteristic feature of the ferritin IRE is that the unpaired C residue is located within an internal loop/bulge.

In iron-starved cells, the IREs become targets of two cytoplasmic iron regulatory proteins, IRP1 and IRP2, which bind to them with high affinity ($K_d \approx 10^{-12}$ M). The IRE/IRP interactions result in stabilization of the otherwise unstable TfR mRNA and in specific translational inhibition of the mRNAs encoding H- and L-ferritin (FIG. 2B). As a result, iron-starved cells increase their capacity to take up transferrin-bound iron by the TfR and minimize its sequestration into ferritin stores. Conversely, in iron-replete cells, IRP1 and IRP2 fail to bind to cognate IREs, permitting TfR mRNA degradation and ferritin mRNA translation. This response inhibits further iron uptake and promotes the storage (and detoxification) of excess iron.

The identification of additional IRE-containing mRNAs showed that the IRE/IRP regulatory system is not confined to the regulation of TfR and ferritin expression. As one would expect, the additional IRE-containing mRNAs[15] primarily encode proteins of iron metabolism, such as the erythroid 5-aminolevulinate synthase (ALAS2), which catalyzes a key reaction for erythroid heme synthesis, and the iron transporters DMT1 and ferroportin. The IRE in the ALAS2 mRNA is located at the 5' UTR and operates as a translational control element.[13] The IRP-mediated inhibition of ALAS2 synthesis in iron-starved erythroid cells is thought to prevent the accumulation of toxic protoporphyrin IX when iron is limiting for heme synthesis. The mRNAs encoding DMT1 and ferroportin contain a single IRE in their 3' or 5' UTR, respectively, which probably accounts for the increased duodenal expression of these transporters in mice fed with an iron-deficient diet.[16,17] The expression of alternatively spliced DMT1 transcripts lacking the IRE should also be noted. The mechanism for IRE-dependent regulation of DMT1 and ferroportin mRNAs is still not clear and may exhibit a cell-type-specific pattern.[18] The functionality of DMT1 IRE appears to require elements of the alternatively spliced upstream exon 1A within DMT1 mRNA.[19]

Interestingly, translation-type IREs have also been identified in the 5' UTR of the mRNAs encoding mammalian mitochondrial aconitase and the insect Ip subunit of succinate dehydrogenase (SDH).[20] Both are iron-containing enzymes catalyzing reactions of the citric acid cycle and do not have an apparent function in the control of iron homeostasis. The iron-dependent regulation of mammalian mitochondrial aconitase and insect SDH via the IRE/IRP system provides regulatory links between iron and energy metabolism. However, the range of iron-dependent translational control in mitochondrial aconitase expression is not as wide as in ferritin.[21-23] The observed quantitative differences may reflect differences in IRE sequences and structure.[24]

IRON REGULATORY PROTEINS, IRP1 AND IRP2

IRP1 and IRP2 are homologous cytoplasmic polypeptides of 889 and 964 amino acids, respectively, and belong to the family of iron-sulfur cluster isomerases.[25–27] Human IRP1 shares 57% sequence identity and 75% similarity with human IRP2. In addition, human IRP1 is 31% identical and 56% similar to porcine mitochondrial aconitase, a well-characterized member of the iron-sulfur cluster isomerases family. By analogy to the known structure of mitochondrial aconitase, both IRP1 and IRP2 are projected to contain three compact domains, linked to a fourth domain by a flexible hinge region (FIG. 3). A notable difference between IRP1 and IRP2 is that the latter contains a cysteine- and proline-rich insertion of 73 amino acids, embedded within domain 1, which is encoded by a unique exon.

IRP1 is expressed ubiquitously and probably this is also the case for IRP2. It was initially proposed that IRP2 exhibits a tissue-specific pattern of expression, but it

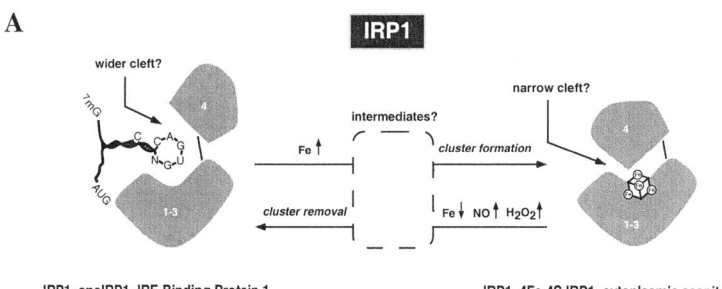

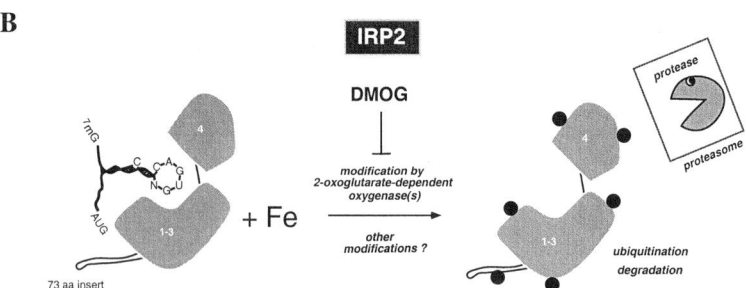

FIGURE 3. (**A**) Regulation of IRP1 by an iron-sulfur cluster switch. In iron-replete cells, IRP1 assembles a cubane 4Fe-4S cluster between domains 1–3 and 4. Iron starvation, NO, or H_2O_2 trigger the switch of 4Fe-4S to apoIRP1, resulting in the conversion of cytosolic aconitase to IRE-binding protein. (**B**) Regulation of IRP2 (depicted with a 73-amino-acid insertion within domain 1) at the level of protein stability. IRP2 is stable in iron-depleted or hypoxic cells. Administration of iron results in IRP2 ubiquitination and degradation by the proteasome. This process can be blocked by dimethyl-oxalyl-glycine (DMOG), an inhibitor of 2-oxoglutarate-dependent oxygenases.

appears now that the abundance of IRP2 in cell lines and tissues was earlier underestimated due to technical constraints.[28] Little is known on the relative contribution of each of the IRPs in the maintenance of cellular iron homeostasis. *In vitro* experiments showed that IRP1 and IRP2 do not differ in their capacity to regulate ferritin mRNA translation[29] and to bind to ferritin and TfR mRNAs.[24,30] However, in murine J774[31] and RAW[32,33] macrophages, the exclusive modulation of IRP2 accounted for the regulation in TfR and ferritin expression, regardless of IRP1 activity. On the other hand, the expression of IRP1$_{C437S}$, a constitutive IRP1 mutant, in human RD4 (rhabdomyosarcoma) cells[34] or in H1299 (lung cancer) cells[35] was sufficient to disrupt iron-dependent regulation of TfR and ferritin. It is conceivable that the regulatory capacity of each of the IRPs may be modulated by cell-specific factors. The study of cells, derived from IRP1$^{-/-}$ and IRP2$^{-/-}$ animals, is expected to shed light on this issue.

IRP1$^{-/-}$ and IRP2$^{-/-}$ mice have been generated in the Rouault lab. The targeted disruption of IRP1 did not yield any obvious phenotype.[28] By contrast, the IRP2$^{-/-}$ mice displayed aberrant iron homeostasis and accumulated iron in the intestinal mucosa and the CNS.[36] The iron overload in specific areas of the brain was associated with the development of a progressive neurodegenerative disorder. These data establish IRP2 as an important regulator of systemic iron metabolism and, furthermore, imply that polymorphisms associated with functional inactivation of IRP2 may also be relevant to human disease. How the loss of IRP2 function leads to iron overload in the brain is currently unclear. It has been speculated that the disruption of iron homeostasis in neurons may be a result of "functional iron depletion" due to overexpression of ferritin and iron sequestration within it in distal axons.[37] This scenario shows some similarities to "neuroferritinopathy", a rare, dominant, adult-onset neurodegenerative disorder, characterized by iron overload in the basal ganglia.[38] The causative defect is a frameshift mutation in L-ferritin gene, predicted to alter the C-terminus of the polypeptide. This may promote the nonreversible storage of metabolically active iron within the mutated ferritin.

REGULATION OF IRP1

IRP1 is regulated by an unusual iron-sulfur cluster switch.[39,40] In iron-replete cells, it assembles a cubane 4Fe-4S cluster, which inhibits its IRE-binding activity. Moreover, this cluster converts IRP1 to a cytosolic aconitase (FIG. 3A), which has a similar catalytic efficiency with its mitochondrial counterpart,[41,42] and may play a role in redox control by promoting the generation of cytosolic NADPH.[43] Thus, IRP1 is a bifunctional protein. In iron-starved cells, the cluster disassembles by a slow mechanism[44] and apoIRP1 acquires IRE-binding activity. This may be a result of a structural rearrangement, providing more space in the cleft for interaction with the IRE.[45,46] The topology of cluster coordination has been mapped by mutational analysis. Three iron atoms bind to cysteine residues in the polypeptide backbone (C437, C503, and C506), while the fourth, Fe$_a$, binds to the solvent (H$_2$O) and interacts with citrate (the aconitase substrate). Phosphomimetic mutations at S138 negatively affect the assembly of the cluster[47] and, in addition, sensitize IRP1 to iron-dependent degradation,[48] suggesting an additional mode of IRP1 regulation at the level of protein stability.

IRP1 was initially appreciated as an intracellular iron sensor, but it soon became clear that it also responds to other stimuli. Exposure of cells to hydrogen peroxide (H_2O_2) or nitric oxide (NO) promotes removal of the cluster and thereby induces IRE-binding activity (reviewed in refs. 26 and 27). The mechanism for cluster removal is not straightforward. While the cluster of IRP1 is generally sensitive to inactivation by reactive oxygen and nitrogen species,[49] which is reflected in loss of aconitase activity, this is not always associated with activation of IRE binding, which requires complete cluster disassembly. Thus, the superoxide anion ($O_2^{\cdot -}$) or peroxynitrite (NOO•) readily diminishes the aconitase activity of IRP1, but fails to convert it into an IRE-binding protein.[50–52] The responses of IRP1 to H_2O_2 and NO are more complex. The H_2O_2-mediated conversion of IRP1 from cytosolic aconitase to IRE-binding protein is a result of a signaling pathway rather than of direct chemical modification of the 4Fe-4S cluster by H_2O_2. First of all, IRP1 activation takes place when intact cells, but not cell extracts, are exposed to H_2O_2.[53,54] By contrast, the treatment of purified 4Fe-4S IRP1 with H_2O_2 yields 3Fe-4S IRP1, which fails to bind to IREs.[42] Importantly, the pathway for H_2O_2-mediated activation of IRP1 is biphasic[44] and can only be reconstituted *in vitro* in the presence of the particulate fraction of permeabilized cells.[55] The latter suggests the involvement of insoluble membrane-associated factors in the pathway.

The mechanism for IRP1 activation by NO is distinct. Exposure of purified IRP1 to NO *in vitro* was recently shown to activate IRE binding,[52] even though in earlier literature this effect was only partial.[56] We have reported that the kinetics of IRP1 activation upon exposure of Ltk⁻ fibroblasts to *S*-nitroso-*N*-acetyl-D,L-penicillamine (SNAP), a donor of NO, and desferrioxamine, an iron chelator, show remarkable similarities.[44] Thus, IRP1 activation requires at least 4 h of continuous treatment with each of these drugs, while the "stress activation" of IRP1 is triggered by a transient pulse with H_2O_2 and occurs within 30–60 min. Moreover, SNAP and desferrioxamine also elicit the activation of IRP2, which does not contain an aconitase-type cluster. These findings are compatible with a model, according to which NO may modulate the LIP and thereby promote responses to iron deficiency. It is also conceivable that a direct, NO-mediated cluster destabilization of IRP1 also accounts at least for a partial activation of IRE binding *in vivo*. The two models are by no means mutually exclusive.[57,58]

Since both H_2O_2 and NO are reactive species generated at high levels by phagocytic cells, their effects on IRP1 also warrant consideration in the context of the inflammatory response. The activation of IRP1 by these stimuli is expected to promote increased cellular uptake from Tf-bound serum iron and may thereby contribute to the shift of extracellular iron to the tissues encountered in conditions such as the anemia of chronic disease.[59] Along these lines, the exposure of B6 fibroblasts to H_2O_2 correlates with an IRP1-dependent activation of TfR expression and an increase in cellular iron uptake.[22] It should, however, be emphasized that, apart from H_2O_2 and NO, phagocytic cells have the capacity to generate a variety of oxidant species, some of which may also have opposing effects on IRP1. Thus, myeloperoxidase-derived hypochlorous acid (HOCl), a potent oxidant and major product of the oxidative burst of phagocytes, antagonizes the H_2O_2-mediated activation of IRP1.[60] This finding illustrates an interesting example of opposing signals driving IRP1-dependent biological responses, by analogy to cytokine signaling networks.

REGULATION OF IRP2

IRP2 is regulated in response to iron and oxygen supply. It is synthesized *de novo* under conditions of iron starvation,[61] remains stable in iron-starved or hypoxic cells,[62] and undergoes proteasomal degradation in iron-replete and normoxic cells.[63] A mechanistic model for the iron-dependent degradation of IRP2, involving the 73-amino-acid insertion, was proposed by Iwai *et al.*[64] and has dominated in the field for many years. According to it, the 73 amino acids define an iron-dependent "degradation domain" and function as an iron sensor. The recognition by the proteasome is mediated by the site-specific oxidation of three critical cysteine residues (C168, C174, and C178), possibly as a result of the direct binding of iron.[65] In support of this view, recent *in vitro* studies with a recombinant "degradation domain" peptide showed iron-dependent generation of oxidized cysteine species.[66] In addition, a two-hybrid screen with the "degradation domain" as a bait, by Iwai and co-workers,[67] resulted in the identification of a RING finger protein with E3 ubiquitin ligase activity, termed HOIL-1 (for heme-oxidized IRP2 ubiquitin ligase-1). However, this work proposed that the signal for IRP2 recognition by HOIL-1 is the binding of heme instead of the oxidation of C168, C174, and C178. Interestingly, a heme-dependent mechanism for IRP2 degradation had also been proposed earlier.[68]

We[69] and others[70] recently presented compelling evidence that the 73-amino-acid insertion is dispensable for the iron-dependent degradation of IRP2. Thus, a triple IRP2 mutant containing C168S, C174S, and C178S substitutions[69] or mutated versions lacking the entire 73-amino-acid domain[69,70] are sensitive to iron-mediated degradation. These results strongly suggest that the "cysteine oxidation model" is irrelevant for the iron-dependent degradation of IRP2 *in vivo*. How then can these findings be reconciled with the previous data and with the recently reported interaction of HOIL-1 with the 73-amino-acid domain? A convincing answer to the first question could be offered by the fact that the degradation of IRP2 exhibits a dose-dependent saturable pattern so that IRP2 expressed at high levels is able to titrate out limiting factors of the degradation machinery.[69] We speculate that previous conclusions were based on results with overexpressed IRP2 mutants. As for the second question, it is conceivable that heme and nonheme iron may trigger distinct pathways for the degradation of IRP2, which possibly involve different domains of the protein.

Pharmacological data suggest that a pathway for iron-mediated degradation of IRP2 requires the activity of 2-oxoglutarate-dependent oxygenases, which are enzymes utilizing iron, oxygen, and ascorbate as cofactors.[71,72] Thus, the substrate analogue dimethyl-oxalyl-glycine (DMOG) can efficiently inhibit the iron-dependent degradation of IRP2 in previously iron-depleted cells, following a pretreatment with the chelator desferrioxamine.[69,70] Interestingly, DMOG fails to inhibit IRP2 degradation in cells not pretreated with desferrioxamine,[69] suggesting that this pathway operates in conjunction with the levels of the LIP and is activated only when previously iron-depleted cells are exposed to iron (FIG. 3B). These findings pave the ground for the identification of *cis*-elements within IRP2 and components of the pathway, including the 2-oxoglutarate-dependent enzymes.

Members of the family of 2-oxoglutarate-dependent oxygenases, and in particular the 4-prolyl-hydroxylases PHD1, PHD2, and PHD3, have recently received a lot of attention because they regulate the oxygen-dependent degradation of the hypoxia inducible factor 1α (HIF-1α)[73–75] and thus function as oxygen sensors. The mechanism involves hydroxylation of P402 and P564 within an oxygen-dependent degradation domain.[76] This modification provides a recognition site for the von Hippel–Lindau tumor suppressor protein (pVHL) component of a ubiquitin ligase complex. The pathways for iron- and oxygen-dependent degradation of IRP2 and HIF-1α share remarkable similarities.[69,70] However, pVHL does not appear to be involved in the regulation of IRP2 as endogenous IRP2 undergoes unimpeded iron-dependent degradation in VHL-deficient cell lines (Wang & Pantopoulos, in preparation).

On a final note, IRP2 also responds to NO. However, there is some confusion in the literature on whether NO actually activates or inhibits IRP2 (discussed in ref. 58). This may be related to opposing effects of different redox species of NO on IRP2. It appears that nitrogen monoxide (the NO• radical) promotes the stabilization of IRP2, possibly indirectly by modulating the LIP,[44] while the nitrosonium cation (NO^+) mediates IRP2 degradation.[33] The complexity of IRP2 (and IRP1) responses to different signals illustrates that the IRE/IRP system integrates diverse regulatory pathways, which link the study of iron metabolism to hypoxia, oxidative stress, and inflammation.

ACKNOWLEDGMENTS

K. Pantopoulos is a scholar of the Canadian Institutes of Health Research (CIHR).

REFERENCES

1. ANDREWS, N.C. 1999. Disorders of iron metabolism. N. Engl. J. Med. **341:** 1986–1995.
2. SIPE, J.C., P. LEE & E. BEUTLER. 2002. Brain iron metabolism and neurodegenerative disorders. Dev. Neurosci. **24:** 188–196.
3. BERG, D. et al. 2001. Brain iron pathways and their relevance to Parkinson's disease. J. Neurochem. **79:** 225–236.
4. BISHOP, G.M. et al. 2002. Iron: a pathological mediator of Alzheimer disease? Dev. Neurosci. **24:** 184–187.
5. ROUAULT, T.A. 2001. Systemic iron metabolism: a review and implications for brain iron metabolism. Pediatr. Neurol. **25:** 130–137.
6. KE, Y. & Z. MING QIAN. 2003. Iron misregulation in the brain: a primary cause of neurodegenerative disorders. Lancet Neurol. **2:** 246–253.
7. PONKA, P. 1997. Tissue-specific regulation of iron metabolism and heme synthesis: distinct control mechanisms in erythroid cells. Blood **89:** 1–25.
8. ANDREWS, N.C. 2000. Iron homeostasis: insights from genetics and animal models. Nat. Rev. Genet. **1:** 208–217.
9. AISEN, P., C. ENNS & M. WESSLING-RESNICK. 2001. Chemistry and biology of eukaryotic iron metabolism. Int. J. Biochem. Cell Biol. **33:** 940–959.
10. PONKA, P., C. BEAUMONT & D.R. RICHARDSON. 1998. Function and regulation of transferrin and ferritin. Semin. Hematol. **35:** 35–54.
11. AISEN, P., M. WESSLING-RESNICK & E.A. LEIBOLD. 1999. Iron metabolism. Curr. Opin. Chem. Biol. **3:** 200–206.
12. KAKHLON, O. & Z.I. CABANTCHIK. 2002. The labile iron pool: characterization, measurement, and participation in cellular processes. Free Radical Biol. Med. **33:** 1037–1046.

13. HENTZE, M.W. & L.C. KÜHN. 1996. Molecular control of vertebrate iron metabolism: mRNA-based regulatory circuits operated by iron, nitric oxide, and oxidative stress. Proc. Natl. Acad. Sci. USA **93:** 8175–8182.
14. ROUAULT, T. & J.B. HARFORD. 2000. Translational control of ferritin synthesis. *In* Translational Control of Gene Expression, pp. 655–670. Cold Spring Harbor Laboratory Press. Cold Spring Harbor, New York.
15. JOHANSSON, H.E. & E.C. THEIL. 2002. Iron-response element (IRE) structure and combinatorial RNA regulation. *In* Molecular and Cellular Iron Transport, pp. 237–253. Dekker. New York/Basel.
16. GUNSHIN, H. *et al.* 1997. Cloning and characterization of a mammalian protein-coupled metal-ion transporter. Nature **388:** 482–488.
17. MCKIE, A.T. *et al.* 2000. A novel duodenal iron-regulated transporter IREG1, implicated in the basolateral transfer of iron to the circulation. Mol. Cell **5:** 299–309.
18. GUNSHIN, H. *et al.* 2001. Iron-dependent regulation of the divalent metal ion transporter. FEBS Lett. **509:** 309–316.
19. HUBERT, N. & M.W. HENTZE. 2002. Previously uncharacterized isoforms of divalent metal transporter (DMT)–1: implications for regulation and cellular function. Proc. Natl. Acad. Sci. USA **99:** 12345–12350.
20. GRAY, N.K. *et al.* 1996. Translational regulation of mammalian and drosophila citric acid cycle enzymes via iron-responsive elements. Proc. Natl. Acad. Sci. USA **93:** 4925–4930.
21. SCHALINSKE, K.L., O.S. CHEN & R.S. EISENSTEIN. 1998. Iron differentially stimulates translation of mitochondrial aconitase and ferritin mRNAs in mammalian cells: implications for iron regulatory proteins as regulators of mitochondrial citrate utilization. J. Biol. Chem. **273:** 3740–3746.
22. CALTAGIRONE, A., G. WEISS & K. PANTOPOULOS. 2001. Modulation of cellular iron metabolism by hydrogen peroxide: effects of H_2O_2 on the expression and function of iron-responsive element–containing mRNAs in B6 fibroblasts. J. Biol. Chem. **276:** 19738–19745.
23. SCHNEIDER, B.D. & E.A. LEIBOLD. 2003. Effects of iron regulatory protein regulation on iron homeostasis during hypoxia. Blood **102:** 3404–3411.
24. KE, Y. *et al.* 1998. Loops and bulge/loops in iron-responsive element isoforms influence iron regulatory protein binding. J. Biol. Chem. **273:** 23637–23640.
25. EISENSTEIN, R.S. 2000. Iron regulatory proteins and the molecular control of mammalian iron metabolism. Annu. Rev. Nutr. **20:** 627–662.
26. PANTOPOULOS, K. & M.W. HENTZE. 2000. Nitric oxide, oxygen radicals, and iron metabolism. *In* Nitric Oxide, pp. 293–313. Academic Press. San Diego.
27. CAIRO, G. & A. PIETRANGELO. 2000. Iron regulatory proteins in pathobiology. Biochem. J. **352:** 241–250.
28. ROUAULT, T.A. 2002. Post-transcriptional regulation of human iron metabolism by iron regulatory proteins. Blood Cells Mol. Dis. **29:** 309–314.
29. KIM, H-Y., R.D. KLAUSNER & T.A. ROUAULT. 1995. Translational repressor activity is equivalent and is quantitatively predicted by *in vitro* RNA binding for two iron-responsive element–binding proteins, IRP1 and IRP2. J. Biol. Chem. **270:** 4983–4986.
30. ERLITZKI, R., J.C. LONG & E.C. THEIL. 2002. Multiple, conserved iron-responsive elements in the 3′-untranslated region of transferrin receptor mRNA enhance binding of iron regulatory protein 2. J. Biol. Chem. **277:** 42579–42587.
31. RECALCATI, S. *et al.* 1998. Nitric oxide–mediated induction of ferritin synthesis in J774 macrophages by inflammatory cytokines: role of selective iron regulatory protein-2 downregulation. Blood **91:** 1059–1066.
32. KIM, S. & P. PONKA. 1999. Control of transferrin receptor expression via nitric oxide–mediated modulation of iron-regulatory protein 2. J. Biol. Chem. **274:** 33035–33042.
33. KIM, S. & P. PONKA. 2002. Nitrogen monoxide–mediated control of ferritin synthesis: implications for macrophage iron homeostasis. Proc. Natl. Acad. Sci. USA **99:** 12214–12219.
34. DERUSSO, P.A. *et al.* 1995. Expression of a constitutive mutant of iron regulatory protein 1 abolishes iron homeostasis in mammalian cells. J. Biol. Chem. **270:** 15451–15454.

35. WANG, J. & K. PANTOPOULOS. 2002. Conditional de-repression of ferritin synthesis in cells expressing a constitutive IRP1 mutant. Mol. Cell. Biol. **22:** 4638–4651.
36. LAVAUTE, T. et al. 2001. Targeted deletion of the gene encoding iron regulatory protein-2 causes misregulation of iron metabolism and neurodegenerative disease in mice. Nat. Genet. **27:** 209–214.
37. ROUAULT, T. A. 2001. Iron on the brain. Nat. Genet. **28:** 299–300.
38. CURTIS, A.R. et al. 2001. Mutation in the gene encoding ferritin light polypeptide causes dominant adult-onset basal ganglia disease. Nat. Genet. **28:** 350–354.
39. HAILE, D.J. et al. 1992. Cellular regulation of the iron-responsive element binding protein: disassembly of the cubane iron-sulfur cluster results in high-affinity RNA binding. Proc. Natl. Acad. Sci. USA **89:** 11735–11739.
40. HAILE, D.J. et al. 1992. Reciprocal control of RNA-binding and aconitase activity in the regulation of the iron-responsive element binding protein: role of the iron-sulfur cluster. Proc. Natl. Acad. Sci. USA **89:** 7536–7540.
41. KENNEDY, M.C. et al. 1992. Purification and characterization of cytosolic aconitase from beef liver and its relationship to the iron-responsive element binding protein. Proc. Natl. Acad. Sci. USA **89:** 11730–11734.
42. BRAZZOLOTTO, X. et al. 1999. Human cytoplasmic aconitase (iron regulatory protein 1) is converted into its [3Fe-4S] form by hydrogen peroxide *in vitro*, but is not activated for iron-responsive element binding. J. Biol. Chem. **274:** 21625–21630.
43. NARAHARI, J. et al. 2000. The aconitase function of iron regulatory protein 1: genetic studies in yeast implicate its role in iron-mediated redox regulation. J. Biol. Chem. **275:** 16227–16234.
44. PANTOPOULOS, K., G. WEISS & M.W. HENTZE. 1996. Nitric oxide and oxidative stress (H_2O_2) control mammalian iron metabolism by different pathways. Mol. Cell. Biol. **16:** 3781–3788.
45. GEGOUT, V. et al. 1999. Ligand-induced structural alterations in human iron regulatory protein-1 revealed by protein footprinting. J. Biol. Chem. **274:** 15052–15058.
46. BRAZZOLOTTO, X. et al. 2002. Structural changes associated with switching activities of the human iron regulatory protein 1. J. Biol. Chem. **277:** 11995–12000.
47. BROWN, N.M. et al. 1998. Novel role of phosphorylation in Fe-S cluster stability revealed by phosphomimetic mutations at Ser-138 of iron regulatory protein 1. Proc. Natl. Acad. Sci. USA **95:** 15235–15240.
48. FILLEBEEN, C. et al. 2003. A phosphomimetic mutation at Ser-138 renders iron regulatory protein 1 sensitive to iron-dependent degradation. Mol. Cell. Biol. **23:** 6973–6981.
49. CAIRO, G. et al. 2002. The iron regulatory proteins: targets and modulators of free radical reactions and oxidative damage. Free Radical Biol. Med. **32:** 1237–1243.
50. GEHRING, N., M.W. HENTZE & K. PANTOPOULOS. 1999. Menadione-induced oxidative stress leads to inactivation of both RNA-binding and aconitase activities of iron regulatory protein-1. J. Biol. Chem. **274:** 6219–6225.
51. KENNEDY, M.C., W.E. ANTHOLINE & H. BEINERT. 1997. An EPR investigation of the products of the reaction of cytosolic and mitochondrial aconitases with nitric oxide. J. Biol. Chem. **272:** 20340–20347.
52. SOUM, E. et al. 2003. Peroxynitrite and nitric oxide differently target the iron-sulfur cluster and amino acid residues of human iron regulatory protein 1. Biochemistry **42:** 7648–7654.
53. PANTOPOULOS, K. & M.W. HENTZE. 1995. Rapid responses to oxidative stress mediated by iron regulatory protein. EMBO J. **14:** 2917–2924.
54. MARTINS, E.A.L., R.L. ROBALINHO & R. MENEGHINI. 1995. Oxidative stress induces activation of a cytosolic protein responsible for control of iron uptake. Arch. Biochem. Biophys. **316:** 128–134.
55. PANTOPOULOS, K. & M.W. HENTZE. 1998. Activation of iron regulatory protein-1 by oxidative stress *in vitro*. Proc. Natl. Acad. Sci. USA **95:** 10559–10563.
56. DRAPIER, J.C. et al. 1993. Biosynthesis of nitric oxide activates iron regulatory factor in macrophages. EMBO J. **12:** 3643–3649.
57. WARDROP, S.L., R.N. WATTS & D.R. RICHARDSON. 2000. Nitrogen monoxide activates iron regulatory protein 1 RNA-binding activity by two possible mechanisms: effect on the [4Fe-4S] cluster and iron mobilization from cells. Biochemistry **39:** 2748–2758.

58. FILLEBEEN, C. & K. PANTOPOULOS. 2002. Redox control of iron regulatory proteins. Redox Rep. **7:** 15–22.
59. WEISS, G. 1999. Iron and anemia of chronic disease. Kidney Int. Suppl. **69:** S12–S17.
60. MÜTZE, S. *et al.* 2003. Myeloperoxidase-derived hypochlorous acid antagonizes the oxidative stress–mediated activation of iron regulatory protein 1. J. Biol. Chem. In press.
61. HENDERSON, B.R. & L.C. KÜHN. 1995. Differential modulation of the RNA-binding proteins IRP1 and IRP2 in response to iron: IRP2 inactivation requires translation of another protein. J. Biol. Chem. **270:** 20509–20515.
62. HANSON, E.S., L.M. FOOT & E.A. LEIBOLD. 1999. Hypoxia post-translationally activates iron-regulatory protein 2. J. Biol. Chem. **274:** 5047–5052.
63. GUO, B. *et al.* 1995. Iron regulates the intracellular degradation of iron regulatory protein 2 by the proteasome. J. Biol. Chem. **270:** 21645–21651.
64. IWAI, K., R.D. KLAUSNER & T.A. ROUAULT. 1995. Requirements for iron-regulated degradation of the RNA binding protein, iron regulatory protein 2. EMBO J. **14:** 5350–5357.
65. IWAI, K. *et al.* 1998. Iron-dependent oxidation, ubiquitination, and degradation of iron regulatory protein 2: implications for degradation of oxidized proteins. Proc. Natl. Acad. Sci. USA **95:** 4924–4928.
66. KANG, D.K. *et al.* 2003. Iron-regulatory protein 2 as iron sensor: iron-dependent oxidative modification of cysteine. J. Biol. Chem. **278:** 14857–14864.
67. YAMANAKA, K. *et al.* 2003. Identification of the ubiquitin-protein ligase that recognizes oxidized IRP2. Nat. Cell Biol. **5:** 336–340.
68. GOESSLING, L.S., D.P. MASCOTTI & R.E. THACH. 1998. Involvement of heme in the degradation of iron-regulatory protein 2. J. Biol. Chem. **273:** 12555–12557.
69. WANG, J. *et al.* 2004. Iron-mediated degradation of IRP2: an unexpected pathway involving a 2-oxoglutarate-dependent oxygenase activity. Mol. Cell. Biol. **24:** 954–965.
70. HANSON, E.S., M.L. RAWLINS & E.A. LEIBOLD. 2003. Oxygen and iron regulation of iron regulatory protein 2. J. Biol. Chem. **278:** 40337–40342.
71. SCHOFIELD, C.J. & Z. ZHANG. 1999. Structural and mechanistic studies on 2-oxoglutarate-dependent oxygenases and related enzymes. Curr. Opin. Struct. Biol. **9:** 722–731.
72. BRUICK, R.K. & S.L. MCKNIGHT. 2001. A conserved family of prolyl-4-hydroxylases that modify HIF. Science **294:** 1337–1340.
73. EPSTEIN, A.C. *et al.* 2001. *C. elegans* EGL-9 and mammalian homologs define a family of dioxygenases that regulate HIF by prolyl hydroxylation. Cell **107:** 43–54.
74. IVAN, M. *et al.* 2001. HIFalpha targeted for VHL-mediated destruction by proline hydroxylation: implications for O_2 sensing. Science **292:** 464–468.
75. JAAKKOLA, P. *et al.* 2001. Targeting of HIF-alpha to the von Hippel–Lindau ubiquitylation complex by O_2-regulated prolyl hydroxylation. Science **292:** 468–472.
76. MASSON, N. *et al.* 2001. Independent function of two destruction domains in hypoxia-inducible factor-alpha chains activated by prolyl hydroxylation. EMBO J. **20:** 5197–5206.

The Metabolism of Neuronal Iron and Its Pathogenic Role in Neurological Disease

Review

TORBEN MOOS[a] AND EVAN H. MORGAN[b]

[a]*Department of Medical Anatomy, University of Copenhagen, Copenhagen, Denmark*
[b]*Department of Physiology, University of Western Australia, Crawley, Western Australia, Australia*

ABSTRACT: Neurons need iron, which is reflected in their expression of the transferrin receptor. The concurrent expression of the ferrous iron transporter, divalent metal transporter I (DMT1), in neurons suggests that the internalization of transferrin is followed by detachment of iron within recycling endosomes and transport into the cytosol via DMT1. To enable DMT1-mediated export of iron from the endosome to the cytosol, ferric iron must be reduced to its ferrous form, which could be mediated by a ferric reductase. The presence of nontransferrin-bound iron in brain extracellular fluids suggests that neurons can also take up iron in a transferrin-free form. Neurons are thought to be devoid of ferritin in many brain regions in which there is an association between iron accumulation and cellular damage, for example, neurons of the substantia nigra pars compacta. The general lack of ferritin together with the prevailing expression of the transferrin receptor indicates that iron acquired by activity of transferrin receptors is directed toward immediate use in relevant metabolic processes, is exported, or is incorporated into complexes other than ferritin. Iron has long been considered to play a significant role in exacerbating degradation processes in brain tissue subjected to acute damage and neurodegenerative disorders. In brain ischemia, the damaging role of iron may depend on the inhibition of detoxifying enzymes responsible for catalyzing the oxidation of ferrous iron. Brain ischemia may also lead to an increase in iron supply to neurons as transferrin receptor expression by brain capillary endothelial cells is increased. Pharmacological blockage of the transferrin receptor/DMT1–mediated uptake could be a target to prevent further iron uptake. In chronic neurodegenerative settings, a deleterious role of iron is suggested since cases of Alzheimer's disease, Parkinson's disease, and Huntington's disease have a significantly higher accumulation of iron in affected regions. Dopaminergic neurons are rich in neuromelanin, shown to be more redox-active in Parkinson's disease cases. Iron-containing inflammatory cells may, however, account for the main portion of iron present in neurodegenerative disorders. More knowledge about iron metabolism in normal and diseased neurons is warranted as this may identify pharmaceutical targets to improve neuronal iron management.

Address for correspondence: Torben Moos, Department of Medical Anatomy, Section B, The Panum Institute, University of Copenhagen, DK-2200 Copenhagen N, Denmark. Voice: +45-35327264; fax: +45-35327252.
t.moos@mai.ku.dk

KEYWORDS: axonal transport; brain; dcytB; divalent metal transporter 1 (DMT1); ferritin; ferroportin; hemorrhage,; hephaestin; iron; iron deficiency; ischemia; Parkinson's disease,; transferrin receptor

INTRODUCTION

Neurons cannot exist without iron. Their need for iron is reflected in several ways. Iron is an essential cofactor for enzymes involved with energy metabolism and synthesis of neurotransmitters, for example, tyrosine hydroxylase and tryptophan hydroxylase.[1-3] The electron transport chain of the mitochondria consists of many iron-containing enzymes in which iron is part of a heme group. Since about 20% of the oxygen consumption of the body relies on mitochondrial respiration in the brain,[3] the neuronal need for iron is obvious. Iron also plays an important role for embryonic, dividing neurons as it is a cofactor for the enzyme ribonucleotide reductase, which is essential for dividing cells.[4] This particular role of iron for neuroectodermal development was recently demonstrated when it was shown that a mutation in the gene encoding the transferrin receptor, which provides transferrin-bound iron to the brain, results in nonviable embryos having severe malformations of the CNS.[5]

The neuronal expression of the transferrin receptor reflects their need for iron. During development, expression of the transferrin receptor in the rat brain is biphasic, being higher in the embryo and the adult brain than in the intervening period of early postnatal life.[6] The adult brain has exclusively nondividing neurons. In contrast, during the initial two postnatal weeks when the CNS develops markedly in terms of extensive myelogenesis, glial cell proliferation, and expansion of the number of synapses, neurons are virtually devoid of detectable levels of transferrin receptors.[7] The transport of iron into the brain is highly upregulated at this developmental stage as verified by radiolabeling[8] and immunohistochemical analyses.[7] Thus, the high iron transport into the brain probably results in sufficient amounts of iron to promote neuronal uptake without significant expression of transferrin receptors. This is because the high availability of iron probably leads to dissociation of iron regulatory proteins (IRPs) from the iron-responsive element (IRE) of the transferrin receptor mRNA, which will lead to a low expression of transferrin receptor mRNA and its protein.[9] Neurons increase their expression of transferrin receptor during iron deficiency, which contrasts with that of astrocytes, oligodendrocytes, and microglia.[7] This different expressional pattern can be regarded as a means of facilitating iron supply to neurons ahead of glial cells in iron-depleted conditions. Indeed, neurons are vulnerable to iron deficiency, which is thought to affect motor and cognitive functions via impaired energy metabolism and transmitter synthesis,[10-12] sometimes even in an irreversible manner.[13]

Recent data indicate that neurons are not only vulnerable to impaired iron metabolism as a result of a reduced iron supply, but also abnormally high cellular iron levels may lead to disordered neuronal function. The identification of mutations in various iron-related proteins as revealed from genetic analyses in patients suffering from neurological diseases[14,15] and targeted deletions of genes encoding proteins involved with cellular iron-handling in mice[16,17] has revealed a tight association between iron accumulation and neuronal damage. Ferrous iron (Fe^{2+}) is capable of forming hydroxyl radicals that are extremely toxic to the neuronal cell membrane.[1,18,19] The concentration of iron in the brain increases with increasing age in rodents and humans.[20,21] Moreover, neuronal iron accumulation is relatively higher in patients suffering from neurodegenerative disorders, including Alzheimer's disease, Parkinson's disease, and Huntington's disease, as compared to age-matched controls,[22–24] which together points toward the idea that a neuronal iron surplus may lead to neurological disease. However, definite proof that iron accumulation can lead to neuronal damage has never been obtained. Another problem is that expression and understanding of the function of proteins involved with physiological iron metabolism in neurons are far from elucidated. This paper reviews neuronal metabolism of iron and critically considers how mishandling of iron may lead to neurological disease.

NEURONAL IRON METABOLISM

Neuronal Iron Uptake

The widespread distribution of neuronal transferrin receptors in the CNS[25] clearly indicates that neurons can acquire iron by means of receptor-mediated uptake of iron-containing transferrin. The concurrent expression of the ferrous iron transporter, divalent metal transporter I (DMT1), in neurons[26–29] suggests that the internalization of transferrin is followed by detachment of iron within recycling endosomes and transport into the cytosol via DMT1. The significance of receptor-mediated internalization of iron-transferrin in the brain was elucidated in a recent study in the brain of the Belgrade (b/b) rat, which has a mutation in the DMT1 molecule.[29] It was found that DMT1 was not expressed in brain capillary endothelial cells of normal or b/b rats. Instead, DMT1 was robustly distributed to neurons, which probably suggests that the low cerebral iron levels in b/b rats are due to impairment of the function of neuronal DMT1.[29] [See FIG. 1.]

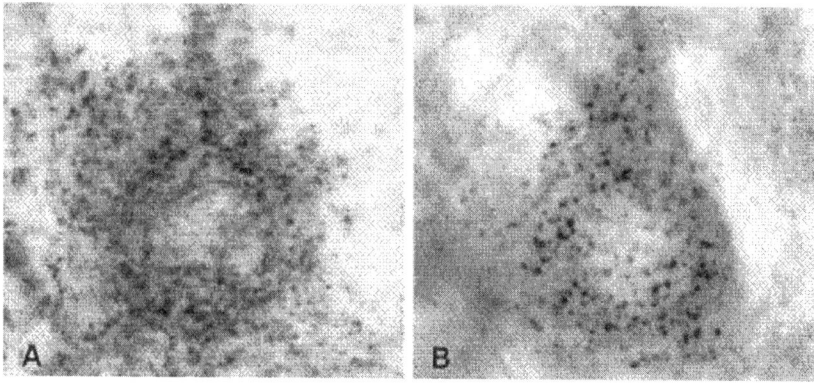

FIGURE 1. (**A**) Transferrin receptor and (**B**) DMT1 in neurons of the red nucleus of the ventral mesencephalon. FIGURE 1B shows a neuron of the Belgrade b/b rat brain known to suffer from iron deficiency, but this condition does not alter the DMT1 immunoreactivity.[29] Both proteins exhibit a dotted immunoreaction product that probably reflects their presence in recycling endosomes. (FIGURE 1B adapted from ref. 29.)

The cellular trafficking of iron once it leaves the endosome is far from elucidated; for example, it is not known which cytosolic molecule(s) binds the iron outside the endosome. To enable DMT1-mediated export of iron from the endosome to the cytosol, the ferric iron (Fe^{3+}) that enters the endosome bound to transferrin must be reduced to its ferrous form. This could be mediated by the ferric reductase present within the endosome or at its membrane. Neurons of the substantia nigra express stromal cell–derived receptor 2,[30] a homologue to the duodenal cytochrome b (dcytB) that exerts ferric reductase activity in duodenal enterocytes.[31] It remains to be shown whether stromal cell–derived receptor 2 serves this function in the brain and whether it is distributed throughout the brain.

The presence of nontransferrin-bound iron in brain extracellular fluids suggests that neurons can also take up iron as transferrin-free iron.[32] This may also apply to astrocytes, oligodendrocytes, and microglia that are devoid of transferrin receptors.[25] Supporting this notion, the brain contains large amounts of the protein SFT (stimulator of Fe transport),[33] which is thought to promote transfer of both nontransferrin-bound iron and transferrin-iron across cell membranes.[34] Iron transported through the blood-brain barrier may occur in the brain interstitial fluid in a low molecular form, for example, as iron-citrate.[32] Iron bound to transferrin in the brain interstitium may also become detached from transferrin due to interaction with other molecules that have the ability of releasing iron from transferrin, for example, ATP and citrate,[35] and are released from astrocytes. The slightly more acidic pH of the brain interstitial fluid (pH 7.3), as compared to that of blood plasma, also favors a displacement of iron from transferrin.[35] It is also possible that a pH much lower than this exists in certain brain microenvironments, such as between brain capillary endothelial cells and the astrocyte end-feet that surround them. The robust expression of transferrin receptors on neurons and the plasticity of these receptors in response to conditions with decreased iron availability[7] nonetheless clearly indicate that neurons can acquire iron via receptor-mediated uptake of transferrin. However, the quantitative importance of the neuronal transferrin receptor for iron uptake during physiological conditions as compared to the uptake of nontransferrin-bound iron remains to be determined.

Proteins Involved with Neuronal Iron Storage

Like other cells of the body, neurons express two different isoforms of ferritin mRNA.[36] The translational activity of these mRNAs is rather limited, however, as the ferritin protein is found neuronally in only a few selected regions, that is, the globus pallidus, medial habenular nucleus, and interpedunculate nucleus.[36] Hence, it is thought that neurons do not contain ferritin-iron in many regions that have been linked to an association between iron accumulation and cellular damage, for example, neurons of the substantia nigra pars compacta.[27] The general lack of ferritin protein, together with the prevailing expression of the transferrin receptor, indicates that iron acquired by activity of transferrin receptors is directed toward immediate use in relevant metabolic processes of the neuron, while excess iron is exported instead of being stored as ferritin. Interestingly, as the exception to the general rule, the medial habenular nucleus expresses significant levels of both ferritin and transferrin receptors,[36,37] which probably indicates that neurons of this nucleus take up iron continuously and partly store it as ferric iron in ferritin. Neurons of the medial habenular nucleus have a major projection to the interpedunculate nucleus of the midbrain. Injection of radiolabeled iron-transferrin into the lateral ventricle leads to receptor-mediated uptake of this complex by the medial habenular neurons, and this is probably followed by axonal transport of iron in anterograde direction toward the interpedunculate nucleus.[38,39]

Iron Export from Neurons

The recent identification of two proteins, ferroportin (MTP1, IREG)[40–42] and hephaestin,[43] has added new information about cellular iron export. Ferroportin is a membrane-bound protein that probably mediates ferrous iron export through the cell membrane. Hephaestin is a copper-containing protein sharing close homology with ceruloplasmin.[46] Both ceruloplasmin and hephaestin exhibit ferrous oxidase activity. Hephaestin is thought to be responsible for oxidation of ferrous iron transported through the basolateral part of duodenal enterocytes, hence allowing for binding of the iron to transferrin in blood plasma.

In the case of the brain, virtually nothing is known about cellular iron export. Recent studies indicate that neurons contain ferroportin,[28,40] which, together with the absence of ferritin in almost all brain nuclei, suggests that neurons export residual iron from the cytosol rather than storing it. The ferroportin protein would export ferrous iron. However, to avoid the toxicity of ferrous iron in the extracellular space in the brain, a concerted action between ferroportin and a ferrous oxidase like hephaestin or ceruloplasmin is predictable. This would lead to neuronal export of ferric iron, which would readily bind to transferrin or low molecular substances such as citrate or ATP in the brain interstitial fluid. So far, reports on the occurrence of hephaestin in the brain are not available, but astrocytes contain ceruloplasmin in a membrane-bound form.[45] Possibly, it could act locally or be released from the astrocytes to influence neuronal iron export, which would indirectly encourage incorporation of ferric iron in apotransferrin for use in the neuronal transferrin receptor. Oxidation of ferrous iron by ceruloplasmin may also explain its beneficial effect on iron uptake on cultured astrocytes.[46] The recent demonstration that mutation of the ceruloplasmin gene, leading to aceruloplasminemia, also causes iron accumulation in some neurons suggests that astrocytic ferrous oxidase activity is involved in iron export from neurons.[16]

Axonal Iron Transport

Long-term survival studies following extracerebral injection of radiolabeled iron into rats showed that iron undergoes a change in distribution with time.[47] Initially, the iron tended to accumulate in regions with high levels of neuronal transferrin receptors. Later, the iron was mainly present in regions containing the projectional fields of such neurons, for example, neurons of the globus pallidus that receive striatal projections, which probably is indicative of anterograde axonal transport of the iron. Several parameters of this putative axonal transport, however, remain unexploited, for example, the oxidation state of iron, the molecules to which it is bound during transport, the transport rate, and the significance of iron in the axonal terminal.

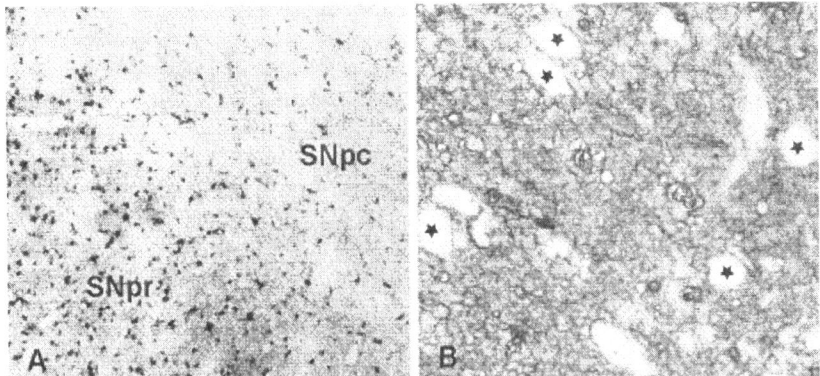

FIGURE 2. (A) Ferritin protein in the ventral mesencephalon of a β2-microglobulin-deficient mouse with circulatory, but not cerebral, iron-overloading due to increased intestinal iron uptake shown at low-power magnification and (B) a case of human Parkinson's disease shown at high power. (A) Note that ferritin is absent from neurons, but present in glial cells, with morphology corresponding to oligodendrocytes and microglia. SNpc: Substantia nigra pars compacta. SNpr: Substantia nigra pars reticulata. (B) This section was subjected to demelanization to remove melanin of the nigral neurons before immunostaining. Neurons are devoid of ferritin immunoreactivity (*stars*). (FIGURE 2A was adapted from ref. 27.)

THE PATHOLOGY OF NEURONAL IRON

Iron accumulation in tissues other than the nervous system leads to gradual cellular damage and loss of function, most clearly illustrated in the case of hemochromatosis. Iron has long been considered to play a significant role in exacerbating degradation processes in brain tissue subjected to acute damage that occurs in response to ischemia, hemorrhage, or trauma. Likewise, the relatively higher amount of iron present in the brains of patients suffering from neurodegenerative disorders is thought to be a key factor in increasing the loss of neurons. The insult of iron-mediated damage arises from the presence of Fe^{2+}, which may generate free radicals in the Fenton reaction. From the perspective of the cell biology of iron, it is necessary to consider how and where ferrous iron may occur in neurons in pathological conditions.

Evidence for Iron Damage in Acute Pathological Settings

In acute diseases occurring in the brain, including ischemia, hemorrhage, and trauma, the contribution of iron to neuronal pathology is likely to be mediated mainly by its release from the heme-containing molecules of the mitochondria in neurons or from ferritin in oligodendrocytes and microglia. In the case of hemorrhage and trauma, however, the main contribution of heme-derived iron comes from hemoglobin of lysing erythrocytes that is taken up by different cell types in the brain, mainly microglia.[48] The enzyme responsible for degradation of heme is heme oxygenase (HO), which occurs in three different isoforms. In pathological conditions, HO-1 is profoundly upregulated in macrophages and microglia.[49–51] Neurons are devoid of HO-1, but they contain HO-2, even in normal conditions, to which the formation of

CO for neurotransmission has been attributed.[51] HO-3 is expressed in the brain, but it appears to be devoid of HO activity (cf. ref. 48). The degradation of heme will lead to formation of Fe^{2+}, which needs instant neutralization to prevent formation of hydroxyl radicals. These radicals are eradicated by enzymes like superoxide dismutase and glutathione peroxidase (cf. ref. 18). Whether formation of free radicals formed by ferrous iron leads to subsequent damage to neurons depends on whether the iron load mediated by heme exceeds the capacity for neutralization of the free radicals. However, as neurons are not believed to take up iron subsequent to degradation of heme derived from red cells (cf. ref. 48), the iron from a hemorrhage is not likely to directly impact on neuronal function. The effect can only be indirect, for example, due to a release of proinflammatory cytokines from microglia involved in the inflammatory process. Interestingly, the process of red cell degradation appears to lead to a permanent regional excess of iron in the brain,[50] which could lead gradually to accumulation in neurons subsequent to its release from inflammatory cells. It should be mentioned, though, that the chemical nature and cellular distribution of iron in the region of a brain hemorrhage are unknown and clearly warrant further examination.

In conditions where ischemia is present, the damaging role of iron probably depends on the inhibition of detoxifying enzymes responsible for catalyzing the oxidation of Fe^{2+}, due to impaired cellular metabolism caused by the lack of oxygen. Moreover, degraded mitochondrial enzymes may release their content of heme iron in the form of Fe^{2+}. If the oxidative enzymes are active, the resulting Fe^{3+} could be stored in ferritin for later use. The iron might also be exported into the brain interstitium and bind to transferrin,[38] or a nontransferrin iron-binding molecule, such as citrate,[33] that could transfer the excess of iron via the ventricular system back to the blood. Neurons do not increase their expression of ferritin protein following ischemia,[53,54] again suggesting that neurons either export unbound iron or retain it in a form other than ferritin. Theoretically, surviving ischemic neurons might increase hephaestin levels and thereby oxidize and export iron, possibly in conjunction with ferroportin. A recent study suggests that ischemia may cause an increase in iron supply to neurons since a period of brain ischemia leads to increased expression of transferrin receptors by brain capillary endothelial cells.[55] This effect is probably due to the increased expression of hypoxia inducing factor (HIF), which is known to increase expression of transferrin receptor mRNA.[56] HIF may also increase neuronal transferrin receptor levels, thus worsening the problem of overcoming a period of ischemia of the neurons. Pharmacological blockage of the transferrin receptor/DMT1–mediated uptake in order to prevent further iron uptake in brain ischemia could be an aim of future research.

Participation of Iron in Chronic Neurodegenerative Disorders

In the case of chronic neurodegenerative diseases, the direct evidence that more iron leads to neurological damage is not settled. There are several clues that together point toward a deleterious role of iron, such as the observation that cases of Alzheimer's disease, Parkinson's disease, and Huntington's disease have a higher accumulation of iron in affected regions as compared to those of age-matched controls.[21–24,57] The cellular distribution of the regional iron accumulation in these disorders is not yet fully determined, though, which has been attributed to a lack of adequate methodology (cf. ref. 57). Many neurons of the aging brain contain Fe^{3+} in contrast to neurons of younger brains.[27,58] This may be due to a reduced capacity to export iron. A clue that neuronal mishandling of iron is associated with neurodegenerative disorders of the human brain can be found in the decreased ratio of ferritin protein to iron in dissected brain regions.[59] The translational activity of ferritin mRNA is regulated by IRPs present in the cytosol that act by binding to IREs on ferritin mRNA in conditions with iron depletion to prevent its translation. IRPs are widely present in the brain, including neurons,[60] but whether the expression or activities of IRPs are modified in neurons of the senescent brain remains to be examined. IRPs are constitutively actively expressed[61] in normal brains and those affected by Parkinson's disease, which is in accord with the absence of ferritin protein in dopaminergic neurons of the substantia nigra.[62] The absence of ferritin in these neurons again suggests that any available iron is either exported or incorporated in other molecules. The dopaminergic neurons are rich in neuromelanin,[63] which was recently shown to be significantly more redox-active in cases of Parkinson's disease compared to age-matched controls.[64] In regard to the higher level of iron in Parkinson's disease, it should be mentioned that the amount that might be bound by neuromelanin represents only 10–20% of the total.[63] Therefore, ferritin of activated microglia may well account for the remaining portion of iron in Parkinson's disease (see below). [See FIG. 2.]

A more direct proof of the involvement of iron in chronic neuronal pathology can be found in studies of patients suffering from disorders in which iron accumulation is high and directly associated with the state of disease, that is, pantothenate kinase associated neurodegeneration (PKAN) (formerly Hallervorden-Spatz disease) and aceruloplasminemia. These disorders are characterized by the excessive iron accumulation in the brain and chronic loss of neurons. Like the more common neurodegenerative disorders mentioned earlier, however, the involvement of iron in PKAN and aceruloplasminemia is thought to be secondary to the primary cause of disease. In PKAN, a mutation in the gene encoding the enzyme pantothenate kinase 2 leads to accumulation of cysteine residues, which in turn, via their affinity for divalent metal ions, may cause accumulation of ferrous iron in neurons.[15,26] In aceruloplasminemia, cells lack the ferroxidase activity of this molecule, which in the brain is expressed as a glycosylphosphatidylinositol-anchored form on membranes of astrocytes.[16,45] Whether the ferroxidase of these cells is also necessary for normal neuronal functioning was not clear until recently when it was shown that targeted deletion of the ceruloplasmin gene leads to iron accumulation and loss of neuronal function, hence clearly suggesting that neurons are affected if astrocytes fail to oxidize ferrous iron in their vicinity.[16] Also, coculturing of neurons with astrocytes was found to prevent the neurodegeneration that occurred in cultures of neurons alone exposed to $FeSO_4$ and other reactive oxygen-generating substances.[65] The recent identification of a mutation of the ferritin light-chain molecule in patients suffering from degeneration of striatal neurons further points toward an association between mismatches in iron handling and neurodegeneration.[14]

Ablation of the gene encoding another iron-related protein, IRP2, also leads to regional iron accumulation and neuronal damage,[17,66] suggesting that impairment of the neuronal iron-handling system has deleterious effects. IRP2 exerts its activity when cells are depleted of iron and operates by inhibiting ferritin mRNA translation and by increasing the stability of transferrin receptor mRNA.[1] Hence, the lack of IRP2 would lead to increased ferritin synthesis. In the mice with IRP2 deletion, there was excessive accumulation of ferritin in axons, a change that preceded axonal degeneration. It was suggested that the excessive synthesis of ferritin led to redistribution of iron in the axons and that the release of iron from the ferritin could cause iron-dependent damage.[17] However, other mechanisms should be considered. For instance, it is possible that the IRP2 knockout mouse fails to express transferrin receptors on brain capillary endothelial cells to a sufficient extent during postnatal development, when there is normally upregulated expression of the receptors on these cells.[9] This would lead to a state of iron deficiency in the brain that would even be worsened at the cellular level by the inability to downregulate ferritin expression. Iron deficiency at this developmental age may impair oligodendrocyte maturation and hence also myelin formation.[67,68] The neuronal abnormalities in the IRP2 knockout mouse might then be a secondary phenomenon due to the long-term effects of impaired myelin formation by oligodendrocytes. Alternatively, the abnormalities of the oligodendrocytes could activate microglia, which in turn could lead to neuronal impairment via the release of proinflammatory cytokines. A chronic inflammatory process in the brain of the IRP2 knockout mouse could lead to increased iron accumulation (see below).

Further evidence for the importance of normal ferritin function in neuronal iron metabolism comes from the newly described inherited disorder, "neuroferritinopathy".[14] In this condition, a mutation of the gene encoding the ferritin light (L) chain leads to increased iron and ferritin deposition, especially in the basal ganglia, and symptoms of extrapyramidal dysfunction. The mechanisms leading to cellular injury are unknown, but may depend on leakage of iron from the mutant protein leading to oxidative damage.[14] Ferritin is highly expressed in oligodendrocytes and microglia,[20,27,36] suggesting that these cell types could provide the iron that affects the neurons.

Many, if not all, of the neurodegenerative disorders are characterized by an accompanying inflammatory reaction that involves reactive microglia. These cells characteristically turn over rapidly as they scavenge dying cells and prevent the subsequent release of potentially hazardous molecules. Microglia contain abundant iron, which probably contributes significantly to the iron accumulation observed in MRI scans of the brain in patients suffering neurodegenerative disorders.[22,23,57] Hence, an often overlooked factor when speaking about iron accumulation and neurodegeneration is that the iron may instead originate from inflammatory cells. Ferritin is not present in neurons of the substantia nigra pars compacta in normal rodents,[27] normal and parkinsonian human autopsies,[62] or dopaminergic neurons of the substantia nigra pars compacta in monkeys at 6 months after exposure to *l*-4-phenyl-1,2,3,6-tetrahydropyridine (MPTP), which is used as a model of Parkinson's disease.[69] However, there is an abundance of reactive, ferritin-positive microglia in the vicinity of dopaminergic neurons in Parkinson's disease cases,[62] which indicates that the iron accumulation in neurodegeneration is due to the accumulation of iron-containing inflammatory cells. Supporting this notion, a recent study that also used MPTP exposure of monkeys demonstrated a profound infiltration of iron-containing microglia in the substantia nigra pars compacta at 5–15 months after MPTP administration, associated significantly with the disappearance of dopaminergic neurons.[70] Iron was not present in the substantia nigra pars compacta until 40% of the dopaminergic neurons had vanished, and it was concluded that iron is not likely to have a primary role in causing nigral cell death, but rather participates by exacerbating the degenerative process once it has started.[71] The latter would most likely be due to the participation of iron-containing microglia that might damage the nigral neurons via their release of proinflammatory cytokines.[62]

CONCLUSIONS

Iron is thought to contribute significantly to the deleterious outcome of acute and chronic lesions in the brain. The role of iron in these conditions is far from resolved, which also applies to the understanding of the neuronal metabolism of iron in physiological conditions. The handling of iron by neurons is thought to involve uptake mediated by the transferrin receptor followed by transport of iron into the cytosol by DMT1, where the iron is directed to incorporation into enzymes or storage in ferritin. There is a substantial lack of information on whether neurons are capable of exporting iron to the cellular exterior. More knowledge about these processes will probably add significantly to the understanding of how iron can cause damage to neurons, and probably also to the identification of putative pharmaceutical targets to improve neuronal iron handling.

ACKNOWLEDGMENTS

The most recent results by the authors were generated by grant support from the Danish Medical Research Council, Dansk Kræftforsknings Fond, Direktør Leo Nielsen & Karen M. Nielsens Legat, Eva og Henry Frænkels Mindefond, Fonden af 17.12.1981, Kong Christian den X.'s Fond, NovoNordisk Fonden, Sigvald og Edith Cecine Kristence Rasmussen, født Poulsen mindelegat (T. Moos), and the Medical Research Fund of Western Australia (E. H. Morgan).

REFERENCES

1. MORGAN, E.H. 1996. Iron metabolism and transport. *In* Hepatology: A Textbook of Liver Disease. Third edition, pp. 526–554. Saunders. Philadelphia.
2. WALKER, B.L., J.W.C. TIONG & W.A. JEFFERIES. 2001. Iron metabolism in mammalian cells. Int. Rev. Cytol. **211:** 241–278.
3. WRIGGELSWORTH, J.M. & H. BAUM. 1988. Iron-dependent enzymes in the brain. *In* Brain Iron: Neurochemical and Behavioral Aspects, pp. 25–66. Taylor & Francis. Philadelphia/London.
4. LASKEY, J., I. WEBB, H.M. SCHULMAN *et al.* 1988. Evidence that transferrin supports cell proliferation by supplying iron for DNA synthesis. Exp. Cell Res. **176:** 87–95.
5. LEVY, J.E., O. JIN, Y. FUJIWARA *et al.* 1999. Transferrin receptor is necessary for development of erythrocytes and the nervous system. Nat. Genet. **21:** 396–399.
6. MOOS, T. & E.H. MORGAN. 2002. A morphological study of the developmentally regulated transport of iron into the brain. Dev. Neurosci. **24:** 99–105.
7. MOOS, T., P.S. OATES & E.H. MORGAN. 1998. Expression of the neuronal transferrin receptor is age dependent and susceptible to iron deficiency. J. Comp. Neurol. **398:** 420–430.
8. TAYLOR, E.M., A. CROWE & E.H. MORGAN. 1991. Transferrin and iron uptake by the brain: effects of altered iron status. J. Neurochem. **57:** 1584–1592.
9. MORGAN, E.H. & T. MOOS. 2002. Mechanism and developmental changes in iron transport across the blood-brain barrier. Dev. Neurosci. **24:** 106–113.

10. ALLEN, R.P., P.B. BARKER, F. WEHRL *et al.* 2001. MRI measurement of brain iron in patients with restless legs syndrome. Neurology **56:** 263–265.
11. DE DEUNGRIA, M., R. RAO, J.D. WOBKEN *et al.* 2000. Perinatal iron deficiency decreases cytochrome c oxidase (CytOx) activity in selected regions of neonatal rat brain. Pediatr. Res. **48:** 169–176.
12. YOUDIM, M.B. & D. BEN-SHACHAR. 1987. Minimal brain damage induced by early iron deficiency: modified dopaminergic neurotransmission. Isr. J. Med. Sci. **23:** 19–25.
13. BEARD, J., K.M. ERIKSON & B.C. JONES. 2003. Neonatal iron deficiency results in irreversible changes in dopamine function in rats. J. Nutr. **133:** 1174–1179.
14. CURTIS, A.R., C. FEY, C.M. MORRIS *et al.* 2001. Mutation in the gene encoding ferritin light polypeptide causes dominant adult-onset basal ganglia disease. Nat. Genet. **28:** 350–354.
15. ZHOU, B., S.K. WESTAWAY, B. LEVINSON *et al.* 2001. A novel pantothenate kinase gene (PANK2) is defective in Hallervorden-Spatz syndrome. Nat. Genet. **28:** 345–349.
16. PATEL, B.N., R.J. DUNN, S.Y. JEONG *et al.* Ceruloplasmin regulates iron levels in the CNS and prevents free radical injury. J. Neurosci. **22:** 6578–6586.
17. LAVAUTE, T., S. SMITH, S. COOPERMAN *et al.* 2001. Targeted deletion of the gene encoding iron regulatory protein-2 causes misregulation of iron metabolism and neurodegenerative disease in mice. Nat. Genet. **27:** 209–214.
18. NAPPI, A.J. & E. VASS. 2000. Iron, metalloenzymes, and cytotoxic reactions. Cell. Mol. Biol. **46:** 637–647.
19. JELLINGER, K.A. 1999. The role of iron in neurodegeneration: prospects for pharmacotherapy of Parkinson's disease. Drugs Aging **14:** 115–140.
20. FOCHT, S.J., B.S. SNYDER, J.L. BEARD *et al.* 1997. Regional distribution of iron, transferrin, ferritin, and oxidatively-modified proteins in young and aged Fischer 344 rat brains. Neuroscience **79:** 255–261.
21. BARTZOKIS, G., M. BECKSON, D.B. HANCE *et al.* 1997. MRI evaluation of age-related increase of brain iron in young adult and older normal males. Magn. Reson. Imag. **15:** 29–35.
22. BARTZOKIS, G., D. SULTZER, J. MINTZ *et al.* 1994. *In vivo* evaluation of brain iron in Alzheimer's disease and normal subjects using MRI. Biol. Psychiatry **35:** 480–487.
23. BARTZOKIS, G., J. CUMMINGS, S. PERLMAN *et al.* 1999. Increased basal ganglia iron levels in Huntington disease. Arch. Neurol. **56:** 569–574.
24. DEXTER, D.T., F.R. WELLS, A.J. LEES *et al.* 1989. Increased nigral iron content and alterations in other metal ions occurring in brain in Parkinson's disease. J. Neurochem. **52:** 1830–1836.
25. MOOS, T. 1996. Immunohistochemical localization of intraneuronal transferrin receptor immunoreactivity in the adult mouse central nervous system. J. Comp. Neurol. **375:** 675–692.
26. GUNSHIN, H., B. MACKENZIE, U.V. BERGER *et al.* 1997. Cloning and characterization of a mammalian proton-coupled metal-ion transporter. Nature **388:** 482–488.
27. MOOS, T., D. TRINDER & E.H. MORGAN. 2000. Cellular distribution of ferric iron, ferritin, transferrin, and divalent metal transporter 1 (DMT1) in substantia nigra and basal ganglia of normal and beta2-microgbulin deficient mouse brain. Cell. Mol. Biol. **46:** 549–561.
28. BURDO, J.R., S.L. MENZIES, I.A. SIMPSON *et al.* Distribution of divalent metal transporter 1 and metal transport protein 1 in the normal and Belgrade rat. J. Neurosci. Res. **66:** 1198–1207.
29. MOOS, T. & E.H. MORGAN. 2003. How does mutation of divalent metal transporter (DMT1) explain the impaired transport of iron into the Belgrade rat brain? J. Neurochem. **88:** 233–245.
30. PONTING, C.P. 2001 Domain homologues of dopamine beta-hydroxylase and ferric reductase: roles for iron metabolism in neurodegenerative disorders? Hum. Mol. Genet. **10:** 1853–1858.
31. MCKIE, A.T., D. BARROW, G.O. LATUNDE-DADA *et al.* 2001. An iron-regulated ferric reductase associated with the absorption of dietary iron. Science **291:** 1755–1759.
32. MOOS, T. & E.H. MORGAN. 1998. Evidence for low molecular weight, non-transferrin-bound iron in rat brain and cerebrospinal fluid. J. Neurosci. Res. **54:** 486–494.

33. KNUTSON, M.D., J.E. LEVY, N.C. ANDREWS *et al.* 2001. Expression of stimulator of Fe transport is not enhanced in Hfe knockout mice. J. Nutr. **131:** 1459–1464.
34. GUTIERREZ, J.A., J. YU, S. RIVERA *et al.* 1997. Functional expression cloning and characterization of SFT, a stimulator of Fe transport. J. Cell Biol. **139:** 895–905.
35. MORGAN, E.H. 1979. Studies on the mechanism of iron release from transferrin. Biochim. Biophys. Acta **580:** 312–326.
36. HANSEN, T.M., H. NIELSEN, N. BERNTH *et al.* 1999. Expression of ferritin protein and subunit mRNAs in normal and iron deficient rat brain. Mol. Brain Res. **65:** 186–197.
37. MOOS, T., P.S. OATES & E.H. MORGAN. 1999. Iron-independent neuronal expression of transferrin receptor mRNA in the rat. Brain Res. Mol. Brain Res. **72:** 231–234.
38. MORRIS, C.M., A.B. KEITH, J.A. EDWARDSON *et al.* 1992. Uptake and distribution of iron and transferrin in the adult rat brain. J. Neurochem. **59:** 300–306.
39. MOOS, T. & E.H. MORGAN. 1998. Kinetics and distribution of [59Fe-125I]transferrin injected into the ventricular system of the rat. Brain Res. **790:** 115–128.
40. ABBOUD, S. & D.J. HAILE. 2000. A novel mammalian iron-regulated protein involved in intracellular iron metabolism. J. Biol. Chem. **275:** 19906–19912.
41. DONOVAN, A., A. BROWNLIE, Y. ZHOU *et al.* 2000. Positional cloning of zebrafish ferroportin 1 identifies a conserved vertebrate iron exporter. Nature **403:** 776–781.
42. MCKIE, A., P. MARCIANI, A. ROLFS *et al.* 2000. A novel duodenal iron-regulated transporter, IREG1, implicated in the basolateral transfer of iron to the circulation. Mol. Cell **5:** 299–309.
43. VULPE, C.D., Y.M. KUO *et al.* 1999. Hephaestin, a ceruloplasmin homologue implicated in intestinal iron transport, is defective in the sla mouse. Nat. Genet. **21:** 195–199.
44. ANDERSON, G.J., D.M. FRAZER *et al.* 2002. The ceruloplasmin homolog hephaestin and the control of intestinal iron absorption. Blood Cells Mol. Dis. **29:** 367–375.
45. PATEL, B.N. & S.A. DAVID. 1997. A novel glycosylphosphatidylinositol-anchored form of ceruloplasmin is expressed by mammalian astrocytes. J. Biol. Chem. **272:** 20185–20190.
46. QIAN, Z.M., Y.K. TSOI, Y. KE *et al.* 2001. Ceruloplasmin promotes iron uptake rather than release in BT325 cells. Exp. Brain Res. **140:** 369–374.
47. DWORK, A.J., G. LAWLER, P.A. ZYBERT *et al.* 1990. An autoradiographic study of the uptake and distribution of iron by the brain of the young rat. Brain Res. **518:** 31–39.
48. WAGNER, K.R., F.R. SHARP, T.D. ARDIZZONE *et al.* 2003. Heme and iron metabolism: role in cerebral hemorrhage. J. Cereb. Blood Flow Metab. **23:** 629–652.
49. TURNER, C.P., M. BERGERON *et al.* 1998. Heme oxygenase-1 is induced in glia throughout brain by subarachnoid hemoglobin. J. Cereb. Blood Flow Metab. **18:** 257–273.
50. HUANG, F.P., G. XI, R.F. KEEP *et al.* 2002. Brain edema after experimental intracerebral hemorrhage: role of hemoglobin degradation products. J. Neurosurg. **96:** 287–293.
51. CHAKRABARTY, A., M.R. EMERSON & S.M. LEVINE. 2003. Heme oxygenase-1 in SJL mice with experimental allergic encephalomyelitis. Mult. Scler. **9:** 372–381.
52. VERMA, A., D.J. HIRSCH, C.E. GLATT *et al.* 1993. Carbon monoxide: a putative neural messenger. Science **259:** 381–384.
53. CHI, S.I., C.K. WANG, J.J. CHEN *et al.* 2000. Differential regulation of H- and L-ferritin messenger RNA subunits, ferritin protein, and iron following focal cerebral ischemia-reperfusion. Neuroscience **100:** 475–484.
54. KONDO, Y., N. OGAWA, M. ASANUMA *et al.* 1995. Regional differences in late-onset iron deposition, ferritin, transferrin, astrocyte proliferation, and microglial activation after transient forebrain ischemia in rat brain. J. Cereb. Blood Flow Metab. **15:** 216–226.
55. OMORI, N., K. MARUYAMA, G. JIN *et al.* 2003. Targeting of post-ischemic cerebral endothelium in rat by liposomes bearing polyethylene glycol–coupled transferrin. Neurol. Res. **25:** 275–279.
56. LOK, C.N. & P. PONKA. 1999. Identification of a hypoxia response element in the transferrin receptor gene. J. Biol. Chem. **274:** 24147–24152.
57. BERG, D., M. GERLACH, M.B. YOUDIM *et al.* 2001. Brain iron pathways and their relevance to Parkinson's disease. J. Neurochem. **79:** 225–236.
58. BENKOVIC, S.A. & J.R. CONNOR. 1993. Ferritin, transferrin, and iron in selected regions of the adult and aged rat brain. J. Comp. Neurol. **338:** 97–113.

59. CONNOR, J.R., B.S. SNYDER, P. AROSIO *et al.* 1995. A quantitative analysis of isoferritins in select regions of aged, parkinsonian, and Alzheimer's diseased brains. J. Neurochem. **65:** 717–724.
60. LEIBOLD, E.A., L.C. GAHRING & S.W. ROGERS. 2001. Immunolocalization of iron regulatory protein expression in the murine central nervous system. Histochem. Cell Biol. **115:** 195–203.
61. FAUCHEUX, B.A., M.E. MARTIN, C. BEAUMONT *et al.* 2003. Neuromelanin associated redox-active iron is increased in the substantia nigra of patients with Parkinson's disease. J. Neurochem. **86:** 1142–1148.
62. MIRZA, B., H. HADBERG, P. THOMSEN *et al.* 2000. The absence of reactive astrocytosis is indicative of a unique inflammatory process in Parkinson's disease. Neuroscience **95:** 425–432.
63. ZECCA, L., D. TAMPELLINI, A. GATTI *et al.* 2002. The neuromelanin of human substantia nigra and its interaction with metals. J. Neural Transm. **109:** 663–672.
64. FAUCHEUX, B.A., M.E. MARTIN, C. BEAUMONT *et al.* 2002. Lack of up-regulation of ferritin is associated with sustained iron regulatory protein-1 binding activity in the substantia nigra of patients with Parkinson's disease. J. Neurochem. **83:** 320–330.
65. TANAKA, J., K. TOKU, B. ZHANG *et al.* 1999. Astrocytes prevent neuronal death induced by reactive oxygen and nitrogen species. Glia **28:** 85–96.
66. HAYFLICK, S.J., S.K. WESTAWAY, B. LEVINSON *et al.* 2003. Genetic, clinical, and radiographic delineation of Hallervorden-Spatz syndrome. N. Engl. J. Med. **348:** 33–40.
67. GRABILL, C., A.C. SILVA, S.S. SMITH *et al.* 2003. MRI detection of ferritin iron overload and associated neuronal pathology in iron regulatory protein-2 knockout mice. Brain Res. **971:** 95–106.
68. OLOYEDE, O.B., A.T. FOLAYAN & A.A. ODUTUGA. 1992. Effects of low-iron status and deficiency of essential fatty acids on some biochemical constituents of rat brain. Biochem. Int. **27:** 913–922.
69. YU, G.S., T.M. STEINKIRCHNER, G.A. RAO *et al.* 1986. Effect of prenatal iron deficiency on myelination in rat pups. Am. J. Pathol. **125:** 620–624.
70. GOTO, K., H. MOCHIZUKI, H. IMAI *et al.* 1996. An immuno-histochemical study of ferritin in 1-methyl-4-phenyl-1,2,3,6-tetrahydropyridine (MPTP)–induced hemiparkinsonian monkeys. Brain Res. **724:** 125–128.
71. HE, Y., P.S. THONG & T. LEE. 2003. Dopaminergic cell death precedes iron elevation in MPTP-injected monkeys. Free Radical Biol. Med. **35:** 540–547.

Fluorescence Detection of Redox-Sensitive Metals in Neuronal Culture

Focus on Iron and Zinc

IAN J. REYNOLDS

Department of Pharmacology, University of Pittsburgh, Pittsburgh, Pennsylvania, USA

ABSTRACT: Detection of neurotoxic metals in the intracellular milieu has made an important contribution to the understanding of the mechanism of metal-induced neuronal injury. Fluorescent, metal-sensitive dyes have proven to be valuable in the measurement of a variety of neurotoxic cations in neurons, and these dyes have provided a number of insights into the relationships between elevations in the cytosolic free-metal concentrations and neuronal death. However, the dyes also have important limitations that can make the interpretation of dye signals difficult. In this review, the characteristics of dyes that can be used to detect both iron and zinc inside neurons, and the methods necessary to distinguish these ions from other intracellular signals, are reviewed. Also provided are examples of the use of the dyes for the redox-sensitive detection of iron and zinc. Finally, the challenges facing the use of these dyes for quantitative determination of changes in intracellular free-ion concentrations are discussed.

KEYWORDS: fluorescence microscopy; intracellular iron; intracellular zinc; fura-2; magfura-2; Newport Green; neurotoxicity

INTRODUCTION

Metals operate in biological systems at a delicate interface between their essential catalytic and structural functions and the havoc that is created by metals that escape their normal control mechanisms. Although the magnitude of metal signals vary considerably, it is fair to generalize that metals functioning in their normal physiological roles are typically present in relatively low concentrations, and that cytotoxic consequences of these same metals occur when the free concentration of the metals is elevated. It follows from this that the goal of techniques intended to provide insight into metal signaling in cell injury must be to selectively report changes in metal ion concentrations within the range of concentration changes associated with injurious stimuli.

There are a variety of methods currently available to monitor changes in metal ion concentrations in cells. The particular focus of this chapter is small molecule

---

Address for correspondence: Ian J. Reynolds, Department of Pharmacology, University of Pittsburgh, W1351 Biomedical Science Tower, Pittsburgh, PA 15261. Voice: 412-648-2134; fax: 412-624-0794.

iannmda@pitt.edu

fluorescent reporters of ion concentrations. These agents, which include fura-2 and fluo-3,[1] were first developed to measure intracellular calcium signals and have proved to be relatively easy to use and quite versatile in the determination of temporal and spatial properties of calcium signaling in a wide range of cell types.[2] The adaptation of these fluorescent reporters to the measurement of redox-active metals is discussed below. There are certain limitations of specificity of these agents, and also practical restrictions in the extent to which such probes can be targeted within cells. The creative use of a series of molecular approaches has provided some alternatives to the use of the fluorescent dyes. For example, combining the calcium recognition properties of calmodulin with fluorescent proteins has generated a series of protein indicators of calcium that can then be targeted to subcellular locations.[3] This concept was recently adapted in relation to the redox-induced release of zinc from a metallothionein-based reporter.[4] In some respects, these molecular methods are a little more difficult to use than the small molecule fluorescent dyes, and this is particularly the case in the study of metal-induced injury to the central nervous system because of the difficulty in introducing genes into neurons. However, their relative sensitivity and selectivity offer some important advantages, as discussed further below.

The fluorescent dyes discussed here are those used to monitor intracellular free iron and zinc. Iron is considered a redox-active metal because of its ready oxidation and reduction, and its participation in a wide range of redox chemistries in the cell. Zinc is not redox active per se. However, as discussed below, zinc is an important endogenous neurotoxin[5–8] and the storage and release of zinc from intracellular sites is regulated by the cellular redox status. Given that oxidant-induced zinc release is neurotoxic,[9] it seems appropriate to consider the fluorescent detection of zinc along with iron under the general heading of redox-active metals.

METAL-SENSITIVE FLUOROPHORES

The first generation of fluorescent ion indicators were based on an adaptation of the calcium chelator EGTA to introduce fluorescence while maintaining the calcium binding properties. The product of this effort was BAPTA, which is fluorescent, but only in the ultraviolet range, which limited its biological utility.[10] A key development in the field was the introduction of a series of indicators that included fura-2, indo-1, and fluo-3.[1] When prepared as the acetoxymethyl (AM) esters, these dyes are cell-permeant,[11] are trapped inside cells by esterase cleavage and the subsequent unmasking of the five charged carboxylic acid residues, and respond to binding of divalent cations with changes in fluorescence intensity, excitation, and/or emission spectra. Although these reagents were designed as calcium indicators, even the earliest studies recognized the potential that the dyes could bind other metal ions,[1] including iron and zinc. These studies also recognized the difference in the interaction of zinc and iron with dyes such as fura-2, where the interaction of zinc resembled that of calcium in producing a shift in the excitation spectrum of fura-2, while iron quenched fura-2 fluorescence. The pioneering work of Tsien and colleagues has been followed by the development of a remarkable range of calcium-sensitive dyes that vary in their calcium affinity, the relative specificity for other ions (such as magnesium[12]), the relative abundance of dyes in intracellular compartments (such as the accumulation of rhod-2 in mitochondria[13]), and their diffusion in the cyto-

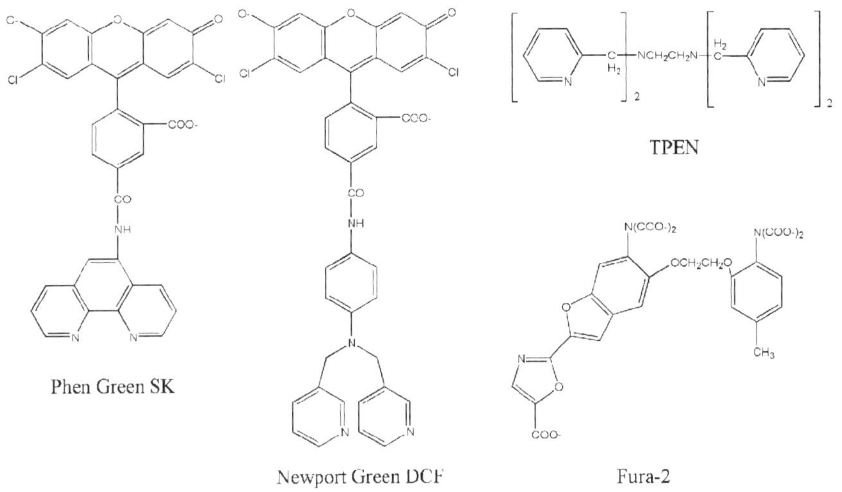

FIGURE 1. Structures of probes used in the fluorescence detection of redox-sensitive metals. Fura-2 is the classic calcium indicator that binds divalent cations using the five carboxylic acid groups. The deesterified form is shown. Fura-2 is sensitive to calcium, zinc, and iron. Shown for comparison are Newport Green DCF, which preferentially binds zinc, and Phen Green SK, which binds iron in preference to zinc. Neither dye binds calcium at physiological concentrations. Also shown is TPEN, a cell-permeable heavy metal chelator. Note that the carboxylic acid groups on Newport Green and Phen Green result in trapping the dye in cells, but do not contribute to ion binding.

plasm (by conjugation with high-molecular-weight molecules such as dextran[14]). Ion binding by the calcium-sensitive dyes is fundamentally a property of divalent cation interaction with the carboxylic acid residues on the dye, and it has proven difficult to unequivocally separate the detection of calcium from other metals with dyes based on this structure.

A separate class of dyes use the coordination of metals with nitrogens rather than carboxylic acids as the basis of metal binding, and here the separation of calcium sensitivity from the sensitivity to other metals is much more robust. Such agents are based on heavy metal chelators like TPEN and include dyes like Newport Green and Phen Green, as well as a number of derivatives[14] (FIG. 1). Although Newport Green has a carboxylic acid group, this is more important for cell retention than zinc binding. These dyes effectively detect zinc, iron, and other species, while being essentially insensitive to biologically relevant concentrations of calcium and magnesium.[15] The selectivity of recognition of different heavy metal species is more limited. Newport Green, for example, will bind zinc, ferrous iron, and ferric iron, and each metal pecies increases fluorescence. The dyes are also sensitive to copper, cadmium, and other metals.[14]

In evaluating the utility of these dyes for detecting redox-active metals, the main considerations are sensitivity and selectivity. Although multiple independent lines of evidence have defined the range of intracellular free calcium and magnesium concentrations, such data are much harder to come by for these less abundant metals.

This makes it difficult to predict what the appropriate affinity should be for a dye to detect relevant ion concentrations. The issue of dye sensitivity in relation to dye affinity is considered in more detail later in this chapter because this issue has some problems that are not fully appreciated. Demonstrating selectivity can also be a problem. When using the calcium-sensitive dyes, it is relatively straightforward to make the distinction between signals derived from calcium or magnesium and other heavy metals. Adding TPEN to cells that show dye responses to a given stimulus reverses heavy metal–mediated dye responses, while having no effect on calcium or magnesium effects. This does not effectively distinguish between various species of heavy metal. Indeed, cell-permeant chelators lack sufficient specificity to make the distinction between copper and zinc, for example. It is also possible to exploit other dye-metal interactions to distinguish responses. Thus, one can distinguish iron from zinc when using fura-2 because iron quenches fura-2 fluorescence, while zinc shifts the excitation wavelengths, as noted earlier.[16] A judicious use of dyes of finite selectivity and sensitivity, chelators with similarly limited selectivity, and a careful examination of the properties of the dye-metal interaction can facilitate the detection of intracellular metal species, but absolute identification of metals using this approach remains challenging.

REDOX-SENSITIVE DETECTION OF INTRACELLULAR IRON

We exploited various properties of iron-sensitive dyes in a recent study that monitored free-iron concentrations in neurons.[16] Much of this volume is focused on the circumstances of iron mobilization and the mechanisms by which iron contributes to cell injury. This section focuses on issues of iron detection in cultured neurons. Our goal in the study was to establish the feasibility of detecting acute changes in cytoplasmic free iron ($[Fe]_i$) in the concentration ranges associated with cell injury. We used the ionophore pyrithione as a means of delivering iron into cells. This agent is selective for heavy metals and does not transport calcium across the plasma membrane, but can effectively deliver ions such as zinc and iron. We evaluated the ability of several different dyes to detect iron delivered by this method, including fura-2, magfura-2, and Phen Green. Each of the dyes was more sensitive to Fe(II) than Fe(III) *in vitro* and in cells, and fura-2 proved to be the most sensitive when loaded into cells (Kress and Reynolds, unpublished observations). Subsequent experiments used fura-2, and we were able to distinguish between dye responses to calcium, zinc, and iron based on the difference in the spectral response to increases in the relevant ion as well as the sensitivity to cell-permeant chelators, as described earlier.

Interestingly, the response of fura-2 to iron was sensitive to an oxidant burden. Thus, exposing Fe(II)-loaded neurons to hydrogen peroxide caused a decrease in the iron-induced quenching of the fluorescence that was consistent with the conversion of Fe(II) to Fe(III) by the peroxide. When a low concentration of peroxide was used, the dequenching was spontaneously reversed, presumably reflecting the ability of the cell to reestablish the normal intracellular reducing environment. We were then able to establish the relative sensitivity of several different cell types to challenge with peroxide on the basis of dequenching of fura-2, as well as the toxic consequences of the iron load in neurons, astrocytes, and oligodendrocytes.[16]

There are several advantages to this approach. Detecting changes in the redox state of an intracellular metal ion is perhaps the most obvious of the advantages. This is likely to be a consequence of the difference in affinity of fura-2 for Fe(II) and Fe(III), such that the differential sensitivity falls within the relevant range for the change in the concentration of the iron species encountered in this set of experiments. These experiments also allowed us to directly determine the effects of intracellular iron chelators such as bipyridyl, TPEN, and desferrioxamine (DFO).[17] Interestingly, DFO was not at all effective when acutely applied to cells, while TPEN and bipyridyl reverse the intracellular iron changes very quickly. The effectiveness of DFO in modifying iron-mediated effects in cells is presumably the result of its endocytosis, but this is clearly a relatively slow process compared to the speed with which the truly cell-permeable chelators work.[17] There are also clear disadvantages to the use of these fluorescent dyes for iron detection. Fundamentally, the method is not that sensitive to iron, which is a reflection of the relatively high dye concentrations in the cell relative to the magnitude of the elevation in $[Fe]_i$ that may typically occur (see below). We were able to detect ionophore-delivered iron, but not changes in $[Fe]_i$ resulting from intracellular acidification, for example (Kress and Reynolds, unpublished observations). Because there is a certain minimum dye concentration necessary for effective fluorescence detection,[15] this sensitivity limit may be difficult to overcome. Nevertheless, the redox sensitivity and the ability to associate $[Fe]_i$ with cellular injury processes illustrate the value of the approach.

MODULATION OF INTRACELLULAR FREE ZINC BY REDOX STATUS

Zinc is a relatively abundant metal in most cells, and in broad terms cells contain similar amounts of iron and zinc. Like iron, cells maintain very low concentrations of free zinc, and it has been estimated that the amount of free zinc is just a few ions per cell.[18] However, it is clear that there are concentrated stores of zinc that are contained within neurotransmitter vesicles[19-21] and that this zinc can be released upon stimulation.[22,23] Further, Koh and colleagues suggested that the entry of vesicular zinc into neurons makes an important contribution to the pathogenesis of ischemic brain injury.[24] It is also clear that zinc applied to neurons in culture is toxic[5,7,25,26] and that the toxicity of zinc is a consequence of zinc entry into neurons.[27-29] The mechanism(s) of zinc toxicity remains a matter of debate, but several studies have suggested that zinc impairs neuronal energy metabolism at the level of glycolysis, the tricarboxylic acid cycle, or the mitochondrion.[30-32]

As noted earlier, zinc is not redox-active in chemical terms. However, it has become clear that the free zinc status is subject to modulation by the redox state of the cell. We encountered this phenomenon in studies originally intended to investigate the regulation of calcium release by sulfhydryl sites on the ryanodine receptor. Exposing neurons to the sulfhydryl oxidizing agent, 2,2'-dithiodipyridine (DTDP), results in neuronal death and small increases in the fluorescence ratio of dyes such as fura-2 and magfura-2. We originally interpreted such changes as increases in calcium in the cytosol, consistent with the hypothesis that oxidation of the ryanodine receptor would activate it and release calcium from intracellular stores.[9] However, studies by Maret and colleagues showed that DTDP could oxidize metallothionein and in doing so greatly reduce the affinity for zinc.[33] Our subsequent studies showed that the DTDP-induced fura-2 and magfura-2 signal could be reversed by TPEN and that TPEN could protect neurons from DTDP-induced neuronal injury.[9] This is consistent with zinc being the metal species being released by DTDP, and the properties of the fura-2 response exclude the possibility that iron is mobilized. However, it is more difficult to be certain that copper is not being mobilized (although copper is certainly less abundant).

Further studies have revealed that a number of stimuli can release a zinc-like metal following oxidation of an intracellular target. These oxidants include nitric oxide[4,34] and phenylarsine oxide.[35] The adaptation of metallothionein into a fluorescence probe using fluorescence resonance energy transfer clearly demonstrates that nitric oxide can liberate zinc from the metallothionein-based fluorophore.[4] In addition, we have recently shown that overexpression of metallothionein in astrocytes both increases the buffering of zinc introduced by an ionophore and also increases the size of the oxidant-releasable zinc pool.[36] All of these studies are consistent with the notion that an increase in oxidant burden can mobilize intracellular zinc (or a closely related metal), and the studies in neurons are consistent with released zinc being neurotoxic. It remains unclear, though, whether the endogenous sulfhydryl-containing zinc binding protein is, in fact, metallothionein. Other zinc-containing proteins, such as protein kinase C, have zinc bound in a way that can be mobilized under physiological conditions,[37] so there are other potential sources of zinc. Studies are currently under way to manipulate endogenous levels of metallothionein in an attempt to address this issue.

QUANTITATIVE CONSIDERATIONS OF INTRACELLULAR METAL DETECTION

The well-earned reputation of fluorescent indicators as effective reporters of changes in intracellular free-ion concentrations is based on their ease of use, their specificity, and their claimed utility for quantitative measurement of intracellular ions. The specificity of the indicators was discussed earlier in this paper, and this section focuses on the question of the determination of free-ion concentrations.

The binding of ions to fluorescent indicators is traditionally modeled on the classic principles of ligand-receptor interactions, whereby the ion represents the ligand and the dye represents the receptor. With the standard assumptions used to simplify the quantitation of drug-receptor interactions, the binding of the ion to dye can be described quite simply by knowing the affinity of the ion for the dye. This is readily determined using buffered solutions of metals that yield known free-ion concentrations and by performing saturation analyses. In practice, such determinations have the ion/buffer concentrations in great excess over that of the fluorescent indicator when saturation parameters are determined in a spectrofluorimeter. The binding of ion by dye can then be related to the change in the fluorescence signal, and a knowledge of the dissociation constant for the dye allows the free-ion concentration to be established by solving the standard binding equation for the free-ligand concentration. The introduction of fura-2 and indo-1 in 1985 was accompanied by the concept of the ratiometric determination of intracellular calcium concentrations.[1] In addition to the advantages of using fluorescence to report ion concentrations, this approach obviated problems associated with alterations in dye concentrations during the experiment (e.g., due to dye leakage or cell swelling) by taking ratios of fluorescence signals at two wavelengths rather than an absolute measure of fluorescence at a single wavelength.

It is apparent, however, that there are some limitations in the extent to which actual free-ion concentrations can be determined using fluorescent dyes. Several factors have now been identified that detract from the basic approach of inferring free-ligand concentrations from dye signals. Relatively early, it was recognized that the intracellular environment in which dyes are used can differ quite appreciably from standard conditions used *in vitro* to characterize dyes. For example, elevated viscosity has been proposed to alter the sensitivity of fura-2 to calcium among several other parameters.[38–40] More recently, we have described a more fundamental problem in the conversion of fluorescence signals to ion concentrations.[15] This issue came to light in the comparison of two dyes, magfura-2 and Newport Green, that bind zinc with affinities that vary by a factor of >100. The two dyes have different fluorescence properties, so signals from the two dyes can be recorded simultaneously. Despite the anticipated difference in sensitivity, the dyes showed identical saturation characteristics when exposed to zinc by the application of zinc together with an ionophore in primary neurons.[15] This led us to reconsider the standard approach used by Grynkiewicz and colleagues.[1] We concluded that the initial assumption about the nature of the ion-dye interactions was flawed. Specifically, the simplifying assumption used in the standard binding equation is that the binding of ligand to receptor does not meaningfully alter the free-ligand concentration. This is a perfectly reasonable assumption for most pharmacological circumstances where there is a large excess of ligand over receptor. However, standard loading procedures for fluorescent dyes result in the accumulation of near-millimolar concentrations of the dyes after they have been trapped by ester cleavage. Under these circumstances, a very substantial amount of the free ion is bound by the dye, and the free concentration decreased, thus undermining the simplifying assumption made by Grynkiewicz *et al*. It remains possible to model the binding of ion to dye under such circumstances, but it is important to note that under the conditions of very high dye concentrations it is the concentration of the dye that drives the binding process and not the affinity of the dye for the ion. Thus, distinguishing dyes by their affinity has very little value when the dye concentration in the cell is very high with respect to the *in vitro* affinity.[15]

There are a number of practical consequences that follow from these conclusions. The effective detection of intracellular cations is often based on the anticipated range of concentration changes in relation to the affinity of the dye. This is a valid approach when the intracellular buffering capacity for the ion of interest is present in excess over the dye concentration. This is probably the case with proton (pH)–sensitive dyes and may be the case for calcium-sensitive dyes too.[41] However, the magnitude of the buffer capacity for zinc and iron is not known and is presumably much smaller than that for calcium. This means that the sensitivity of the reporting system is likely to be driven by dye concentration as much as affinity. Thus, dye concentrations matter, and things that change dye concentrations will alter apparent responses. It is also difficult to decrease dye concentrations to the point where this is no longer a factor in the measurements because the dye signal is lost before the dye concentrations become sufficiently low.[15] Under these circumstances, it remains quite feasible to detect changes in the free concentration of redox-sensitive metals in neurons, but it is difficult to quantitatively interpret such signals. The effective calibration of such responses requires a determination of the concentration of the dye in addition to knowing the affinity of the dye and an estimate of the endogenous buffer capacity of the ion of interest. Two out of three of these parameters are very difficult to obtain experimentally, and this provides a limit to the extent to which the dye signals can be calibrated.

In summary, there are a number of tools available for the fluorescence detection of redox-sensitive metals in neurons. The use of these tools has allowed the characterization of changes in the intracellular free-metal concentrations that are associated with neuronal injury, particularly in cell-culture systems. This has provided valuable information about the relationship between ion movement, elevation in intracellular ion concentrations, and cell death. The studies are limited by problems in specificity of some of the measurement approaches (which are experimentally addressable) as well as difficulties in calibration of the dye signals (which are more difficult to deal with). Nevertheless, these approaches have considerable value for the characterization of redox-sensitive metals and neuronal injury.

REFERENCES

1. GRYNKIEWICZ, G., M. POENIE & R.Y. TSIEN. 1985. A new generation of Ca^{2+} indicators with greatly improved fluorescence properties. J. Biol. Chem. **260**: 3440–3450.
2. KEITH, C.H., R. RATAN, F.R. MAXFIELD et al. 1985. Local cytoplasmic calcium gradients in living mitotic cells. Nature **316**: 848–850.
3. MIYAWAKI, A., J. LLOPIS, R. HEIM et al. 1997. Fluorescent indicators for Ca^{2+} based on green fluorescent proteins and calmodulin. Nature **388**: 882–887.
4. PEARCE, L.L., R.E. GANDLEY, W. HAN et al. 2000. Role of metallothionein in nitric oxide signaling as revealed by a green fluorescent fusion protein. Proc. Natl. Acad. Sci. USA **97**: 477–482.
5. CHOI, D.W., M. YOKOYAMA & J. KOH. 1988. Zinc neurotoxicity in cortical cell culture. Neuroscience **24**: 67–79.
6. KOH, J-Y. & D.W. CHOI. 1988. Zinc alters excitatory amino acid neurotoxicity in cortical neurons. J. Neurosci. **8**: 2164–2171.
7. DUNCAN, M.W., A.M. MARINI, R. WATTERS et al. 1992. Zinc, a neurotoxin to cultured neurons, contaminates cycad flour prepared by traditional Guamanian methods. J. Neurosci. **12**: 1523–1537.
8. WEISS, J.H., D.M. HARTLEY, J. KOH & D.W. CHOI. 1993. AMPA receptor activation potentiates zinc neurotoxicity. Neuron **10**: 43–49.
9. AIZENMAN, E., A.K. STOUT, K.A. HARTNETT et al. 2000. Induction of neuronal apoptosis by thiol oxidation: putative role of intracellular zinc release. J. Neurochem. **75**: 1878–1888.
10. TSIEN, R.Y. 1980. New calcium indicators and buffers with high selectivity against magnesium and protons: design, synthesis, and properties of prototype structures. Biochemistry **19**: 2396–2404.
11. TSIEN, R.Y. 1981. A non-disruptive technique for loading calcium buffers into cells. Nature **290**: 527–528.
12. RAJU, B., E. MURPHY, L.A. LEVY et al. 1989. A fluorescent indicator for measuring cytosolic free magnesium. Am. J. Physiol. **256**: C540–C548.
13. SIMPSON, P.B. & J.T. RUSSELL. 1996. Mitochondria support inositol 1,4,5-trisphosphate-mediated Ca^{2+} waves in cultured oligodendrocytes. J. Biol. Chem. **271**: 33493–33501.
14. HAUGLAND, R.P. 2002. Handbook of Fluorescent Probes and Research Products. Molecular Probes. Eugene, OR.
15. DINELEY, K.E., L.M. MALAIYANDI & I.J. REYNOLDS. 2002. A reevaluation of neuronal zinc measurements: artifacts associated with high intracellular dye concentration. Mol. Pharmacol. **62**: 618–627.

16. KRESS, G.J., K.E. DINELEY & I.J. REYNOLDS. 2002. The relationship between intracellular free iron and cell injury in cultured neurons, astrocytes, and oligodendrocytes. J. Neurosci. **22:** 5848–5855.
17. ZANNINELLI, G., H. GLICKSTEIN, W. BREUER *et al.* 1997. Chelation and mobilization of cellular iron by different classes of chelators. Mol. Pharmacol. **51:** 842–852.
18. OUTTEN, C.E. & T.V. O'HALLORAN. 2001. Femtomolar sensitivity of metalloregulatory proteins controlling zinc homeostasis. Science **292:** 2488–2492.
19. CRAWFORD, I.L. & J.D. CONNOR. 1972. Zinc in maturing rat brain: hippocampal concentration and localization. J. Neurochem. **19:** 1451–1458.
20. PEREZ-CLAUSELL, J. & G. DANSCHER. 1985. Intravesicular localization of zinc in rat telencephalic boutons: a histochemical study. Brain Res. **337:** 91–98.
21. BUDDE, T., A. MINTA, J.A. WHITE & A.R. KAY. 1997. Imaging free zinc in synaptic terminals in live hippocampal slices. Neuroscience **79:** 347–358.
22. ASSAF, S.Y. & S.H. CHUNG. 1984. Release of endogenous Zn^{2+} from brain tissue during activity. Nature **308:** 734–736.
23. HOWELL, G.A., M.G. WELCH & C.J. FREDERICKSON. 1984. Stimulation-induced uptake and release of zinc in hippocampal slices. Nature **308:** 736–738.
24. KOH, J-Y., S.W. SUH, B.J. GWAG *et al.* 1997. The role of zinc in selective neuronal death after transient global cerebral ischemia. Science **272:** 1013–1016.
25. KOH, J-Y. & D.W. CHOI. 1994. Zinc toxicity on cultured cortical neurons: involvement of *N*-methyl-D-aspartate receptors. Neuroscience **60:** 1049–1057.
26. MANEV, H., E. KHARLAMOV, T. UZ *et al.* 1997. Characterization of zinc-induced neuronal death in primary cultures of rat cerebellar granule cells. Exp. Neurol. **146:** 171–178.
27. CANZONIERO, L.M.T., D.M. TURETSKY & D.W. CHOI. 1999. Measurement of intracellular free zinc concentrations accompanying zinc-induced neuronal death. J. Neurosci. **19:** D1–D6.
28. CHENG, C. & I.J. REYNOLDS. 1998. Calcium-sensitive fluorescent dyes can report increases in intracellular free zinc concentration in cultured forebrain neurons. J. Neurochem. **71:** 2401–2410.
29. SENSI, S., L.M.T. CANZONIERO, S.P. YU *et al.* 1997. Measurement of intracellular free zinc in living cortical neurons: routes of entry. J. Neurosci. **17:** 9554–9564.
30. SHELINE, C.T., M.M. BEHRENS & D.W. CHOI. 2000. Zinc-induced cortical neuronal death: contribution of energy failure attributable to loss of NAD+ and inhibition of glycolysis. J. Neurosci. **20:** 3139–3146.
31. BROWN, A.M., B.S. KRISTAL, M.S. EFFRON *et al.* 2000. Zn^{2+} inhibits alpha-ketoglutarate-stimulated mitochondrial respiration and the isolated alpha-ketoglutarate dehydrogenase complex. J. Biol. Chem. **275:** 13441–13447.
32. DINELEY, K.E., T.V. VOTYAKOVA & I.J. REYNOLDS. 2003. Zinc inhibition of cellular energy production: implications for mitochondria and neurodegeneration. J. Neurochem. **85:** 563–570.
33. MARET, W. & B.L. VALLEE. 1998. Thiolate ligands in metallothionein confer redox activity on zinc clusters. Proc. Natl. Acad. Sci. USA **95:** 3478–3482.
34. ST. CROIX, C.M., K.J. WASSERLOOS, K.E. DINELEY *et al.* 2002. Nitric oxide–induced changes in intracellular zinc homeostasis are mediated by metallothionein/thionein. Am. J. Physiol. Lung Cell. Mol. Physiol. **282:** L185–L192.
35. SENSI, S.L., D. TON-THAT & J.H. WEISS. 2002. Mitochondrial sequestration and Ca(2+)-dependent release of cytosolic Zn(2+) loads in cortical neurons. Neurobiol. Dis. **10:** 100–108.
36. MALAIYANDI, L.M., K.E. DINELEY & I.J. REYNOLDS. 2004. Divergent consequences arise from metallothionein overexpression in astrocytes: zinc buffering and oxidant-induced zinc release. Glia **45:** 346–353.
37. KNAPP, L.T. & E. KLANN. 2002. Potentiation of hippocampal synaptic transmission by superoxide requires the oxidative activation of protein kinase C. J. Neurosci. **22:** 674–683.
38. BLATTER, L.A. & W.G. WIER. 1990. Intracellular diffusion, binding, and compartmentalization of the fluorescent calcium indicators indo-1 and fura-2. Biophys. J. **58:** 1491–1499.

39. POENIE, M. 1990. Alteration of intracellular fura-2 fluorescence by viscosity: a simple correction. Cell Calcium **11:** 85–91.
40. LATTANZIO, F.A. & D.K. BARTSCHAT. 1991. The effect of pH on rate constants, ion selectivity, and thermodynamic properties of fluorescent calcium and magnesium indicators. Biochem. Biophys. Res. Commun. **177:** 184–191.
41. NEHER, E. 1995. The use of fura-2 for estimating Ca buffers and Ca fluxes. Neuropharmacology **34:** 1423–1442.

Metal-Catalyzed Disruption of Membrane Protein and Lipid Signaling in the Pathogenesis of Neurodegenerative Disorders

MARK P. MATTSON

Laboratory of Neurosciences, National Institute on Aging Intramural Research Program, and Department of Neuroscience, Johns Hopkins University School of Medicine, Baltimore, Maryland, USA

ABSTRACT: Membrane lipid peroxidation and oxidative modification of various membrane and associated proteins (e.g., receptors, ion transporters and channels, and signal transduction and cytoskeletal proteins) occur in a range of neurodegenerative disorders. This membrane-associated oxidative stress (MAOS) is promoted by redox-active metals, most notably iron and copper. The mechanisms whereby different genetic and environmental factors initiate MAOS in specific neurological disorders are being elucidated. In Alzheimer's disease (AD), the amyloid β-peptide generates reactive oxygen species and induces MAOS, resulting in disruption of cellular calcium homeostasis. In Parkinson's disease (PD), mitochondrial toxins and perturbed ubiquitin-dependent proteolysis may impair ATP production and increase oxyradical production and MAOS. The inheritance of polyglutamine-expanded huntingtin may promote neuronal degeneration in Huntington's disease (HD), in part, by increasing MAOS. Increased MAOS occurs in amyotrophic lateral sclerosis (ALS) as the result of genetic abnormalities (e.g., Cu/Zn–superoxide dismutase mutations) or exposure to environmental toxins. Levels of iron are increased in vulnerable neuronal populations in AD and PD, and dietary and pharmacological manipulations of iron and copper modify the course of the disease in mouse models of AD and PD in ways that suggest a role for these metals in disease pathogenesis. An increasing number of pharmacological and dietary interventions are being identified that can suppress MAOS and neuronal damage and improve functional outcome in animal models of AD, PD, HD, and ALS. Novel preventative and therapeutic approaches for neurodegenerative disorders are emerging from basic research on the molecular and cellular actions of metals and MAOS in neural cells.

KEYWORDS: Alzheimer's disease; apoptosis; copper; Huntington's disease; iron; Parkinson's disease

Address for correspondence: Mark P. Mattson, Laboratory of Neurosciences, National Institute on Aging Intramural Research Program, 5600 Nathan Shock Drive, Baltimore, MD 21224. Voice: 410-558-8463; fax: 410-558-8465.
    mattsonm@grc.nia.nih.gov

INTRODUCTION

Different neurodegenerative disorders are distinguished by the specific neuronal populations affected, certain histological features, and behavioral symptoms. Neurons in the hippocampus, entorhinal cortex, amygdala, basal forebrain, and some regions of cerebral cortex degenerate in Alzheimer's disease (AD), accounting for the cognitive impairment and emotional problems in this disease.[1] The neurons most drastically affected in Parkinson's disease (PD) are those located in the substantia nigra that use dopamine as a neurotransmitter; degeneration of these dopaminergic neurons results in the inability of PD patients to control their body movements.[2] Huntington's disease (HD) involves degeneration of neurons in the striatum and cortex, with patients often exhibiting involuntary motor functions.[3] Lower and upper motor neurons degenerate in amyotrophic lateral sclerosis (ALS), resulting in progressive paralysis and death.[4] Despite the differences in the manifestations of these different neurodegenerative disorders, research has revealed shared aspects of the neurodegenerative mechanisms. In each disorder, neurons are subjected to increased oxidative stress, metabolic compromise, and perturbed ion homeostasis.[5–12]

During the past decade, major advances have been made in identifying genes that cause, or affect the risk of, neurodegenerative disorders (TABLE 1). Whereas HD is an inherited disease caused by trinucleotide expansions in the gene encoding huntingtin,[13] only a small percentage of cases of AD, PD, and ALS are inherited. Three genes have been identified in which mutations are linked to early-onset autosomal dominant AD: the amyloid precursor protein (APP), presenilin-1 (PS1), and presenilin-2 (PS2).[14] Some cases of familial PD are caused by mutations in α-synuclein, DJ-1, or parkin.[15] Mutations in Cu/Zn-SOD and alsin are responsible for some cases of ALS.[16] The identification of the genetic defects that cause these diseases has led to the development of novel animal models of the diseases and to relatively rapid advances in understanding why neurons become dysfunctional and die in humans with these diseases. In addition to disease-causing genes, genetic factors that either increase or decrease disease risk are being identified. One example of a disease-modifying gene is apolipoprotein E; individuals with the E4 isoform are at increased risk of AD.[17]

Most cases of AD, PD, and ALS have no clear genetic basis and thus many studies are aimed at identifying environmental factors that may determine whether or not an individual develops the disease (TABLE 1). Changes that occur in cells during the

TABLE 1. Examples of genetic and environmental factors in neurodegenerative disorders

Disorder	Genes	Environment
AD	APP, presenilins, ApoE	Calories, physical and mental inactivity
PD	α-Synuclein, parkin, DJ-1	Toxins, iron, calories
HD	Huntingtin	Calories, iron
ALS	Cu/Zn-SOD, alsin	Toxins
Stroke	Notch-3, ApoE	Calories, lipids, smoking

aging process play an important role because disease risk increases sharply after middle age. Dietary factors may influence the risk of neurodegenerative disorders. For example, studies of animal models of AD, PD, and HD have shown that dietary restriction (decreased meal size and/or frequency) can protect neurons, delay disease onset, and increase survival,[18–21] although dietary restriction was not beneficial in a mouse model of ALS.[22] Dietary metals may also influence the risk of neurodegenerative disorders. For example, a high intake of iron and/or elevated serum iron levels have been associated with increased risk of AD and PD,[23,24] and iron increases the vulnerability of neurons to dysfunction and death in animal and cell culture models of AD and PD.[25,26]

MEMBRANE-ASSOCIATED OXIDATIVE STRESS (MAOS) IN NEURODEGENERATIVE DISORDERS

The evidence that neurons are subjected to increased oxidative stress in aging and neurodegenerative disorders is extensive and will not be reviewed here. Instead, we describe results of our studies of cell culture and animal models that provide evidence for a pivotal role for MAOS in the dysfunction of synapses and the death of neurons in neurodegenerative disorders. These studies suggest a sequence of events resulting in perturbed membrane protein function and lipid metabolism, which initially impairs the functions of synapses and later triggers the death of neurons.

MAOS in AD

Analyses of postmortem brain tissue samples have documented increased oxidative damage to proteins, DNA, and lipids in AD patients compared to age-matched neurologically normal control subjects. For example, levels of protein carbonyls, 8-oxoguanine, and lipid peroxidation products were greater in samples from AD patients (for review, see ref. 27). Levels of oxidative damage are particularly high in the environment of amyloid plaques and in neurons with neurofibrillary tangles. There appears to be a major increase in the amount of lipid peroxidation that occurs in the membranes of brain cells in AD because the concentration of the amount of the lipid peroxidation product, 4-hydroxynonenal (HNE), is elevated in the cerebrospinal fluid of patients compared to controls.[28] Levels of iron are increased in degenerating neurons in AD,[29] suggesting a contribution of this metal to the MAOS in AD. Transgenic mice that express a familial AD APP mutation develop progressive deposition of Aβ in the hippocampus and cerebral cortex, and the Aβ deposition is associated with increased levels of oxidative stress.[30] Mice expressing a PS1 mutation exhibit enhanced membrane lipid peroxidation when exposed to a neurotoxin that induces hippocampal damage and memory impairment.[31,32]

When exposed to Aβ, cultured neurons, and synapses isolated from the brains of rats, lipid peroxidation occurs.[25,33–35] Our data[36] and those of others[37] suggest that Aβ itself generates ROS (hydrogen peroxide and hydroxyl radical) by a chemical reaction that requires oxygen and trace amounts of Fe^{2+} or Cu^+. The generation of ROS by Aβ is associated with, and may promote, aggregation of the peptide (FIG. 1). Low amounts of Fe^{2+} greatly enhance Aβ-induced lipid peroxidation and neuro-

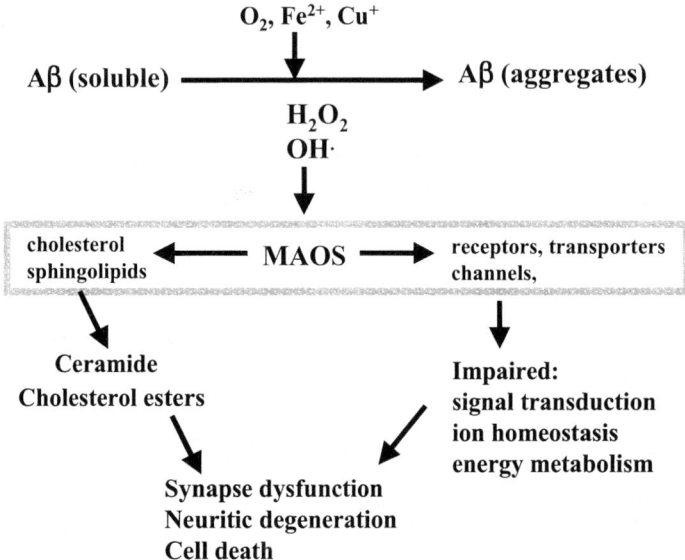

FIGURE 1. Mechanisms whereby Fe^{2+} and Cu^+ promote amyloid β-peptide (Aβ) aggregation and neurotoxicity in AD. See references 5, 9, 27, and 101 for discussion of these mechanisms.

TABLE 2. Examples of membrane-associated molecules modified by oxidative stress in models of neurodegenerative disorders

Proteins	Lipids
Na+/K+-ATPase[34,35,38]	Cholesterol[42,67,99,100]
Ca^{2+}-ATPase[38]	Unsaturated fatty acids[33]
Glucose transporter[39,61]	Sphingomyelin[42,67]
Glutamate transporter[41,61]	Ceramides[42,67]
Gq11[40]	
NMDA receptor[97]	
Ca^{2+} channel[98]	
Apolipoprotein E[12,60]	

toxicity.[25] We discovered that levels of HNE are increased when cultured neurons or cortical synaptosomes are exposed to Aβ.[34,35] The HNE modifies several membrane proteins including ion-translocating ATPases, glucose transporters, glutamate transporters, and GTP-binding proteins.[38-41] The production of HNE was suppressed and neuronal viability preserved when neurons were treated with vitamin E and other antioxidants that inhibit membrane lipid peroxidation.[25,39] Moreover, treatment of neurons with a membrane-permeant form of glutathione, which scavenges HNE, protected them from being killed by Aβ.[25,39]

Membrane lipids are altered in AD and may contribute to synaptic dysfunction and neuronal cell death (TABLE 2). Mass spectrometry analyses of various lipids in membrane samples from vulnerable and nonvulnerable brain regions of AD patients and control subjects revealed significant increases in levels of long-chain ceramides and free cholesterol in the patients.[42] The magnitude of the elevations of ceramide and cholesterol levels was positively associated with disease severity. These abnormalities in membrane lipids may be the result of increased MAOS because exposure of cultured neurons to Aβ resulted in increased levels of long-chain ceramides and free cholesterol in the neurons.[42] Ceramide accumulation can trigger apoptosis. When cultured hippocampal neurons were treated with a drug called ISP-1 that inhibits ceramide production, they were more resistant to being killed by Aβ and HNE.[42] Collectively, these findings suggest an important role for ceramide production resulting from MAOS in the pathogenesis of AD.

MAOS in PD

Dopaminergic neurons in the substantia nigra contain higher amounts of iron than most other cells in the brain, and these neurons accumulate even greater amounts of

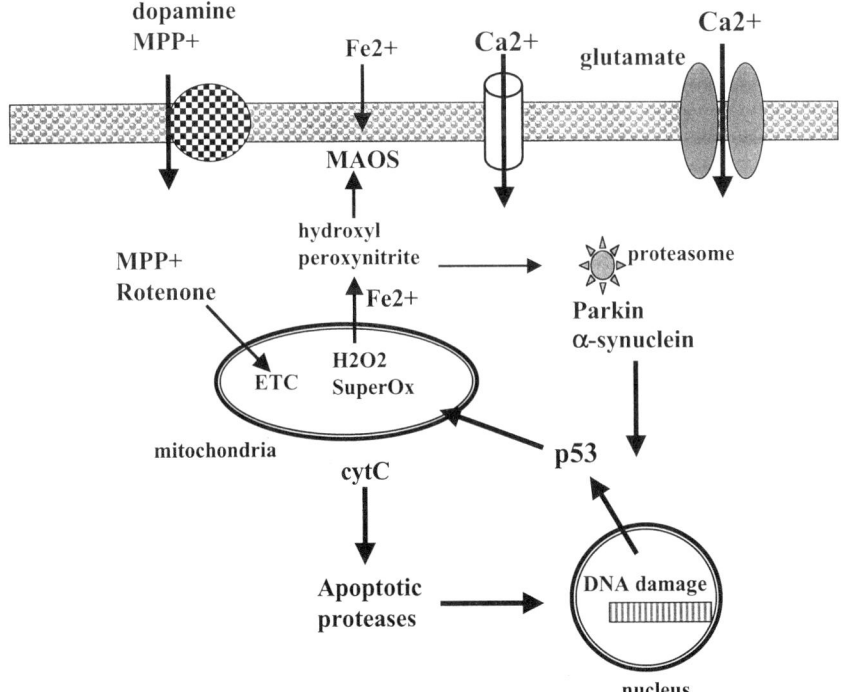

FIGURE 2. Possible mechanisms responsible for degeneration of dopaminergic neurons in PD. Terms: cytC, cytochrome C; ETC, electron transport chain; MPP+, 1-methyl-4-phenyl-pyridinium ion; SuperOx, superoxide anion radical. See references 49–52 for discussion of these mechanisms.

iron in PD patients.[43–45] Analyses of brain tissue samples have revealed evidence for MAOS in PD, with levels of HNE being increased in dopaminergic neurons.[46] Toxins that inhibit mitochondrial complex I cause PD-like pathology and motor dysfunction in mice, monkeys, and humans.[47,48] These toxins, which include MPTP and rotenone, increase the production of superoxide and peroxynitrite, and induce membrane lipid peroxidation in neurons[49,50] (FIG. 2). Several different antioxidants have been reported to protect dopaminergic neurons from being damaged by the toxins,[51] suggesting a pivotal role for MAOS in these toxin-based models of PD.

We recently performed experiments aimed at establishing the effects of dietary iron availability on the pathogenic process and functional outcome in a mouse model of PD.[52] Mice were fed diets containing low (4 ppm), normal (48 ppm), or high (400 ppm) amounts of iron for 6 weeks and were then administered MPTP. In mice fed the normal iron diet, MPTP reduced striatal dopamine levels and impaired motor behavior. Low dietary iron increased serum total iron binding capacity and provided protection against the adverse effects of MPTP on motor function.[52] Iron supplementation alone significantly impaired dopamine synthesis and motor behavior, and was lethal when combined with MPTP treatment. Dietary iron also affected striatal levels of HNE, sphingomyelin C16:0, and ceramide C16:0, each of which have been linked to neuronal damage and death. In studies of cultured neurons, iron and the active form of MPTP, MPP+, acted synergistically to induce the formation of mitochondrial reactive oxygen species and the proapoptotic protein p53, resulting in reduced neuronal viability. Collectively, these findings suggest that dietary iron can exacerbate MAOS, resulting in dysfunction and death of dopaminergic neurons in an animal model of PD.

MAOS in HD

Neurons in the striatum (caudate and putamen) and cerebral cortex are the most severely affected in HD. Associated with the neurodegenerative process is evidence of increased oxidative stress, including membrane lipid peroxidation.[53] Data suggest that iron levels are increased in the striatum of HD patients.[54] The antioxidant creatine significantly retarded the neurodegenerative process and increased the survival of huntingtin mutant mice.[55] In our studies, when huntingtin mutant mice are maintained on a dietary restriction regimen (intermittent fasting), the neurodegenerative process is attenuated and their survival time is increased by more than 10%.[21] Although it is not known if dietary restriction suppresses MAOS in HD, this is a possibility because it has been shown that dietary restriction reduces circulating free iron levels[56] and that dietary restriction can protect neurons against insults that induce MAOS.[57] It is not yet known how mutant huntingtin promotes oxidative stress in neurons, but some data suggest the mechanism involves perturbed signal transduction and/or protein aggregation.[58] A possible specific mechanism involves impairment of brain-derived neurotrophic factor (BDNF) gene transcription by aggregates of mutant huntingtin, resulting in loss of neuroprotective BDNF signaling (FIG. 3). It will be interesting to determine the possible roles of iron and copper in the pathogenesis of HD. Because excellent mouse models of HD are now available, it will be possible to test the effects of dietary iron and copper, and of chelators of these metals, on disease pathogenesis.

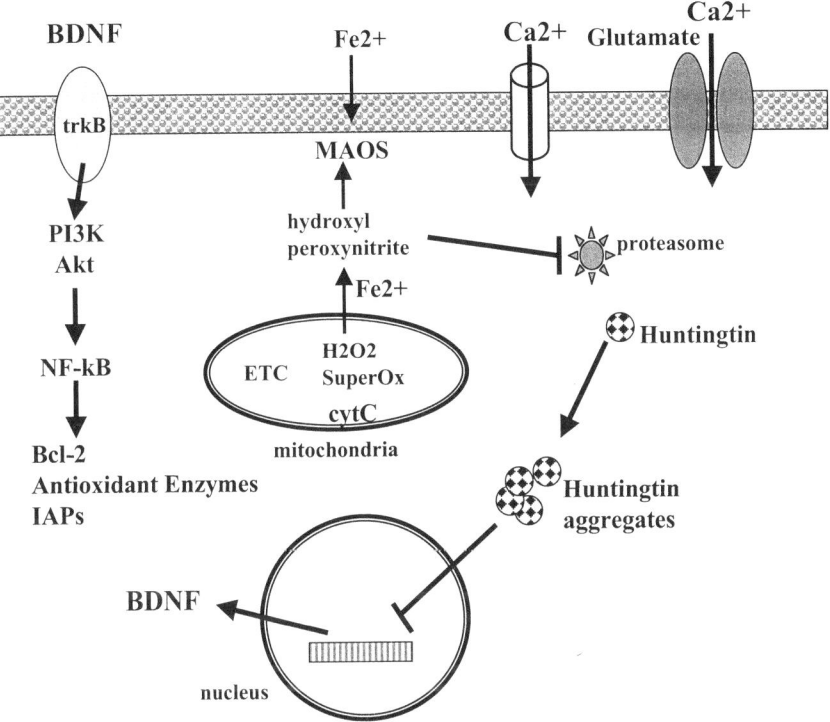

FIGURE 3. Possible mechanisms whereby mutant huntingtin causes the degeneration of striatal and cortical neurons in HD. Terms: BDNF, brain-derived neurotrophic factor; cytC, cytochrome C; ETC, electron transport chain; IAP, inhibitor of apoptosis proteins; SuperOx, superoxide anion radical. See references 10, 21, and 58 for discussion of these mechanisms.

MAOS in ALS

We showed that levels of free and protein-bound HNE are greatly increased in spinal cord tissue and cerebrospinal fluid from ALS patients compared to levels in control subjects.[12,59] Several proteins were modified by HNE including apolipoprotein E, a protein with the ability to protect neurons against MAOS by scavenging HNE.[60] Exposure of cultured motor neurons to HNE resulted in depletion of choline acetyltransferase, a key enzyme in acetylcholine synthesis, and ultimately killed the neurons.[61] Analyses of spinal cord tissue from Cu/Zn-SOD mutant mice revealed increased HNE levels in association with degenerating motor neurons. Others have shown that antioxidants, including vitamin E, can significantly suppress the neurodegenerative process and increase the survival of Cu/Zn-SOD mutant mice.[62] MAOS may promote excitotoxicity in motor neurons, a form of cell death that data suggest occurs in ALS.[63]

With regard to the role of MAOS in pathogenesis of inherited ALS, data suggest that Cu/Zn-SOD mutations alter the conformation of the protein in a way that increases the access of the copper to hydrogen peroxide, thereby resulting in the production of hydroxyl radical,[64] a potent inducer of MAOS that can disrupt ion homeostasis in motor neurons.[65] Environmental toxins, including metals, have been implicated in ALS, with the evidence being particularly strong for a population in Guam that suffered an unusually high incidence of a form of ALS that appears not to have a genetic cause.[66]

Membrane lipids are altered in ALS and may contribute to the degeneration of motor neurons in this disease. Mass spectrometry analyses of various lipids in membrane samples of spinal cord tissue from ALS patients and control subjects revealed significant increases in levels of long-chain ceramides and cholesterol esters in the patients.[67] Similar changes in these lipids occurred in Cu/Zn-SOD mutant mice and the magnitude of the elevations of ceramide and cholesterol levels was positively associated with disease severity. These abnormalities in membrane lipids may be the result of increased MAOS because exposure of cultured motor neurons to oxidative insults resulted in increased levels of long-chain ceramides and cholesterol esters in the neurons.[67] When cultured motor neurons were treated with ISP-1 to inhibit ceramide production, they were more resistant to being killed by oxidative insults.[67] Collectively, these findings suggest an important role for ceramide production resulting from MAOS in the pathogenesis of ALS.

MAOS in Stroke

A stroke occurs when a blood vessel in the brain becomes occluded or ruptured, resulting in a severe reduction in the supply of oxygen and nutrients to the cells normally perfused by the affected blood vessel. Neurons that rely totally upon the affected blood vessel rapidly undergo necrosis after the stroke, whereas neurons in the surrounding region that receives blood from other vessels as well may die by apoptosis or they may survive. Rodent models of stroke, particularly those involving occlusion of the middle cerebral artery, have provided insight into the cascade of events that results in death of neurons in the ischemic penumbra.[68] A key role for MAOS in ischemic neuronal death is suggested by studies showing that antioxidants that suppress MAOS are effective in decreasing neuronal loss and improving functional outcome in rodent models of stroke.[69,70] Iron status has been linked to stroke,[71,72] and it was reported that the iron chelator, deferoxamine, was effective in reducing focal ischemic brain injury in a rodent model of stroke.[73]

Implications for Dietary and Pharmacological Interventions

High dietary iron intake is associated with an increasing number of prominent disorders, including cardiovascular disease and stroke.[74,75] The evidence that iron contributes to the pathogenesis of several different neurodegenerative conditions, some being described above, suggests that a reduction in iron intake may be an approach for reducing the risk of neurodegenerative disorders. Agents that chelate iron and/or copper have been shown to be beneficial in animal models of AD and PD,[76–78] and a clinical trial of a copper-chelating agent in AD patients is in progress.[79] A clue that metal chelators may be useful in AD patients came from a report that deferoxamine significantly improved cognitive function in AD patients.[80] It may also be possible to modulate the amounts or activities of the various proteins in the blood and brain that bind and transport iron and copper (FIG. 4).

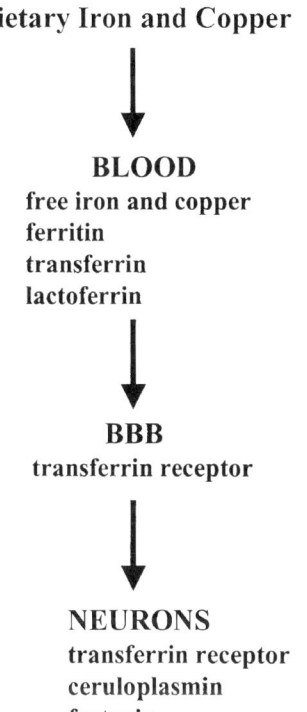

FIGURE 4. Examples of proteins that bind and transport iron and copper in the blood and the brain. Terms: BBB, blood-brain barrier; DMT-1, divalent metal transporter-1.

A second approach is to counteract the ability of iron and copper to promote MAOS. Vitamin E and estrogens are both quite effective at suppressing membrane lipid peroxidation and protecting neurons in experimental models of neurodegenerative disorders.[25,81–83] Additional antioxidants that suppress MAOS and that have proven to be neuroprotective in cell culture and/or animal models include coenzyme Q10, nordihydroguaiaretic acid.[84,85] Many novel antioxidants are being identified as being effective in reducing neuronal damage in cell culture and animal models of neurodegenerative disorders. Medicinal chemistry developed around such molecules promises to reveal agents with improved efficacy in suppressing MAOS and with improved ability to enter the central nervous system.

Overeating is a major risk factor for several prominent age-related diseases in humans, including cardiovascular disease and stroke, diabetes, and certain types of cancer. When rats or mice are maintained on a dietary restriction regimen (every-other-day fasting), we have found that neurons in their brains are more resistant to degeneration in models relevant to AD, PD, HD, and stroke.[86,87] We have gained insight into the cellular and molecular mechanisms responsible for the neuroprotective effects of DR. Neurons in the brains of rats or mice maintained on DR exhibit increased levels of BDNF and increased amounts of several stress resistance protein chaperones.[19,88–90] We had previously shown that BDNF can enhance antioxidant defenses in neurons and can protect them against oxidative insults, including insults such as Fe^{2+} that induce MAOS.[91]

Recent research has provided evidence that dietary folate serves an important neuroprotective function and that low dietary intake of folate is associated with elevated levels of homocysteine and increased risk of stroke, AD, and PD.[92] Homocysteine can cause DNA damage and can induce MAOS,[93,94] and dietary folate deficiency elevated homocysteine and caused degeneration of hippocampal neurons in APP mutant transgenic mice, an animal model of AD.[95] Dietary folate deficiency and homocysteine also endangered oligodendrocytes and promoted damage to white matter in PS1 mutant mice, another model relevant to AD.[96] Folate deficiency and elevated homocysteine levels have also been linked to the pathogenesis of PD.[92] Folic acid supplementation is therefore recommended as a means of reducing one's risk of neurodegenerative disorders.

REFERENCES

1. WISCHIK, C.M., C.R. HARRINGTON, E.B. MUKAETOVA-LADINSKA *et al.* 1992. Molecular characterization and measurement of Alzheimer's disease pathology: implications for genetic and environmental aetiology. Ciba Found. Symp. **169:** 268–293.
2. LOTHARIUS, J. & P. BRUNDIN. 2002. Pathogenesis of Parkinson's disease: dopamine, vesicles, and alpha-synuclein. Nat. Rev. Neurosci. **3:** 932–942.
3. REINER, A., R.L. ALBIN, K.D. ANDERSON *et al.* 1988. Differential loss of striatal projection neurons in Huntington's disease. Proc. Natl. Acad. Sci. USA **85:** 5733–5737.
4. HAVERKAMP, L.I., V. APPEL & S.H. APPEL. 1995. Natural history of amyotrophic lateral sclerosis in a database population: validation of a scoring system and a model for survival prediction. Brain **118:** 707–719.
5. MATTSON, M.P., S.W. BARGER, B. CHENG *et al.* 1993. β-Amyloid precursor protein metabolites and loss of neuronal calcium homeostasis in Alzheimer's disease. Trends Neurosci. **16:** 409–415.
6. MARK, R.J., E.M. BLANC & M.P. MATTSON. 1996. Amyloid β-peptide and oxidative cell injury in Alzheimer's disease. Mol. Neurobiol. **12:** 211–224.
7. CHAN, S.L., K. FURUKAWA & M.P. MATTSON. 2002. Presenilins and APP in neuritic and synaptic plasticity: implications for the pathogenesis of Alzheimer's disease. Neuromol. Med. **2:** 167–196.
8. JENNER, P. 2003. Oxidative stress in Parkinson's disease. Ann. Neurol. **53:** S26–S36.
9. MATTSON, M.P., W.A. PEDERSEN, W. DUAN *et al.* 1999. Cellular and molecular mechanisms underlying perturbed energy metabolism and neuronal degeneration in Alzheimer's and Parkinson's diseases. Ann. N.Y. Acad. Sci. **893:** 154–175.
10. PETERSEN, A. & P. BRUNDIN. 2002. Huntington's disease: the mystery unfolds? Int. Rev. Neurobiol. **53:** 315–339.
11. DIB, M. 2003. Amyotrophic lateral sclerosis: progress and prospects for treatment. Drugs **63:** 289–310.
12. PEDERSEN, W.A., W. FU, J.N. KELLER *et al.* 1998. Protein modification by the lipid peroxidation product 4-hydroxynonenal in spinal cord tissue of ALS patients. Ann. Neurol. **44:** 819–824.
13. YOUNG, A.B. 2003. Huntingtin in health and disease. J. Clin. Invest. **111:** 299–302.
14. ROGAEVA, E. 2002. The solved and unsolved mysteries of the genetics of early-onset Alzheimer's disease. Neuromol. Med. **2:** 1–10.
15. COOKSON, M.R. 2003. Pathways to parkinsonism. Neuron **37:** 7–10.
16. CUDKOWICZ, M.E., D. MCKENNA-YASEK, P.E. SAPP *et al.* 1997. Epidemiology of mutations in superoxide dismutase in amyotrophic lateral sclerosis. Ann. Neurol. **41:** 210–221.
17. SMITH, J.D. 2000. Apolipoprotein E4: an allele associated with many diseases. Ann. Med. **32:** 118–127.

18. BRUCE-KELLER, A.J., G. UMBERGER, R. MCFALL & M.P. MATTSON. 1999. Food restriction reduces brain damage and improves behavioral outcome following excitotoxic and metabolic insults. Ann. Neurol. **45:** 8–15.
19. DUAN, W. & M.P. MATTSON. 1999. Dietary restriction and 2-deoxyglucose administration improve behavioral outcome and reduce degeneration of dopaminergic neurons in models of Parkinson's disease. J. Neurosci. Res. **57:** 195–206.
20. ZHU, H., Q. GUO & M.P. MATTSON. 1999. Dietary restriction protects hippocampal neurons against the death-promoting action of a presenilin-1 mutation. Brain Res. **842:** 224–229.
21. DUAN, W., Z. GUO, H. JIANG et al. 2003. Dietary restriction normalizes glucose metabolism and BDNF levels, slows disease progression, and increases survival in huntingtin mutant mice. Proc. Natl. Acad. Sci. USA **100:** 2911–2916.
22. PEDERSEN, W.A. & M.P. MATTSON. 1999. No benefit of dietary restriction on disease onset or progression in ALS Cu/Zn-SOD mutant mice. Brain Res. **833:** 117–120.
23. KENNARD, M.L., H. FELDMAN, T. YAMADA & W.A. JEFFRIES. 1996. Serum levels of the iron binding protein p97 are elevated in Alzheimer's disease. Nat. Med. **2:** 1230–1235.
24. LOGROSCINO, G., K. MARDER, J. GRAZIANO et al. 1997. Altered systemic iron metabolism in Parkinson's disease. Neurology **49:** 714–717.
25. GOODMAN, Y. & M.P. MATTSON. 1994. Secreted forms of beta-amyloid precursor protein protect hippocampal neurons against amyloid beta-peptide–induced oxidative injury. Exp. Neurol. **128:** 1–12.
26. YOUDIM, M.B., D. BEN-SHACHER & P. RIEDERER. 1993. The possible role of iron in the etiopathology of Parkinson's disease. Mov. Disord. **8:** 1–12.
27. MARKESBERY, W.R. & J.M. CARNEY. 1999. Oxidative alterations in Alzheimer's disease. Brain Pathol. **9:** 133–146.
28. LOVELL, M.A., W.D. EHMANN, M.P. MATTSON & W.R. MARKESBERY. 1997. Elevated 4-hydroxynonenal in ventricular fluid in Alzheimer's disease. Neurobiol. Aging **18:** 457–461.
29. CONNOR, J.R., S.L. MENZIES, S.M. ST. MARTIN et al. 1992. A histochemical study of iron, transferrin, and ferritin in Alzheimer's diseased brains. J. Neurosci. Res. **31:** 75–83.
30. LIM, G.P., T. CHU, F. YANG et al. 2001. The curry spice curcumin reduces oxidative damage and amyloid pathology in an Alzheimer transgenic mouse. J. Neurosci. **21:** 8370–8377.
31. MATTSON, M.P., H. ZHU, J. YU & M.S. KINDY. 2000. Presenilin-1 mutation increases neuronal vulnerability to focal ischemia *in vivo*, and to hypoxia and glucose deprivation in cell culture: involvement of perturbed calcium homeostasis. J. Neurosci. **20:** 1358–1364.
32. LAFONTAINE, M.A., M.P. MATTSON & D.A. BUTTERFIELD. 2002. Oxidative stress in synaptosomal proteins from mutant presenilin-1 knock-in mice: implications for familial Alzheimer's disease. Neurochem. Res. **27:** 417–421.
33. BUTTERFIELD, D.A., K. HENSLEY, M. HARRIS et al. 1994. β-Amyloid peptide free radical fragments initiate synaptosomal lipoperoxidation in a sequence-specific fashion: implications to Alzheimer's disease. Biochem. Biophys. Res. Commun. **200:** 710–715.
34. KELLER, J.N., R.J. MARK, A.J. BRUCE et al. 1997. 4-Hydroxynonenal, an aldehydic product of membrane lipid peroxidation, impairs glutamate transport and mitochondrial function in synaptosomes. Neuroscience **80:** 685–696.
35. MARK, R.J., M.A. LOVELL, W.R. MARKESBERY et al. 1997. A role for 4-hydroxynonenal in disruption of ion homeostasis and neuronal death induced by amyloid β-peptide. J. Neurochem. **68:** 255–264.
36. HENSLEY, K., J.M. CARNEY, M.P. MATTSON et al. 1994. A model for β-amyloid aggregation and neurotoxicity based on free radical generation by the peptide: relevance to Alzheimer's disease. Proc. Natl. Acad. Sci. USA **91:** 3270–3274.
37. HUANG, X., C.S. ATWOOD, M.A. HARTHSHOM et al. 1999. The A beta peptide of Alzheimer's disease directly produces hydrogen peroxide through metal ion reduction. Biochemistry **38:** 7609–7616.
38. MARK, R.J., K. HENSLEY, D.A. BUTTERFIELD & M.P. MATTSON. 1995. Amyloid β-peptide impairs ion-motive ATPase activities: evidence for a role in loss of neuronal Ca^{2+} homeostasis and cell death. J. Neurosci. **15:** 6239–6249.

39. MARK, R.J., Z. PANG, J.W. GEDDES *et al.* 1997. Amyloid β-peptide impairs glucose uptake in hippocampal and cortical neurons: involvement of membrane lipid peroxidation. J. Neurosci. **17:** 1046–1054.
40. BLANC, E.M., J.F. KELLY, R.J. MARK & M.P. MATTSON. 1997. 4-Hydroxynonenal, an aldehydic product of lipid peroxidation, impairs signal transduction associated with muscarinic acetylcholine and metabotropic glutamate receptors: possible action on $G\alpha_{q/11}$. J. Neurochem. **69:** 570–580.
41. BLANC, E.M., J.N. KELLER, S. FERNANDEZ & M.P. MATTSON. 1998. 4-Hydroxynonenal, a lipid peroxidation product, inhibits glutamate transport in astrocytes. Glia **22:** 149–160.
42. CUTLER, R.G., J. KELLEY, K. STORIE *et al.* 2004. Involvement of oxidative stress–induced abnormalities in ceramide and cholesterol metabolism in brain aging and Alzheimer's disease. Proc. Natl. Acad. Sci. USA. In press.
43. DEXTER, D.T., F.R. WELLS, F. AGID *et al.* 1987. Increased nigral iron content in postmortem parkinsonian brain. Lancet **2:** 1219–1220.
44. SOFIC, E., P. RIEDERER, H. HEINSEN *et al.* 1988. Increased iron (III) and total iron content in postmortem substantia nigra of parkinsonian brain. J. Neural Transm. **74:** 199–205.
45. GORELL, J.M., R.J. ORDIDGE, G.G. BROWN *et al.* 1995. Increased iron-related MRI contrast in the substantia nigra in Parkinson's disease. Neurology **45:** 1138–1143.
46. YORITAKA, A., N. HATTORI, K. UCHIDA *et al.* 1996. Immunohistochemical detection of 4-hydroxynonenal protein adducts in Parkinson disease. Proc. Natl. Acad. Sci. USA **93:** 2696–2701.
47. BETARBET, R., T.B. SHERER, G. MACKENZIE *et al.* 2000. Chronic systemic pesticide exposure reproduces features of Parkinson's disease. Nat. Neurosci. **3:** 1301–1306.
48. SCHNEIDER, J.S., A. YUWILER & C.H. MARKHAM. 1987. Selective loss of subpopulations of ventral mesencephalic dopaminergic neurons in the monkey following exposure to MPTP. Brain Res. **411:** 144–150.
49. DUAN, W., Z. ZHANG, D.M. GASH & M.P. MATTSON. 1999. Participation of prostate apoptosis response-4 in degeneration of dopaminergic neurons in models of Parkinson's disease. Ann. Neurol. **46:** 587–597.
50. DUAN, W., Z. ZHU, B. LADENHEIM *et al.* 2002. p53 inhibitors preserve dopamine neurons and motor function in experimental parkinsonism. Ann. Neurol. **52:** 597–606.
51. JENNER, P. & C.W. OLANOW. 1998. Understanding cell death in Parkinson's disease. Ann. Neurol. **44:** S72–S84.
52. LEVENSON, C.W., W. DUAN, R.G. CUTLER *et al.* 2003. Dietary iron availability influences motor function and damage to dopaminergic neurons in a mouse model of Parkinson's disease. J. Neurochem. Submitted.
53. PEREZ-SEVERIANO, F., C. RIOS & J. SEGOVIA. 2000. Striatal oxidative damage parallels the expression of a neurological phenotype in mice transgenic for the mutation of Huntington's disease. Brain Res. **862:** 234–237.
54. BARTZOKIS, G., J. CUMMINGS, S. PERLMAN *et al.* 1999. Increased basal ganglia iron levels in Huntington disease. Arch. Neurol. **56:** 569–574.
55. ANDREASSEN, O.A., A. DEDEOGLU, R.J. FERRANTE *et al.* 2001. Creatine increases survival and delays motor symptoms in a transgenic animal model of Huntington's disease. Neurobiol. Dis. **8:** 479–491.
56. KRETSCH, M.J., A.K. FONG, M.W. GREEN & H.L. JOHNSON. 1998. Cognitive function, iron status, and hemoglobin concentration in obese dieting women. Eur. J. Clin. Nutr. **52:** 512–518.
57. GUO, Z., A. ERSOZ, D.A. BUTTERFIELD & M.P. MATTSON. 2000. Beneficial effects of dietary restriction on cerebral cortical synaptic terminals: preservation of glucose transport and mitochondrial function after exposure to amyloid β-peptide and oxidative and metabolic insults. J. Neurochem. **75:** 314–320.
58. MATTSON, M.P. 2002. Accomplices to neuronal death. Nature **415:** 377–379.
59. SMITH, R.G., Y.K. HENRY, M.P. MATTSON & S.H. APPEL. 1998. Presence of 4-hydroxynonenal in cerebrospinal fluid of patients with sporadic amyotrophic lateral sclerosis. Ann. Neurol. **44:** 696–699.
60. PEDERSEN, W.A., S.L. CHAN & M.P. MATTSON. 2000. A mechanism for the neuroprotective effect of apolipoprotein E: isoform-specific modification by the lipid peroxidation product 4-hydroxynonenal. J. Neurochem. **74:** 1426–1433.

61. PEDERSEN, W.A., N. CASHMAN & M.P. MATTSON. 1999. The lipid peroxidation product 4-hydroxynonenal impairs glutamate and glucose transport and choline acetyltransferase activity in NSC-19 motor neuron cells. Exp. Neurol. **155:** 1–10.
62. GURNEY, M.E., F.B. CUTTING, P. ZHAI et al. 1996. Benefit of vitamin E, riluzole, and gabapentin in a transgenic model of familial amyotrophic lateral sclerosis. Ann. Neurol. **39:** 147–157.
63. PEDERSEN, W.A., H. LUO, W. FU et al. 2000. Evidence that Par-4 participates in motor neuron death in amyotrophic lateral sclerosis. FASEB J. **14:** 913–924.
64. WIEDAU-PAZOS, M., J.J. GOTO, S. RABIZADEH et al. 1996. Altered reactivity of superoxide dismutase in familial amyotrophic lateral sclerosis. Science **271:** 515–518.
65. KRUMAN, I.I., W.A. PEDERSEN, J.E. SPRINGER & M.P. MATTSON. 1999. ALS-linked Cu/Zn-SOD mutation increases vulnerability of motor neurons to excitotoxicity by a mechanism involving increased oxidative stress and perturbed calcium homeostasis. Exp. Neurol. **160:** 28–39.
66. ROMAN, G.C. 1996. Neuroepidemiology of amyotrophic lateral sclerosis: clues to aetiology and pathogenesis. J. Neurol. Neurosurg. Psychiatry **61:** 131–137.
67. CUTLER, R.G., W.A. PEDERSEN, S. CAMANDOLA et al. 2002. Evidence that accumulation of ceramides and cholesterol esters mediates oxidative stress–induced death of motor neurons in ALS. Ann. Neurol. **52:** 448–457.
68. MATTSON, M.P. 2000. Apoptosis in neurodegenerative disorders. Nat. Rev. Mol. Cell Biol. **1:** 120–129.
69. KELLER, J.N., M.S. KINDY, F.W. HOLTSBERG et al. 1998. Mitochondrial manganese superoxide dismutase prevents neural apoptosis and reduces ischemic brain injury: suppression of peroxynitrite production, lipid peroxidation, and mitochondrial dysfunction. J. Neurosci. **18:** 687–697.
70. YU, Z.F., A.J. BRUCE-KELLER, Y. GOODMAN & M.P. MATTSON. 1998. Uric acid protects neurons against excitotoxic and metabolic insults in cell culture, and against focal ischemic brain injury in vivo. J. Neurosci. Res. **53:** 613–625.
71. SOROND, F.A. & R.R. RATAN. 2000. Ironing-out mechanisms of neuronal injury under hypoxic-ischemic conditions and potential role of iron chelators as neuroprotective agents. Antioxid. Redox Signal **2:** 421–436.
72. DAVALOS, A., J. CASTILLO, J. MARRUGAT et al. 2000. Body iron stores and early neurologic deterioration in acute cerebral infarction. Neurology **54:** 1568–1574.
73. PALMER, C., R.L. ROBERTS & C. BERO. 1994. Deferoxamine posttreatment reduces ischemic brain injury in neonatal rats. Stroke **25:** 1039–1045.
74. SEMPOS, C.T. & A.C. LOOKER. 1999. Iron status and the risk of coronary heart disease. Nutr. Metab. Cardiovasc. Dis. **9:** 294–303.
75. GARIBALLA, S.E. 2000. Nutritional factors in stroke. Br. J. Nutr. **84:** 5–17.
76. CHERNY, R.A., C.S. ATWOOD, M.E. XILINAS et al. 2001. Treatment with a copper-zinc chelator markedly and rapidly inhibits beta-amyloid accumulation in Alzheimer's disease transgenic mice. Neuron **30:** 665–676.
77. YOUDIM, M.B., M. GASSEN, A. GROSS et al. 2000. Iron chelating, antioxidant, and cytoprotective properties of dopamine receptor agonist; apomorphine. J. Neural Transm. Suppl. **58:** 83–96.
78. KAUR, D. et al. 2003. Genetic or pharmacological iron chelation prevents MPTP-induced neurotoxicity in vivo: a novel therapy for Parkinson's disease. Neuron **37:** 899–909.
79. BUSH, A.I. 2003. The metallobiology of Alzheimer's disease. Trends Neurosci. **26:** 207–214.
80. CRAPPER MCLACHLAN, D.R., A.J. DALTON, T.P. KRUCK et al. 1991. Intramuscular desferrioxamine in patients with Alzheimer's disease. Lancet **337:** 1304–1308.
81. HARA, H., H. KATO & K. KOGURE. 1990. Protective effect of alpha-tocopherol on ischemic neuronal damage in the gerbil hippocampus. Brain Res. **510:** 335–338.
82. GOODMAN, Y., A.J. BRUCE, B. CHENG & M.P. MATTSON. 1996. Estrogens attenuate and corticosterone exacerbates excitotoxicity, oxidative injury, and amyloid beta-peptide toxicity in hippocampal neurons. J. Neurochem. **66:** 1836–1844.
83. KELLER, J.N. & M.P. MATTSON. 1997. 17β-Estradiol attenuates oxidative impairment of synaptic Na^+/K^+-ATPase activity, glucose transport, and glutamate transport induced by amyloid β-peptide and iron. J. Neurosci. Res. **50:** 522–530.

84. GOODMAN, Y., M.R. STEINER, S.M. STEINER & M.P. MATTSON. 1994. Nordihydroguaiaretic acid protects hippocampal neurons against amyloid β-peptide toxicity, and attenuates free radical and calcium accumulation. Brain Res. **654:** 171–176.
85. SHULTS, C.W., D. OAKES & K. KIEBURTZ. 2002. Effects of coenzyme Q10 in early Parkinson disease: evidence of slowing of the functional decline. Arch. Neurol. **59:** 1541–1550.
86. PROLLA, T.A. & M.P. MATTSON. 2001. Molecular mechanisms of brain aging and neurodegenerative disorders: lessons from dietary restriction. Trends Neurosci. **24:** S21–S31.
87. MATTSON, M.P., S.L. CHAN & W. DUAN. 2002. Modification of brain aging and neurodegenerative disorders by genes, diet, and behavior. Physiol. Rev. **82:** 637–672.
88. YU, Z.F. & M.P. MATTSON. 1999. Dietary restriction and 2-deoxyglucose administration reduce focal ischemic brain damage and improve behavioral outcome: evidence for a preconditioning mechanism. J. Neurosci. Res. **57:** 830–839.
89. DUAN, W., Z. GUO & M.P. MATTSON. 2001. Brain-derived neurotrophic factor mediates an excitoprotective effect of dietary restriction in mice. J. Neurochem. **76:** 619–626.
90. LEE, J., W. DUAN & M.P. MATTSON. 2002. Evidence that brain-derived neurotrophic factor is required for basal neurogenesis and mediates, in part, the enhancement of neurogenesis by dietary restriction in the hippocampus of adult mice. J. Neurochem. **82:** 1367–1375.
91. MATTSON, M.P., M.A. LOVELL, K. FURUKAWA & W.R. MARKESBERY. 1995. Neurotrophic factors attenuate glutamate-induced accumulation of peroxides, elevation of $[Ca^{2+}]_i$ and neurotoxicity, and increase antioxidant enzyme activities in hippocampal neurons. J. Neurochem., **65:** 1740–1751.
92. MATTSON, M.P. & T.B. SHEA. 2003. Folate and homocysteine metabolism in neural plasticity and neurodegenerative disorders. Trends Neurosci. **26:** 137–146.
93. HO, P.I. et al. 2001. Homocysteine potentiates beta-amyloid neurotoxicity: role of oxidative stress. J. Neurochem. **78:** 249–253.
94. KRUMAN, I., S.L. CHAN, C. CULMSEE et al. 2000. Homocysteine elicits a DNA damage response in neurons resulting in apoptosis and hypersensitivity to excitotoxicity. J. Neurosci. **20:** 6920–6926.
95. KRUMAN, I.I., T.S. KUMARAVEL, A. LOHANI et al. 2002. Folic acid deficiency and homocysteine impair DNA repair in hippocampal neurons and sensitize them to amyloid toxicity in experimental models of Alzheimer's disease. J. Neurosci. **22:** 1752–1762.
96. PAK, K., S.L. CHAN & M.P. MATTSON. 2003. Homocysteine and folate deficiency sensitize oligodendrocytes to the cell death-promoting effects of a presenilin-1 mutation and amyloid beta-peptide. Neuromol. Med. **3:** 119–127.
97. LU, C., S.L. CHAN, N. HAUGHEY et al. 2001. Selective and biphasic effect of the membrane lipid peroxidation product 4-hydroxy-2,3-nonenal on N-methyl-D-aspartate channels. J. Neurochem. **78:** 577–589.
98. LU, C., S.L. CHAN, W. FU & M.P. MATTSON. 2002. The lipid peroxidation product 4-hydroxynonenal facilitates opening of voltage-dependent Ca^{2+} channels in neurons by increasing protein tyrosine phosphorylation. J. Biol. Chem. **277:** 24368–24375.
99. KELLER, J.N., K.B. HANNI, S.P. GABBITA et al. 1999. Oxidized lipoproteins increase reactive oxygen species formation in microglia and astrocyte cell lines. Brain Res. **830:** 10–15.
100. KELLER, J.N., K.B. HANNI, W.A. PEDERSEN et al. 1999. Opposing actions of native and oxidized lipoprotein on motor neuron-like cells. Exp. Neurol. **157:** 202–210.
101. MATTSON, M.P. 1998. Modification of ion homeostasis by lipid peroxidation: roles in neuronal degeneration and adaptive plasticity. Trends Neurosci. **21:** 53–57.

Iron, Atherosclerosis, and Neurodegeneration

A Key Role for Cholesterol in Promoting Iron-Dependent Oxidative Damage?

WEI-YI ONG[a] AND BARRY HALLIWELL[b]

Departments of [a]Anatomy and [b]Biochemistry, National University of Singapore, Singapore 119260

> ABSTRACT: This article reviews the roles and interactions of iron, atherosclerosis, and neurodegeneration. It highlights the importance of cholesterol in promoting iron-dependent oxidative damage. An intriguing possibility is that hypercholesterolemia can increase brain iron load and both the aggregation of beta-amyloid and the ability of iron on plaques to catalyze oxidative damage. This could explain why hypercholesterolemia is a risk factor for Alzheimer's disease. Further work is necessary to study the mechanism of increased iron transport across the blood brain barrier in atherosclerosis.
>
> KEYWORDS: Alzheimer's disease; atherosclerosis; brain; cholesterol; iron; hypercholesterolemia; statin; transferrin

INTRODUCTION

Iron is a key component of many proteins essential for brain metabolism, including cytochromes a, b, and c; cytochrome oxidase; and the iron-sulfur complexes of the mitochondrial electron transport chain. Iron is also a cofactor for tyrosine hydroxylase and tryptophan hydroxylase, enzymes involved in synthesis of neurotransmitters;[1] succinate dehydrogenase; and aconitase of the TCA cycle.[2] The highest levels of iron are found in the motor system, with the globus pallidus, substantia nigra pars reticulata, red nucleus, and myelinated fibers of the putamen showing the highest staining reactivity. Iron is predominantly accumulated in glial cells.[3,4] Cerebral contusion, cortical laceration, intracerebral hematoma formation, and hemorrhagic cortical infarction cause extravasation of red blood cells, followed by hemolysis, decompartmentalization of iron,[5] and possible increased incidence of epilepsy.[6,7]

Iron is transported in the bloodstream largely in the form of transferrin-bound iron. The binding of transferrin to transferrin receptors on brain endothelial cells is important in the transport of iron across the blood brain barrier, and the majority of iron transport in the brain occurs as the result of receptor-mediated endocytosis of iron-transferrin by capillary endothelial cells, followed by release of iron from transferrin within the cell, recycling of transferrin to the blood, and transport of iron into

---

Address for correspondence: Dr. Wei-Yi Ong, Department of Anatomy, National University of Singapore, Singapore 119260. Voice: +65-68743662; fax: +65-67787643.
antongwy@nus.edu.sg

the brain.[8] In addition to transferrin, metal transporters including IREG (MTP1) and divalent metal transporter-1 (DMT1, also known as Nramp2) are important in iron transport across the blood brain interface. DMT1 is an iron influx transporter, whereas IREG is an iron efflux transporter. Both DMT1[9–11] and IREG[11] have been localized to astrocytes, and astrocytic end-feet around blood vessels, suggesting that these cells also play a role in iron transport across the blood brain interface.

Cholesterol is an integral component of cell membranes and is also used for steroid synthesis. It is transported in the bloodstream as free cholesterol and cholesterol esters within various lipoproteins. Low density lipoprotein (LDL) provides cholesterol to the tissues. In the liver and most extrahepatic tissues, LDL binds to its receptor at the cell surface and is recruited into clathrin-coated endocytotic vesicles. After endocytosis, LDL enters the endosomal/lysosomal system, where cholesteryl ester, a major lipid found in LDL, is hydrolyzed by the enzyme acid lipase.[12] The cholesterol that is formed by the hydrolytic action of acid lipase is transported from the endosome/lysosome to the plasma membrane or to the endoplasmic reticulum for reesterification. Cholesterol leaves as well as enters cells. The cholesterol that exits cells is absorbed into high density lipoprotein (HDL). The enzyme β-hydroxyl-β-methylglutaryl-CoA (HMG-CoA) reductase is the rate-limiting enzyme in cholesterol synthesis.[12]

Atherosclerosis is associated with high lipid content in the blood, particular LDL. This extremely widespread disease predisposes to myocardial infarction, cerebral thrombosis, and other serious illness. The plasma concentrations of LDL and HDL as well as of total cholesterol need to be considered when evaluating patients. Individuals with elevated LDL have a higher incidence of atherosclerosis and its complications, whereas individuals with elevated HDL have a lower incidence.[12]

The central nervous system accounts for only 2% of the whole body mass, but contains almost a quarter of the unesterified cholesterol present in the whole individual. This sterol is largely present in two pools comprising the cholesterol in the plasma membranes of glial cells and neurons and the cholesterol present in the specialized membranes of myelin. From 0.02% (human) to 0.4% (mouse) of the cholesterol in these pools turns over each day so that the absolute flux of sterol across the brain is only approximately 0.9% as rapid as the turnover of cholesterol in the whole body.[13] The input of cholesterol into the central nervous system comes almost entirely from *in situ* synthesis, and there is currently little evidence for the net transfer of sterol from the plasma into the brain of the fetus, newborn, or adult. In the steady state in the adult, an equivalent amount of cholesterol must move out of the brain and this output is partly accounted for by the formation and excretion of 24*S*-hydroxycholesterol. Cholesterol from damaged or dying neurons is converted to 24*S*-hydroxycholesterol by cholesterol 24-hydroxylase. The oxysterol is subsequently transferred across the blood brain barrier, transported to the liver by LDL, and excreted as bile acids. Most of plasma 24*S*-hydroxycholesterol is derived from brain cholesterol catabolism.[14] An increased synthesis of cholesterol occurs in neurons, after excitotoxic injury.[15] Indirect evidence also suggests that there is a large turnover of cholesterol between neurons and glial cells in the central nervous system.[13]

There is accumulating evidence that hypertension or hypercholesterolemia could cause damage to endothelial cells. An increase in plasma von Willebrand factor and vascular endothelial growth factor and lowered soluble endothelial growth factor receptor, signifying increased endothelial cell damage, has been observed in hypertension and vascular disease.[16–18] These indices are particularly abnormal in high-risk hypertensives, with three or more clinical risk factors, compared to a low-risk group. After 6 months of intensive cardiovascular risk factor management (including antihypertensive/hypercholesterolemia treatment), both von Willebrand factor and vascular endothelial growth factor were reduced. In addition, soluble endothelial growth factor receptor levels decreased with increased cardiovascular risk and were increased after pharmacological management. These data suggest abnormal angiogenesis, as well as endothelial damage in hypertension, which can be beneficially improved with pharmacologic treatment and cardiovascular risk factor management.[18]

IRON AND ATHEROSCLEROSIS

Iron and copper ions appear important in the oxidative modification of LDL, even though it is not known exactly how LDL is oxidized in the subendothelial space.[19] Advanced human atherosclerotic lesions contain redox-active metal ions that can promote LDL oxidation.[20] Incubation with the iron chelator deferoxamine or pretreatment of LDL-treated endothelial cells with deferoxamine suppressed ferritin-induced intracellular oxidation of LDL.[21] Oxidative modification of LDL by metal ions results in numerous changes to the lipoprotein particles that are potentially pro-atherogenic[22] and can ultimately result in uncontrolled uptake of these lipoproteins by macrophages and other cells.[23] Most of the cell types present in the intima of vessel walls can promote the oxidation of LDL *in vitro*. The presence of transition metals appears to be an absolute requirement for cell-mediated oxidation of lipoproteins,[24] and there is always the possibility that these effects are artifacts due to the presence of added or contaminating metal ions in cell culture media.[25]

The mechanisms by which cells oxidize LDL in the presence of transition metals are not entirely clear. It appears, however, that the cells can accelerate ongoing metal-catalyzed LDL oxidation. This ability is (at least in part) due to redox cycling of transition metals by the action of $O_2^{\bullet-}$ and of transplasma membrane electron transport systems.[26,27] Lipid peroxidation of fatty acid chains leads to the formation of an array of different products with diverse and powerful biological activities, among which are a variety of different aldehydes.[28] The major products of lipid peroxidation, lipid hydroperoxides,[29] undergo fission in the presence of transition metals, giving rise to the formation of peroxyl and alkoxyl radicals as well as short-chain, unesterified aldehydes,[29,30] or a second class of aldehydes still esterified to the parent lipid, called core aldehydes.[31,32] These reactive aldehydes are cytotoxic. In addition to the fatty acid chains, the cholesterol moiety of cholesterol esters may also be oxidized, yielding cholesterol oxidation products, or oxysterols. The latter may be generated within tissues by nonenzymatic oxidation of cholesterol by reactive oxygen species, including peroxyl or alkoxyl radicals derived from lipid peroxidation,[33,34] or by enzymatic catalysis. Oxysterols such as 7β-hydroxycholesterol and 7-ketocholesterol have been shown to induce damage to the cells of the vascular wall[35] and to monocytic cells,[36] and may be important mediators of the toxic and inflammatory processes in atherosclerosis.[37]

As well as being present in advanced lesions,[20] iron may also be involved in the early stages of atherosclerosis. Iron deposition in the endothelium and in the media is closely associated with the progression of atherosclerosis.[38] A sevenfold increase in iron and an average of nearly twofold increase in phosphorus in the atherosclerotic lesion has been observed in aortae from hypercholesterolemic rabbits, compared with healthy tissue, using the sensitive method of nuclear microscopy.[39,40] The finding of iron accumulation in vessel walls of atherosclerosis is consistent with magnetic resonance imaging studies, which showed accumulation of supramagnetic iron oxide particles in atherosclerotic plaques. Intravenously administered supramagnetic iron particles accumulate in atherosclerotic plaques of hypercholesterolemic rabbits and produced a pronounced focal signal loss in the vessel wall on magnetic resonance images.[41] The histologic results of these experiments indicate that the iron particles are taken up by endothelial cells along the aortic wall. This finding was interpreted as a regional increase in endothelial iron permeability and endothelial dysfunction. Interestingly, iron particles were also observed in the subendothelial intima.[41] As in hypercholesterolemic rabbits, pronounced signal loss was also observed in the wall of the aorta and pelvic arteries seen in part of an elderly patient population after intravenous supramagnetic iron oxide administration, indicating that the supramagnetic iron oxide particles also accumulated in human atherosclerotic plaques.[42]

Attempts to influence atherosclerosis in hypercholesterolemic rabbits by manipulating iron levels have yielded mixed results. Iron supplementation with iron dextran has been reported to result in a lowering of plasma cholesterol levels and a decrease in the rate of development of atherosclerosis, whereas iron depletion by phlebotomy had no significant effect.[43] These observations are somewhat contrary to reports that weekly bleeding decreased iron uptake into the artery wall and delayed the onset of atherogenesis in hypercholesterolemic rabbits,[40] as well as the finding of a trend to a reduction in the progression of atherosclerosis lesions in hypercholesterolemic rabbits treated with an iron chelator, desferrioxamine.[44] A key question may be to what extent these various manipulations have affected the iron content of the vessel wall, and what proportion of iron present is redox-active rather than stored in inactive forms.[45]

IRON, CHOLESTEROL, AND THE BRAIN

It is possible that the increased permeability to iron, such as that reported for endothelial cells in the aorta or other larger vessels in hypercholesterolemia, might also occur in endothelial cells of brain blood vessels. The blood vessels in the cerebral cortex are mostly capillaries one to two cells thick,[46,47] and any iron that has crossed the endothelial layer would find itself in contact with astrocytic end-feet or the neuropil.

This possibility, of increased passage of iron into the brain parenchyma in hypercholesterolemia, is consistent with several studies, which showed increased lipid peroxidation in the brain in hypercholesterolemic rabbits. Brain malondialdehyde in rabbits treated with cholesterol was higher than those fed a normal diet.[48] Rabbits fed with an atherogenic diet containing 1% cholesterol appeared to show increased lipid peroxidation in tissues including the brain, as measured by thiobarbituric acid reactive substances (TBARS). The increase in TBARS was reduced by dietary supplementation with olive oil.[49] Cholesterol feeding has also been shown to cause a significant increase in the lipid peroxide concentrations of plasma, erythrocytes, liver, and brain as measured by increased amounts of TBARS, in a recent study.[50] A link between ApoE deficiency, hypercholesterolemia, and increased brain lipid peroxidation has also been reported. ApoE-deficient mice develop hypercholesterolemia on a normal diet, and the brains of these mice showed an age-dependent increase in

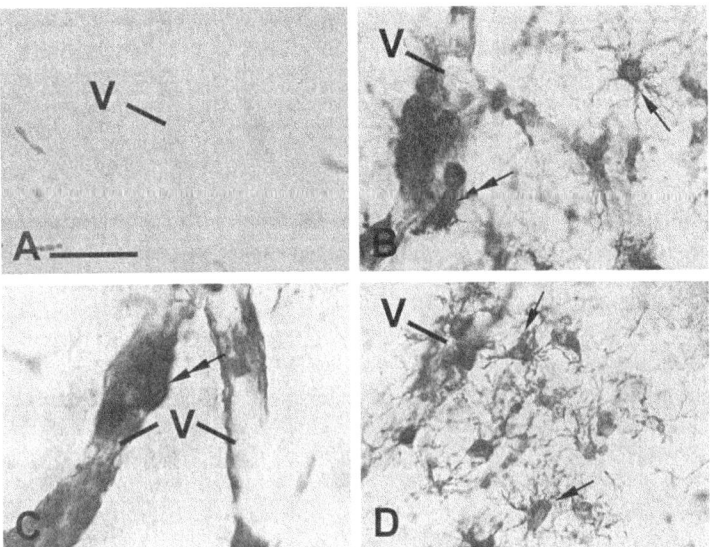

FIGURE 1. Sections of the frontal cortex of a rabbit fed with normal diet (A) or rabbits fed with diet containing 1% cholesterol (B for 8 weeks; C,D for 12 weeks). The sections were reacted with Perls' stain for ferric iron. (A) Very few iron-containing cells are present around blood vessels (V) in the normal rabbit. (B,C,D) Large numbers of Perls' stained cells are observed around blood vessels (V) in the cholesterol-fed rabbit. These include cells with few processes that are closely adherent to the vessels, characteristic of "perivascular cells" or perivascular macrophages (B,C: double-arrows), and cells in the neuropil with thin and straight processes, characteristic of oligodendrocytes (B,D: arrows). The dense iron staining in oligodendrocytes is likely to be a reflection of increased iron levels in the extracellular space in the hypercholesterolemic rabbits. For quantitative details, see ref. 57. Scale: 40 µm.

levels of lipid peroxidation products, isoprostane F2α-III and isoprostane F2α-VI, compared with wild-type C57Bl/6 mice.[51] Isoprostanes are far more reliable markers of lipid peroxidation than such parameters as TBARS.[52,53] In humans, blood cholesterol levels have been reported to be significantly higher in men, but not in women, with an E4 allele as opposed to those without the allele.[54]

In a recent study, we examined some of the possible effects of hypercholesterolemia on the brain, using the cholesterol-fed New Zealand White rabbit as a model. This is a widely used model to study atherosclerosis, but effects of a high cholesterol diet on the brains of these animals have not been examined previously. An increase in the number of iron-positive cells, identified as oligodendrocytes, was observed in the brain parenchyma, in rabbits treated with a high cholesterol diet for 8 weeks. At this time, no neuronal death was observed, indicating that the increased iron did not occur secondary to neuronal injury, as is known to happen after kainate-induced neurodegeneration.[55,56] No heme oxygenase-1 or bilirubin immunoreactivity was also observed in the brains in these rabbits, indicating that the iron accumulation did not occur as a consequence of increased breakdown of heme.[57] [See FIG. 1.]

How can this rise in iron be explained? We postulate that increased cholesterol could result in increased endocytosis of iron in endothelial cells and increased transport across the blood brain interface in the hypercholesterolemic rabbits. Endocytosis occurs by clathrin-independent as well as clathrin-dependent mechanisms.[58–60] Clathrin-dependent endocytosis is a better defined process and is responsible for the rapid uptake of, for example, hormones, growth factors, and transferrin. The interaction of the molecules involved in this process has been investigated both *in vivo* and *in vitro*, resulting in characterization of a number of important proteins such as clathrin, adaptors, and dynamin.[61] Less is known about clathrin-independent endocytosis, but different forms seem to exist; for instance, both dynamin-dependent and dynamin-independent mechanisms have been reported.[62–65] Caveolae are nonclathrin-containing microinvaginations of the cell plasma membrane involved in cell transport and metabolism as well as signal transduction. These functions depend on the presence of integral proteins called caveolins. Caveolin-1 has been localized to rat and human brain microvessels.[66]

In response to free cholesterol, the gene expression of caveolin is upregulated by the sterol-responsive elements from the caveolin promoter.[67,68] On the other hand, treatment by a statin, atorvastatin, resulted in reduced caveolin abundance in cultured endothelial cells *in vitro*.[69] Caveolin-1 is essential for caveolae formation in adipocytes, endothelial cells, fibroblasts, and epithelial cells.[70] It has been postulated that the increased caveolin expression in hypercholesterolemia may be responsible for the increased number of caveolae, leading to increased transendothelial transport of molecules including LDL.[71] Although it is possible that an increased caveolin-1 expression and possible increase in number of caveolae might account for some of the increased iron transport into the brain in hypercholesterolemia, it seems unlikely to account for the bulk of the increase since transferrin receptors are present in clathrin-coated vesicles,[72] whereas caveolae are nonclathrin microinvaginations.

In addition to caveolin-1-dependent caveolae formation, cholesterol also plays a role in clathrin-dependent endocytosis. The role of cholesterol in endocytosis of transferrin was investigated in HEp-2 and other cell lines by using methyl-β-cyclodextrin (MCD) to selectively extract cholesterol from the plasma membrane.[73] MCD treatment strongly inhibited endocytosis of transferrin and EGF, whereas endocytosis of ricin was less affected. The inhibition of transferrin endocytosis was completely reversible. On removal of MCD, it was restored by continued incubation of the cells, even in serum-free medium. Interestingly, the recovery in serum-free medium was inhibited by addition of lovastatin, which prevents cholesterol synthesis, but endocytosis recovered when a water-soluble form of cholesterol was added together with lovastatin. Although clathrin-independent (and caveolae-independent) endocytosis still operates after removal of cholesterol, these results indicate that cholesterol is essential for the formation of clathrin-coated endocytic vesicles.[73] It is postulated that this clathrin-dependent transferrin endocytosis could be increased by an increase in cholesterol, leading to increased transport of iron across endothelial cells. Further work is necessary to investigate this possibility.

ATHEROSCLEROSIS, CHOLESTEROL, AND NEURODEGENERATION

Atherosclerosis reduces the resilience of large arteries and can induce hypertension. In turn, sustained hypertension worsens atherosclerosis, seemingly driving it into the walls of small branch arteries (0.5 mm or less). The coats of the vessels become impregnated in a hyaline-lipid material (lipohyalinosis). The segment so affected may weaken and allow the formation of a small dissecting aneurysm, which some neuropathologists hold responsible for hypertensive stroke. Lipohyalinosis may also result in thrombosis of small penetrating arteries in the brain.[74] The atheromatous process in brain arteries runs parallel to, but is somewhat less severe than that in the aorta, heart, and lower limbs. As with coronary and peripheral atherosclerosis, individuals with low blood levels of HDL and high levels of LDL cholesterol are particularly disposed to cerebral atherosclerosis.[75]

HMG-CoA reductase inhibitors (statins) have been shown to result in significant reductions in risk of ischemic stroke among patients with established vascular disease. The Long-term Intervention with Pravastatin in Ischemic Disease (LIPID) study reported a 19% relative risk reduction in strokes in the pravastatin-treated group.[76] These observations are supported by meta-analyses demonstrating that statin therapy lowers stroke risk by ~30%.[77,78] In two recent trials, the Myocardial Ischemia Reduction with Aggressive Cholesterol Lowering (MIRACL)[79,80] and the Heart Protection Study,[81] further strong prospective evidence was provided that statin therapy represents a powerful means of reducing stroke incidence. Statins may exert a plaque-stabilizing effect in aorta and carotid arteries similar to that observed in coronary arteries,[82] thereby preventing plaque disruption and thromboembolism.[83,84]

In addition to prevention of stroke, statins have been reported to influence the outcome of established strokes.[85,86] Treatment with mevastatin reduced infarct size and improved neurological deficits in a dose- and time-dependent manner.[85] There could be several ways in which statins could exert a beneficial effect in the stroke: (1) statins might upregulate brain endothelial nitric oxide synthase, thus increasing blood flow and improving neurologic outcome;[85,87] (2) statins might have anti-inflammatory actions, including reducing interleukin-1 and tumor necrosis factor production, thereby inhibiting the consequences of stroke;[88] (3) statins may enhance brain plasticity by increasing vascular endothelial growth factor, angiogenesis, endogenous cell proliferation, and neurogenesis, thus improving functional outcome after stroke;[86] (4) statins may exert their beneficial effect by inhibiting increased brain cholesterol synthesis after neuronal injury. Increased synthesis of cholesterol and increased formation of the toxic cholesterol oxidation product, 7-ketocholesterol, have been shown in degenerating hippocampal neurons after kainate-induced excitotoxic injury.[15] The use of statins may prevent such an increase of endogenously synthesized cholesterol and may lead to a reduction in the formation of cholesterol oxidation products. The usefulness of statins to reduce brain cholesterol synthesis is shown by the ability of these drugs to lower plasma levels of a product of brain cholesterol metabolism, 24S-hydroxycholesterol.[14]

Atherosclerotic (multi-infarct) dementia or vascular dementia refers to an impairment of intellectual function due to multiple infarcts. A history of one or more strokes can usually be elicited, but the degree of focal neurologic deficit may be small in comparison to the cognitive impairment and personality changes.[74] It has been suggested that many forms of vascular dementia might be preventable, with good control of vascular risk factors in middle age.[89] A recent study has shown significant improvement of cerebral vasomotor reactivity by statin therapy in patients with cerebral small-vessel disease.[90]

Alzheimer's disease is a progressive, neurodegenerative disorder characterized by senile plaques and neurofibrillary components. Beta-amyloid is a principal component of senile plaques and is thought to be central to the pathogenesis of the disease. The Alzheimer's disease brain is under significant oxidative stress,[91] and the beta-amyloid peptide is known to cause oxidative stress *in vitro*. Recent evidence suggests that the toxic species in beta-amyloid toxicity is prefibrillar beta-amyloid or oligomeric beta-amyloid, and not the beta-amyloid fibril.[92] Experimental and clinical studies suggest that there is a pathophysiological relation between amyloid and cholesterol levels. Elevated amyloid-42 levels and the 4 allele of apolipoprotein E (ApoE4) are risk factors for Alzheimer's disease.[93] In addition, ApoE4 has been correlated with increased risk for atherosclerosis[94] and amyloid plaque formation.[95] In transgenic mice overexpressing amyloid precursor protein, cholesterol levels inversely regulate amyloid production and Alzheimer's pathology in the mouse brain. ApoE plays an important role in the central nervous system as a cholesterol transport protein and a chaperone to promote conversion of beta-amyloid to an insoluble form.[95] There is evidence of increased turnover of cholesterol in the brain in Alzheimer's disease. Elevated concentrations of 24*S*-hydroxycholesterol were observed in the cerebrospinal fluid of Alzheimer's disease patients. These findings could indicate increased cholesterol turnover in the central nervous system during neurodegeneration. The observed influence of the apolipoprotein E epsilon 4 (ApoE4) allele on cerebrospinal fluid 24*S*-hydroxycholesterol concentrations with a gene-dosage effect also suggests the existence of a link between ApoE and brain cholesterol metabolism. The elevation of 24*S*-hydroxycholesterol appears to occur early in the disease process since patients with mild cognitive impairment also had increased CSF concentrations of this compound.[96] In addition to ApoE, recent evidence shows that metal ions, especially Zn, Cu, and Fe, are also important in oligomerization and toxicity of beta-amyloid. Compounds that interfere with metal-ion binding to beta-amyloid dissolve brain deposits *in vitro*, and one such compound, clioquinol, inhibits beta-amyloid deposition in the Tg2576 mouse model for Alzheimer's disease and might be useful clinically.[97] Iron bound to plaques may be redox-active and able to catalyze oxidative damage.[91,97]

In addition to the effects of brain cholesterol, there is also evidence that plasma cholesterol levels may be related to Alzheimer's disease. Epidemiologic studies show that patients with elevated serum cholesterol have an increased incidence of Alzheimer's disease.[98,99] Recently, even mild hypercholesterolemia has been shown to be an early risk factor for the development of Alzheimer's amyloid pathology.[100] Autopsy cases of patients older than 40 years were reviewed and correlated with cholesterolemia and presence or absence of amyloid deposition (amyloid-positive vs. amyloid-negative subjects) and cholesterolemia and amyloid load. Cholesterolemia correlates with presence of amyloid deposition in the youngest subjects (40 to 55 years) with early amyloid deposition (diffuse type of senile plaques). In this group, increases in cholesterolemia from 181 to 200 almost tripled the odds for developing amyloid, independent of the ApoE isoform.[100] Since very little cholesterol is thought to cross the blood brain barrier,[13] this suggests that changes in the blood vessels in atherosclerosis might affect the progression of Alzheimer's disease. These changes could possibly involve increased iron transport across endothelial cells, as described after hypercholesterolemia.[57]

Statins have been shown to have beneficial effects on Alzheimer's disease.[101,102] The mechanism by which statins could exert these beneficial effects is being investigated and includes studies on the relationship between cholesterol and beta-amyloid formation,[103–108] or studies on the reduction of oxysterols after statin treatment.[14] It is probably significant that a hydrophilic statin that crosses the blood brain barrier poorly (pravastatin) has been reported to be as effective as one that crosses the blood brain barrier readily (lovastatin) in decreasing the risk of Alzheimer's disease.[102] The finding that atorvastatin (which crosses the blood brain barrier poorly) was able to markedly attenuate beta-amyloid deposition in the PSAPP transgenic mouse model of Alzheimer's amyloidosis[109] also suggests that it is not necessary for the statins to cross the blood brain barrier to influence beta-amyloid deposition. In addition, lowering of blood cholesterol by means other than inhibition of HMG-CoA reductase appears to have a beneficial effect on brain beta-amyloid. Thus, treatment with the cholesterol-lowering drug, BM15.766, which inhibits 7-dehydrocholesterol reductase, the enzyme catalyzing the last step of cholesterol biosynthesis, was shown to reduce plasma cholesterol, brain beta-amyloid peptides, and beta-amyloid load in transgenic mice exhibiting an Alzheimer's beta-amyloid phenotype.[110]

CONCLUSIONS

The above observations suggest that a significant proportion of the beneficial effect of statins in Alzheimer's disease is mediated by reductions in serum cholesterol level, with probably additional effects on brain cholesterol synthesis, in the case of statins that cross the blood brain barrier. The mechanism by which this could be accomplished is not known. However, an intriguing possibility, in view of the reported effects of metal ions on beta-amyloid aggregation[97] and our finding of increased iron staining in the brains of hypercholesterolemic rabbits,[57] is that the reduction in hypercholesterolemia could have reduced brain iron load and both the aggregation of beta-amyloid and the ability of iron on plaques to catalyze oxidative damage. Further work is necessary to study the mechanism of increased iron transport across the blood brain barrier in atherosclerosis.

ACKNOWLEDGMENTS

This work was supported by the National University of Singapore Academic Research Fund/Singapore Totalizator Board/Office of Life Sciences (Grant R-183-000-072-112/650/432).

REFERENCES

1. CHEN, Q., J.L. BEARD & B.C. JONES. 1995. Abnormal rat brain monoamine metabolism in iron deficiency anemia. J. Nutr. Biochem. **5:** 486–493.
2. CONNOR, J.R., S.L. MENZIES, J.R. BURDO & P.J. BOYER. 2001. Iron and iron management proteins in neurobiology. Pediatr. Neurol. **25:** 118–129.
3. CONNOR, J.R., B.S. SNYDER, J.L. BEARD *et al.* 1992. Regional distribution of iron and iron-regulatory proteins in the brain in aging and Alzheimer's disease. J. Neurosci. Res. **31:** 327–335.
4. MORRIS, C.M., J.M. CANDY, A.E. OAKLEY *et al.* 1992. Histochemical distribution of non-haem iron in the human brain. Acta Anat. **144:** 235–257.
5. WAGNER, K.R., F.R. SHARP, T.D. ARDIZZONE *et al.* 2003. Heme and iron metabolism: role in cerebral hemorrhage. J. Cereb. Blood Flow Metab. **23:** 629–652.

6. WILLMORE, L.J. 1990. Post-traumatic epilepsy: cellular mechanisms and implications for treatment. Epilepsia **31:** S67–S73.
7. ONG, W.Y., B.K.C. ONG, A.A. FAROOQUI *et al.* 2003. Iron and epilepsy. *In* Metal Ions and Neurodegenerative Disorders, pp. 365–398. World Scientific. Singapore.
8. MOOS, T. & E.H. MORGAN. 2000. Transferrin and transferrin receptor function in brain barrier systems. Cell. Mol. Neurobiol. **20:** 77–95.
9. WANG, X.S., W.Y. ONG & J.R. CONNOR. 2001. A light and electron microscopic study of the iron transporter protein DMT-1 in the monkey cerebral neocortex and hippocampus. J. Neurocytol. **30:** 353–360.
10. WANG, X.S., W.Y. ONG & J.R. CONNOR. 2002. A light and electron microscopic study of divalent metal transporter-1 distribution in the rat hippocampus, after kainate-induced neuronal injury. Exp. Neurol. **177:** 193–201.
11. JEONG, S.Y. & S. DAVID. 2003. Glycosylphosphatidylinositol-anchored ceruloplasmin is required for iron efflux from cells in the central nervous system. J. Biol. Chem. **278:** 27144–27148.
12. GANONG, W.F. 2001. Review of Medical Physiology. Twentieth edition. McGraw–Hill. New York.
13. DIETSCHY, J.M. & S.D. TURLEY. 2001. Cholesterol metabolism in the brain. Curr. Opin. Lipidol. **12:** 105–112.
14. VEGA, G.L., M.F. WEINER, A.M. LIPTON *et al.* 2003. Reduction in levels of 24S-hydroxycholesterol by statin treatment in patients with Alzheimer disease. Arch. Neurol. **60:** 510–515.
15. ONG, W.Y., E.W.S. GOH, X.R. LU *et al.* 2003. Increase in cholesterol and cholesterol oxidation products, and role of cholesterol oxidation products in kainate-induced neuronal injury. Brain Pathol. **13:** 250–262.
16. BLANN, A.D., T. NAQVI, M. WAITE & C.N. MCCOLLUM. 1993. Von Willebrand factor and endothelial damage in essential hypertension. J. Hum. Hypertens. **7:** 107–111.
17. BELGORE, F.M., A.D. BLANN, F.L. LI-SAW-HEE *et al.* 2001. Plasma levels of vascular endothelial growth factor and its soluble receptor (SFlt-1) in essential hypertension. Am. J. Cardiol. **87:** 805–807.
18. FELMEDEN, D.C., C.G. SPENCER, F.M. BELGORE *et al.* 2003. Endothelial damage and angiogenesis in hypertensive patients: relationship to cardiovascular risk factors and risk factor management. Am. J. Hypertens. **16:** 11–20.
19. CHISOLM, G.M. & D. STEINBERG. 2000. The oxidative modification hypothesis of atherogenesis: an overview. Free Radical Biol. Med. **28:** 1815–1826.
20. SMITH, C., M.J. MITCHINSON, O.I. ARUOMA & B. HALLIWELL. 1992. Stimulation of lipid peroxidation and hydroxyl-radical generation by the contents of human atherosclerotic lesions. Biochem. J. **286:** 901–905.
21. SATOH, T. & O. TOKUNAGA. 2002. Intracellular oxidative modification of low density lipoprotein by endothelial cells. Virchows Arch. **440:** 410–417.
22. STEINBERG, D., S. PARTHASARATHY, T.E. CAREW *et al.* 1989. Beyond cholesterol: modifications of low-density lipoprotein that increase its atherogenicity. N. Engl. J. Med. **320:** 915–924.
23. GOLDSTEIN, J.L. & M.S. BROWN. 1977. The low-density lipoprotein pathway and its relation to atherosclerosis. Annu. Rev. Biochem. **46:** 897–930.
24. CHAIT, A. & J.W. HEINECKE. 1994. Lipoprotein modification: cellular mechanisms. Curr. Opin. Lipidol. **5:** 365–370.
25. HALLIWELL, B. 2003. Oxidative stress in cell culture: an under-appreciated problem? FEBS Lett. **540:** 3–6.
26. CRANE, F.L., I.L. SUN, M.G. CLARK *et al.* 1985. Transplasma-membrane redox systems in growth and development. Biochim. Biophys. Acta **811:** 233–264.
27. GARNER, B., D. VAN REYK, R.T. DEAN & W. JESSUP. 1997. Direct copper reduction by macrophages: its role in low density lipoprotein oxidation. J. Biol. Chem. **272:** 6927–6935.
28. WITZ, G. 1989. Biological interaction of α,β-unsaturated aldehydes. Free Radical Biol. Med. **7:** 333–349.
29. TERAO, J. 1990. Reactions of lipid hydroperoxides. *In* Membrane Lipid Oxidation, pp. 219–238. CRC Press. Boca Raton, FL.

30. GIROTTI, A.W. 1998. Lipid hydroperoxide generation, turnover, and effector action in biological systems. J. Lipid Res. **39:** 1529–1542.
31. ITABE, H. 1998. Oxidized phospholipids as a new landmark in atherosclerosis. Prog. Lipid Res. **37:** 181–207.
32. KARTEN, B., H. BOECHZELT, P.M. ABUJA et al. 1999. Macrophage-enhanced formation of cholesteryl ester–core aldehydes during oxidation of low density lipoprotein. J. Lipid Res. **40:** 1240–1253.
33. SMITH, L.L. 1981. Cholesterol Autoxidation. Plenum. New York
34. SMITH, L.L. & B.H. JOHNSON. 1989. Biological activites of oxysterols. Free Radical Biol. Med. **7:** 285–332.
35. LIZARD, G., S. GUELDRY, O. SORDET et al. 1998. Glutathione is implied in the control of 7-ketocholesterol-induced apoptosis, which is associated with radical oxygen species production. FASEB J. **12:** 1651–1663.
36. CLARE, K., S.J. HARDWICK, K.L.H. CARPENTER et al. 1995. Toxicity of oxysterols to human monocytes-macrophages. Atherosclerosis **118:** 67–75.
37. LEONARDUZZI G., B. SOTTERO & G. POLI. 2002. Oxidized products of cholesterol: dietary and metabolic origin, and proatherosclerotic effects (review). J. Nutr. Biochem. **13:** 700–710.
38. FERRARI, M., G.S. WERNER, J. RIEBER et al. 2001. No influence of hemochromatosis-related gene mutations on restenosis rate in a retrospective study of 137 patients after coronary stent implantation. Int. J. Cardiovasc. Intervent. **4:** 181–186.
39. THONG, P.S., M. SELLEY & F. WATT. 1996. Elemental changes in atherosclerotic lesions using nuclear microscopy. Cell. Mol. Biol. **42:** 103–110.
40. PONRAJ, D., J. MAKJANIC, P.S. THONG et al. 1999. The onset of atherosclerotic lesion formation in hypercholesterolemic rabbit is delayed by iron depletion. FEBS Lett. **459:** 218–222.
41. SCHMITZ, S.A., S.E. COUPLAND, R. GUST et al. 2000. Superparamagnetic iron oxide–enhanced MRI of atherosclerotic plaques in Watanabe hereditable hyperlipidemic rabbits. Invest. Radiol. **35:** 460–471.
42. SCHMITZ, S.A., M. TAUPITZ, S. WAGNER et al. 2001. Magnetic resonance imaging of atherosclerotic plaques using superparamagnetic iron oxide particles. J. Magn. Reson. Imaging **14:** 355–361.
43. DABBAGH, A.J., G.T. SHWAERY, J.F. KEANEY, JR. & B. FREI. 1997. Effect of iron overload and iron deficiency on atherosclerosis in the hypercholesterolemic rabbit. Arterioscler. Thromb. Vasc. Biol. **17:** 2638–2645.
44. REN, M., F. WATT, B.H. TAN & B. HALLIWELL. 2003. Correlation of iron and zinc levels with lesion depth in newly formed atherosclerotic lesions. Free Radical Biol. Med. **34:** 746–752.
45. HALLIWELL, B. & J.M. GUTTERIDGE. 1990. Role of free radicals and catalytic metal ions in human disease: an overview. Methods Enzymol. **186:** 1–85.
46. ZHANG, H.F., W.Y. ONG & S.K. LEONG. 1997. Ultrastructural changes in cerebral cortical blood vessels of immature and adult rats after a single episode of neonatal hypoxia. J. Hirnforsch. **38:** 81–97.
47. ZHANG, H.F., W.Y. ONG, S.K. LEONG & L.J. GAREY. 1997. Ultrastructural characteristics of blood vessels in the infant and adult human cerebral cortex. Histol. Histopathol. **12:** 85–97.
48. AYDEMIR, E.O., C. DUMAN, H.A. CELIK et al. 2000. Effects of defibrotide on aorta and brain malondialdehyde and antioxidants in cholesterol-induced atherosclerotic rabbits. Int. J. Clin. Lab. Res. **30:** 101–107.
49. DE LA CRUZ, J.P., L. QUINTERO, M.A. VILLALOBOS & F. SANCHEZ DE LA CUESTA. 2000. Lipid peroxidation and glutathione system in hyperlipidemic rabbits: influence of olive oil administration. Biochim. Biophys. Acta **1485:** 36–44.
50. GOKKUSU, C. & T. MOSTAFAZADEH. 2003. Changes of oxidative stress in various tissues by long-term administration of vitamin E in hypercholesterolemic rats. Clin. Chim. Acta **328:** 155–161.
51. PRACTICO, D., J. ROKACH & R.K. TANGIRALA. 1999. Brains of aged apolipoprotein E–deficient mice have increased levels of F2-isoprostanes, *in vivo* markers of lipid peroxidation. J. Neurochem. **73:** 736–741.

52. MORROW, J.D., B. FREI, A.W. LONGMIRE et al. 1995. Increase in circulating products of lipid peroxidation (F2-isoprostanes) in smokers. N. Engl. J. Med. **332:** 1198–1203.
53. PRACTICO, D., R. TANGIRALA, D. RADAR et al. 1998. Vitamin E suppresses isoprostane generation *in vivo* and reduces atherosclerosis in apoE-deficient mice. Nat. Med. **4:** 1189–1192.
54. DUPUY, A.M., E. MAS, K. RITCHIE et al. 2001. The relationship between apolipoprotein E4 and lipid metabolism is impaired in Alzheimer's disease. Gerontology **47:** 213–218.
55. ONG, W.Y., M.Q. REN, J. MAKJANIC et al. 1999. A nuclear microscopic study of elemental changes in the rat hippocampus after kainate-induced neuronal injury. J. Neurochem. **72:** 1574–1579.
56. WANG, X.S., W.Y. ONG & J.R. CONNOR. 2002. Increase in ferric and ferrous iron in the rat hippocampus with time after kainate-induced excitotoxic injury. Exp. Brain Res. **143:** 137–148.
57. ONG, W.Y., B. TAN, N. PAN et al. 2004. Increased iron staining in the cerebral cortex of cholesterol fed rabbits. Mech. Ageing Dev. In press.
58. VAN DEURS, B., O.W. PETERSEN, S. OLSNES & K. SANDVIG. 1989. The ways of endocytosis. Int. Rev. Cytol. **117:** 131–177.
59. SANDVIG, K. & B. VAN DEURS. 1994. Endocytosis without clathrin. Trends Cell Biol. **4:** 275–277.
60. SANDVIG, K. & B. VAN DEURS. 1996. Endocytosis, intracellular transport, and cytotoxic action of shiga toxin and ricin. Am. Physiol. Soc. **76:** 949–965.
61. SCHMID, S.L. 1997. Clathrin-coated vesicle formation and protein sorting: an integrated process. Annu. Rev. Biochem. **6:** 511–548.
62. DAMKE, H., T. BABA, D.E. WARNOCK & S.L. SCHMID. 1994. Induction of mutant dynamin specifically blocks endocytic coated vesicle formation. J. Cell Biol. **127:** 915–934.
63. DAMKE, H., T. BABA, A.M. VAN DER BLIEK & S.L. SCHMID. 1995. Clathrin-independent pinocytosis is induced in cells overexpressing a temperature-sensitive mutant of dynamin. J. Cell Biol. **131:** 69–80.
64. OH, P., D.P. MCINTOSH & J.E. SCHNITZER. 1998. Dynamin at the neck of caveolae mediates their budding to form transport vesicles by GTP-driven fission from the plasma membrane of endothelium. J. Cell Biol. **141:** 101–114.
65. HENLEY, J.R., E.W. KRUEGER, B.J. OSWALD & M.A. MCNIVEN. 1998. Dynamin-mediated internalization of caveolae. J. Cell Biol. **141:** 85–99.
66. VIRGINTINO, D., D. ROBERTSON, M. ERREDE et al. 2002. Expression of caveolin-1 in human brain microvessels. Neuroscience **115:** 145–152.
67. FIELDING, C.J., A. BIST & P.E. FIELDING. 1997. Caveolin mRNA levels are up-regulated by free cholesterol and down-regulated by oxysterols in fibroblast monolayers. Proc. Natl. Acad. Sci. USA **94:** 3753–3758.
68. BIST, A., P.E. FIELDING & C.J. FIELDING. 1997. Two sterol regulatory element–like sequences mediate up-regulation of caveolin gene transcription in response to low density lipoprotein free cholesterol. Proc. Natl. Acad. Sci. USA **94:** 10693–10698.
69. BROUET, A., P. SONVEAUX, C. DESSY et al. 2001. Hsp90 and caveolin are key targets for the proangiogenic nitric oxide–mediated effects of statins. Circ. Res. **89:** 866–873.
70. DRAB, M., P. VERKADE, M. ELGER et al. 2001. Loss of caveolae, vascular dysfunction, and pulmonary defects in caveolin-1 gene-disrupted mice. Science **293:** 2449–2452.
71. SIMIONESCU, M., A. GAFENCU & F. ANTOHE. 2002. Transcytosis of plasma macromolecules in endothelial cells: a cell biological survey. Microsc. Res. Tech. **57:** 269–288.
72. DAUTRY-VARSAT, A. 1986. Receptor-mediated endocytosis: the intracellular journey of transferrin and its receptor. Biochimie **68:** 375–381.
73. RODAL, S.K., G. SKRETTING, O. GARRED et al. 1999. Extraction of cholesterol with methyl-beta-cyclodextrin perturbs formation of clathrin-coated endocytic vesicles. Mol. Biol. Cell **10:** 961–974.
74. ADAMS, R., M. VICTOR & A.H. ROPPER. 1997. Principles of Neurology. Sixth edition. McGraw–Hill. New York.
75. NUBIOLA, A.R., L. MASANA, S. MASDEU & J. RUBIES-PRAT. 1981. High-density lipoprotein cholesterol in cerebrovascular disease. Arch. Neurol. **38:** 468.
76. WHITE, H.D., R.J. SIMES, N.E. ANDERSON et al. 2000. Pravastatin therapy and the risk of stroke. N. Engl. J. Med. **343:** 317–326.

77. BLAUW, G.J., A.M. LAGAAY, A.H. SMELT & R.G. WESTENDORP. 1997. Stroke, statins, and cholesterol: a meta-analysis of randomized, placebo-controlled, double-blind trials with HMG-CoA reductase inhibitors. Stroke **28:** 946–950.
78. CROUSE, J.R., III, R.P. BYINGTON, H.M. HOEN & C.D. FURBERG. 1997. Reductase inhibitor monotherapy and stroke prevention. Arch. Intern. Med. **157:** 1305–1310.
79. SCHWARTZ, G.G., A.G. OLSSON, M.D. EZEKOWITZ *et al.* 2001. Effects of atorvastatin on early recurrent ischemic events in acute coronary syndromes—the MIRACL study: a randomized controlled trial. JAMA **285:** 1711–1718.
80. WATERS, D.D., G.G. SCHWARTZ, A.G. OLSSON *et al.* 2002. Effects of atorvastatin on stroke in patients with unstable angina on non-Q-wave myocardial infarction: a Myocardial Ischemia Reduction with Aggressive Cholesterol Lowering (MIRACL) substudy. Circulation **106:** 1690–1695.
81. COLLINS, R., J. ARMITAGE, S. PARISH *et al.* 2003. MRC/BHF Heart Protection Study of cholesterol-lowering with simvastatin in 5963 people with diabetes: a randomised placebo-controlled trial. Lancet **361:** 2005–2016.
82. DELANTY, N. & C.J. VAUGHAN. 1997. Vascular effects of statins in stroke. Stroke **28:** 2315–2320.
83. AMARENCO, P., A. COHEN, C. TZOURIO *et al.* 1994. Atherosclerotic disease of the aortic arch and the risk of ischemic stroke. N. Engl. J. Med. **331:** 1474–1479.
84. DRESSLER, F.A., W.R. CRAIG, R. CASTELLO & A.J. LABOVITZ. 1998. Mobile aortic atheroma and systemic emboli: efficacy of anticoagulation and influence of plaque morphology on recurrent stroke. J. Am. Coll. Cardiol. **31:** 134–138.
85. AMIN-HANJANI, S., N.E. STAGLIANO, M. YAMADA *et al.* 2001. Mevastatin, an HMG-CoA reductase inhibitor, reduces stroke damage and upregulates endothelial nitric oxide synthase in mice. Stroke **32:** 980–986.
86. CHEN, J., Z.G. ZHANG, Y. LI *et al.* 2003. Statins induce angiogenesis, neurogenesis, and synaptogenesis after stroke. Ann. Neurol. **53:** 743–751.
87. VAUGHAN, C.J. 2003. Prevention of stroke and dementia with statins: effects beyond lipid lowering. Am. J. Cardiol. **91:** 23B–29B.
88. BALDUINI, W., E. MAZZONI, S. CARLONI *et al.* 2003. Prophylactic but not delayed administration of simvastatin protects against long-lasting cognitive and morphological consequences of neonatal hypoxic-ischemic brain injury, reduces interleukin-1β and tumor necrosis factor-α mRNA induction, and does not affect endothelial nitric oxide synthase expression. Stroke **34:** 2007–2012.
89. ERKINJUNTTI, T. & K. ROCKWOOD. 2003. Vascular dementia. Semin. Clin. Neuropsychiatry **8:** 37–45.
90. STERZER, P., F. MEINTZSCHEL, A. ROSLER *et al.* 2001. Pravastatin improves cerebral vasomotor reactivity in patients with subcortical small-vessel disease. Stroke **32:** 2817–2820.
91. BUTTERFIELD, D.A. 2002. Amyloid beta-peptide (1–42)–induced oxidative stress and neurotoxicity: implications for neurodegeneration in Alzheimer's disease brain—a review. Free Radical Res. **36:** 1307–1313.
92. DRAKE, J., C.D. LINK & D.A. BUTTERFIELD. 2003. Oxidative stress precedes fibrillar deposition of Alzheimer's disease amyloid beta-peptide (1–42) in a transgenic *Caenorhabditis elegans* model. Neurobiol. Aging **24:** 415–420.
93. CORDER, E.H., A.M. SAUNDERS, W.J. STRITTMATTER *et al.* 1993. Gene dose of apolipoprotein E type 4 allele and the risk of Alzheimer's disease in late onset families. Science **261:** 921–923.
94. HOFMAN, A., A. OTT, M.M. BRETELER *et al.* 1997. Atherosclerosis, apolipoprotein E, and prevalence of dementia and Alzheimer's disease in the Rotterdam Study. Lancet **349:** 151–154.
95. BALES, K.R., T. VERINA, R.C. DODEL *et al.* 1997. Lack of apolipoprotein E dramatically reduces amyloid beta-peptide deposition. Nat. Genet. **17:** 263–264.
96. PAPASSOTIROPOULOS, A., D. LUTJOHANN, M. BAGLI *et al.* 2002. 24S-Hydroxycholesterol in cerebrospinal fluid is elevated in early stages of dementia. J. Psychiatr. Res. **36:** 27–32.
97. BUSH, A.I. 2003. The metallobiology of Alzheimer's disease. Trends Neurosci. **26:** 207–214.

98. JARVIK, G.P., E.M. WIJSMAN, W.A. KUKULL *et al.* 1995. Interactions of apolipoprotein E genotype, total cholesterol level, age, and sex in prediction of Alzheimer's disease: a case-control study. Neurology **45**: 1092–1096.
99. NOTKOLA, I.L., R. SULKAVA, J. PEKKANEN *et al.* 1998. Serum total cholesterol, apolipoprotein E epsilon 4 allele, and Alzheimer's disease. Neuroepidemiology **17**: 14–20.
100. PAPPOLLA, M.A., T.K. BRYANT-THOMAS, D. HERBERT *et al.* 2003. Mild hypercholesterolemia is an early risk factor for the development of Alzheimer amyloid pathology. Neurology **61**: 199–205.
101. JICK, H., G.L. ZORNBERG, S.S. JICK *et al.* 2000. Statins and the risk of dementia. Lancet **356**: 1627–1631.
102. WOLOZIN, B., W. KELLMAN, P. RUOSSEAU *et al.* 2000. Decreased prevalence of Alzheimer disease associated with 3-hydroxy-3-methyglutaryl coenzyme A reductase inhibitors. Arch. Neurol. **57**: 1439–1443.
103. SPARKS, D.L., S.W. SCHEFF, J.C. HUNSAKER III *et al.* 1994. Induction of Alzheimer-like beta-amyloid immunoreactivity in the brains of rabbits with dietary cholesterol. Exp. Neurol. **126**: 88–94.
104. BODOVITZ, S. & W.L. KLEIN. 1996. Cholesterol modulates alpha-secretase cleavage of amyloid precursor protein. J. Biol. Chem. **271**: 4436–4440.
105. HOWLAND, D.S., S.P. TRUSKO, M.J. SAVAGE *et al.* 1998. Modulation of secreted beta-amyloid precursor protein and amyloid beta-peptide in brain by cholesterol. J. Biol. Chem. **273**: 16576–16582.
106. SIMONS, M., P. KELLER, B. DE STROOPER *et al.* 1998. Cholesterol depletion inhibits the generation of beta-amyloid in hippocampal neurons. Proc. Natl. Acad. Sci. USA **95**: 6460–6464.
107. REFOLO, L.M., B. MALESTER, J. LAFRANCOIS *et al.* 2000. Hypercholesterolemia accelerates the Alzheimer's amyloid pathology in a transgenic mouse model. Neurobiol. Dis. **7**: 321–331.
108. FASSBENDER, K., M. SIMONS, C. BERGMANN *et al.* 2001. Simvastatin strongly reduces levels of Alzheimer's disease beta-amyloid peptides A beta 42 and A beta 40 *in vitro* and *in vivo*. Proc. Natl. Acad. Sci. USA **98**: 5856–5861.
109. PENTACESKA, S.S., S. DEROSA, V. OLM *et al.* 2002. Statin therapy for Alzheimer's disease: will it work? J. Mol. Neurosci. **19**: 155–161.
110. REFOLO, L.M., M.A. PAPPOLLA, J. LAFRANCOIS *et al.* 2001. A cholesterol-lowering drug reduces beta-amyloid pathology in a transgenic mouse model of Alzheimer's disease. Neurobiol. Dis. **8**: 890–899.

Severity of Neurodegeneration Correlates with Compromise of Iron Metabolism in Mice with Iron Regulatory Protein Deficiencies

SOPHIA R. SMITH,[a] SHARON COOPERMAN,[a] TIM LAVAUTE,[a,b] NANCY TRESSER,[c] MANIK GHOSH,[a] ESTHER MEYRON-HOLTZ,[a] WILLIAM LAND,[a] HAYDEN OLLIVIERRE,[a] BERNARD JORTNER,[d] ROBERT SWITZER III,[e] ALBEE MESSING,[f] AND TRACEY A. ROUAULT[a,g]

[a]*National Institute of Child Health and Human Development, Cell Biology and Metabolism Branch, Bethesda, Maryland, USA*

[d]*Laboratory for Neurotoxicity Studies, Virginia Tech, Blacksburg, Virginia, USA*

[e]*NeuroScience Associates, Knoxville, Tennessee, USA*

[f]*Department of Pathobiological Sciences and Waisman Center, University of Wisconsin-Madison, Madison, Wisconsin, USA*

> ABSTRACT: In mammals, iron regulatory proteins 1 and 2 (IRP1 and IRP2) posttranscriptionally regulate expression of several iron metabolism proteins including ferritin and transferrin receptor. Genetically engineered mice that lack IRP2, but have the normal complement of IRP1, develop adult-onset neurodegenerative disease associated with inappropriately high expression of ferritin in degenerating neurons. Here, we report that mice that are homozygous for a targeted deletion of IRP2 and heterozygous for a targeted deletion of IRP1 (IRP1+/– IRP2–/–) develop a much more severe form of neurodegeneration, characterized by widespread axonopathy and eventually by subtle vacuolization in several areas, particularly in the substantia nigra. Axonopathy develops in white matter tracts in which marked increases in ferric iron and ferritin expression are detected. Axonal degeneration is significant and widespread before evidence for abnormalities or loss of neuronal cell bodies can be detected. Ultimately, neuronal cell bodies degenerate in the substantia nigra and some other vulnerable areas, microglia are activated, and vacuoles appear. Mice manifest gait and motor impairment at stages when axonopathy is pronounced, but neuronal cell body loss is minimal. These observations suggest that therapeutic strategies that aim to revitalize neurons by treatment with neurotrophic factors may be of value in IRP2–/– and IRP1+/– IRP2–/– mouse models of neurodegeneration.
>
> KEYWORDS: iron; neurodegeneration; IRP; substantia nigra; axonopathy; ferritin

[b]Present address: Tim LaVaute, University of Wisconsin, Madison, WI.
[c]Formerly of Neuroimmunology Branch, National Institute of Neurologic Disease and Stroke, Bethesda, MD.
[g]Address for correspondence: Dr. Tracey A. Rouault, NIH/NICHD/CBMB, Building 18T, Room 101, Bethesda, MD 20892. Voice: 301-496-7060; fax: 301-402-0078.
trou@helix.nih.gov

Iron is an essential element for function of the central nervous system since it participates in a variety of critical metabolic processes including oxygen transport, electron transport, and DNA synthesis. In the CNS, iron is required for biosynthesis of several neurotransmitters[1,2] and for the function of numerous enzymes required for oxidative phosphorylation, including iron-sulfur proteins and cytochromes. In normal animals, iron reproducibly accumulates in characteristic brain areas as animals mature.[3]

Accumulation of excess iron is hypothesized to play a role in pathogenesis of some neurodegenerative diseases. Increased iron has frequently been noted in brain regions that develop pathology in common diseases such as Parkinson's disease[4] and Alzheimer's disease,[5] as well as in rare conditions such as NBIA 1 disease (neuronal brain iron accumulation type 1, formerly known as Hallervorden-Spatz syndrome),[6] autosomal recessive juvenile parkinsonism,[7] multiple systems atrophy,[8] and Friedreich's ataxia.[9]

Recently, we reported that genetic ablation of iron regulatory protein 2 (IRP2–/–) in mice results in misregulation of iron metabolism in several tissues and is associated with adult-onset neurodegenerative disease.[10] IRP2 is one of two iron regulatory proteins (IRPs) responsible for regulation of iron homeostasis in mammalian cells (reviewed in refs. 11 and 12). Both IRPs are cytosolic proteins that bind to RNA stem-loop motifs known as IREs found in the 5′ UTR of ferritin H and L subunit transcripts, in the 3′ UTR of the transferrin receptor (TfR), and in several other iron metabolism gene transcripts. Each of the two IRPs, IRP1 and IRP2, binds to IREs when cells are depleted of iron. When IRPs bind to ferritin transcripts, they interfere with translation and decrease synthesis of ferritin protein; in contrast, when they bind to the TfR transcript, they protect the mRNA from degradation, leading to increased synthesis of TfR. The two IRPs are 58% homologous in sequence, but they have different mechanisms for sensing iron levels.[11,12]

Although both IRPs are expressed in the central nervous system, neurodegenerative disease does not develop in IRP1–/– animals (Meyron-Holtz et al., *EMBO J.*, in press). The unexpected absence of a neurological phenotype in IRP1–/– mice raised the possibility that IRP1 does not contribute to posttranscriptional regulation of iron metabolism in the brain. However, here we report that animals that lack both copies of IRP2 and one copy of IRP1 (IRP1+/– IRP2–/–) develop a form of neurodegenerative disease that is much more severe than that of IRP1+/+ IRP2–/– mice. In these mice, additional loss of one IRP1 allele further compromises regulation of iron metabolism in several tissues including brain, indicating that IRP1 contributes to iron homeostasis in multiple tissues, but its role is apparent only when the mouse lacks sufficient IRP2 for regulation. In this report, we focus on characterizing the phenotype and pathology of IRP1+/– IRP2–/– mice, an important new animal model of adult-onset neurodegeneration.

MATERIALS AND METHODS

Targeted Deletion of IRP1 and IRP2

Embryonic stem cells were transfected with constructs, chimeric mice were bred, and Southern analyses were performed on mouse tail cuts as previously described.[10]

IRP1+/− IRP2−/− animals were bred by mating IRP1+/− IRP2+/− animals and genotyping offspring.

Chemistries and Blood Work

Serum chemistries and complete blood counts were performed by the Veterinary Research Pathology Branch of the NIH.
Serum ferritins were measured as previously described.[10]

Western Analysis of Mouse Tissues

Mouse tissues were minced in Triton buffer containing 25 mM Tris, 40 mM KCl, 0.1 mM DTT, 1 mM AEBSF, 10 µg/mL leupeptin, 1 tablet/50 mL complete, and 1% Triton. Homogenized tissues were quickly centrifuged to remove large debris. Protein concentrations were measured using Bradford assay. Equal amounts of proteins from the tissues of three different genotypes were loaded on a 13% SDS-PAGE (for ferritin) and an 8% gel (for TfR), and detection of ferritin and TfR was as previously described.[10]

Histopathology

Age- and sex-matched WT, IRP2−/−, and IRP1+/− IRP2+/− mice between the ages of 4 and 12 months were anesthetized with pentobarbital IP and transcardially perfused with 4% paraformaldehyde/PBS.[10] Tissues were removed by dissection and postfixed for 24 h at 4°C, followed by embedding in gelatin and staining of sections using H&E, GFAP, Perls' DAB,[13] or amino-cupric silver stain[14] as previously described.[10] Alternatively, tissues of perfused animals were cryoprotected in 30% sucrose/PBS overnight. Forty-µm cryostat sections were processed for detection of microglial activation using rat anti-mouse Mac 1 antibody (1:200; Serotec) followed by biotinylated goat anti-rat Ig secondary antibody (1:200, Serotec).

Tyrosine hydroxylase immunohistochemistry was performed on paraffin-embedded sagittal sections of mice of each genotype using mouse anti-tyrosine hydroxylase antibodies (Zymed) at a final concentration of 1 µg/mL.

Ferritin immunohistochemistry was performed as previously described.[10]

For high-resolution light microscopy and transmission electron microscopy studies, deeply anesthetized animals were transcardially perfused with 4% paraformaldehyde and 0.2% glutaraldehyde. Multiple regions of brain and spinal cord were postfixed in osmium tetroxide, embedded in Polybed epoxy® resin, sectioned at 1-µm thickness, stained using toluidine blue and safranin, and examined by light microscopy. Selected tissue blocks from this material were used for ultrastructural analyses. These were thin-sectioned, with the sections mounted on copper grids and stained with lead citrate and uranyl acetate. These specimens were examined by transmission electron microscopy.

Forelimb and hindlimb grip strength was assessed by hang-test using a custom-made grip as previously described.[10] Statistical analysis was performed using an unpaired, Student's *t* test.

RESULTS

IRP1+/– IRP2–/– Animals Develop Early-Onset Neurodegeneration

We generated mice with the genotype IRP1+/– IRP2–/– through breeding and compared them to IRP2–/– and WT mice. The loss of one copy of IRP1 significantly worsened the neurodegeneration of IRP2–/– mice. Symptoms including failure to support the abdomen and slow uncoordinated gait were easily observed at 4 months of age in IRP1+/– IRP2–/– animals, whereas these symptoms did not appear until many months later in IRP2–/– mice. To assess the neurologic impairment quantitatively, we employed the hang-test, in which the length of time that a mouse can hang on to an inverted wire screen is measured.[10] Notably, at 1 year of age, the IRP1+/– IRP2–/– mice were able to hang on for only 10.7 ± 9.3 s ($n = 17$), whereas IRP2–/– mice could hang on for an average of 33.9 ± 25.1 s ($n = 22$) and WT mice could usually hang on for a full minute (57.1 ± 6.7 s) ($n = 34$).

A. TfR Western

Liver
Forebrain

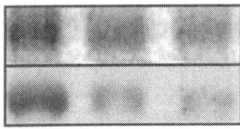

IRP1+/+ IRP1+/+ IRP1+/–
IRP2+/+ IRP2–/– IRP2–/–

B. Ferritin Western

Liver
Forebrain

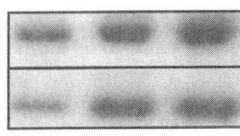

IRP1+/+ IRP1+/+ IRP1+/–
IRP2+/+ IRP2–/– IRP2–/–

FIGURE 1. Misregulation of ferritin and TfR expression in tissues of WT, IRP2–/–, and IRP1+/– IRP2–/– mice. Total protein lysates were prepared from the mouse tissues indicated and subjected to Western analysis using the antibodies to TfR (**A**) and to ferritin (**B**). Results shown are representative examples from experiments that have been performed on multiple mice of the indicated genotypes.

Ferritin and Transferrin Receptor Expression Are Abnormal in IRP1+/– IRP2–/– Mice

Since a major role of IRPs is to appropriately repress ferritin synthesis, we compared ferritin expression levels in lysates of multiple tissues from WT, IRP2–/–, and IRP1+/– IRP2–/– animals. Notably, ferritin levels in the forebrain and liver were higher in IRP1+/– IRP2–/– mice than in the IRP2–/– mice, indicating that loss of one IRP1 allele further impairs the compromised ability of IRP2–/– mice to repress ferritin translation (FIG. 1A). Serum ferritin was markedly increased in IRP1+/– IRP2–/– mice compared to IRP2–/– mice; serum ferritin of IRP1+/– IRP2–/– was 3000 ± 1900 ng/mL, whereas serum ferritin of IRP2–/– mice was 1800 ± 640 ng/mL, compared to WT serum ferritins of 740 ± 16 ng/mL. Transferrin receptor expression was decreased in the forebrain and liver of IRP2–/– and IRP1+/– IRP2–/– mice compared to WT (FIG. 1B), with slightly lower expression apparent in the IRP1+/– IRP2–/– mice.

To assess for compromise of tissue functions, we autopsied animals and evaluated histopathology of multiple tissues. In addition, we measured serum chemistries and complete blood counts of IRP1+/– IRP2–/– animals and WT animals. Despite abnormal iron metabolism in the liver, liver function tests including ALT, GST, LDH, and bilirubin were normal. BUN, a potential indicator of renal dysfunction, was also normal. Serum phosphorus was elevated in several IRP1+/– IRP2–/– animals that were agonal. The most noteworthy difference in blood work was that IRP1+/– IRP2–/– animals were anemic. Hemoglobin for WT animals was 13.32 ± 0.982 ($n = 6$), whereas serum hemoglobin was 9.86 ± 0.303 ($n = 6, p < .0005$). In addition, MCV was 50.35 ± 0.923 in WT vs. $38.1 \pm 1.835, p < .0005$. These values are consistent with an iron-deficiency anemia.

Ferric Iron Accumulation Colocalizes with Axonopathy in IRP2–/– and IRP1+/– IRP2–/– Mice

In order to examine pathology in the brain, we evaluated coronal sections of WT, IRP2–/–, and IRP1+/– IRP2–/– genotypes using the enhanced Perls' DAB stain for the detection of iron[13] and the amino-cupric silver stain[14] for the detection of axonal degeneration. As shown in FIGURES 2a–c, ferric iron accumulation in cerebellar white matter was substantially increased in IRP1+/– IRP2–/– mice compared to IRP2–/– animals. Moreover, comparable increases in silver impregnation of cerebellar white matter were seen in the IRP1+/– IRP2–/– animal compared to the IRP2–/– animal (FIGS. 2d–f).

Iron accumulation was observed in distinctive and reproducible regions of white and gray matter throughout the brain (summarized in TABLE 1). Usually when iron accumulation was observed in white matter tracts, evidence of axonal degeneration was also found, as evidenced by black threadlike areas of silver impregnation. Widespread axonal degeneration was also revealed throughout the brain by detection of myelin dense bodies characteristic of axonal degeneration, as illustrated in FIGURE 3. When iron accumulation was detected in gray matter areas such as the interpeduncular nucleus (FIG. 4) and mammillary body (not shown), there was rarely evidence of silver impregnation of cell bodies. However, when the projections of intact iron-laden neuronal cell bodies were examined, there was often evidence of axonal

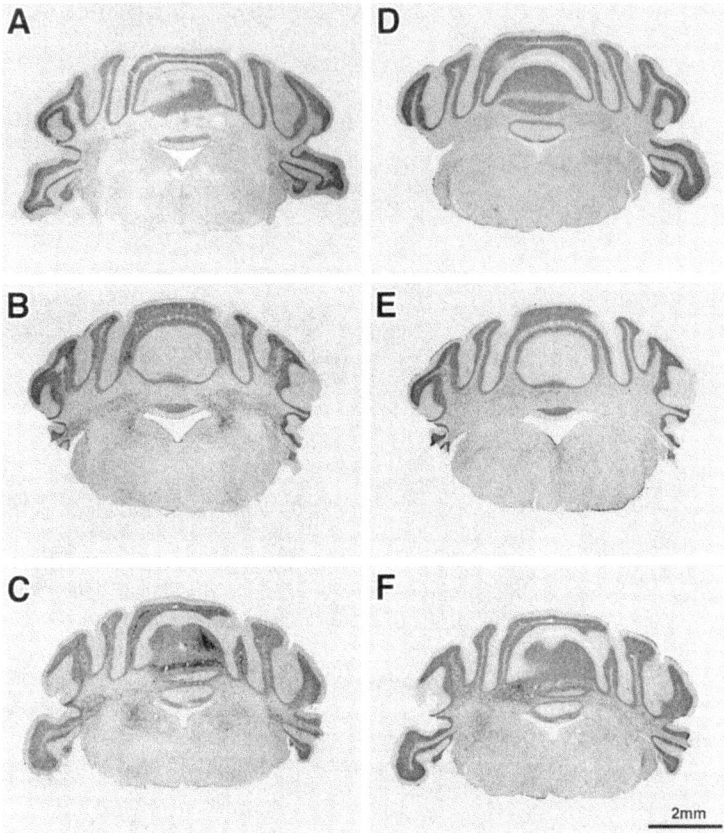

FIGURE 2. Iron accumulation and axonal degeneration colocalize and reveal an increase in axonal degeneration in the IRP1+/− IRP2−/− as compared to the WT (**A**) and IRP2−/− alone. In coronal sections through the metencephalon, ferric iron was detected as a brown precipitate with Perls' DAB stain in white matter of the IRP2−/− animal (**B**) and in the IRP1+/− IRP2−/− animal, where increased iron relative to the IRP2−/− animal was detected. Nuclei are counterstained with methyl green to show general morphology (**C**). Silver impregnation indicative of axonal degeneration was prominent in areas that showed substantial iron accumulation, particularly at the tips of the cerebellar folia (**D–F**). Nuclei are counterstained with neutral red to show morphology. [Figure shown in black and white.]

degeneration. For example, the mammillothalamic tract projects from cell bodies in the mammillary body that are heavily laden with iron, but silver does not impregnate the cell bodies and is discernible only in axons. When the time course of disease development was analyzed by comparing 4-, 9-, 12-, and 28-month-old mice, iron accumulation in white matter tracts preceded evidence of silver degeneration (data not shown).

TABLE 1. Iron accumulation in discrete tracts and nuclei of the brain correlates with areas of degeneration detected by the amino-cupric silver degeneration stain

Region	Enhanced Prussian Blue	Amino cupric
Telencephalon		
Anterior commissure	++	+
Caudate-putamen	+++	++
Hippocampus	++	++
Corpus callosum	-	-
Frontal cortex	-	-
CN II	+	+++
Diencephalon		
Fasciculus retroflexus	+++	-
Mamillary body	++	-
Mamillothalamic tract	-	++
Fornix	+++	+
Thalamus	++	++
Medial lemniscus	++	++
Medial longitudinal fasciculus	++	++
Posterior commissure	++	-
Mesencephalon		
Inferior and superior colliculi	+++	+++
Substantia nigra	++	++
CN III	+	+
CN IV	+	+
Interpeduncular nucleus	+++	-
Superior cerebellar peduncle	++	++
Nigrostriatal tract	++	+++
Metencephalon		
Cerebellum	+++	+++
Middle cerebellar peduncle	++	++
Inferior cerebellar peduncle	++	++
CN V	+	+
CN VI	+	+
CN VII	+	+
CN VIII	++	+
Myencephalon		
Nucleus gracilis	++	++

NOTE: Tracts in which significant iron staining was observed were evaluated to assess the corresponding level of silver impregnation, and conversely those with significant silver impregnation were examined for presence of iron. The nigrostriatal and cerebellar white matter tract silver impregnation was most intense and was used as the standard for +++ staining. The intense iron staining of the caudate putamen and cerebellar white matter was used as the standard for +++ iron staining. + was used to reflect faint, but clearly detectable iron staining or silver impregnation, and ++ was used to reflect easily detected silver or iron staining. As indicated, numerous areas of the brain were iron-loaded and showed silver impregnation. However, some areas including frontal cortex, internal capsule, and corpus callosum were completely spared.

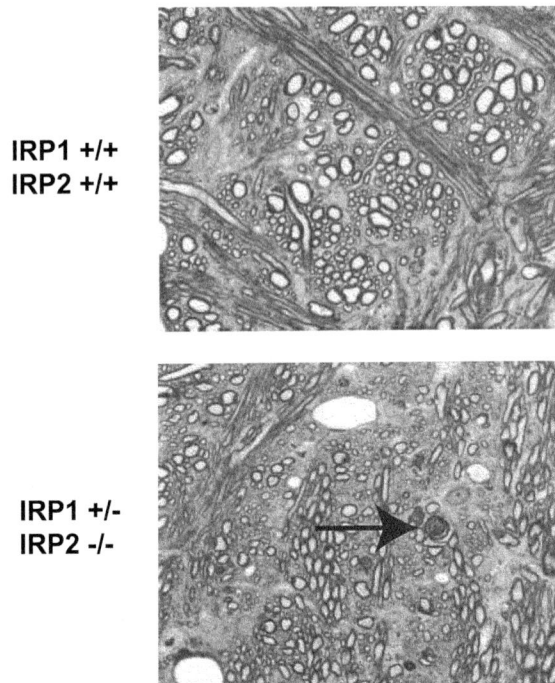

FIGURE 3. Axonopathy is also manifested by widespread presence of myelin dense bodies. Myelin dense bodies were detected throughout the brain of IRP1+/− IRP2−/− mice, but not in WT mice. Sections through the pons of epoxy-embedded, toluidine blue–stained brains of WT (*top*) and IRP1+/− IRP2−/− animals (*bottom*) are shown. Numerous myelin dense bodies are present, the largest of which is marked with an *arrow*.

Iron Overload, Degeneration, and Vacuolization Occur in the Substantia Nigra

The substantia nigra was very iron-rich in IRP2−/− and IRP1+/− IRP2−/− mice (FIG. 5a; left-hand panel) compared to WT controls. In addition, the nigrostriatal tract showed impressive silver degeneration, and some silver impregnation was detected in cell bodies in the substantia nigra pars compacta (SNPC) of IRP2−/− and IRP1+/− IRP2−/− mice (not shown). Careful morphologic evaluation of the SNPC on toluidine blue–stained sections of epoxy-embedded brains revealed that neuronal cell bodies were misshapen (FIG. 5a; right-hand panel). There was blebbing of the plasma membrane, loss of integrity of the nuclear membrane, and formation of numerous vacuoles ranging in size from approximately 5 to 50 microns in diameter that were most markedly increased in the IRP1+/− IRP2−/− mice. In FIGURE 5a, several typical vacuoles filled with debris are seen. In some instances, neuronal remnants were identifiable in these vacuoles, which we have previously detected in IRP2−/− mice using MRI.[15] Notably, although TUNEL assays were negative, assays

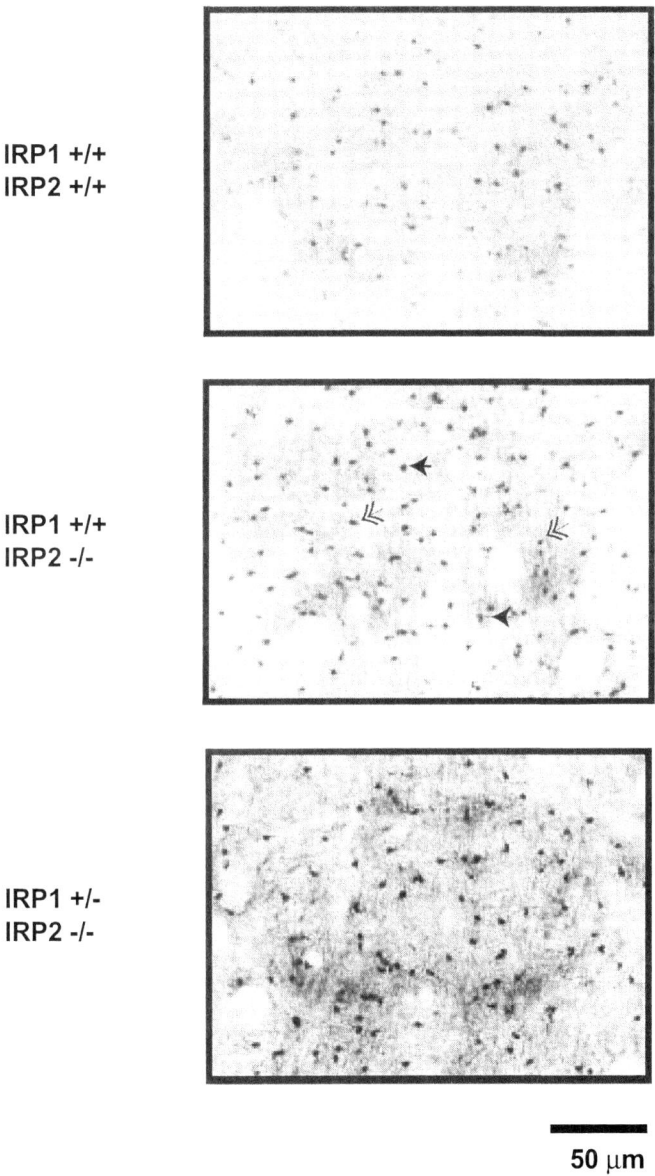

FIGURE 4. Iron accumulates in neuronal cell bodies and oligodendrocytes of the interpeduncular nucleus. Iron accumulation detected by Perls' DAB stain is present in neuronal cell bodies (*solid arrow*), oligodendrocytes (*double arrow*), and their processes. Iron accumulation is markedly increased in IRP1+/− IRP2−/− mice compared to IRP2−/− and to WT.

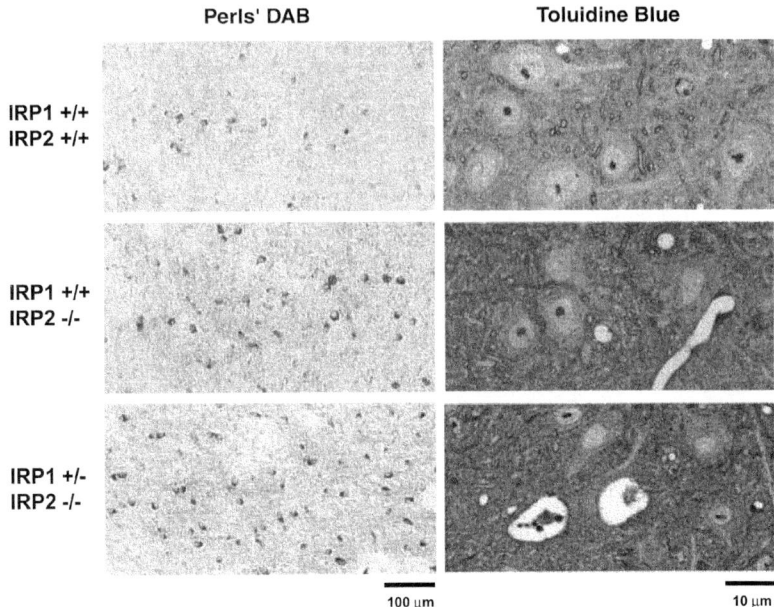

FIGURE 5a. Iron accumulation in the substantia nigra correlates with neuronal degeneration and the appearance of vacuolation. Iron accumulation is greatest in the substantia nigra of IRP1+/− IRP2−/− mice. Neurons of abnormal size and shape are detectable in substantia nigra of toluidine blue stains of epoxy-embedded brains in IRP2−/− mice, and more markedly in IRP1+/− IRP2−/− mice. Several characteristic vacuoles that contain debris are present in the IRP1+/− IRP2−/− toluidine blue stain.

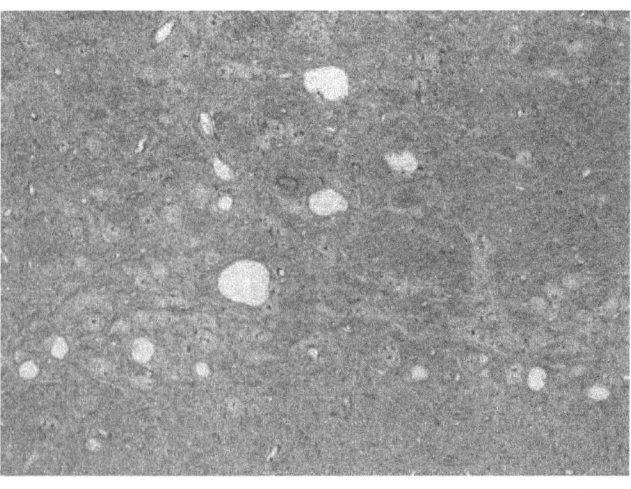

FIGURE 5b. A low-power view of the substantia nigra pars compacta of IRP1+/− IRP2−/− reveals multiple vacuoles ranging up to 30 μm in diameter. Many of the vacuoles are the approximate size of neuronal cell bodies.

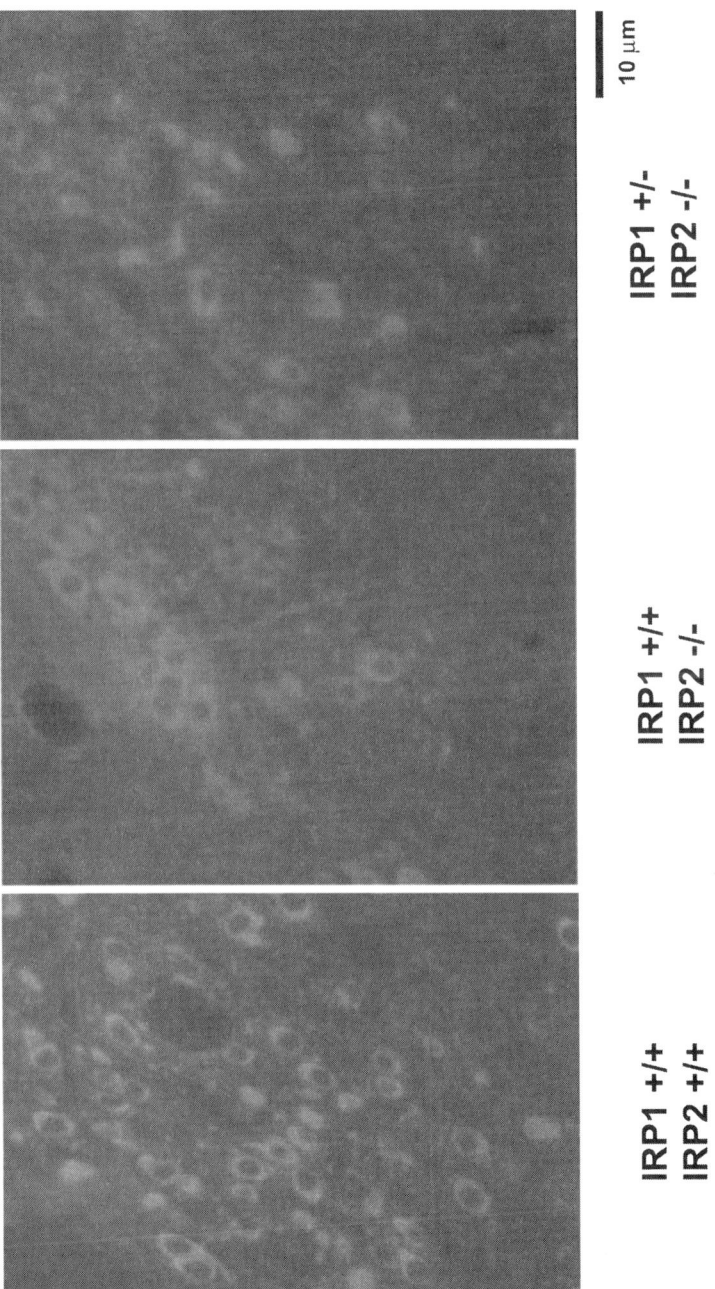

FIGURE 6. Decreased numbers of tyrosine hydroxylase–positive cells are detected in IRP2–/– and IRP1+/– IRP2–/– mice. Multiple sagittal sections were stained for tyrosine hydroxylase as described in MATERIALS AND METHODS and matched for anatomical location. Fewer tyrosine hydroxylase–positive cells were consistently detected in IRP2–/– and IRP1+/– IRP2–/– mice.

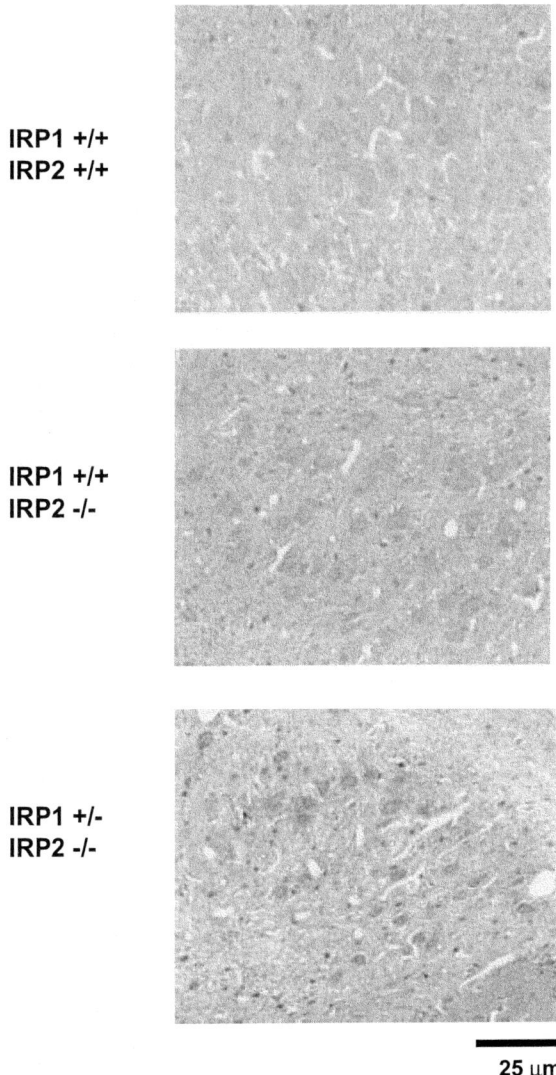

FIGURE 7. Increased ferritin is detected in neuronal cell bodies in the deep cerebellar nuclei of IRP2–/– and IRP1+/– IRP2–/– mice. Perfused and paraffin-embedded sagittal brain sections were stained with ferritin antibody as described in MATERIALS AND METHODS. Increased ferritin expression compared to WT was noted in the neuronal cell bodies in multiple brain areas (see TABLE 1). Shown here is increased ferritin expression in the deep cerebellar nuclei of IRP2–/– mice, a finding that was even more pronounced in the IRP1+/– IRP2–/– mice.

for activation of caspase III were lightly positive in the substantia nigra and in several other regions of the brain (data not shown). We counted the vacuoles in the SNPC of WT vs. IRP2–/– vs. IRP1+/– IRP2–/– mice and discovered that vacuole formation was much more pronounced in the IRP1+/– IRP2–/– mice than in the IRP2–/– mice, yielding a spongiform appearance in low-power views of the SN (FIG. 5b). In three 9-month-old IRP1+/– IRP2–/– mice, we counted an average of 17.8 ± 5.1 vacuoles per single section of SN; in contrast, in two age- and sex-matched IRP2–/– mice, we counted 8.5 ± 2.5 vacuoles per SN, and comparable vacuoles were not detected in two similarly processed age- and sex-matched WT

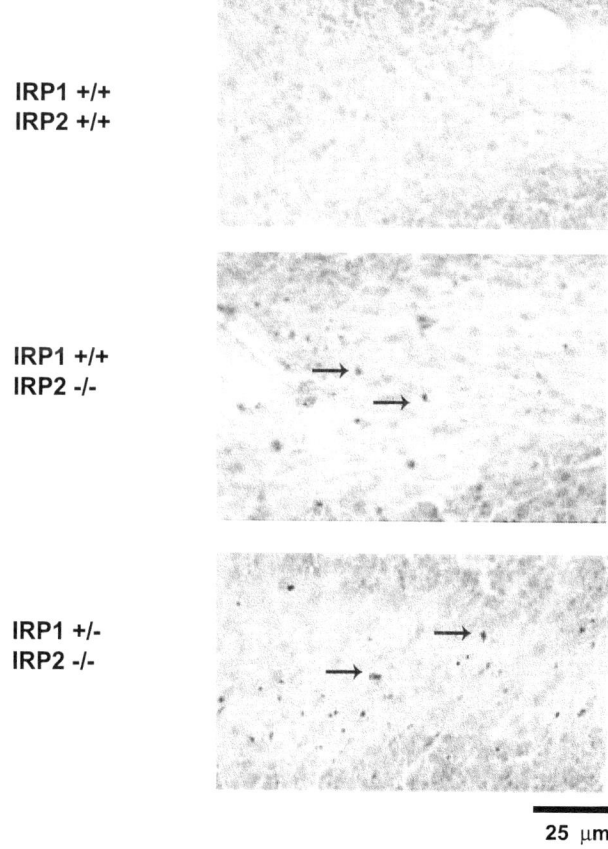

FIGURE 8. Markedly increased numbers of ubiquitin inclusions are present in cerebellar white matter of IRP1+/– IRP2–/– compared to IRP2–/– mice. Ubiquitin inclusions were detected in coronal sections of cerebellar white matter using anti-ubiquitin antibodies followed by avidin-biotin complex immunoperoxidase staining. Ubiquitin inclusions appear as intense brown spots (*arrows*), and nuclei of cerebellar granule cells are counterstained with methyl green (*appears as light gray in figure*).

mice. Immunohistochemistry for tyrosine hydroxylase confirmed that dopaminergic neurons were reduced in number in IRP1+/− IRP2−/− animals (FIG. 6).

Increased Ferritin Expression Is Detected within Neuronal Cell Bodies by Immunohistochemistry

In most areas that showed increased ferric iron, increased ferritin was detected by immunohistochemistry. Ferritin in neuronal cell bodies of the deep cerebellar nuclei increased markedly in IRP1+/− IRP2−/− mice compared to IRP2−/− animals, as

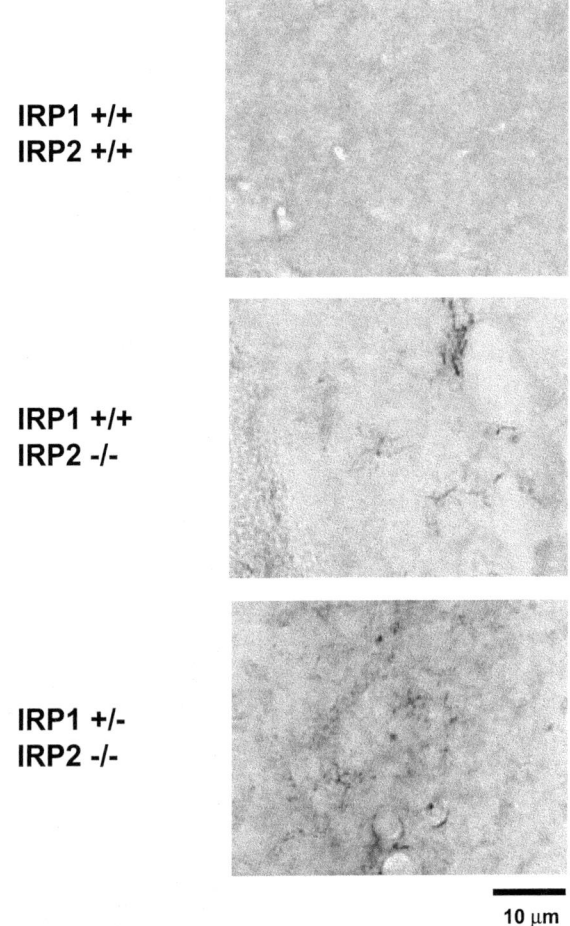

FIGURE 9. Foci of activated microglia are detected in IRP2−/− and IRP1+/− IRP2−/− mice. Using antibodies to activated microglia, frozen sections of brain were stained, and numerous activated microglia were found in IRP2−/− and IRP1+/− IRP2−/− mice.

indicated in FIGURE 7, and neuronal ferric iron increased comparably (data not shown). These observations are consistent with the interpretation that much of the increased ferric iron is sequestered within ferritin.[10]

Ubiquitin Inclusions Are Present in White Matter

A common feature of neurodegenerative diseases is the presence of ubiquitin-positive inclusions. Using anti-ubiquitin antibodies, we detected ubiquitin inclusions in IRP2–/– and IRP1+/– IRP2–/– mice. Notably, the number of ubiquitin inclusions in IRP1+/– IRP2–/– mice was markedly increased over the number found in IRP2–/– mice (FIG. 8).

Activated Microglia Are Prevalent in Brains of IRP2–/– and IRP1+/– IRP2–/– Animals

To pursue possible mechanisms that could lead to the formation of vacuoles observed in SNPC and to a lesser degree in the superior colliculus and pons (not shown), we looked for evidence of microglial or astroglial activation. Foci of activated microglia were found in some of the affected regions of the brain of IRP2–/– and IRP1+/– IRP2–/– mice. Activated microglia were distinctly increased in mice that were IRP1+/– IRP2–/– compared to mice that lacked IRP2 alone (FIG. 9). We did not find evidence for activation of astrocytes using the GFAP stain (not shown).

DISCUSSION

Mice with compromised iron regulatory protein function develop adult-onset neurodegenerative disease characterized by widespread axonopathy and later by vacuolization in discrete and reproducible areas of the brain, most prominently in the substantia nigra. The results of breeding between IRP1–/– and IRP2–/– mice indicate that IRP1 and IRP2 are partially redundant in function. Loss of one copy of IRP1 further exacerbates the neurodegenerative disease of IRP2–/– mice. This result is very interesting because IRP1–/– mice do not develop neurodegenerative disease, and we have discovered that IRP1 contributes minimally to normal iron regulation in WT mice (Meyron-Holtz et al., submitted). The pronounced worsening of neurodegenerative disease in IRP1+/– IRP2–/– mice both clinically and pathologically compared to IRP2–/– mice correlates with increased compromise of the ability of cells in many tissues of IRP1+/– IRP2–/– mice to appropriately repress ferritin synthesis and increase stability of the transferrin receptor transcript.

In the central nervous system of IRP2–/– and IRP1+/– IRP2–/– mice, inappropriate ferritin expression is associated with ferric iron accumulation in discrete tracts and nuclei throughout the brain. Iron accumulation is substantial in numerous specific white matter tracts, where iron is present in invaginations of oligodendrocyte cytosol into the axonal space as well as in the axons themselves (Sheng et al., in preparation). White matter tracts with significant iron accumulation degenerate, as indicated by the amino-cupric silver degeneration stain (FIG. 2) and presence of myelin dense bodies detected in these regions on toluidine blue–stained epoxy-embedded tissue (FIG. 3). In contrast, silver impregnation generally spares most neuronal cell bodies. However, morphologic abnormalities can be detected in neuronal somata in several locations (FIGS. 5–7).[10] In the SN, neurons that accumulated iron and ferritin appeared to shrink, and occasionally the nucleolus could often be detected as a remnant in a developing vacuole (FIG. 5). Although the morphology of these pyknotic cells was not typical for apoptotic cells,[16] we found immunohistochemical evidence for activation of caspase III in the SN and to a lesser extent in several other locations. Therefore, the contribution of apoptosis to neurodegeneration is not yet clear.

In the SN, we found multiple vacuoles, many of which conformed to the size and shape of neuronal cell bodies. Using GFAP staining, we did not find a marked increase in astrocyte activity compared to WT controls. However, stains for activated microglia revealed multiple foci of activated microglia in regions of the brain that were degenerating. The presence of activated microglia may help to explain the genesis of the vacuoles in our mice. We suggest that microglia may remove remnants of shrunken pyknotic neurons and that absence of astroglial activation explains why vacuoles persist in affected regions of the brain. The degree of vacuolization is minimal compared to prion disease[17] and the recently described mahoganoid neurodegeneration model.[18] Thus, we have not yet discovered any good counterparts to our model in the literature. Electron microscopic analysis of these vacuoles indicates that these vacuoles are not membrane-bound, and the debris contents of the vacuole are not distinctive (not shown).

Our results do not yet allow us to determine why axonopathy develops in IRP2–/– and IRP1+/– IRP2–/– mice. The consequences of loss of IRP2 function include increased ferritin expression, decreased transferrin receptor expression,[10] and perhaps other as yet uncharacterized abnormalities in regulation of gene expression. We have previously proposed that axonal degeneration may be one of the earliest events in disease progression in IRP2–/– mice,[19] and our data support a similar interpretation for IRP1+/– IRP2–/– mice. The axonal degeneration takes place in tracts that are clearly loaded with ferritin and ferric iron. In white and gray matter areas that are iron-overloaded as judged by staining for ferric iron and ferritin, we find that oligodendrocytes are extremely loaded with ferritin iron (FIG. 3) and this includes not only the cytosol in the cell body of the oligodendrocyte, but also invaginations from the inner myelin sheath into the region of the axon (Sheng et al., in preparation). The fact that only those oligodendrocytes that are in close physical contact with iron-overloaded neurons develop ferritin iron overload could mean that the oligodendrocytes acquire iron from the adjacent neuronal cell body or axon. Notably, iron overload is not detected in oligodendrocytes of the frontal cortex, a region of the brain in which iron overload is not detected in neurons in our models. Since oligodendrocytes are generally regarded as a relatively homogeneous cell type, it seems most likely that the iron overload observed in distinct regions of the brain is secondary to some feature of the immediate environment. However, it will be important in the future to culture oligodendrocytes and various types of neurons from our animals to distinguish cells that are programmed to make primary errors in regulation of iron metabolism as a consequence of loss of IRP activity from those in which iron overload is determined by their location in the central nervous system.

Although overexpression of ferritin in our mouse models correlates well with development of neurodegeneration, these results conflict somewhat with those of a recently reported study in which ferritin overexpression in the SN protected animals from the effects of MPTP injection.[20] However, it is also possible that our mice would be similarly protected from a single toxic insult from MPTP injection early in the course of disease. Similarly, it is possible that the transgenic mice that overexpress ferritin in the SN may have late-term pathology that has not yet been appreciated.

Although greater misregulation of ferritin and transferrin receptor expression is appreciable in multiple tissues of the IRP1+/– IRP2–/– mouse, it is interesting to note that the only systemic abnormality is an anemia in IRP1+/– IRP2–/– mice. We do not detect abnormalities of the liver (this paper), kidney (unpublished observations), and small intestine[10] as judged by serum tests and pathologic examination in IRP1+/– IRP2–/– mice, as was true for the IRP2–/– mice. These results underscore a conundrum that we previously encountered in IRP2–/– mice, namely that neurodegeneration occurs when other tissues affected by the same metabolic compromise exhibit normal function and histology. The subset of neurons that show evidence of misregulation of iron metabolism develop an axonopathy, whereas neurons in the cortex that do not accumulate iron and ferritin are spared.

It is not clear why some neurons are very adversely affected by a fundamental mistake in iron metabolism, whereas other tissues are spared. One possibility is that it takes time for damage to accrue, and injuries accumulate slowly in postmitotic neurons. However, vulnerability is unlikely to be determined solely by the expected life span of the cell; cardiomyocytes are also postmitotic cells, but there is no evidence for cardiomyocyte pathology in IRP2–/– or IRP1+/– IRP2–/– mice (not shown). Another possibility is that neurons are unusually vulnerable to mistakes in iron metabolism because they are highly polarized and iron redistributes to the axon and myelin sheath inappropriately in IRP2–/–[19] and IRP1+/– IRP2–/– mice (this paper).

A logical prediction from the observed exacerbation of misregulation of iron metabolism that occurs in IRP1+/– IRP2–/– mice is that IRP1–/– IRP2–/– mice would have an extremely severe form of neurodegenerative disease. However, we are unable to evaluate IRP1–/– IRP2–/– mice because complete loss of expression of both IRPs is embryonic lethal, with death occurring at the blastocyst stage (Smith and Rouault, in preparation). These results underscore our conclusion that both IRP1 and IRP2 contribute to normal posttranscriptional iron regulation in mammalian cells. IRP1–/– IRP2+/– animals look very similar to IRP1+/+ IRP2+/– animals[10] in that they have later-onset neurodegenerative disease that is less severe.

The presence in our mouse models of a pronounced axonopathy that is generally not associated with pathology and loss of neuronal cell bodies is interesting for studies of pathophysiology and therapeutics. In some instances, it is now recognized that axonopathy is preceded by nerve terminal injury.[21] A common form of axonopathy occurs in a process referred to as "dying back", but there is no evidence that dying back plays a role in our mice. Axonal degeneration that does not involve a classical "dying back" response has been described in association with axonal inclusions in motor neuron, Parkinson's, and Huntington's diseases.[22] If much of the disability of our mice results from compromised function of neurons that are alive, but impaired, as our models suggest, then therapeutic interventions could include delivery of neurotrophic factors such as NGF[22] or GDNF.[23] Neurotrophic factors could potentially revitalize function of impaired living neurons and reverse clinical symptomatology. IRP1+/– IRP2–/– mice may thus provide a model in which therapeutic interventions relevant to human disease can be tested.

ACKNOWLEDGMENTS

We thank the Veterinary Resources Program of the NIH for measuring complete blood counts and serum chemistries. This work was supported by the Intramural Program of the National Institute of Child Health and Human Development, with generous help from the Lookout Fund.

REFERENCES

1. PONTING, C.P. 2001. Domain homologues of dopamine beta-hydroxylase and ferric reductase: roles for iron metabolism in neurodegenerative disorders? Hum. Mol. Genet. **10:** 1853–1858.
2. CONNOR, J.R., G. PAVLICK, D. KARLI et al. 1995. A histochemical study of iron-positive cells in the developing rat brain. J. Comp. Neurol. **355:** 111–123.
3. HILL, J.M. & R.C. SWITZER III. 1984. The regional distribution and cellular localization of iron in the rat brain. Neuroscience **11:** 595–603.
4. BERG, D., M. GERLACH, M.B. YOUDIM et al. 2001. Brain iron pathways and their relevance to Parkinson's disease. J. Neurochem. **79:** 225–236.
5. SMITH, M.A., P.L. HARRIS, L.M. SAYRE & G. PERRY. 1997. Iron accumulation in Alzheimer disease is a source of redox-generated free radicals. Proc. Natl. Acad. Sci. USA **94:** 9866–9868.
6. HAYFLICK, S.J. 2003. Pantothenate kinase–associated neurodegeneration (formerly Hallervorden-Spatz syndrome). J. Neurol. Sci. **207:** 106–107.
7. TAKANASHI, M., H. MOCHIZUKI, K. YOKOMIZO et al. 2001. Iron accumulation in the substantia nigra of autosomal recessive juvenile parkinsonism. ARJP **7:** 311–314.
8. DICKSON, D.W., W. LIN, W.K. LIU & S.H. YEN. 1999. Multiple system atrophy: a sporadic synucleinopathy. Brain Pathol. **9:** 721–732.
9. WILSON, R.B. 2003. Frataxin and frataxin deficiency in Friedreich's ataxia. J. Neurol. Sci. **207:** 103–105.
10. LAVAUTE, T., S. SMITH, S. COOPERMAN et al. 2001. Targeted deletion of iron regulatory protein 2 causes misregulation of iron metabolism and neurodegenerative disease in mice. Nat. Genet. **27:** 209–214.
11. ROUAULT, T. & R. KLAUSNER. 1997. Regulation of iron metabolism in eukaryotes. Curr. Top. Cell. Regul. **35:** 1–19.
12. SCHNEIDER, B.D. & E.A. LEIBOLD. 2000. Regulation of mammalian iron homeostasis. Curr. Opin. Clin. Nutr. Metab. Care **3:** 267–273.
13. FRANCOIS, C., J. NGUYEN-LEGROS & G. PERCHERON. 1981. Topographical and cytological localization of iron in rat and monkey brains. Brain Res. **215:** 317–322.
14. DE OLMOS, J.S., C.A. BELTRAMINO & S. DE OLMOS DE LORENZO. 1994. Use of an amino-cupric-silver technique for the detection of early and semiacute neuronal degeneration caused by neurotoxicants, hypoxia, and physical trauma. Neurotoxicol. Teratol. **16:** 545–561.
15. GRABILL, C., A.C. SILVA, S.S. SMITH et al. 2003. MRI detection of ferritin iron overload and associated neuronal pathology in iron regulatory protein-2 knockout mice. Brain Res. **971:** 95–106.
16. MARTIN, L.J., N.A. AL-ABDULLA, A.M. BRAMBRINK et al. 1998. Neurodegeneration in excitotoxicity, global cerebral ischemia, and target deprivation: a perspective on the contributions of apoptosis and necrosis. Brain Res. Bull. **46:** 281–309.
17. ARMSTRONG, R.A., P.L. LANTOS & N.J. CAIRNS. 2002. Spatial patterns of the vacuolation in subcortical white matter in sporadic Creutzfeldt-Jakob disease (sCJD). Clin. Neuropathol. **21:** 284–288.
18. HE, L., X.Y. LU, A.F. JOLLY et al. 2003. Spongiform degeneration in mahoganoid mutant mice. Science **299:** 710–712.
19. ROUAULT, T.A. 2001. Iron on the brain. Nat. Genet. **28:** 299–300.

20. KAUR, D., F. YANTIRI, S. RAJAGOPALAN et al. 2003. Genetic or pharmacological iron chelation prevents MPTP-induced neurotoxicity in vivo: a novel therapy for Parkinson's disease. Neuron **37:** 899–909.
21. LOPACHIN, R.M., C.D. BALABAN & J.F. ROSS. 2003. Acrylamide axonopathy revisited. Toxicol. Appl. Pharmacol. **188:** 135–153.
22. RAFF, M.C., A.V. WHITMORE & J.T. FINN. 2002. Axonal self-destruction and neurodegeneration. Science **296:** 868–871.
23. GILL, S.S., N.K. PATEL, G.R. HOTTON et al. 2003. Direct brain infusion of glial cell line–derived neurotrophic factor in Parkinson disease. Nat. Med. **9:** 589–595.

Heme Oxygenase-1: Transducer of Pathological Brain Iron Sequestration under Oxidative Stress

HYMAN M. SCHIPPER

Center for Neurotranslational Research, Lady Davis Institute for Medical Research, Sir Mortimer B. Davis Jewish General Hospital, Department of Neurology and Neurosurgery, and Department of Medicine (Geriatrics), McGill University, Montreal, Quebec, Canada

ABSTRACT: Mechanisms responsible for the pathological deposition of redox-active brain iron in human neurological disorders remain incompletely understood. Heme oxygenase-1 (HO-1) is a 32-kDa stress protein that degrades heme to biliverdin, free iron, and carbon monoxide. In this chapter, we review evidence that (1) HO-1 is overexpressed in CNS tissues affected by Alzheimer's disease (AD), Parkinson's disease (PD), multiple sclerosis (MS), and other degenerative and nondegenerative CNS diseases; (2) the pro-oxidant effects of dopamine, hydrogen peroxide, β-amyloid, and proinflammatory cytokines stimulate HO-1 expression in some of these conditions; and (3) upregulation of HO-1 in astrocytes exacerbates intracellular oxidative stress and promotes sequestration of nontransferrin-derived iron by the mitochondrial compartment. A model is presented implicating glial HO-1 induction as a "final common pathway" leading to pathological iron sequestration and mitochondrial insufficiency in a host of human CNS disorders.

KEYWORDS: Alzheimer's disease; amyloid; astrocyte; cytokines; dopamine; heme oxygenase-1; iron; mitochondria; multiple sclerosis; oxidative stress; Parkinson's disease

IRON DEPOSITION AND THE CNS

Derangements in iron homeostasis and pathological deposition of this redox-active metal in brain have been implicated in a host of adult and pediatric conditions representing virtually every major category of neurological affliction: neurodegenerative (Alzheimer's disease, Parkinson's disease, progressive supranuclear palsy), metabolic (PANK-2 deficiency, aceruloplasminemia), immunologic (multiple sclerosis), ischemic (cerebral infarction), hemorrhagic (cerebral hematoma), traumatic (cerebral contusion), and infectious (HIV-1 encephalitis). Given iron's propensity to generate

Address for correspondence: Hyman M. Schipper, Center for Neurotranslational Research, Lady Davis Institute for Medical Research, Sir Mortimer B. Davis Jewish General Hospital, Department of Neurology and Neurosurgery, McGill University, 3755 Côte Ste-Catherine Road, Montreal, Quebec H3T 1E2, Canada. Voice: 514-340-8260; fax: 514-340-7502.
hyman.schipper@mcgill.ca

Ann. N.Y. Acad. Sci. 1012: 84–93 (2004). © 2004 New York Academy of Sciences.
doi: 10.1196/annals.1306.007

cytotoxic free radicals (Fenton chemistry), the advent of effective neuroprotection for many of these conditions may be facilitated by a thorough understanding of the pathophysiological mechanisms responsible for the abnormal brain metal profiles. While many of the mechanisms hypothesized will likely prove unique to specific disorders, others may constitute "final common pathways" shared by clusters of conditions that inevitably give rise to homologous patterns of aberrant brain metal deposition. In this chapter, evidence will be reviewed implicating induction of the enzyme, heme oxygenase-1, in astrocytes as one such common pathway to pathological brain iron sequestration and oxidative mitochondrial injury in Alzheimer's disease (AD), Parkinson's disease (PD), multiple sclerosis (MS), and other human neurological disorders. The model developed will also serve to unite several consistent neuropathological features of these conditions, namely, oxidative stress, iron mobilization, and mitochondrial insufficiency,[1] into a single "lesion" devolving on the action of HO-1. The primary literature attesting to central iron overload in these CNS disorders and other potential mechanisms mediating the abnormal accumulation of brain metals will be addressed elsewhere in this volume of the *Annals* and will not be covered here.

HEME OXYGENASE-1: REGULATION AND PHYSIOLOGY

Heme catabolism in a wide range of species is mediated by the heme oxygenase family of enzymes (E.C. 1:14:99:3; heme-hydrogen donor:oxygen oxidoreductase). Heme oxygenases are located within the endoplasmic reticulum, where they serve, in concert with NADPH cytochrome P450 reductase, to oxidize heme to biliverdin, free ferrous iron, and carbon monoxide (CO). Biliverdin is metabolized further to the bile pigment, bilirubin, by action of biliverdin reductase.[2] Mammalian cells express at least two isoforms of heme oxygenase, HO-1 (a.k.a. heat shock protein *32*) and HO-2. A third protein, HO-3,[3] may be a retrotransposition of the HO-2 gene unique to rats.[4] HO-1 and HO-2 are encoded by distinct genes and exhibit significant differences with regard to molecular weight, electrophoretic mobility, susceptibility to proteolysis, tissue distribution, regulation, and antigenicity. Substrate and cofactor specificities are, however, identical for the two isoforms.[5] Whereas HO-2 protein is widely distributed throughout the rodent neuraxis,[6] basal HO-1 expression in the normal brain is confined to small groups of scattered neurons and neuroglia.[7]

In humans, the *ho-1* gene is located on chromosome 22q12 and contains 4 introns and 5 exons. A 500-bp promoter, a proximal enhancer, and 2 or more distal enhancers occur in the regulatory region of the mammalian *ho-1* gene.[5] The latter exhibits AP-1, AP-2, nuclear factor kappa B (NFκB), and HIF-1 binding sites, as well as heat shock consensus (HSE) sequences, metal response elements (MtRE, CdRE), and antioxidant response elements (ARE). These response elements render the *ho-1* gene highly inducible by a wide array of pro-oxidant and inflammatory stimuli including heme, β-amyloid, dopamine, H_2O_2, UV light, transition metals, prostaglandins, Th1 cytokines, and lipopolysaccharide.[5,8] Furthermore, susceptibility to transcriptional suppression by glucocorticoids is mediated by a 56-bp sequence (STAT-3 acute-phase response factor binding site) located within the *ho-1* promoter.[9] A PEST [proline (P)–glutamic acid (E)–serine (S)–threonine (T)] sequence at the carboxy-terminus predisposes the HO-1 peptide to rapid degradation.[10] In mammalian cells, the half-lives of HO-1 mRNA and protein are in the range of 3 h and 15–21 h, respectively.[5]

The products of the heme oxygenase reaction, biliverdin (and its derivative, bilirubin), iron, and CO are all biologically active molecules. In the face of oxidative challenge, induction of HO-1 may protect cells by augmenting the breakdown of pro-oxidant heme to the radical-scavenging bile pigments, biliverdin and bilirubin.[7,11–14] Furthermore, biliverdin, bilirubin, and CO have been shown to have potent immunomodulatory properties. For example, in animal models, HO-1 upregulation attenuates inflammation of the cornea, pleura, and renal glomerulus; suppresses cardiac and renal xenograft rejection; and decreases mortality following systemic lipopolysaccharide exposure.[15,16] In some tissues and under certain experimental conditions, the induction of HO-1 may actually promote rather than protect against cellular injury. The intracellular liberation of potentially cytotoxic free ferrous iron and CO may account for these "sinister" effects of HO-1. In many tissues, coinduction of ferritin provides a measure of cytoprotection by providing a "sink" for the excess intracellular iron. Under some circumstances, however, heme-derived iron and CO may exacerbate intracellular oxidative stress and cellular injury by promoting free radical generation within the mitochondrial compartment.[17,18]

THE "JANUS" FACES OF HO-1 IN NEURAL INJURY

As described in the previous section, stressor-induced HO-1 expression may confer cytoprotection or, conversely, contribute to cellular injury. Perhaps nowhere has this disparate behavior of HO-1 engendered more controversy than in disorders of the nervous system. While there exists ample evidence for HO-1-mediated neuroprotection in various whole animal and tissue culture models of CNS injury and disease, a growing literature attesting to the neuroendangering aspects of HO-1 activity is also at hand. Details of this debate are beyond the scope of this article and are reviewed elsewhere.[19] Suffice it to say that these conflicting positions should not be viewed as mutually exclusive; the intensity and chronicity of HO-1 induction and the chemistry of the local redox microenvironment may determine whether the antioxidant benefits of a diminished heme:bilirubin ratio or oxidative damage accruing from intracellular mobilization of iron/CO are predominant.[20,21]

HO-1: TRANSDUCER OF MITOCHONDRIAL IRON TRAPPING IN "STRESSED" ASTROGLIA

In this section, we review evidence that HO-1 overexpression in astrocytes perturbs cellular redox homeostasis and promotes mitochondrial iron sequestration and oxidative injury. In the remainder of the paper, an argument will be developed implicating glial HO-1 expression as a common pathway leading to pathological iron deposition and oxidative mitochondrial damage in a host of human neurological disorders.

Stimulation of Glial HO-1 Expression and Mitochondrial Iron Sequestration

Cysteamine (CSH; 880 µM), dopamine (DA; 1 µM), tumor necrosis factor-α (TNFα; 20 ng/mL), interleukin-1 (IL-1β; 20 ng/mL), and β-amyloid$_{40/42}$ (3–15 µM) upregulate HO-1 mRNA, protein, and/or activity levels in cultured neonatal rat astroglia within 3–12 h of treatment. Within 3–6 days of exposure to these stimuli, sequestration of nontransferrin-derived ^{59}Fe (or ^{55}Fe) by the mitochondrial compartment is significantly augmented in these cells.[22–24] These treatments had no appreciable effect on mitochondrial trapping of diferric transferrin–derived iron,[22–24] an observation commensurate with the fact that brain iron deposition under various pathological conditions *in situ* may occur independently of transferrin and its receptor.[8] The effects of the aforementioned stimuli on HO-1 expression and mitochondrial iron sequestration could be mimicked by hydrogen peroxide (300–500 µM) or menadione (100 µM) administration, and cotreatment with potent antioxidants (ascorbate, melatonin, or *trans*-resveratrol) attenuated the HO-1 response to CSH, DA, and the proinflammatory cytokines in these cells.[22–24] Thus, oxidative stress is a likely common mechanism mediating glial *ho-1* gene induction in these experimental paradigms.

Pivotal Role of HO-1 in Mitochondrial Iron Trapping

Coadministration of tin mesoporphyrin (SnMP; 1 µM), a competitive inhibitor of heme oxygenase activity, or dexamethasone (DEX; 50 µg/mL), a transcriptional suppressor of the *ho-1* gene, significantly attenuated mitochondrial iron sequestration in cultured astrocytes exposed to DA, β-amyloid, TNFα, or IL-1β. Similarly, administration of SnMP or DEX abolished the pathological accumulation of mitochondrial ^{55}Fe observed in rat astroglia engineered to overexpress the human *ho-1* gene by transient transfection.[22–24] These findings suggest that upregulation of HO-1 is a critical event in the cascade leading to excessive mitochondrial iron deposition in oxidatively challenged astroglia.

HO-1 Perpetuates Oxidative Stress in Cultured Astroglia

Treatment with ascorbate, melatonin, or resveratrol blocks the late, compensatory induction of the manganese superoxide dismutase gene in astrocytes transiently transfected with human HO-1 cDNA.[18] In addition, preliminary data suggest that levels of protein carbonyls, isoprostanes, and 8-OHdG are significantly increased in mitochondrial fractions derived from astrocytes transfected with hHO-1 relative to sham-transfected cultures (author's unpublished results). Taken together, these findings strongly suggest that HO-1 overexpression, at least in astroglia, exacerbates intracellular oxidative stress. Treatment with cyclosporin A or trifluoperazine, potent inhibitors of the mitochondrial permeability transition pore, curtails mitochondrial iron trapping in hHO-1 transfected glia and cells exposed to DA, TNFα, or IL-1β.[22,24] Conceivably, intracellular oxidative stress accruing from HO-1 activity promotes pore opening[25,26] and influx of cytosolic iron to the mitochondrial matrix. The pathological accumulation of mitochondrial iron and other transition metals in astroglia overexpressing HO-1 may sensitize nearby neuronal elements to oxidative injury and thereby contribute directly to local neurodegenerative processes.[27,28] As such, glial HO-1 activation may represent a legitimate target for pharmacological manipulation, bearing in mind the caveats implied above (see THE "JANUS" FACES OF HO-1 IN NEURAL INJURY). Further analysis of glial HO-1 as a potential therapeutic target is justified on account of its augmented expression in various human neurological disorders, as described in the following section.

HO-1 EXPRESSION IN HUMAN CNS DISORDERS

Alzheimer's Disease

In AD brain, HO-1 protein colocalizes to neurons, GFAP-positive astrocytes, ependymocytes, neurofibrillary tangles, senile plaques, corpora amylacea, and some vascular smooth muscle and endothelial cells.[29,30] HO-1 immunoreactivity is greatly enhanced in neurons of the AD temporal cortex and hippocampus relative to corresponding tissues derived from nondemented controls matched for age and postmortem interval.[30] An *in situ* hybridization study also revealed increased HO-1 mRNA levels in AD-affected brain tissues.[31] We observed that 86% of GFAP-positive astrocytes residing within the AD hippocampus exhibited HO-1 immunoreactivity, whereas the fraction of hippocampal astroglia expressing HO-1 in normal control tissue was in the range of only 6–7%. Similarly, Western blots of protein extracts derived from AD hippocampus and temporal cortex revealed intense HO-1 bands, whereas the latter were faint or absent in normal control preparations.[30] As described above (see HO-1: TRANSDUCER OF MITOCHONDRIAL IRON TRAPPING IN "STRESSED" ASTROGLIA), oxidative stress resulting from excessive amyloid burden or the elaboration of proinflammatory cytokines may be responsible for the induction of HO-1 in the Alzheimer-diseased cerebral cortex and hippocampus. On the basis of the model presented earlier (see HO-1: TRANSDUCER OF MITOCHONDRIAL IRON TRAPPING IN "STRESSED" ASTROGLIA), we submit that the dysregulation of iron homeostasis and mitochondriopathy observed in AD brain[1,30] may be a consequence of sustained HO-1 overproduction in the affected tissues.

A curious recent observation was the relative downregulation of HO-1 protein in AD choroid plexus epithelial cells.[32] The latter may explain the apparent suppression of HO-1 concentrations previously documented in AD CSF relative to control values.[33] The subnormal levels of HO-1 protein in AD blood[33] and choroid plexus[32] may be due to the presence of a circulating HO-1 suppressor factor in this disease that is currently under investigation.[34,35]

Parkinson's Disease

In both PD and normal brain, moderate HO-1 immunoreactivity was observed in dopaminergic neurons of the substantia nigra.[36] In the PD samples, diseased dopaminergic neurons could be readily identified by the presence of cytoplasmic Lewy bodies that were prominently decorated with HO-1 staining.[36,37] The proportion of GFAP-positive astroglia expressing HO-1 in the PD nigra was significantly greater (77.1%) than that computed in age-matched control subjects (18.7%). Percentages of GFAP-positive astroglia coexpressing HO-1 in other subcortical nuclei, such as the caudate, putamen, and globus pallidus, were relatively low and not substantially different between PD and control specimens.[36] Dopamine released from dying nigrostriatal neurons, dopamine-derived hydrogen peroxide (see HO-1: TRANSDUCER OF MITOCHONDRIAL IRON TRAPPING IN "STRESSED" ASTROGLIA), catechol-generated semiquinone radicals, and endogenous MPTP-like neurotoxins are all plausible inducers of astroglial HO-1 in the PD nigra.[22,24] Augmentation of glial HO-1 activity may, in turn, promote the transferrin receptor–independent accumulation of iron and mitochondrial electron transport (complex I) deficits consistently reported in the basal ganglia of PD subjects.[1,28]

Multiple Sclerosis

Proinflammatory cytokines, oxidative stress, and mitochondrial iron deposition have been implicated in the pathogenesis of MS and experimental autoimmune encephalomyelitis (EAE), an animal model of MS.[24,38] In a neuropathological survey,[24] the percentage of GFAP-positive astrocytes expressing HO-1 in spinal cord plaques derived from MS patients (57.3%) was noted to be significantly greater than that observed in the spinal white matter of normal subjects (15.4%). In MS, glial HO-1 overexpression may be secondary to the enhanced release of TNFα, IL-1β (see HO-1: TRANSDUCER OF MITOCHONDRIAL IRON TRAPPING IN "STRESSED" ASTROGLIA), or myelin basic protein[39] within the affected tissues. Upregulation of HO-1, in turn, may amplify intracellular oxidative stress and give rise to mitochondrial iron deposits reported in this condition (see HO-1: TRANSDUCER OF MITOCHONDRIAL IRON TRAPPING IN "STRESSED" ASTROGLIA).

Prominent HO-1 induction also occurs in the CNS of rodents with EAE (see chapter by S. LeVine in this volume). In a recent study, administration of tin-protoporphyrin-IX, a competitive inhibitor of heme oxygenase activity, ameliorated behavioral deficits, weight loss, and indices of neural oxidative stress in female SJL mice with EAE.[40] These findings conflict with an earlier report[41] wherein worsening of disease was witnessed following metalloporphyrin suppression of heme oxygenase activity in Lewis rats with EAE. These interesting studies support an active role for HO-1 in the pathophysiology of EAE. They also signal further caution against facile extrapolation from one disease or disease model to another when contemplating manipulation of HO-1 as a therapeutic strategy for human neurological disorders.

Other CNS Disorders

Robust HO-1 overexpression has been documented in postmortem brain tissue procured from patients with cerebrovascular ischemia,[42] progressive supranuclear palsy, Pick's disease, corticobasal ganglionic degeneration,[43] and cerebral malaria.[44] The pathological profiles of ischemic brain injury and many of the human neurodegenerations also feature abnormal iron mobilization, oxidative molecular damage, and bioenergetic failure[1] that we submit may be due, at least in part, to the antecedent induction of HO-1.

SUMMARY AND CONCLUSIONS

First, the *ho-1* gene is exquisitely sensitive to oxidative stress and, in astrocytes, is rapidly upregulated following exposure to the pro-oxidant effects of hydrogen peroxide, dopamine, β-amyloid, and Th1 cytokines. Free ferrous iron and CO derived from the HO-1-mediated degradation of heme exacerbates intracellular oxidative stress even after initiating insults may have dissipated. The intraglial oxidative stress promotes opening of mitochondrial permeability transition pores, thereby allowing iron ions or low-molecular-weight iron chelates to accumulate within the mitochondrial matrix. With time, many of the effete, iron-laden mitochondria in these "stressed" astroglia undergo morphological transformation to Gomori-positive granules and corpora amylacea, inclusions that classically accumulate in aging and degenerating neural tissues[45] (FIG. 1). In spite of these profound changes, the astro-

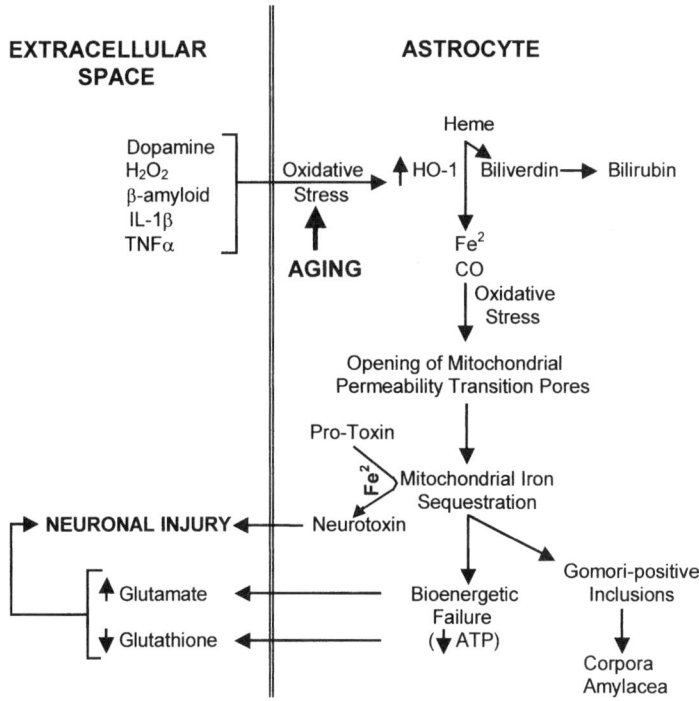

FIGURE 1. Putative role of astroglial HO-1 in pathological iron deposition and mitochondrial insufficiency in human neurodegenerative and inflammatory disorders.

cytes tend to remain viable because (i) they are endowed with potent antioxidant defenses in comparison with other neural cell types, (ii) they rapidly elaborate a number of heat shock proteins known to confer cytoprotection, and (iii) in contradistinction to neurons and oligodendroglia, their capacity to shift to robust anaerobic metabolism may allow them to sacrifice a significant proportion of their mitochondria with relative impunity.[22]

Second, prolonged survival of iron-laden astrocytes *in situ* may have important ramifications for the pathological and clinical progression of neurodegenerative diseases. Using electron spin resonance spectroscopy, we demonstrated that the glial mitochondrial iron behaves as a pseudo-peroxidase activity that promotes the oxidation of dopamine and other catechols to potentially neurotoxic ortho-semiquinone radicals.[28] The redox-active glial iron also facilitates bioactivation of the protoxin, MPTP, to the dopaminergic neurotoxin, MPP+, in the face of monoamine oxidase blockade.[46] We observed that neuronlike PC12 cells grown on a substratum of astrocytes replete with mitochondrial iron are far more susceptible to dopamine/H_2O_2-related killing than PC12 cells cocultured with control, "iron-poor" astroglia.[27] Taken together, these findings suggest that the progressive increase in glial mitochondrial iron that has been documented in subcortical regions of the aging rodent and human brain may enhance the susceptibility of the latter to parkinsonism and other free radical-related neurodegenerative disorders[28] (FIG. 1).

Third and last, HO-1 protein is overexpressed in astrocytes and other cells indigenous to CNS tissues affected by AD, PD, other aging-related neurodegenerations, MS, ischemia, and malaria. Numbers of neuroglia immunoreactive for HO-1 also correlate positively with aging in the normal human cerebral cortex and hippocampus.[47] These observations, in conjunction with the aforementioned *in vitro* findings, suggest that the aberrant deposition of (nontransferrin) iron, redox pathophysiology, and mitochondrial insufficiency characterizing many, if not all, of these disparate conditions may represent immediate downstream effects of glial HO-1 induction in the afflicted tissues (FIG. 1). This is not to imply that one should immediately move to suppress brain HO-1 in these patients; as discussed above, HO-1 is a "double-edged sword" that, under certain circumstances, may confer neuroprotection. Further investigation will be required to resolve these important issues. It may also prove interesting to see whether HO-1 contributes to pathological brain iron deposition in subjects with PANK-2 deficiency (formerly Hallervorden-Spatz disease), aceruloplasminemia, Friedreich's ataxia, sideroblastic anemia with ataxia, and neuroferritinopathy above and beyond the dysregulation of metal homeostasis incurred by the primary genetic defects.

ACKNOWLEDGMENTS

The skillful secretarial assistance of Lucia Badolato is greatly appreciated. The author's work described herein was supported by grants from the Canadian Institutes of Health Research, the Fonds de la recherche en santé du Québec, and the Alzheimer's Association (United States).

REFERENCES

1. BEAL, M.F. 1996. Mitochondria, free radicals, and neurodegeneration. Curr. Opin. Neurobiol. **6:** 661–666.
2. MAINES, M.D. 2000. The heme oxygenase system and its functions in the brain. Cell. Mol. Biol. (Noisy-Le-Grand) **46:** 573–585.
3. MCCOUBREY, W.K., JR., T.J. HUANG & M.D. MAINES. 1997. Isolation and characterization of a cDNA from the rat brain that encodes hemoprotein heme oxygenase-3. Eur. J. Biochem. **247:** 725–732.
4. SCAPAGNINI, G. *et al.* 2002. Gene expression profiles of heme oxygenase isoforms in the rat brain. Brain Res. **954:** 51–59.
5. DENNERY, P.A. 2000. Regulation and role of heme oxygenase in oxidative injury. Curr. Top. Cell. Regul. **36:** 181–199.
6. VERMA, A. *et al.* 1993. Carbon monoxide: a putative neural messenger. Science **259:** 381–384.
7. BARANANO, D.E. & S.H. SNYDER. 2001. Neural roles for heme oxygenase: contrasts to nitric oxide synthase. Proc. Natl. Acad. Sci. USA **98:** 10996–11002.
8. SCHIPPER, H.M. 2000. Heme oxygenase-1: role in brain aging and neurodegeneration. Exp. Gerontol. **35:** 821–830.
9. LAVROVSKY, Y., G.S. DRUMMOND & N.G. ABRAHAM. 1996. Downregulation of the human heme oxygenase gene by glucocorticoids and identification of 56b regulatory elements. Biochem. Biophys. Res. Commun. **218:** 759–765.
10. DWYER, B.E. *et al.* 1992. Heme oxygenase is a heat shock protein and PEST protein in rat astroglial cells. Glia **5:** 300–305.
11. STOCKER, R. *et al.* 1987. Bilirubin is an antioxidant of possible physiological importance. Science **235:** 1043–1046.

12. NAKAGAMI, T. *et al.* 1993. A beneficial role of bile pigments as an endogenous tissue protector: anti-complement effects of biliverdin and conjugated bilirubin. Biochim. Biophys. Acta **1158:** 189–193.
13. LLESUY, S.F. & M.L. TOMARO. 1994. Heme oxygenase and oxidative stress: evidence of involvement of bilirubin as physiological protector against oxidative damage. Biochim. Biophys. Acta **1223:** 9–14.
14. DORE, S. *et al.* 1999. Bilirubin, formed by activation of heme oxygenase-2, protects neurons against oxidative stress injury. Proc. Natl. Acad. Sci. USA **96:** 2445–2450.
15. WILLIS, D. *et al.* 1996. Heme oxygenase: a novel target for the modulation of the inflammatory response. Nat. Med. **2:** 87–90.
16. TOSAKI, A. & D.K. DAS. 2002. The role of heme oxygenase signaling in various disorders. Mol. Cell. Biochem. **232:** 149–157.
17. ZHANG, J. & C.A. PIANTADOSI. 1992. Mitochondrial oxidative stress after carbon monoxide hypoxia in the rat brain. J. Clin. Invest. **90:** 1193–1199.
18. FRANKEL, D., K. MEHINDATE & H.M. SCHIPPER. 2000. Role of heme oxygenase-1 in the regulation of manganese superoxide dismutase gene expression in oxidatively-challenged astroglia. J. Cell. Physiol. **185:** 80–86.
19. MAWAL, Y., B.D. KRAVITZ & H.M. SCHIPPER. 2002. Heme oxygenase-1 and Alzheimer disease. *In* Heme Oxygenase in Biology and Medicine, pp. 145–155. Kluwer Academic/Plenum. New York.
20. GALBRAITH, R. 1999. Heme oxygenase: who needs it? Proc. Soc. Exp. Biol. Med. **222:** 299–305.
21. SUTTNER, D.M. & P.A. DENNERY. 1999. Reversal of HO-1 related cytoprotection with increased expression is due to reactive iron. FASEB J. **13:** 1800–1809.
22. SCHIPPER, H.M. 1999. Glial HO-1 expression, iron deposition, and oxidative stress in neurodegenerative diseases. Neurotox. Res. **1:** 57–70.
23. HAM, D. & H.M. SCHIPPER. 2000. Heme oxygenase-1 induction and mitochondrial iron sequestration in astroglia exposed to amyloid peptides. Cell. Mol. Biol. (Noisy-Le-Grand) **46:** 587–596.
24. MEHINDATE, K. *et al.* 2001. Proinflammatory cytokines promote glial heme oxygenase-1 expression and mitochondrial iron deposition: implications for multiple sclerosis. J. Neurochem. **77:** 1386–1395.
25. PETRONILLI, V. *et al.* 1993. Physiological effectors modify voltage sensing by the cyclosporin A–sensitive permeability transition pore of mitochondria. J. Biol. Chem. **268:** 21939–21945.
26. BERNARDI, P. 1996. The permeability transition pore: control points of a cyclosporin A–sensitive mitochondrial channel involved in cell death. Biochim. Biophys. Acta **1275:** 5–9.
27. FRANKEL, D. & H.M. SCHIPPER. 1999. Cysteamine pretreatment of the astroglial substratum (mitochondrial iron sequestration) enhances PC12 cell vulnerability to oxidative injury. Exp. Neurol. **160:** 376–385.
28. SCHIPPER, H.M. 2001. Mitochondrial iron deposition in aging astroglia: mechanisms and disease implications. *In* Mitochondrial Ubiquinone (Coenzyme Q): Biochemical, Functional, Medical, and Therapeutic Aspects in Human Health and Disease, pp. 267–280. Prominent Press. Scottsdale, AZ.
29. SMITH, M.A. *et al.* 1994. Heme oxygenase-1 is associated with the neurofibrillary pathology of Alzheimer's disease. Am. J. Pathol. **145:** 42–47.
30. SCHIPPER, H.M., S. CISSE & E.G. STOPA. 1995. Expression of heme oxygenase-1 in the senescent and Alzheimer-diseased brain. Ann. Neurol. **37:** 758–768.
31. PREMKUMAR, D.R. *et al.* 1995. Induction of heme oxygenase-1 mRNA and protein in neocortex and cerebral vessels in Alzheimer's disease. J. Neurochem. **65:** 1399–1402.
32. ANTHONY, S.G. *et al.* 2003. Stress protein expression in the Alzheimer-diseased choroid plexus. J. Alzheimer's Dis. **5:** 171–177.
33. SCHIPPER, H.M. *et al.* 2000. Evaluation of heme oxygenase-1 as a systemic biological marker of sporadic AD. Neurology **54:** 1297–1304.
34. BERLIN, D., A. LIBERMAN & H.M. SCHIPPER. 2002. Partial characterization of a heme oxygenase-1 suppressor in Alzheimer plasma. Ann. Neurol. **52**(suppl. 1): S30.

35. KRAVITZ, S. *et al.* 2002. Heme oxygenase-1 suppressor activity in Alzheimer plasma. Ann. Neurol. **52**(suppl. 1): S31.
36. SCHIPPER, H.M., A. LIBERMAN & E.G. STOPA. 1998. Neural heme oxygenase-1 expression in idiopathic Parkinson's disease. Exp. Neurol. **150**: 60–68.
37. CASTELLANI, R. *et al.* 1996. Glycoxidation and oxidative stress in Parkinson disease and diffuse Lewy body disease. Brain Res. **737**: 195–200.
38. LEVINE, S.M. 1997. Iron deposits in multiple sclerosis and Alzheimer's disease brains. Brain Res. **760**: 298–303.
39. BUSINARO, R. *et al.* 2002. Myelin basic protein induces heme oxygenase 1 in human astroglial cells. Glia **37**: 83–88.
40. CHAKRABARTY, A., M.R. EMERSON & S.M. LEVINE. 2003. Heme oxygenase-1 in SJL mice with experimental allergic encephalomyelitis. Multiple Sclerosis **9**: 372–381.
41. LIU, Y. *et al.* 2001. Heme oxygenase-1 plays an important protective role in experimental autoimmune encephalomyelitis. Neuroreport **12**: 1841–1845.
42. BESCHORNER, R. *et al.* 2000. Long-term expression of heme oxygenase-1 (HO-1, HSP-32) following focal cerebral infarctions and traumatic brain injury in humans. Acta Neuropathol. (Berlin) **100**: 377–384.
43. CASTELLANI, R. *et al.* 1995. Evidence for oxidative stress in Pick disease and corticobasal degeneration. Brain Res. **696**: 268–271.
44. SCHLUESENER, H.J. & K. SEID. 2000. Heme oxygenase-1 in lesions of rat experimental autoimmune encephalomyelitis and neuritis. J. Neuroimmunol. **110**: 114–120.
45. MYDLARSKI, M.B. 1998. Astrocyte granulogenesis and the cellular stress response. *In* Astrocytes in Brain Aging and Neurodegeneration, pp. 207–234. R. G. Landes. Austin, TX.
46. DI MONTE, D.A. *et al.* 1995. Iron-mediated bioactivation of 1-methyl-4-phenyl-1,2,3,6-tetrahydropyridine (MPTP) in glial cultures. Glia **15**: 203–206.
47. HIROSE, W., K. IKEMATSU & R. TSUDA. 2003. Age-associated increases in heme oxygenase-1 and ferritin immunoreactivity in the autopsied brain. Leg. Med. (Tokyo) **5**(suppl.): S360–S366.

Nutritional Iron Deprivation Attenuates Kainate-Induced Neurotoxicity in Rats: Implications for Involvement of Iron in Neurodegeneration

S. SHOHAM[a] AND M. B. H. YOUDIM[b]

[a]Research Department, Herzog Hospital, Jerusalem, Israel

[b]Eve Topf and National Parkinson Foundation Centers for Neurodegenerative Diseases, Department of Pharmacology, Technion–Israel Institute of Technology, Haifa, Israel

ABSTRACT: There is evidence suggesting that oxidative stress contributes to kainate neurotoxicity. Since iron promotes oxidative stress, the present study explores how change in nutritional iron content modulates kainate-induced neurotoxicity. Rats received an iron-deficient diet (ID) from 22 days of age for 4 weeks. One control group received the same diet supplemented with iron and another control group received standard rodent diet. Cellular damage after subcutaneous kainate (10 mg/kg) was assessed by silver impregnation and gliosis by staining microglia. ID reduced cellular damage in piriform and entorhinal cortex, in thalamus, and in hippocampal layers CA1–3. ID also attenuated gliosis, except in the hippocampal CA1 layer. Given involvement of zinc in hippocampal neurotransmission and in oxidative stress, we tested for a possible interaction of nutritional iron with nutritional zinc. Rats were made iron-deficient and then assigned to supplementation with iron, zinc, or iron + zinc. Controls were continued on ID diet. After 2 weeks, rats were treated with kainate. Iron supplementation abolished the protective effect of ID in piriform and entorhinal cortex. In hippocampal CA1 and dorsal thalamus, neither iron nor zinc supplementation alone abolished the protective effect of ID against cellular damage. Iron + zinc supplementation abolished ID protection in dorsal thalamus, but not in reuniens nucleus. Kainate-induced gliosis in CA1 remained unaffected by nutritional treatments. Thus, in piriform and entorhinal cortex, nutritional iron has a major impact on cellular damage and gliosis. In hippocampal CA1, gliosis may associate with synaptic plasticity not modulated by nutritional iron, while cellular damage is sensitive to nutritional iron and zinc.

KEYWORDS: kainate; hippocampus; piriform cortex; thalamus; oxidative stress; neurodegeneration; iron; zinc; epilepsy; microglia

Address for correspondence: Dr. Shai Shoham, Research Department, Herzog Hospital, Jerusalem 91351, Israel. Voice: +972-2-5316860; fax: +972-2-6536035.
sshoham@md2.huji.ac.il

INTRODUCTION

In human temporal lobe epilepsy (TLE), recurrent seizures represent repetitive and synchronous firing of large populations of neurons in distinct anatomical brain regions.[1] Several lines of evidence suggest that seizure-induced damage occurs in brain regions, which are active during seizures, including the temporal cortex, amygdala, hippocampus, and thalamus.[2]

Systemic injection of kainate (KA), an excitotoxic glutamate receptor agonist, to rats has been shown to mimic human TLE.[2] In this model, there is a cascade of events: in the first hours, there is intense neural activation,[2,3] shifts in extracellular pH,[4] edema, and transient breakdown of the blood-brain barrier.[5] Within the first 3 days, apoptosis develops[6] and, 5–7 days after KA injection, neurodegeneration is manifested in cell shrinkage, neuronal loss, and microgliosis in piriform cortex, hippocampus, amygdala, and thalamus.[5]

There is a large body of evidence to support the involvement of oxidative stress in the KA model.[7–9] This evidence includes increased hydroxyl radical generation, depletion of reduced glutathione (GSH), lipid peroxidation, and prevention of KA-induced neurode generation by radical scavengers.[10–14]

Chelatable (free) iron promotes oxidative stress by Fenton chemistry[15] and has been implicated to play a pivotal role in several neurodegenerative diseases, including Parkinson's disease, Alzheimer's disease, amyotrophic lateral sclerosis, Hallervorden-Spatz disease, Huntington's disease, and Friedreich's ataxia.[16–20] Indeed, the mechanism of neurotoxicity of the parkinsonian neurotoxins, 6-hydroxydopamine (6-OHDA) and N-methyl-4-phenyl-1,2,3,6-tetrahydropyridine (MPTP), involves misregulation of iron metabolism in substantia nigra identical to what has been observed in parkinsonian substantia nigra.[20] Indeed, pretreatment with iron chelators, desferal and VK-28, is neuroprotective in both parkinsonian models.[21–24] Furthermore, nutritional iron deficiency has recently been shown to be neuroprotective against 6-OHDA neurotoxicity of substantia nigra.[25] Similar to substantia nigra, there is a high concentration of iron in the hippocampal CA3 region,[26] where substantial damage occurs in the KA model.[2] The neurotoxicity of KA is also associated with increased levels of iron divalent metal transporter-1 and iron levels in the rat hippocampus.[27–29] Moreover, KA-induced seizures involve intense glutamatergic neurotransmission,[2] and microinjection of glutamate analogues induces iron accumulation.[30–32] Thus, the present investigation focuses on the modulation of KA neurotoxicity by brain iron.

Two major approaches may be considered to study modulation of KA neurotoxicity by iron: iron chelation and iron deprivation. Most iron chelators such as desferal have almost no penetration through the blood-brain barrier and must be injected into brain,[21] which confounds the trauma involved in intracerebral injection with the trauma of KA-induced processes. In contrast, iron deprivation in rats, from 3 weeks of age for 4 weeks, has been extensively investigated. Iron deprivation has been shown to lower brain levels by 30–40% without affecting most brain biochemical and neurotransmitter metabolism pathways.[33–35] The iron-dependent brain biochemistry is more resistant to nutritional iron deficiency than systemic organs such as the liver.[35] Moreover, in preliminary work, we have not found cell pathology by iron deprivation per se and have verified that anatomical organization of the hippocampus and seizures are not altered by iron deprivation.[36,37] Furthermore, in gerbils, nutritional iron deficiency protects against stroke-induced brain edema,[38] thus demonstrating the ability of nutritional iron deprivation to protect from pathology.

In the present investigation, rats were fed a synthetic iron-deficient (ID) diet and compared to animals receiving the same diet supplemented with iron to the level available in standard rodent diet and to animals given standard rodent diet (as food pellets). After 4 weeks of dietary treatment, all rats received KA injection.

Since zinc has been suggested to be a factor in KA pathology[2,39,40] and to contribute to oxidative stress,[41] we examined a possible interaction of iron and zinc in contribution to KA-induced neurotoxicity. Thus, after 4 weeks of iron deprivation, iron-deprived rats were randomized to four groups that received a diet supplemented with iron, zinc, or iron + zinc, or were continued on an ID diet for 2 weeks and then treated with KA. A preliminary report of this study was made.[17]

METHODS

Animals and Nutritional Treatments

Male albino Sprague Dawley rats, 28 days of age, were purchased from the colonies of Levenstein Yoqne'am, Israel, and housed in polycarbonate cages provided with tops, feeders, and bottle caps, all made of stainless steel (Tecniplast, Italy). Room temperature was maintained at $22 \pm 2°C$ and lights were on for 12 h (16:00 to 04:00). An ethical committee of the Hebrew University in Jerusalem approved the study.

One group of animals received an iron-deficient diet (group ID) with an iron content of <5 ppm. The diet consisted of a milk-powder mixture of ingredients, modified from McCall *et al.*,[42] and is detailed elsewhere.[43] From previous work by our group[35] and by others,[16] it is known that such an ID diet, given to rats from 22 to 28 days of age for 4–5 weeks, results in reduction of up to 30–35% iron content in several brain regions, including caudate, cortex, cerebellum, and hippocampus.[33–35,44] Due to these previous works, biochemical analyses of brain iron were not performed in the present investigation.

One control group (CD) received the same ID diet, to which iron was added to the level of iron content in the standard laboratory diet (245 mg/kg diet). In this group, iron was given as ferric ammonium sulfate (Sigma F1668).

A second control group (SD) received a standard rodent diet ("Nutrilab" pellets, diet no. 1605, from Mahmad, Be'er Sheva, Israel, with an iron content of 245 mg/kg). The nutritional treatments just mentioned were given from 22 days of age for 4 weeks. Then, rats received subcutaneous KA injection, at a dose of 10 mg/kg, and were sacrificed 7 days later.

For studies on iron replenishment and on the contribution of zinc to KA-induced neurotoxicity, rats were made iron-deficient for the same period as just described. The synthetic ID diet contained 85 mg/kg zinc, the same as the SD rats. Then, ID rats were assigned to the following nutritional treatments for 2 weeks: One group received iron replenishment (ID + Fe, identical to the CD diet described earlier). One group received zinc supplementation (ID + Zn) of 12 mg/kg diet (without iron replenishment). One group received both iron replenishment and zinc supplementation (ID + Zn + Fe). A control group continued to receive an ID diet for 2 weeks.

Then, rats received a subcutaneous KA injection at 10 mg/kg and were sacrificed 7 days later.

KA Injection and Behavioral Observations

KA, at a dose of 10 mg/kg and a concentration of 10 mg/mL in 0.02 M phosphate-buffered saline (PBS), pH 7.2, was injected subcutaneously in the dorsal neck region. Behavioral observations were made and the rating system of Sperk et al.[45] was used: 0, normal behavior; 1, no overt seizures, occurrence of "wet dog shakes", staring, and strong immobility; 2, strongly increased rate of "wet dog shakes", rare and mild limbic seizures affecting the forelimbs and head; 3, "wet dog shakes", mild to severe intermittent seizures (affecting the whole body with rearing, falling over); 4, prolonged generalized, severe seizures with the same symptoms as in 3; 5, generalized seizures with early death. Rats that received a score of 3 or 4 throughout the third and fourth hours following KA injection were included in the present study.

Blood Sampling and Hematological Status

Blood was sampled just before transcardial perfusion with paraformaldehyde: rats were injected with Pental (pentobarbitone sodium, 200 mg/mL) at 0.3 mL/100 g body weight. The chest cavity was opened and blood was drawn from the heart for analysis of serum iron and glucose. Biochemical analyses of iron and glucose were performed with a DuPont Dimension AR (Wilmington, DE) automatic analyzer. TABLE 1 shows that, after 4 weeks on the ID diet, rats have serum iron values of 4.0 ± 1.8 µmol/L, as compared with controls (CD) with values of 26.0 ± 8.4 µmol/L.

Tissue Preparation (Including Preparation for Sulfide/Silver Stain)

After a blood sample was taken, the following sequence of transcardial perfusion was performed: 100 mL of 0.02 M PBS, pH 7.4, containing heparin at 5 U/mL followed by ice-cold 220 mL of 4% paraformaldehyde in 0.1 M PBS containing 4% sucrose. Brains were cut into coronal blocks and immersed in the same fixative for 2 h at 2–8°C. Subsequently, brain blocks were kept in 10% sucrose in 0.1 M PBS, pH 7.4, until sectioning in a cryostat. Floating sections, 30 µm thick, were collected for immunohistochemistry and silver impregnation and kept in a cryopreservation buffer at −20°C until staining. The cryopreservation buffer included polyvinylpyrrolidone (Sigma, PVP-40) and ethylene glycol (ICN, Ohio, Cat. No. 151089) in 0.1 M potassium acetate, pH 6.5. In addition, sections, 15 µm thick, were thaw-mounted onto silane-coated slides (SuperFrost Plus, Menzel Glaser, Germany) for cresyl violet and sulfide/silver staining. Slides for sulfide/silver staining were kept in the same cryopreservation buffer as for immunohistochemistry.

Silver Impregnation Stain

For silver impregnation staining, floating sections were processed using a kit (PK301 from FD Neurotechnologies, Ellicott City, MD). The principle of the staining is that degenerating neurons become argyrophilic.[46] Degenerating neurons and neural processes are stained blue-black, whereas nondegenerating neurons are stained yellow (an example is given in FIG. 1). Five sections were sampled from each rat.

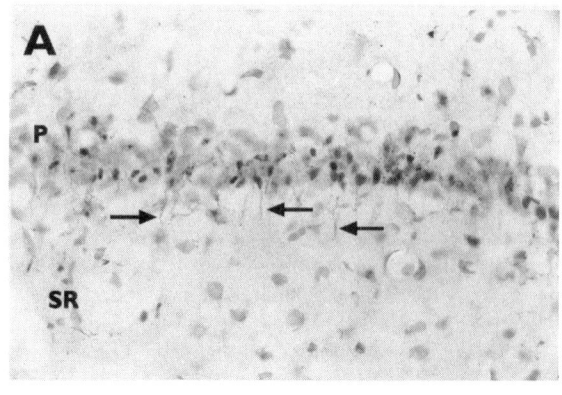

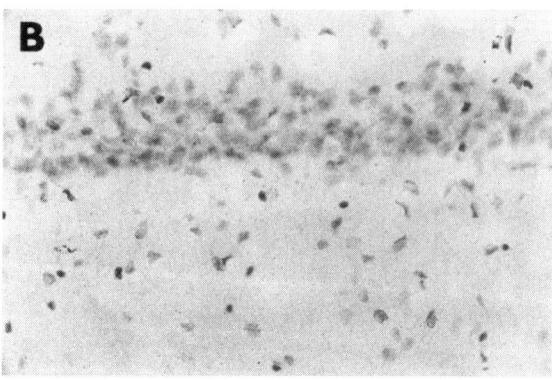

FIGURE 1. Detection of cellular damage by silver impregnation staining in the hippocampal CA1 region. **(A)** A section from a kainate-treated rat. There are darkly stained cells in the pyramidal cell layer (P) with silver-positive processes (*arrows*) that project into the stratum radiatum layer (SR). Other cells in this photograph are stained yellow, which appear as *light gray*. **(B)** A section from a naive rat. Calibration bar: 50 µm.

Immunohistochemistry

Microglia were identified using a monoclonal antibody (clone OX42, Cedarlane, Canada) directed against rat complement receptor 3 (CR3) at a dilution of 1:500.

Floating sections were rinsed twice in 0.02 M PBS and then endogenous peroxidase activity was quenched by incubation with 0.15% hydrogen peroxide for 25 min. Sections were incubated with the primary antiserum for 1 h at room temperature and then refrigerated at 2–8°C overnight. Sections were then incubated with biotinylated goat antimouse (Sigma B0529) 1:200 for 1 h and with Extravidin-peroxidase (Sigma E2886) 1:100 for 45 min. The final color reaction included diaminobenzidine 0.0125%, nickel ammonium sulfate 0.05%, and hydrogen peroxide 0.0015% for 2 min.[47]

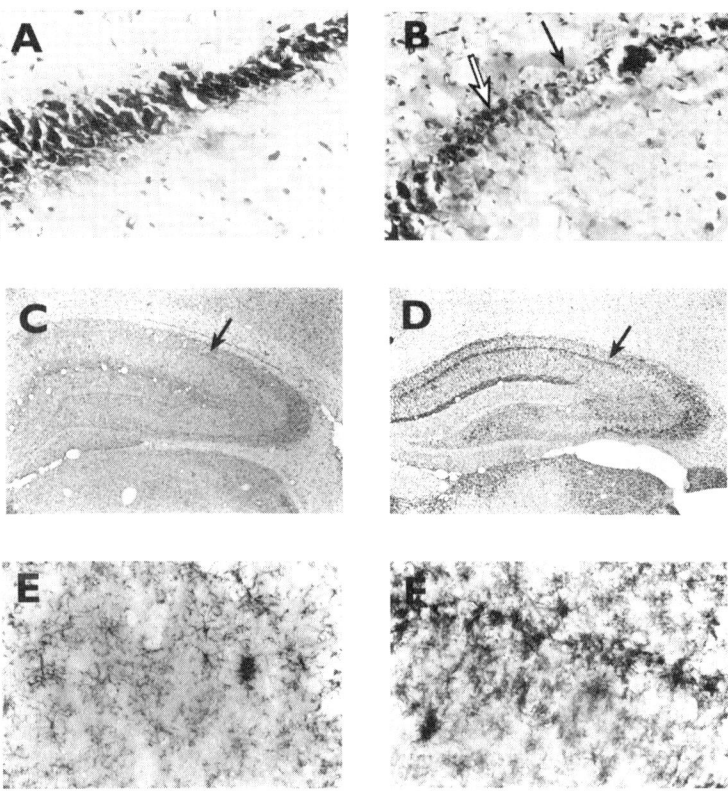

FIGURE 2. (A, B) Detection of cellular damage with cresyl violet stain and **(C–F)** detection of gliosis by immunohistochemical staining of microglia. The left column (A, C, E) shows photographs from a naive rat and the right column (B, D, F) shows photographs from a kainate-treated rat. In (B), cresyl violet stain demonstrates damaged segments of the CA1 pyramidal layer. The *black arrow* points at shrunken cells, whereas the *white arrow* points at a region with only cell debris. In (C) and (E), immunostaining of glia demonstrates that, in the naive condition, the CA1 (*arrow*) is largely devoid of reactive microglia. (D) In kainate-treated rats, the CA1 (*arrow*) was lined up with reactive microglia. (F) Multiple reactive microglia line the pyramidal cell layer [magnification from (D)]. Calibration bar: 45 μm (A, B, E, F); 180 μm (C, D).

Quantification of Neurotoxicity

All microscopic quantitative analyses were performed using a Seescan Sonata image analysis system (Seescan, Cambridge, United Kingdom).

Six coronal brain sections were sampled from each rat: three at the range of 0.48–1.30 mm posterior to bregma ("anterior piriform cortex") and three at the range of 2.80–3.60 mm posterior to bregma ("posterior piriform cortex").[48] "Cellular damage" was defined based on both cresyl violet and silver impregnation staining. Silver impregnation identified degenerating cells (FIG. 1), which by cresyl violet staining were not always recognized as damaged. Cresyl violet staining visualized cell shrinkage loss (FIG. 2) and general tissue deterioration. "Gliosis" was defined as the appearance of multiple microglia and microglial fibers spread over the region.

In all brain regions, cellular damage and gliosis formed discrete and continuous territories, which did not always overlap. Thus, cellular damage and gliosis were quantified separately. In each brain region, the affected areas were measured and summed. The area with cellular damage or gliosis was divided by the total area of the brain region in that brain section. This ratio is presented in the figures as percentage of brain region and is used for analyses of variance as described later. In hippocampus, both cellular damage and gliosis appeared in irregular segments of the pyramidal layer (a damaged segment is illustrated in FIG. 2). In each brain section containing the hippocampus, the lengths of damaged/gliotic segments of the pyramidal layer were summed and the percentage of this sum out of the full length of the given layer (CA1 or CA3) was calculated.

Statistical Analyses

Analysis of variance was used to test the main effect of nutritional treatment, and subsequent post hoc Newman-Keuls tests were employed for specific comparisons between treatments. Since the comparison was of the percentage of brain region affected and since under some treatments damage covered almost the entire brain region or a very small percentage of the brain region, the 2* arcsinus [square root (Xi)] transformation was used to improve the resolution between treatments and stabilize the variance. One ANOVA compared three groups: ID, CD, and SD rats. A second ANOVA compared five groups: ID, ID + Fe, ID + Zn, ID + Fe + Zn, and SD. Pearson's R correlation coefficient was employed to examine relationships between damage/microgliosis and blood levels of iron.

RESULTS

After 4 weeks of iron deprivation, ID rats had lower serum iron levels compared to controls (TABLE 1). Although they had lower body weight, ID rats appeared generally healthy and maintained glucose regulation similar to controls (TABLE 1). In another group of ID rats, after 4 weeks of iron deprivation and 2 additional weeks of nutritional iron supplementation, serum iron levels were elevated (TABLE 2).

In general, KA-induced neurotoxicity was observed in piriform and entorhinal cortex, thalamus (dorsal and reuniens), and hippocampus (CA1–3 layer). The impact of nutritional iron manipulations varied between brain regions. Therefore, results are presented separately for each brain region.

TABLE 1. Body weight and blood biochemistry

	ID (n = 6)	CD (n = 6)	SD (n = 6)	F test	p value
Body weight (g)	152.0 ± 12.9	232.4 ± 11.7	223.0 ± 9.7	13.0	0.0001
Glucose (mmol/L)	8.0 ± 0.5	7.5 ± 0.9	7.4 ± 0.7	0.34	N.S.
Iron (µmol/L)	4.0 ± 1.8	26.0 ± 8.4	19.0 ± 2.8	4.96	0.022

NOTE: Weanling rats were deprived of iron for 4 weeks (ID) and compared to two control groups: one that received the same synthetic diet supplemented with iron (CD) and one that received standard rodent diet (SD). The F test and p values are from ANOVA. Post hoc Newman-Keuls tests revealed that rats given SD had greater body weight and more serum iron compared to ID rats ($p < .01$). Iron-supplemented rats (CD) also had greater body weight and more serum iron compared to ID rats ($p < .01$). Serum glucose did not differ among treatments.

TABLE 2. Body weight and blood biochemistry in the study on the role of zinc

	ID (n = 6)	ID + Fe (n = 8)	ID + Zn (n = 9)	ID + Fe + Zn (n = 6)	F test	p value
Body weight (g)	225.0 ± 15.1	219.0 ± 12.8	233.0 ± 11.6	253.0 ± 5.0	1.6	N.S.
Glucose (mmol/L)	6.6 ± 0.6	7.3 ± 0.5	6.9 ± 0.6	7.0 ± 0.3	0.3	N.S.
Iron (µmol/L)	8.0 ± 1.1	26.5 ± 5.1	10.5 ± 3.1	27.7 ± 4.6	6.6	0.0021

NOTE: Weanling rats were deprived of iron for 4 weeks. Then, they were assigned to the following treatments for 2 weeks: one group continued to receive the iron-deficient diet (ID); one group received the same diet, but supplemented with iron (ID + Fe); one group received the same diet, but supplemented with zinc (ID + Zn); and one group received the same diet supplemented with both iron and zinc (ID + Fe + Zn). The F test and p values are from ANOVA. Post hoc Newman-Keuls tests revealed that iron-supplemented groups (CD and ID + Fe + Zn) had more serum iron compared to the other groups (ID and ID + Zn) ($p < .02$).

Piriform and Entorhinal Cortex

Effects of Iron Deprivation

The main effect of dietary treatment of cellular damage was significant in the piriform cortex with $F(2,15) = 19.6$, $p < .001$, and in entorhinal cortex with $F(2,15) = 24.9$, $p < .0001$. In both regions, post hoc tests revealed that ID reduced cellular damage significantly compared to both CD and SD control groups, $p < .02$ (FIGS. 3 and 4).

In both CD and SD rats, gliosis was more extensive than cellular damage (FIG. 4B compared to FIG. 4A). The main effect of dietary treatment on gliosis was significant in piriform cortex, with $F(2,15) = 25.5$, $p < .0001$, and in entorhinal cortex with $F(2,15) = 11.6$, $p < .0001$. In both piriform and entorhinal cortical regions, post hoc tests revealed that ID reduced gliosis significantly compared to both CD and SD control groups, $p < .05$ (FIG. 4B). Serum iron correlated positively with cellular damage ($R = .53$, $p < .05$) and with gliosis ($R = .55$, $p < .05$).

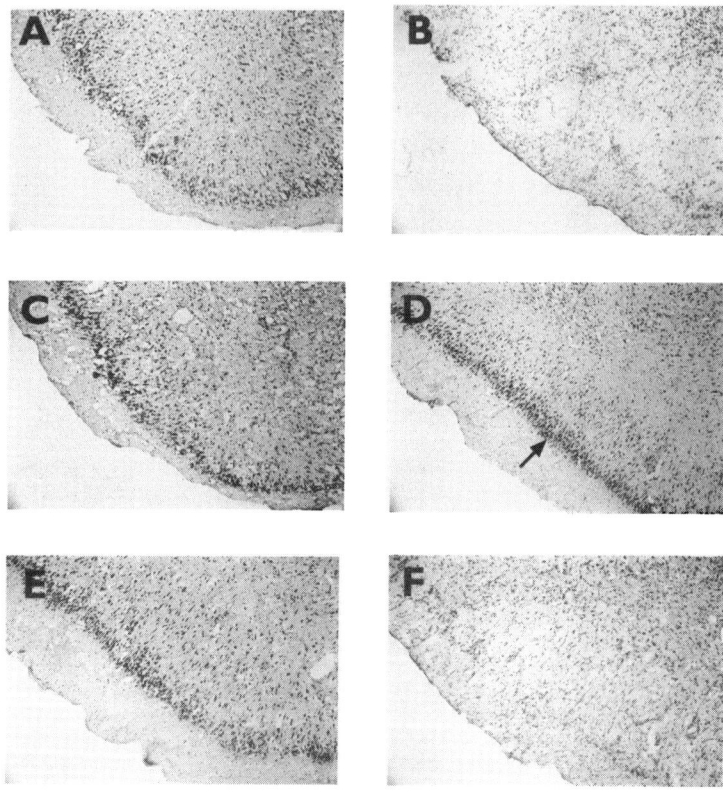

FIGURE 3. Protection of piriform cortex demonstrated with cresyl violet staining. The left column **(A, C, E)** represents control, noninjected rats. The right column **(B, D, F)** represents kainate-treated rats. The distribution pattern of cells in control rats given the standard rodent diet (SD in A) was similar to that of rats given an iron-deficient diet [ID in (C)] and to that of rats given an iron-supplemented diet [CD in (E)]. However, after kainate treatment, there was cellular damage in the SD (B) and in the CD (F) groups, whereas ID rats were protected (D) [*arrow* in (D) points at the surviving layer]. Calibration bar: 250 µm.

Effects of Iron and Zinc Supplementation to ID Rats

In the piriform cortex, the main effect of dietary treatment on cellular damage was significant with $F(4,30) = 6.9, p < .001$, and in the entorhinal cortex with $F(4,30) = 9.2, p < .001$ (FIG. 4C). Post hoc comparisons revealed that zinc supplementation to ID rats did not abolish the protection conferred by ID, but iron supplementation did, increasing both cellular damage ($p < .05$, FIG. 4C) and gliosis ($p < .01$, FIG. 4D). Iron supplementation brought the level of cellular damage to that observed in SD rats.

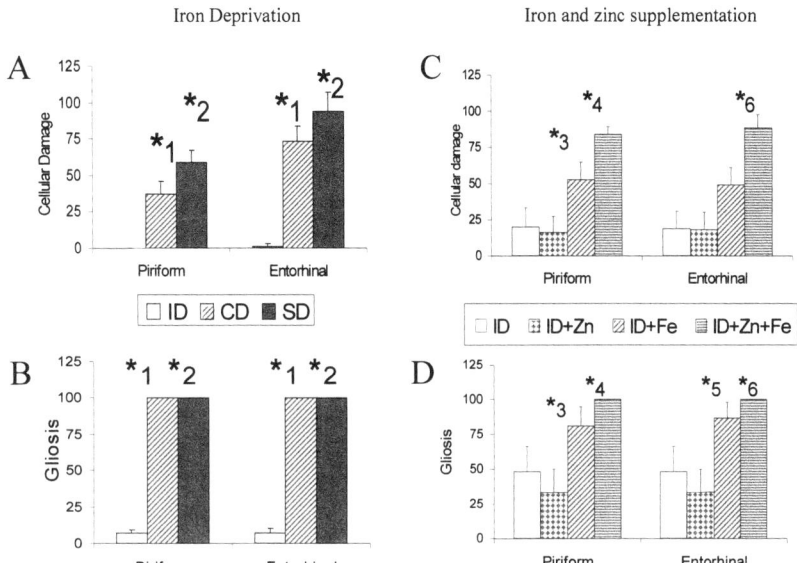

FIGURE 4. Effects of nutritional treatments on piriform and entorhinal cortex. **(A)** ID attenuated cellular damage and **(B)** gliosis compared to CD (*1, $p < .05$) and to SD (*2, $p < .05$). In piriform cortex, ID + Fe increased cellular damage **(C)** and gliosis **(D)** compared to ID and to ID + Zn (*3, $p < .05$) and brought the damage to a level that was not different from that of SD. Cellular damage was amplified significantly by ID + Zn + Fe compared to ID + Fe (C, *4). In entorhinal cortex, ID + Fe increased gliosis (D, *5), but not cellular damage (C), compared to ID and ID + Zn. In entorhinal cortex, ID + Zn + Fe increased cellular damage (C) above ID and ID + Zn (*6) to the level of damage of SD.

Cellular damage was further increased by combined iron and zinc supplementation (FIG. 4C).

Compared to ID, iron supplementation increased gliosis (FIG. 4D), but not cellular damage in entorhinal cortex (FIG. 4C). However, combined iron and zinc supplementation (ID + Fe + Zn) increased cellular damage in entorhinal cortex and brought it to the level of cellular damage observed in SD rats (FIG. 4C).

Thalamus

Effects of Iron Deprivation

KA-induced neurotoxicity was observed in the dorsal thalamus, including the paraventricular, dorsal, intermediodorsal, and centromedian nuclei (FIG. 5). Neurotoxicity also was observed in the ventral thalamus in the reuniens nucleus (FIG. 5). In the dorsal thalamic group, the main effect of nutritional treatment on cellular damage was significant with $F(2,15) = 5.87, p < .02$; in the reuniens, the main effect was significant with $F(2,15) = 17.3, p < .001$ (FIG. 6A). Post hoc comparisons revealed that ID reduced cellular damage compared to both CD and SD controls (FIG. 6A).

The main effect of nutritional treatment on gliosis was significant in the dorsal thalamus with $F(2,15) = 3.7$, $p < .05$, and in the reuniens with $F(2,15) = 3.4$, $p < .05$ (FIG. 6B). Post hoc comparisons revealed that ID reduced gliosis compared to both CD and SD controls ($p < .05$, FIG. 6B). Serum iron correlated positively with cellular damage ($R = .4$, $p < .05$) in the dorsal thalamus, but not in reuniens.

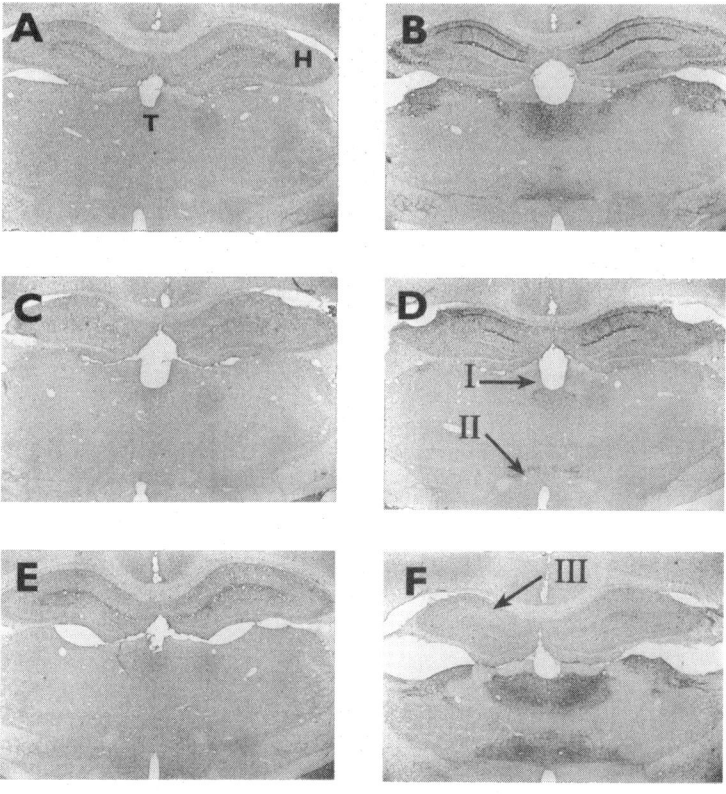

FIGURE 5. Distribution of gliosis in the thalamus and hippocampus. The left column **(A, C, E)** represents control noninjected rats. The right column **(B, D, F)** represents kainate-treated rats. In noninjected rats, there was diffuse staining in the thalamus [T in (A)], which was comparable in all experimental groups: SD in (A), ID in (C), and CD in (E). In kainate-injected rats, gliosis overlapped discrete subregions of the dorsal and ventral medial thalamus. In ID rats (D), kainate-induced gliosis was mild in the dorsal nuclei (*arrow* I) and in reuniens and gelatinous nuclei (*arrow* II). In the hippocampus of kainate-treated SD rats (B), most subregions were gliotic; in ID rats (D) and CD rats (F), the gliosis, especially in the dentate gyrus, was mild [*arrow* III in (F)] compared to SD rats (B). Calibration bar: 650 μm.

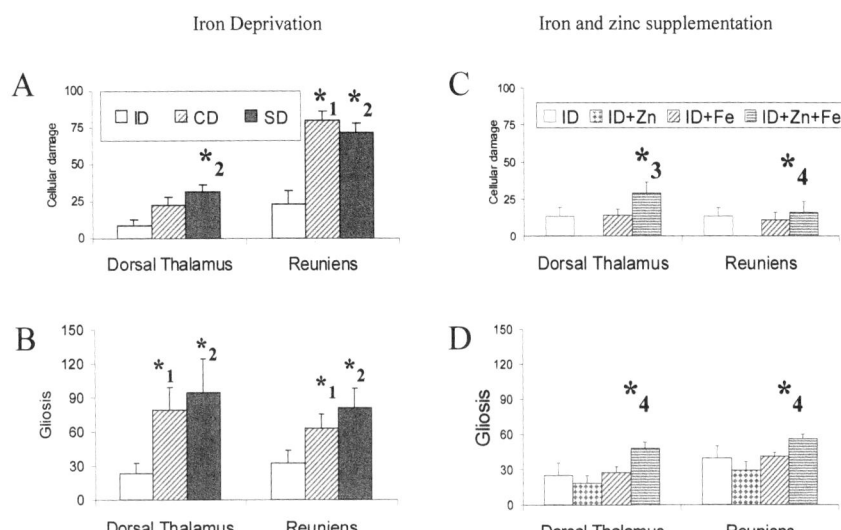

FIGURE 6. Effects of nutritional treatments on thalamic dorsal and reuniens nuclei. **(A)** ID attenuated cellular damage in the dorsal thalamus compared to SD (*2, $p < .05$) and in reuniens compared to both CD (*1, $p < .05$) and SD (*2, $p < .05$). **(B)** In both dorsal thalamus and reuniens, ID attenuated gliosis compared to both CD (*1, $p < .05$) and SD (*2, $p < .05$). **(C)** In both dorsal thalamus and reuniens, ID + Fe failed to increase cellular damage above ID level. ID + Zn treatment not only did not increase cellular damage, but was consistently associated with reduced cellular damage. In dorsal thalamus, ID + Fe + Zn increased cellular damage significantly (*3, $p < .05$) above ID to the level of SD. However, in reuniens, ID + Fe + Zn failed to increase cellular damage. **(D)** In both dorsal thalamus and reuniens, all nutritional treatments failed to reverse the effects of ID on gliosis and remained significantly below the level of gliosis in SD rats (*4, $p < .05$).

Effects of Iron and Zinc Supplementation to ID Rats

The main effect of dietary treatment on cellular damage was significant in dorsal thalamus with $F(4,30) = 2.84$, $p < .04$, and in reuniens with $F(4,30) = 14.8$, $p < .0001$ (FIG. 6C). In the dorsal thalamus, post hoc comparisons revealed that iron replenishment did not abolish the protection conferred by ID. However, iron + zinc supplementation abolished the protection conferred by ID and elevated the damage induced by KA to a level significantly above that of iron-supplemented ($p < .07$) and zinc-supplemented ($p < .03$) rats (FIG. 6C). Combined iron + zinc supplementation increased KA-induced cellular damage to the level observed in SD rats.

The main effect of dietary treatment on gliosis was significant in dorsal thalamus with $F(4,30) = 7.1$, $p < .0005$. However, neither iron nor zinc supplementation abolished the protection conferred by ID. By contrast, iron + zinc supplementation increased gliosis significantly compared to all other nutritional treatment groups ($p < .006$) in dorsal thalamus (FIG. 6D). Nevertheless, even iron + zinc supplementation did not bring gliosis up to the level observed in SD rats (FIG. 6D) ($p < .05$).

In thalamic reuniens nucleus, no nutritional treatment abolished the effect of ID on cellular damage (FIG. 6C) or gliosis (FIG. 6D). Thus, for KA-induced cellular damage in iron-supplemented and iron + zinc–supplemented rats, cellular damage and gliosis were lower than cellular damage and gliosis observed in SD rats, based on ANOVA with $F(4,30) = 4.7$, $p < .005$ (FIGS. 6C and 6D) (post hoc, $p < .05$).

Hippocampus

Effects of Iron Deprivation

The main effect of dietary treatment on cellular damage was significant in the CA1 region, with $F(2,15) = 4.27$, $p < .04$, and in the CA3 region, with $F(2,15) = 4.9$, $p < .03$ (FIG. 7A). Post hoc tests revealed that ID reduced cellular damage significantly in the CA1 and CA3 regions ($p < .05$). In the CA1 region, CD controls had less damage compared to SD rats ($p < .05$, FIG. 7A). In CA3, damage in the CD group was intermediate between SD and ID (FIG. 7A). No cellular damage was detected in the dentate gyrus (DG) in SD rats.

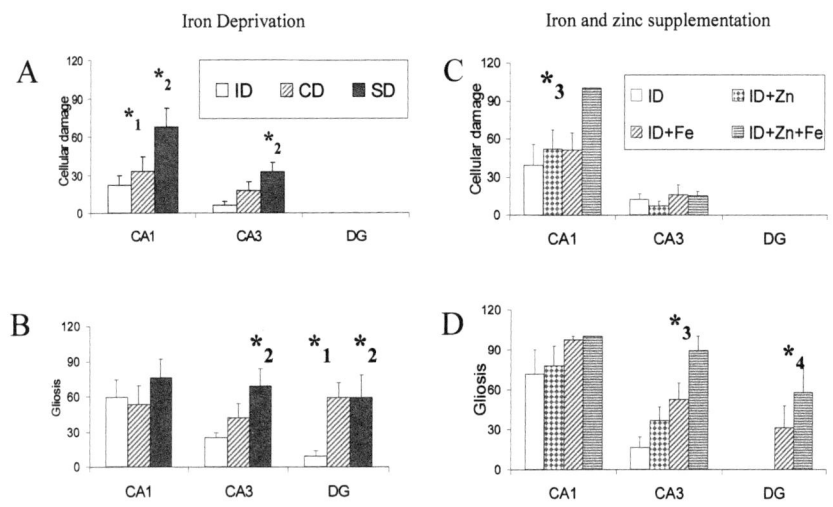

FIGURE 7. Effects of nutritional treatments on hippocampus. **(A)** In CA1, ID reduced cellular damage compared to both CD (*1, $p < .05$) and SD (*2, $p < .05$); in CA3, ID attenuated cellular damage compared to SD (*2, $p < .05$). **(B)** In CA1, ID failed to attenuate gliosis compared to either CD or SD; in CA3, ID reduced gliosis compared to SD (*2, $p < .05$); in dentate gyrus, there was no cellular damage, but ID attenuated gliosis compared to both CD (*1, $p < .05$) and to SD (*2, $p < .05$). **(C)** In both CA1 and CA3, ID + Fe failed to increase cellular damage. ID + Zn also failed, but ID + Fe + Zn increased cellular damage (*3, $p < .005$). **(D)** In CA1, gliosis was not altered by nutritional treatments. In CA3, ID + Fe increased gliosis to the level of SD. ID + Fe + Zn increased gliosis even further (*3, $p < .01$). In dentate gyrus, ID + Fe and ID + Fe + Zn increased gliosis significantly compared to ID (*4, $p < .05$) to the level of SD.

The main effect of dietary treatment on gliosis was significant in the CA3 region with $F(2,15) = 3.9, p < .05$, and in DG with $F(2,15) = 4.5, p < .03$ (FIG. 7B). Post hoc tests revealed that ID attenuated gliosis in both CA3 and dentate regions ($p < .05$). In CA1, ID did not attenuate gliosis. The extent of gliosis was similar in CA1 of both SD and CD rats.

In sum, in hippocampus, there were separate spatial patterns of cellular damage and gliosis. Cellular damage was absent in DG. There was a trend of increased cellular damage along the hippocampal neuroanatomical pathway from DG to CA3 to CA1 (FIG. 7A). Gliosis was present in all hippocampal subregions to a similar extent (FIG. 7B). The impact of ID on gliosis followed the pattern where the internal hippocampal pathway was greatest in DG, was less in CA3, and had no impact in CA1 (FIG. 7B).

Effects of Iron and Zinc Supplementation on ID Rats

The effect of dietary treatment on cellular damage in the CA1 region was significant with $F(4,30) = 2.7, p < .05$. In the CA3 region, the effect was not significant, and cellular damage was not detected in the DG (FIG. 7C). Post hoc tests revealed that, in CA1, neither iron supplementation (ID + Fe) nor zinc supplementation (ID + Zn) abolished the protection conferred by ID. However, iron supplementation combined with zinc supplementation (ID + Fe + Zn) abolished the protection conferred by ID (FIG. 7C).

Gliosis in the CA1 region was not attenuated by ID or by any other dietary treatment (FIG. 7D). However, the effect of dietary treatment on gliosis was significant in the CA3 region, with $F(4,30) = 5.7, p < .02$, and in the DG, with $F(4,30) = 4.9, p < .004$. In the CA3 region, post hoc comparisons revealed that iron supplementation (ID + Fe) and iron + zinc supplementation (ID + Fe + Zn) increased gliosis compared to ID ($p < .05$) to a level not different from SD (FIG. 7D). Furthermore, post hoc comparisons revealed that, in the DG, iron supplementation (ID + Fe) and iron + zinc supplementation (ID + Fe + Zn) increased gliosis compared to ID ($p < .04$, FIG. 7D). Serum iron levels correlated significantly with gliosis in the hippocampal DG ($R < .49, p < 0.05$).

In sum, in hippocampus, there were regional differences between the impacts of nutritional treatment on cellular damage vs. gliosis. While no cellular damage was observed in DG, gliosis was induced by KA and was sensitive to nutritional treatment (FIGS. 7C and 7D). In CA3, while cellular damage was not affected by nutritional treatment, gliosis was (FIGS. 7C and 7D). In CA1, while cellular damage was affected by nutritional treatment, gliosis was not (FIGS. 7C and 7D).

DISCUSSION

In the present study, KA induced cellular damage and gliosis in the piriform and entorhinal cortex, thalamus, and hippocampus similar to patterns reported in previous studies.[2,3,5] The present study demonstrates that nutritional ID attenuates KA-induced neurotoxicity.

Thus, it could be predicted that iron supplementation following iron deprivation would abolish the protection conferred by ID. However, the present study revealed brain region–specific patterns in the response to iron supplementation, which included dissociations between cellular damage and gliosis. Since both cellular damage and gliosis have been linked to oxidative stress, there must be involvement of additional factors besides iron availability and oxidative stress. Since zinc has long been considered important in modulation of KA-induced neurotoxicity,[2,40] the present study examined the impact of iron and zinc supplementation on ID rats. In the following, the effects of dietary treatments are discussed with respect to each brain region.

Piriform and Entorhinal Cortex

The protective effects of ID and the finding that iron supplementation to ID rats abolished the protection by ID suggest an important contribution of nutritional iron to KA-induced neurotoxicity in piriform and entorhinal cortex. The mechanism of iron involvement possibly involves a combination of edema and oxidative stress. During KA-induced seizures, the piriform and entorhinal cortex become swollen, resulting in rupture of small blood vessels (edema), which in turn results in exposure of cortical tissue to blood hemoglobin.[5,49] Iron bound in hemoglobin may be released due to alterations in extracellular pH that occur during seizures.[4] Previous work has demonstrated that iron released from hemoglobin can be toxic to cortical cells.[49] Oxidative stress was reported in piriform cortex following KA injection.[50–52]

Zinc added to iron supplementation amplified KA-induced neurotoxicity. This may be related to the transport of zinc and its release in piriform cortex.[53] Zinc can cause oxidative neuronal necrosis in cortical cells.[41] On the other hand, it should be noted that, in the present study, zinc supplementation alone did not abolish the protection conferred by ID. Thus, there is a critical role for iron in KA-induced neurotoxicity in piriform cortex.

Thalamus

In dorsal thalamus and in the thalamic reuniens nucleus, ID reduced cellular damage and gliosis. However, iron supplementation failed to abolish the protection conferred by ID. The failure of iron supplementation in thalamus stands in contrast to the effectiveness of iron supplementation in the piriform and entorhinal cortex previously discussed. Earlier studies found that KA-induced edema was less common in thalamus than in piriform cortex.[5] This suggests that less blood iron might penetrate thalamus associated with edema to contribute to neurotoxicity as in piriform cortex.

Combined iron and zinc supplementation increased cellular damage significantly in dorsal thalamus, but came short of matching the extent of damage seen in control rats receiving the standard rodent diet (SD). Zinc is released in the thalamus associated with hippocampothalamic glutamatergic neurotransmission.[54] However, combined iron and zinc supplementation had no effect in the thalamic reuniens nucleus. Possibly, propagation of excitotoxic neurotransmission to thalamus depends on hippocampal output during seizures, which would depend on how severely the hippocampus is affected by KA. It should be noted that zinc supplementation alone appeared to reduce cellular damage in both dorsal thalamus and reuniens. Thus, besides interaction with iron that promotes cellular damage, this suggests that zinc participates in other processes that may inhibit cellular damage,[55,56] accounting for the small net effect of combined iron and zinc supplementation in thalamus.

Hippocampus

ID reduced cellular damage in the CA1–CA3 hippocampal fields. However, iron supplementation alone failed to abolish the protection conferred by ID, suggesting that additional factors are involved. Combined iron and zinc supplementation abolished the protection by ID in CA1, but not in CA3, suggesting that zinc is one additional factor in the contribution of iron to cellular damage in specific regions.

These findings may be interpreted in terms of hippocampal neurocircuitry. There is evidence that KA-induced seizures begin with excitation of the hippocampal DG.[2,3] Excited DG granule cells then transmit excitatory messages through the mossy fiber (MF) system and release glutamate at their terminal field, which is the hippocampal CA3 region. Pyramidal neurons in CA3 then release glutamate in CA1. DG cells being the initiators of the excitatory cascade are apparently the least at risk for degeneration, but CA3 and CA1 are at risk. Differential vulnerability of CA1 vs. CA3 may depend on their position in the neurocircuitry of seizures and the availability of iron.

Interruption of signal transduction in the MF system attenuates KA-induced neurotoxicity in CA3.[57] The CA3 field accumulates iron.[26] Iron is stored in ferritin by oligodendrocytes and microglia along the MF terminal region.[58] In a previous study, we found that 4 weeks of iron deprivation in juvenile rats, as done in the present study, retards the normal development of cellular iron storage in the hippocampal MF system.[58] This may reduce the release of free iron in the hippocampus in the course of excitatory stimulation,[7] and hence reduce oxidative stress. In addition, retardation of hippocampal MF development may be protective since Stafstrom *et al.*[59] have shown that KA neurotoxicity rapidly increases in severity with age in juvenile rats.

Although iron supplementation for 2 weeks does not restore cellular iron storage in the MF system, combined iron + zinc supplementation does.[58] This may explain why in the present study iron supplementation did not abolish protection by ID in CA1, whereas combined iron and zinc supplementation did. Possibly, combined iron and zinc supplementation normalizes both iron storage and neurotransmission functions in the MF system. Consequently, the excitatory output from CA3 to CA1 is restored, and hence the increased neurotoxicity to CA1. The suggestion that iron storage is involved in vulnerability to KA is supported by recent studies showing accumulation of iron in hippocampus following KA-induced seizures.[27,28,31]

There is still the question of why the CA3 region remained protected in ID rats that received combined iron and zinc supplementation. Possibly, neurotoxicity in CA3 depends on several reciprocal connections with brain regions besides the DG. The entorhinal cortex is one major source of excitatory input to DG. The entorhinal cortex itself is damaged early following KA injection (6–12 h) due to edema, whereas apoptotic neurodegeneration in CA3 is slower, developing gradually over several days.[6] Thus, early edema-associated damage may remove some excitatory inputs to CA3, thus attenuating apoptotic events in CA3, but not in CA1.

Furthermore, some protective mechanisms besides reduced iron levels may be operating in ID rats. ID rats eat less and their body weight is lower than age-matched controls. Food restriction has neuroprotective actions[60] and one of the reported effects of food restriction is reduction in iron accumulation and oxidative stress.[61,62] In the present study, however, ID rats supplemented with iron, zinc, or iron + zinc had similar body weight and yet different neurotoxicity patterns. Thus, protective mechanisms secondary to dietary restriction do not account for the present findings.

Effects of ID on Gliosis

ID reduced KA-induced gliosis in all regions, except for hippocampal CA1. Furthermore, in hippocampal DG, KA induced gliosis, but not cellular damage. In the hippocampal CA3 field, combined iron + zinc restored gliosis, but did not abolish the protection from cellular damage conferred by ID.

It is possibe that "gliosis" should not be viewed as a unitary process, but rather as multiple processes, some of which are sensitive to iron and some of which are not. Reactive microglia are apparently involved in causation of cellular damage,[63] but reactive microglia are also involved in synaptic plasticity and reorganization.[64] Iron may be involved in both types of microglial action since iron participates in generation of redox-sensitive transcription factor NF-kB[65,66] and inflammatory cytotoxic cytokines such as tumor necrosis factor (TNF) alpha and interleukin 1 and 6,[67] which affect neural plasticity and reactive microgliosis.[68] Zinc also may be involved since microglia are sensitive to glutamate via N-methyl-D-aspartate (NMDA) receptors.[69] It is possible that the addition of zinc modulates functional states of microglia via zinc modulation of NMDA receptor function.[70] Furthermore, there is evidence that microglia secrete factors that enhance neuronal excitability in the hippocampus.[71]

SUMMARY

The present findings showing protection by iron deprivation corroborate evidence that iron-induced oxidative stress has a role in KA-induced neurotoxicity.[7,9,52] Furthermore, the present study joins a series of studies in which nutritional iron deprivation and iron chelation with desferrioxamine reduce ischemia/reperfusion-induced edema in gerbil brains[38,72,73] and retard 6-OHDA-induced dopamine loss[21,23] and MPTP-induced degeneration.[22,24] In all cases, there is evidence for reduction of oxidative stress by reduction of available free iron.

The pathology of neurodegenerative diseases (Parkinson's disease, Alzheimer's disease, Huntington's chorea, amyotrophic lateral sclerosis) is associated with proliferation of reactive microglia (gliosis), accumulation of iron, and biochemical evidence for oxidative stress processes.[17,19,66,74-76] The mechanistic relationship between these phenomena is not yet understood, but evidence is now accumulating for iron metabolism misregulation involving mutated iron metabolism genes.[20] The present study exposes several cases of association and dissociation between iron and gliosis that warrant further analysis of iron's contribution to gliosis.

It is of interest to investigate to what extent iron and zinc may contribute to the long-term synaptic reorganization, which is believed to lead to recurrent seizures after KA treatment[77,78] and possibly in human TLE.[79] Thus, it is possible that the evidence from the present study on the role of iron in TLE will lead to novel pharmacological approaches in a syndrome that is rather resistant to currently available drugs.[80]

ACKNOWLEDGMENTS

This work was funded by a grant from the bureau of the chief scientist of the Israeli Ministry of Health, Israel Center for Psychopharmacology, Jerusalem, as well as by the National Parkinson Foundation (Miami, FL), the Stein Foundation (Philadelphia, PA), and the Golding and Friedman Parkinson Research Funds (Technion, Haifa).

REFERENCES

1. COLDER, B.W. et al. 1996. Neuronal synchrony in relation to burst discharge in epileptic human temporal lobes. J. Neurophysiol. **75:** 2496–2508.
2. BEN-ARI, Y. 1985. Limbic seizure and brain damage produced by kainic acid: mechanisms and relevance to human temporal lobe epilepsy. Neuroscience **14:** 375–403.
3. LOTHMAN, E.W. & R.C. COLLINS. 1981. Kainic acid induced limbic seizures: metabolic, behavioral, electroencephalographic, and neuropathological correlates. Brain Res. **218:** 299–318.
4. CHESLER, M. & K. KAILA. 1992. Modulation of pH by neuronal activity. TINS **15:** 396–402.
5. LASSMAN, H. et al. 1984. The role of brain edema in epileptic brain damage induced by systemic kainic acid injection. Neuroscience **13:** 691–704.
6. CHARRIAUT-MARLANGUE, D., A. AGOUN-ZOUAOUI, Y. REPRESA & Y. BEN-ARI. 1996. Apoptotic features of selective neuronal death in ischemia, epilepsy, and gp120 toxicity. TINS **19:** 109–114.
7. COYLE, J.T. & P. PUTTFARCKEN. 1993. Oxidative stress, glutamate, and neurodegenerative disorders. Science **262:** 689–695.
8. HIRATA, H. & J.L. CADET. 1997. Kainate-induced hippocampal DNA damage is attenuated in superoxide dismutase transgenic mice. Mol. Brain Res. **48:** 145–148.
9. KONDO, T. et al. 1997. DNA fragmentation and prolonged expression of c-fos, c-jun, and hsp70 in kainic acid–induced neuronal cell death in transgenic mice overexpressing human CuZn–superoxide dismutase. J. Cereb. Blood Flow Metab. **17:** 241–256.
10. CHEN, S.T. & J.I. CHUANG. 1999. The antioxidant melatonin reduces cortical neuronal death after intrastriatal injection of kainate in the rat. Exp. Brain Res. **124:** 241–247.
11. LOCKHART, B. et al. 2001. Protective effect of the antioxidant 6-ethoxy-2,2-pentamethylen-1,2-dihydroquinoline (S 33113) in models of cerebral neurodegeneration. Eur. J. Pharmacol. **416:** 59–68.
12. KIM, H.C. et al. 2000. Phenidone prevents kainate-induced neurotoxicity via antioxidant mechanisms. Brain Res. **874:** 15–23.
13. CECCON, M. et al. 2000. Intracellular glutathione levels determine cerebellar granule neuron sensitivity to excitotoxic injury by kainic acid. Brain Res. **862:** 83–89.
14. SKAPER, S.D. et al. 1999. Excitotoxicity, oxidative stress, and the neuroprotective potential of melatonin. Ann. N.Y. Acad. Sci. **890:** 107–118.
15. HALLIWELL, B.J. & J.M.C. GUTTERIDGE. 1984. Oxygen toxicity, oxygen radicals, transition metals, and disease. Biochem. J. **219:** 1–14.
16. BEARD, J.L., J.R. CONNOR & B.C. JONES. 1993. Brain iron: location and function. Prog. Food Nutr. Sci. **17:** 183–221.
17. SHOHAM, S. & M.B.H. YOUDIM. 2000. Iron involvement in neural damage and microgliosis in models of neurodegenerative diseases. Cell. Mol. Biol. (Noisy-Le-Grand) **46:** 743–760.
18. YOUDIM, M.B.H., D. BEN-SHACHAR & P. RIEDERER. 1990. The role of monoamine oxidase, iron-melatonin interaction, and intracellular calcium in Parkinson's disease. J. Neural Transm. Suppl. **32:** 239–248.

19. YOUDIM, M.B.H. & P. RIEDERER. 1993. The role of iron in senescence of dopaminergic neurons in Parkinson's disease. J. Neural Transm. Suppl. **40:** 57–67.
20. YOUDIM, M.B.H. & P. RIEDERER. 2003. Iron in normal and pathological brain. *In* Encyclopedia of Neuroscience. Elsevier. Amsterdam. In press.
21. BEN-SHACHAR, D., G. ESHEL, J.P.M. FINBERG & M.B.H. YOUDIM. 1991. The iron chelator desferrioxamine (desferal) retards 6-hydroxydopamine-induced degeneration of nigrostriatal neurons. J. Neurochem. **56:** 1441–1444.
22. KAUR, D. *et al.* 2003. Genetic or pharmacological iron chelation prevents MPTP-induced neurotoxicity *in vivo*: a novel therapy for Parkinson's disease. Neuron **37:** 899–909.
23. BEN-SHACHAR, D. *et al.* 2003. Neuroprotection by a novel brain permeable iron chelator, VK-28, against 6-hydroxydopamine lesion in rats. Neuropharmacology. In press.
24. LAN, J. & D.H. JIANG. 1997. Excessive iron accumulation in the brain: a possible potential risk of neurodegeneration in Parkinson's disease. J. Neural Transm. **104:** 649–660.
25. LEVENSON, C.W. *et al.* 2003. Effect of dietary iron on motor behaviour and neuronal death in an experimental model of parkinsonism. J. Am. Aging Assoc. In press.
26. HILL, J.M. & C. SWITZER III. 1984. The regional distribution and cellular localization of iron in the rat brain. Neuroscience **11:** 596–603.
27. WANG, X.S., W.Y. ONG & J.R. CONNOR. 2002. Increase in ferric and ferrous iron in the rat hippocampus with time after kainate-induced excitotoxic injury. Exp. Brain Res. **143:** 137–148.
28. WANG, X.S., W.Y. ONG & J.R. CONNOR. 2002. A light and electron microscopic study of divalent metal transporter-1 distribution in the rat hippocampus, after kainate-induced neuronal injury. Exp. Neurol. **177:** 193–201.
29. WANG, X.S., W.Y. ONG & J.R. CONNOR. 2003. Quinacrine attenuates increases in divalent metal transporter-1 and iron levels in the rat hippocampus, after kainate-induced neuronal injury. Neuroscience **120:** 21–29.
30. SHOHAM, S., E. WERTMAN & P. EBSTEIN. 1992. Iron accumulation in the rat basal ganglia after excitatory amino acid injections—dissociation from neuronal loss. Exp. Neurol. **118:** 227–241.
31. ONG, W.Y. *et al.* 1999. A nuclear microscopic study of elemental changes in the rat hippocampus after kainite-induced neuronal injury. J. Neurochem. **72:** 1574–1579.
32. MATSUOKA, Y. *et al.* 1998. Kainic acid induction of heme oxygenase *in vivo* and *in vitro*. Neuroscience **85:** 1223–1233.
33. DALLMAN, P.R., M.N. SIIMES & E.C. MANIES. 1975. Brain iron: persistent deficiency following short term iron deprivation in the young rat. Br. J. Haematol. **31:** 209–215.
34. YOUDIM, M.B. & A.R. GREEN. 1978. Iron deficiency and neurotransmitter synthesis and function. Proc. Nutr. Soc. **37:** 173–179.
35. YOUDIM, M.B.H., D. BEN-SHACHAR & S. YEHUDA. 1989. Putative biological mechanisms of the effect of iron deficiency on brain biochemistry and behavior. Am. J. Clin. Nutr. **50:** 607–617.
36. YOUDIM, M.B. *et al.* 1980. The effects of iron deficiency on brain biogenic monoamine biochemistry and function in rats. Neuropharmacology **19:** 259–267.
37. SHOHAM, S., Y. GLINKA, Z. TANNE & M.B.H. YOUDIM. 1996. Brain iron: function and dysfunction in relation to cognitive processes. *In* Iron Nutrition in Health and Disease, pp. 205–218. Libbey. London.
38. TERADA, L.S., I.R. WILLINGHAM & J.E. REPINE. 1992. Iron and stroke. *In* Iron and Human Disease, pp. 313–327. CRC Press. Boca Raton, FL.
39. CHOI, D.W., M. YOKOYAMA & J. KOH. 1988. Zinc neurotoxicity in cortical cell culture. Neuroscience **24:** 67–79.
40. FREDERICKSON, C.J., M.D. HERNANDEZ & J.F. MCGINTY. 1989. Translocation of zinc may contribute to seizure-induced death of neurons. Brain Res. **480:** 317–321.
41. KIM, E.Y. *et al.* 1999. Zn2+ entry produces oxidative neuronal necrosis in cortical cell cultures. Eur. J. Neurosci. **11:** 327–334.
42. MCCALL, M.G. *et al.* 1960. Studies in iron metabolism. I. The experimental production of iron deficiency in the growing rat. Br. J. Nutr. **16:** 297–304.

43. TANNE, Z. et al. 1994. Ultrastructural and cytochemical changes in the heart of iron-deficient rats. Biochem. Pharmacol. **47:** 1759–1766.
44. BEN-SHACHAR, D., R. ASHKENAZI & M.B. YOUDIM. 1986. Long-term consequence of early iron-deficiency on dopaminergic neurotransmission in rats. Int. J. Dev. Neurosci. **4:** 81–88.
45. SPERK, G. et al. 1985. Kainic acid–induced seizures: dose-relationship of behavioural, neurochemical, and histopathological changes. Brain Res. **338:** 289–295.
46. GALLYAS, F., F.H. GULDNER, G. ZOLTAGY & R. WOLFF. 1990. Golgi-like demonstration of "dark" neurons with an argyrophil III method for experimental neuropathology. Acta Neuropathol. **79:** 620–628.
47. SHOHAM, S. & R.P. EBSTEIN. 1997. The distribution of beta amyloid precursor protein in rat cortex after kainate-induced seizures. Exp. Neurol. **147:** 361–376.
48. PAXINOS, G. & C. WATSON. 1986. The Rat Brain in Stereotaxic Coordinates. Academic Press. San Diego.
49. REGAN, R.F. & S.S. PANTER. 1996. Hemoglobin potentiates excitotoxic injury in cortical cell culture. Neurotrauma **13:** 223–231.
50. BRUCE, A.J. & M. BAUDRY. 1995. Oxygen free radicals in rat limbic structures after kainate-induced seizures. Free Radical Biol. Med. **18:** 993–1002.
51. CANDELARIO-JALIL, E. et al. 2001. Selective vulnerability to kainate-induced oxidative damage in different rat brain regions. J. Appl. Toxicol. **21:** 403–407.
52. LAYTON, M.E. & T.L. PAZDERNIK. 1999. Reactive oxidant species in piriform cortex extracellular fluid during seizures induced by systemic kainic acid in rats. J. Mol. Neurosci. **13:** 63–68.
53. TAKEDA, A., M. OHNUMA, J. SAWASHITA & S. OKADA. 1997. Zinc transport in the rat olfactory system. Neurosci. Lett. **225:** 69–71.
54. MENGUAL, E., C. CASANOVAS-AGUILAR, J. PEREZ-CLAUSELL & J.M. GIMENEZ-AMAYA. 2001. Thalamic distribution of zinc-rich terminal fields and neurons of origin in the rat. Neuroscience **102:** 863–884.
55. EIMERL, S. & M. SCHRAMM. 1993. Potentiation of ^{45}Ca uptake and acute toxicity mediated by the N-methyl-D-aspartate receptor: the effect of metal binding agents and transition metal ions. J. Neurochem. **61:** 518–525.
56. MATSUSHITA, K. et al. 1996. Effect of systemic zinc administration on delayed neuronal death in the gerbil hippocampus. Brain Res. **743:** 362–365.
57. OKAZAKI, M.M. & J.V. NADLER. 1988. Protective effects of mossy fiber lesions against kainic acid–induced seizures and neuronal degeneration. Neuroscience **26:** 763–781.
58. SHOHAM, S. & M.B.H. YOUDIM. 2002. The effects of iron deficiency and iron and zinc supplementation on rat hippocampus ferritin. J. Neural Transm. **109:** 1241–1256.
59. STAFSTROM, C.E., J.L. THOMPSON & G.L. HOLMES. 1992. Kainic acid seizures in the developing brain: status epilepticus and spontaneous recurrent seizures. Dev. Brain Res. **65:** 227–236.
60. MATTSON, M.P. 2000. Neuroprotective signaling and the aging brain: take away my food and let me run. Brain Res. **886:** 47–53.
61. COOK, C.I. & B.P. YU. 1998. Iron accumulation in aging: modulation by dietary restriction. Mech. Ageing Dev. **102:** 1–13.
62. SOHAL, R.S. & R. WEINDRUCH. 1996. Oxidative stress, caloric restriction, and aging. Science **273:** 59–63.
63. BOJE, K.M. & P.K. ARORA. 1992. Microglial-produced nitric oxide and reactive nitrogen oxides mediate neuronal cell death. Brain Res. **587:** 250–256.
64. EYUPOGLU, I.Y., I. BECHMANN & R. NITSCH. 2003. Modification of microglia function protects from lesion-induced neuronal alterations and promotes sprouting in the hippocampus. FASEB J. **17:** 1110–1111.
65. LIN, M. et al. 1997. Role of iron in NF-kB activation and cytokine gene expression by rat hepatic macrophages. Am. J. Physiol. **272:** G1355–G1364.
66. YOUDIM, M.B., E. GRUNBLATT & S. MANDEL. 1999. The pivotal role of iron in NF-kappa B activation and nigrostriatal dopaminergic neurodegeneration: prospects for neuroprotection in Parkinson's disease with iron chelators. Ann. N.Y. Acad. Sci. **890:** 7–25.
67. O'NEILL, L.A.J. & C. KALTSCMIDT. 1997. NF-kB: a crucial transcription factor for glial and neuronal cell function. TINS **20:** 252–258.

68. MILLER, W.J., D.G. MACGREGOR & T.W. STONE. 1994. Time course of purine protection against kainate-induced increase in hippocampal [3H]-PK11195 binding. Brain Res. Bull. **34:** 133–136.
69. TIKKA, T.M. & J.E. KOISTINAHO. 2001. Minocycline provides neuroprotection against N-methyl-D-aspartate neurotoxicity by inhibiting microglia. J. Immunol. **166:** 7527–7533.
70. XIE, X., R.C. HIDER & T.G. SMART. 1994. Modulation of GABA-mediated synaptic transmission by endogenous zinc in the immature rat hippocampus *in vitro*. J. Physiol. **478**(pt. 1): 75–86.
71. HEGG, C.C. & S.A. THAYER. 1999. Monocytic cells secrete factors that evoke excitatory synaptic activity in rat hippocampal cultures. Eur. J. Pharmacol. **385:** 231–237.
72. PATT, A. *et al.* 1988. Xanthine oxidase–derived hydrogen peroxide and iron contribute to ischemia-reperfusion-induced edema in gerbil brain. J. Clin. Invest. **81:** 1556–1564.
73. PATT, A. *et al.* 1990. Iron depletion or chelation reduces ischemia-reperfusion-induced edema in gerbil brains. J. Pediatr. Surg. **25:** 224–236.
74. CONNOR, J.R. 1992. Proteins of iron regulation in the brain in Alzheimer's disease. *In* Iron and Human Disease, pp. 366–395. CRC Press. Boca Raton, FL.
75. JELLINGER, K. *et al.* 1990. Brain iron and ferritin in Parkinson's and Alzheimer's diseases. J. Neural Transm. Parkinson's Dis. Dementia Sect. **2:** 327–340.
76. THOMPSON, K.J., S. SHOHAM & J.R. CONNOR. 2001. Iron and neurodegenerative disorders. Brain Res. Bull. **55:** 155–164.
77. CRONIN, J., A. OBENHAUS, C.R. HOUSER & F.E. DUDEK. 1992. Electrophysiology of dentate granule cells after kainate-induced synaptic reorganization of the mossy fibres. Brain Res. **573:** 305–310.
78. SLOVITER, R.S. 1992. Possible functional consequences of synaptic reorganization in the dentate gyrus of kainate-treated rats. Neurosci. Lett. **137:** 91–96.
79. BABB, T.L. *et al.* 1991. Synaptic reorganization by mossy fibres in human epileptic fascia dentata. Neuroscience **42:** 351–363.
80. OJEMANN, G.A. 1997. Treatment of temporal lobe epilepsy. Annu. Rev. Med. **48:** 317–328.

Manganese Neurotoxicity

ALLISON W. DOBSON, KEITH M. ERIKSON, AND MICHAEL ASCHNER

Department of Physiology and Pharmacology, and Interdisciplinary Program in Neuroscience, Wake Forest University School of Medicine, Winston-Salem, North Carolina, USA

ABSTRACT: Manganese is an essential trace element and it is required for many ubiquitous enzymatic reactions. While manganese deficiency rarely occurs in humans, manganese toxicity is known to occur in certain occupational settings through inhalation of manganese-containing dust. The brain is particularly susceptible to this excess manganese, and accumulation there can cause a neurodegenerative disorder known as manganism. Characteristics of this disease are described as Parkinson-like symptoms. The similarities between the two disorders can be partially explained by the fact that the basal ganglia accumulate most of the excess manganese compared with other brain regions in manganism, and dysfunction in the basal ganglia is also the etiology of Parkinson's disease. It has been proposed that populations already at heightened risk for neurodegeneration may also be more susceptible to manganese neurotoxicity, which highlights the importance of investigating the human health effects of using the controversial compound, methylcyclopentadienyl manganese tricarbonyl (MMT), in gasoline to increase octane. The mechanisms by which increased manganese levels can cause neuronal dysfunction and death are yet to be elucidated. However, oxidative stress generated through mitochondrial perturbation may be a key event in the demise of the affected central nervous system cells. Our studies with primary astrocyte cultures have revealed that they are a critical component in the battery of defenses against manganese-induced neurotoxicity. Additionally, evidence for the role of oxidative stress in the progression of manganism is reviewed here.

KEYWORDS: manganese; neurotoxicity; manganism; oxidative stress; reactive oxygen species

INTRODUCTION

Manganese is a trace element that is essential in the diet of all animals. While this metal is inhaled from the atmosphere, diet is normally a far greater source of human exposure to manganese. Because there are homeostatic systems of regulation for absorption and excretion of manganese in the body, the levels found in tissues are usually very stable, regardless of intake levels. However, manganese can accumulate in certain brain regions following elevated exposures, and manganese-induced neuro-

Address for correspondence: Michael Aschner, Ph.D., Department of Physiology and Pharmacology, Wake Forest University School of Medicine, Medical Center Boulevard, Winston-Salem, NC 27157-1083. Voice: 336-716-8530; fax: 336-716-8501.
maschner@wfubmc.edu

toxicity can ensue. The symptomatic cases of this neurotoxicity are known as manganism, and clinically this presents with a Parkinson-like motor dysfunction. Little is known about the detailed mechanisms of manganese neurotoxicity. Studies on manganese transport and other biochemical end points are reviewed here. In particular, evidence for the involvement of oxidative stress in the progression of manganism is discussed.

ESSENTIALITY

Manganese is found in all body tissues as it is essential for many ubiquitous enzymatic reactions, including synthesis of amino acids, lipids, proteins, and carbohydrates.[1] Also particularly noteworthy for neurotoxicity studies is the requirement for manganese in the reactions catalyzed by arginase, glutamine synthetase, phosphoenolpyruvate decarboxylase, and manganese-dependent superoxide dismutase, to name a few.[2] The National Academies' Institute of Medicine has set dietary reference intakes in their 2002 report.[3] The report sets an adequate intake (AI) level for manganese at 2.3 mg per day for men and 1.8 mg per day for women. The tolerable upper intake level (UL) is set at 11 mg for adults. Studies have demonstrated that the female gastrointestinal tract is more efficient at absorbing manganese than in men.[4] Iron availability may be related to this difference in absorbance. Furthermore, adjustments in the manganese requirement were made in consideration of pregnancy (2.0 mg/day), lactation (2.6 mg/day), and the developmental stages of childhood (range of 0.003–2.2 mg/day, depending on age and sex).[3]

Manganese deficiency can cause a wide range of problems, including impaired growth, skeletal defects, reduced fertility, birth defects, abnormal glucose tolerance, and altered lipid and carbohydrate metabolism.[5,6] However, these effects were observed in lab animals and this deficiency is not clinically recognized in humans.

SOURCES OF MANGANESE

Most adults have a daily intake of manganese below 5 mg manganese/kg, with a reported range of 0.9 to 10 mg manganese per day.[1,7] Grains, tea, and green leafy vegetables contain the highest amounts of manganese in the normal adult male diet as reported in the Total Diet Study.[8] The manganese content of human milk has been found to vary with stages of lactation.[9–11] There are also reports from both rat and human studies of much higher manganese absorption in the neonatal period.[12–15] This evidence is consistent with the higher manganese levels believed to be required for brain development at early stages. Infant formulas tend to have more manganese than human milk, and this has been a cause for some concern.[16,17]

Individuals receiving total parenteral nutrition (TPN) make up a subpopulation that is at even higher risk for manganese toxicity. The TPN solution can be formulated to contain manganese, but sometimes the manganese is found as an unintended contaminant.[18,19] Because the normal regulating mechanisms for manganese metabolism are bypassed (i.e., the gut), 100% of the manganese in the TPN solution enters the body as compared to approximately 5% of that taken orally. There have been reported intoxications from TPN solutions containing 0.1 mg Mn/day. The symptoms and manganese measurements were consistent with other forms of manganese toxicity; withdrawal from the TPN alleviates symptoms.[20,21]

Airborne manganese can exist as fumes, aerosols, or suspended particulate matter.[7] This manganese "dust" can be inhaled and deposited in parts of the upper or lower respiratory tract, where the manganese can then be absorbed into the bloodstream. The levels of manganese in the air vary, depending on the industries nearby, wind erosion, and other factors. Ferroalloy production, iron and steel foundries, and combustion emissions from power plants and coke ovens make significant contributions to the concentration of manganese in air.[22] The average levels reported by ATSDR[7] for urban and nonurban air are 33 and 5 ng Mn/m^3, respectively. Nevertheless, the average daily Mn intake from ambient air is estimated to be <2 μg Mn/day.[23,24]

The gasoline additive, methylcyclopentadienyl manganese tricarbonyl (MMT), is a somewhat controversial source of additional airborne manganese. This compound may be used as a replacement for lead as an antiknock agent, and the debate surrounding its use in the United States was recently reviewed in *Science*.[25] Upon combustion in automobile engines, MMT yields a complex mixture of phosphate, sulfate, and oxide forms of manganese. It has been used in Canada for over 10 years, and studies of the Canadian cities with the most traffic have shown the air manganese content to be near or below the current inhalation (RfC) reference concentration for inhalable manganese, which is 0.05 μg Mn/m^3 as set by the United States Environmental Protection Agency.[24,26–28]

ABSORPTION AND TRANSPORT

Only about 1–5% of the manganese ingested by humans is absorbed into the body by the gastrointestinal tract under normal conditions.[4,29,30] This value is reportedly higher when measurements are taken less than 24 h postingestion, but similar studies in animals indicate that much of the manganese that is retained for shorter time periods is localized to the liver and intestinal tract and eliminated through biliary excretion.[31] As such, it would not reach the brain or other systemic tissues in significant amounts.

The molecular details of oral manganese absorption are not well understood. There is one line of evidence suggesting an active transport process,[32] and another group has demonstrated absorption through a simple passive diffusion-like process.[33] Furthermore, there are many other factors that have been found to affect manganese absorption, including dietary manganese levels,[1,29,34–36] dietary levels of various minerals,[37–39] age and developmental state of the individual,[13–15,40] and especially iron status. There seems to be an inverse relationship between body iron stores and manganese absorption, perhaps due to competition for transport machinery such as DMT-1 (divalent metal transporter, also known as DCT-1 or nramp-2).[41,42] Several studies have demonstrated that iron deficiency increases transport of orally administered manganese into the body as well as delivery to the brain.[1,43–45]

Absorption of manganese via the lungs has only recently been investigated and it seems to depend largely on particle solubility. Whereas $MnCl_2$, which is a soluble salt, is quickly taken into the bloodstream, insoluble MnO_2 given at similar doses was very slowly absorbed and at much lower overall levels.[46] This report also showed that the soluble salt was more readily delivered to the brain. More recently,

Dorman et al.[47] also showed that inhaled $MnSO_4$ was cleared from the lung faster than the less soluble phosphate or tetroxide manganese compounds, and transport into the brain and other tissues reflected this pattern based on particle solubility as well.

Blood manganese is largely bound to β-globulin and albumin (~80%), and a small percentage of trivalent (3+) manganese is found complexed to transferrin.[48–50] Nevertheless, because of the large number of unoccupied binding sites, transferrin has been implicated as a potential transport system for manganese to traverse the blood-brain barrier and other membranes.[51] No other specific transporters are known for blood manganese. Typical serum concentrations of manganese are in the range of 0.8–2.1 μg Mn/L. Neonates generally have the highest levels, a decreasing trend is observed through the first year, and adults have the lowest serum manganese content (reviewed in refs. 52 and 53).

Distribution of manganese to the body tissues is fairly homogeneous. Increased concentration of manganese is found in tissues rich in mitochondria and pigmentation. Bone, liver, pancreas, and kidney tend to have higher manganese levels than other tissues.[54,55] Liver especially accumulates manganese after high exposures, and most absorbed manganese is excreted in bile. Liver disease, therefore, is a risk factor for increased accumulation of manganese in the brain.[56–58]

Transport of manganese into the central nervous system (CNS) has been directly investigated in a limited number of studies. Together, these reports implicate three sites of manganese entry into the brain. The cerebral capillaries, the cerebrospinal fluid (CSF; via choroid plexus transport), and the olfactory nerve are all potential locations of manganese import.[59–61] Acute bolus intravenous injections of large amounts of manganese leads to a saturable transferrin-independent transport across the blood-brain barrier via either active or passive processes.[59,62] The choroid plexus, the site of CSF production, is where ^{54}Mn first appears in rodent brain after bolus injection into the circulation.[63,64] However, at relevant manganese exposure levels, the capillary endothelium seems to represent the route that is physiologically most germane to manganese entry into the CNS. Furthermore, the likeliest modes of transport are by transferrin/transferrin receptor and DMT-1.

Iron, manganese, and other metals are able to be complexed and carried by some of the same transporters. Transferrin/transferrin receptor and DMT-1, especially, are thought to transport both of these metals, with iron being far more prevalent under normal circumstances. Evidence from Suarez and Eriksson[65] and Aschner and Gannon[66] strongly suggests transport of trivalent manganese complexed to transferrin into the brain capillary endothelium. As such, the high concentration of transferrin receptors in the nucleus accumbens and caudate putamen, which provide efferent fibers to areas rich in manganese (ventral pallidum, globus pallidus, and substantia nigra), is consistent with transferrin-mediated manganese transport.

The role of DMT-1 in brain manganese transport is currently an area of intense investigation. It has been suggested that much of the manganese that gains access to the CNS does so via DMT-1 in brain endothelium. Absorption of manganese in the gut is thought to be mediated by DMT-1.[52] Studies of the Belgrade rat, which carries a mutation in the DMT-1 gene, show that (in addition to frank deficiency in uptake of iron) the homozygote demonstrates lower uptake of radiolabeled manganese than the heterozygote.[67] The Belgrade rat may also be a good model for dissecting the mechanisms of manganese transport into the brain. Additional experiments to elucidate the role of DMT-1 in manganese transport into rat brain endothelial cells are under way.

It has been well documented that xenobiotics can travel directly to the brain via the olfactory system.[68] Axonal transport of manganese has also been conclusively demonstrated.[61,63,69–72] Delivery of inhaled manganese is likely through direct intra-axonal transport,[73] and it has been reported in rat, mouse, and freshwater pike after intranasal instillation.[69,73,74] Additionally, Dorman et al.[75] have studied inhalation of various manganese-containing particulates and also found delivery along the olfactory route. However, the significance of the contribution of this pathway toward manganese toxicity is not yet clear. The striatum and other nonolfactory brain structures do not seem to accumulate much manganese through this route.[61,75,76] Further, there are substantial physiological differences known between human and rodent nasal and brain anatomy that complicate interpretation of comparative studies.[77]

Manganese toxicity studies have revealed that distribution of the metal to the various brain regions is not homogeneous and may even differ across species. Magnetic resonance imaging (MRI) techniques show that, in exposed humans and macaque monkeys, manganese concentrations are highest in striatum, globus pallidus, and substantia nigra.[21,78–80] Mixed results have been observed in rodent studies, however. Brenneman and coworkers[81] have reported that rat striatum and globus pallidus do not preferentially accumulate manganese after excess exposure. However, a very recent study in our laboratory showed that, after dietary iron deprivation, manganese accumulated in globus pallidus, hippocampus, and substantia nigra of rat brain.[45] This suggests that iron deficiency in humans might also lead to a higher tendency toward manganese accumulation in brain regions normally rich in iron. Previous work has demonstrated higher absorbance and accumulation in the brain in iron-deficient animals.[45]

TOXICITY

Inhalation of particulate manganese is the most recognized occupational risk for human toxicity. This manganese dust in various forms irritates lungs of humans and animals, causing an inflammatory response,[82] as do many other particulates. Studies thus far have not clarified whether this response is specific to manganese or is representative of the general reaction to inhaled particulates. A recent report shows that oxidative stress in lungs and heart is observed after 5-h inhalation exposure to concentrated ambient particles containing a mixture of metals including Mn.[83] This suggests that the lung inflammation may be a general response to inhaled metal particulates. Nevertheless, there are significant neurological effects specific to manganese particulate inhalation. Impotence and loss of libido have been reported in manganese-exposed workers,[7] but the later-stage neurological effects are the most compelling cause for concern about manganese exposure.

Chronic exposure to high levels of inhalable manganese (>1–5 mg Mn/m^3) is the most frequently observed cause of manganese-induced neurotoxicity.[7,84,85] Ingestion of very large amounts of manganese in water from contaminated wells has also been reported to cause neurotoxicity.[86,87] However, confounding variables in both of these studies make the interpretation questionable. Additionally, Vieregge and coworkers[88] studied human populations chronically consuming high levels of manganese and they were unable to confirm adverse health effects even after 10 years of exposure.

The disorder known as manganism is strongly associated with elevated levels of manganese in the brain. Specifically, structures of the basal ganglia—caudate putamen, globus pallidus, substantia nigra, and subthalamic nuclei, all of which contain substantial levels of nonheme iron—represent regions of highest manganese concentration.[85] The earliest symptoms associated with abnormal manganese accumulation are psychiatric. Compulsive or violent behavior, emotional instability, and hallucinations are characteristic, and patients may also suffer from fatigue, headache, muscle cramps, loss of appetite, apathy, insomnia, and diminished libido. The most severe forms of manganism present with prolonged muscle contractions (known as dystonia), decreased muscle movement (known as hypokinesia), rigidity, and muscle tremors. The physical traits of this disorder thus resemble Parkinson's disease, but there are distinguishing features.[78,85,89] While generalized bradykinesia and rigidity are

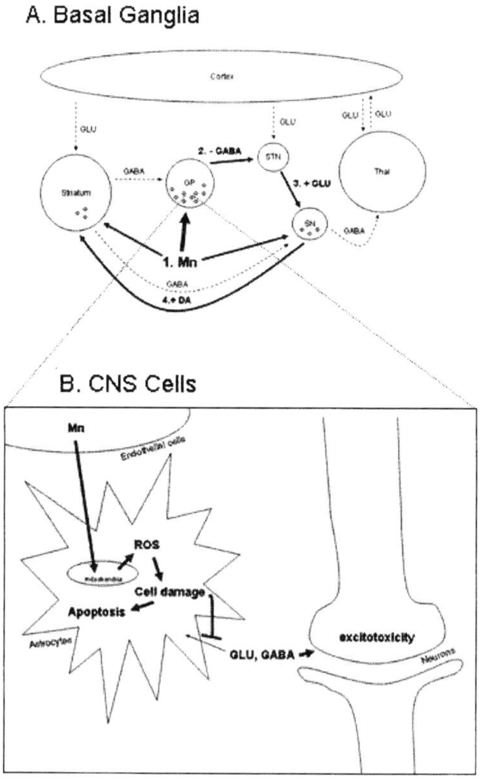

FIGURE 1. Manganese-induced neurotoxicity. **(A)** Regions of the basal ganglia are affected by manganese accumulation: (1) Manganese preferentially accumulates in globus pallidus (GP), leading to (2), decreased GABA innervation into subthalamic nuclei (STN), which leads to (3), unregulated glutamate (GLU) input into the substantia nigra (SN) and (4), increased dopamine input into the striatum. **(B)** Manganese induces oxidative stress through mitochondrial perturbation, leading to astrocytic dysfunction and imbalanced extracellular neurochemistry. (Adapted from ref. 107.)

found in both syndromes, the dystonia of manganism is a neurological sign attributed to damage to the globus pallidus[78] and is only minimally observed in Parkinson's patients. Other features of manganism that differ from parkinsonism were noted in a comprehensive survey of patients with these disorders, and they include less frequent resting tremor, a propensity to fall backward, little or no sustained response to levodopa therapy, and normal fluorodopa uptake, as observed by positron emission tomography (PET).[78,85]

Glutamate from cortical neurons along with γ-amino-butyric acid (GABA) and dopamine from other basal ganglia structures all influence striatal control of motor activity.[90] In Parkinson's disease, the nigrostriatal pathway is affected due to demise of dopaminergic neurons in the substantia nigra. Based on these observations, a model has been proposed in which the etiological damage in manganism is likely to occur to the output pathways downstream of the nigrostriatal dopaminergic pathway (FIG. 1A).[78,85,91]

Animal studies are providing insight into the hypothesis that the developing brain and nervous system may be substantially more sensitive to accumulation of manganese than adult brain. A recent study by Dorman et al.[92] found that rat neonates chronically exposed to oral manganese chloride ($MnCl_2$) accumulate more brain manganese than adult rats given equal doses (0, 25, or 50 mg/kg body wt/day). This evidence may suggest increased susceptibility for the developing nervous system, but an alternative explanation is that it could reflect the increased requirement for manganese in the developing CNS.[52] Additionally, there are species differences to consider. The literature available on manganese toxicity in nonhuman primates reports data similar to those obtained from humans with manganism.[52,78] However, rodent studies have yielded mixed findings concerning regional brain manganese distribution and neurochemical and neuropathological responses to manganese exposure.[81,92–94] Further, the behavioral changes observed in manganese-poisoned humans are not replicable in rodents, which further confounds interpretation of results from those studies in assessing the consequences of human exposure.

MANGANESE-INDUCED OXIDATIVE STRESS

Oxidative stress and its effects on mitochondrial energy metabolism have lately been implicated in a wide range of pathological processes, and especially in neurodegenerative conditions such as Parkinson's or Alzheimer's disease.[95] Furthermore, the intense investigation surrounding the free radical theory of aging is leading many scientists to believe that aging mitochondria are the primary culprits.[96] They are more susceptible to oxidative damage and less efficient at repairing this damage than young mitochondria. Indeed, there is good evidence that in multicellular organisms, such as *C. elegans*, oxidative stress is an important factor in limiting life span.[97] Witholt et al.[98] recently investigated increased risk to manganese-induced damage using a preparkinsonian rat model treated with low cumulative doses of manganese. They report exacerbation of both neurochemical and motor function changes in the senescent group. A previous report showed that exposure of neurons to MMT resulted in rapid increases in reactive oxygen species followed by mitochondrially induced apoptosis.[99,100] However, it is noteworthy that combustion of MMT in cars yields various manganese salts, the most abundant being phosphate and sulfate.[23,24]

Oxygen radicals can damage components of the electron transport and oxidative phosphorylation machinery, and this leads to generation of more reactive oxygen species (namely, superoxide). The new radicals exacerbate the damage and a "downward spiral" ensues.[101] In this scenario, cells are ultimately subjected to energy failure as ATP production declines. The membrane potential is lost as the mitochondria undergo permeability transition, which then leads to cell death.[102] This mitochondrial dysfunction coincides with decreased cerebral metabolic rates in Alzheimer's disease, Parkinson's disease, Huntington's disease, and other neurodegenerative disorders.[102,103] Whether the mitochondrial demise has a causal role or appears as a secondary effect in these disorders is still a subject of intense debate. Albin et al.[104] reviewed a variety of basal ganglia toxicants and concluded that the probable mechanism of action for almost all known basal ganglia neurotoxins is inhibition of mitochondrial function. Studies of this interrelationship are clouded by the fact that mitochondrial function declines as a normal part of the aging process, and age itself is a risk factor for these neurodegenerative diseases. Altogether, the literature seems to point to a strong association between aging, mitochondrial impairment, and oxidative stress.

Optimal brain function is dependent upon cross talk between multiple cell types. In particular, astrocytes produce trophic factors, regulate neurotransmitter and ion concentrations, and remove toxins and debris from the extracellular space around the neurons. Therefore, oxidative impairment of astrocytic functions has the potential to indirectly induce and/or exacerbate neuronal dysfunction.[105] Specifically, removal of the neurotransmitters, GABA, glutamate, and dopamine, from the extracellular fluid can be altered by manganese treatment.[45,106–108] Neurons vicinal to the affected astrocytes are then potentially made susceptible to excitotoxicity or other downstream dysfunction because of the imbalanced extracellular neurochemistry (see FIG. 1B).

On the subcellular level, manganese is most concentrated in mitochondria.[109] However, the overall percentage of manganese found in the mitochondria of specific brain regions did not increase after manganese exposure in neonatal rats,[81] which indicates that there is not additional selective uptake into this organelle at higher manganese levels. Nevertheless, decreased complex I activity, increased oxidative damage, and altered activities of antioxidant defense enzymes have been demonstrated in Parkinson's disease.[95] This supports a growing body of literature on oxidative stress in neurodegeneration.

Gavin et al.[110] showed evidence suggesting that the ATPase complex is inhibited at very low levels of mitochondrial manganese and that complex I is inhibited only at higher manganese concentrations. In another study, treatment of striatal neurons with manganese showed dose-dependent losses of mitochondrial membrane potential and complex II activity.[111] Collectively, these results indicate that manganese may trigger apoptotic-like neuronal death secondary to mitochondrial dysfunction. However, it is possible that necrosis may be involved to some extent as Roth et al.[112] found that caspases were not involved in manganese-induced neuronal death.

Zwingmann and colleagues recently reported that neurons treated for 5 days with $MnCl_2$ are extremely susceptible to oxidative stress and energy failure through the resulting mitochondrial dysfunction,[113] whereas astrocytes fare slightly better after the same treatment. When the cells were cocultured, comparative NMR data showed "disturbed astrocytic function and a failure of astrocytes to provide neurons with substrates for energy and neurotransmitter metabolism, leading to deterioration of neuronal antioxidant capacity (decreased glutathione levels) and energy metabolism". These results are consistent with previous reports from our lab and others demonstrating the important role of astrocytes in effectively buffering the extracellular environment to protect the more sensitive neurons. It has also been reported in many cases that astrocytes have higher levels of glutathione and some other antioxidant defenses than neurons.[114,115]

A final factor in manganese toxicity is the oxidation state of the metal. It has been shown that trivalent manganese is more effective at inhibiting complex I,[116-118] but the divalent form is by far the predominant species within cells and is largely bound to ATP.[118] Nevertheless, manganese in any state will spontaneously give rise to infinitesimal amounts of trivalent manganese, and HaMai et al.[119] demonstrated that trivalent manganese, even at trace amounts, can cause formation of reactive oxygen species. Interestingly, the mitochondria also paradoxically rely heavily on manganese for antioxidant protection as it is the critical cofactor for the important superoxide dismutase enzyme specific to this organelle. In fact, mice lacking the mitochondrial isoform of SOD have a mean life span of 8 days, whereas mice deficient in cytosolic or extracellular SODs have a very benign phenotype.[97]

The mechanisms of manganese toxicity in the brain are slowly being elucidated. At present, the preponderance of evidence indicates that oxidative stress and mitochondria play major roles in the manganese-induced degenerative processes that lead to dysfunction in the basal ganglia. It is well established that astrocytes are critical in defending the more sensitive neurons, and we will continue to study their role in preventing degenerative mechanisms. Furthermore, there is insufficient data to determine whether any subpopulations are in fact more susceptible to manganese-induced neurodegeneration. These issues clearly warrant future investigations.

ACKNOWLEDGMENTS

This work was supported by NIH Grants NIEHS 10563 (to M. Aschner) and NIH T32-ES-07331 Multidisciplinary Training in Molecular Toxicology (to A. W. Dobson) and by the RJR–Leon Golberg Fellowship in Pharmacology and Toxicology (to K. M. Erikson).

REFERENCES

1. FINLEY, J.W. & C.D. DAVIS. 1999. Manganese deficiency and toxicity: are high or low dietary amounts of manganese cause for concern? Biofactors **10:** 15–24.
2. TAKEDA, A. 2003. Manganese action in brain function. Brain Res. Brain Res. Rev. **41:** 79–87.
3. NAS. 2002. Dietary reference intakes for vitamin A, vitamin K, arsenic, boron, chromium, copper, iodine, iron, manganese, molybdenum, nickel, silicon, vanadium, and zinc. Available at www.nap.edu/books/0309072794/html/.
4. FINLEY, J.W., P.E. JOHNSON & L.K. JOHNSON. 1994. Sex affects manganese absorption and retention by humans from a diet adequate in manganese. Am. J. Clin. Nutr. **60:** 949–955.
5. FREELAND-GRAVES, J. & C. LLANES. 1994. Models to study manganese deficiency. *In* Manganese in Health and Disease, pp. 115–120. CRC Press. Boca Raton, FL.
6. KEEN, C.L. *et al.* 1999. Nutritional aspects of manganese from experimental studies. Neurotoxicology **20:** 213–223.

7. AGENCY FOR TOXIC SUBSTANCES AND DISEASE REGISTRY. 2000. Toxicological profile for manganese. U.S. Department of Health and Human Services Public Health Service. Available at www.atsdr.cdc.gov/toxprofiles/tp151.html/.
8. PENNINGTON, J.A. & S.A. SHOEN. 1996. Total Diet Study: estimated dietary intakes of nutritional elements, 1982–1991. Int. J. Vitam. Nutr. Res. **66:** 350–362.
9. CASEY, C.E., K.M. HAMBIDGE & M.C. NEVILLE. 1985. Studies in human lactation: zinc, copper, manganese, and chromium in human milk in the first month of lactation. Am. J. Clin. Nutr. **41:** 1193–1200.
10. STASTNY, D., R.S. VOGEL & M.F. PICCIANO. 1984. Manganese intake and serum manganese concentration of human milk-fed and formula-fed infants. Am. J. Clin. Nutr. **39:** 872–878.
11. VAUGHAN, L.A., C.W. WEBER & S.R. KEMBERLING. 1979. Longitudinal changes in the mineral content of human milk. Am. J. Clin. Nutr. **32:** 2301–2306.
12. GRUDEN, N. 1977. Suppression of transduodenal manganese transport by milk diet supplemented with iron. Nutr. Metab. **21:** 305–309.
13. KEEN, C.L., J.G. BELL & B. LONNERDAL. 1986. The effect of age on manganese uptake and retention from milk and infant formulas in rats. J. Nutr. **116:** 395–402.
14. ZLOTKIN, S.H., S. ATKINSON & G. LOCKITCH. 1995. Trace elements in nutrition for premature infants. Clin. Perinatol. **22:** 223–240.
15. DORNER, K. *et al.* 1989. Longitudinal manganese and copper balances in young infants and preterm infants fed on breast-milk and adapted cow's milk formulas. Br. J. Nutr. **61:** 559–572.
16. KRACHLER, M. & E. ROSSIPAL. 2000. Concentrations of trace elements in extensively hydrolysed infant formulae and their estimated daily intakes. Ann. Nutr. Metab. **44:** 68–74.
17. LONNERDAL, B. 1994. Nutritional aspects of soy formula. Acta Paediatr. Suppl. **402:** 105–108.
18. HAMBIDGE, K.M. *et al.* 1989. Plasma manganese concentrations in infants and children receiving parenteral nutrition. J. Parenter. Enteral Nutr. **13:** 168–171.
19. KURKUS, J., N.W. ALCOCK & M.E. SHILS. 1984. Manganese content of large-volume parenteral solutions and of nutrient additives. J. Parenter. Enteral Nutr. **8:** 254–257.
20. BERTINET, D.B. *et al.* 2000. Brain manganese deposition and blood levels in patients undergoing home parenteral nutrition. J. Parenter. Enter. Nutr. **24:** 223–227.
21. KAFRISTA, Y. *et al.* 1998. Long-term outcome of brain manganese deposition in patients on home parenteral nutrition. Arch. Dis. Child. **79:** 263–265.
22. LIOY, P.J. 1983. Air pollution emission profiles of toxic and trace elements from energy related sources: status and needs. Neurotoxicology **4:** 103–112.
23. LYNAM, D.R. *et al.* 1999. Environmental effects and exposures to manganese from use of methylcyclopenta-dienyl manganese tricarbonyl (MMT) in gasoline. Neurotoxicology **20:** 145–150.
24. ZAYED, J. *et al.* 1999. Airborne manganese particulates and methylcyclopentadienyl manganese tricarbonyl (MMT) at selected outdoor sites in Montreal. Neurotoxicology **20:** 151–157.
25. KAISER, J. 2003. Manganese: a high-octane dispute. Science **300:** 926–928.
26. CLAYTON, C.A. *et al.* 1999. Estimating distributions of long-term particulate matter and manganese exposures for residents of Toronto, Canada. Atmos. Environ. **33:** 2515–2526.
27. LORANGER, S. & J. ZAYED. 1997. Environmental contamination and human exposure to airborne total and respirable manganese in Montreal. J. Air Waste Manag. Assoc. **47:** 983–989.
28. PELLIZARI, E.D. *et al.* 1999. Particulate matter and manganese exposures in Toronto, Canada. Atmos. Environ. **33:** 721–734.
29. DAVIS, C.D., L. ZECH & J.L. GREGER. 1993. Manganese metabolism in rats: an improved methodology for assessing gut endogenous losses. Proc. Soc. Exp. Biol. Med. **202:** 103–108.
30. DAVIDSSON, L. *et al.* 1988. Intrinsic and extrinsic labeling for studies of manganese absorption in humans. J. Nutr. **118:** 1517–1521.
31. BALLATORI, N., E. MILES & T.W. CLARKSON. 1987. Homeostatic control of manganese excretion in the neonatal rat. Am. J. Physiol. **252:** R842–R847.

32. GARCIA-ARANDA, J.A., R.A. WAPNIR & F. LIFSHITZ. 1983. In vivo intestinal absorption of manganese in the rat. J. Nutr. **113:** 2601–2607.
33. BELL, J.G., C.L. KEEN & B. LONNERDAL. 1989. Higher retention of manganese in suckling than in adult rats is not due to maturational differences in manganese uptake by rat small intestine. J. Toxicol. Environ. Health **26:** 387–398.
34. BRITTON, A.A. & G.C. COTZIAS. 1966. Dependence of manganese turnover on intake. Am. J. Physiol. **211:** 203–206.
35. DORMAN, D.C. et al. 2001. Influence of dietary manganese on the pharmacokinetics of inhaled manganese sulfate in male CD rats. Toxicol. Sci. **60:** 242–251.
36. MALECKI, E.A. et al. 1996. Biliary manganese excretion in conscious rats is affected by acute and chronic manganese intake, but not by dietary fat. J. Nutr. **126:** 489–498.
37. DAVIDSSON, L. et al. 1991. The effect of individual dietary components on manganese absorption in humans. Am. J. Clin. Nutr. **54:** 1065–1070.
38. LAI, J.C. et al. 1999. Manganese mineral interactions in brain. Neurotoxicology **20:** 433–444.
39. PLANELLS, E. et al. 2000. Effect of magnesium deficiency on enterocyte Ca, Fe, Cu, Zn, Mn, and Se content. J. Physiol. Biochem. **56:** 217–222.
40. GRUDEN, N. 1977. Suppression of transduodenal manganese transport by milk diet supplemented with iron. Nutr. Metab. **21:** 305–309.
41. DAVIS, C.D., E.A. MALECKI & J.L. GREGER. 1992. Interactions among dietary manganese, heme iron, and nonheme iron in women. Am. J. Clin. Nutr. **56:** 926–932.
42. GUNSHIN, H. et al. 1997. Cloning and characterization of a mammalian proton-coupled metal-ion transporter. Nature **388:** 482–488.
43. CHANDRA, S.V. & G.S. SHUKLA. 1976. Role of iron deficiency in inducing susceptibility to manganese toxicity. Arch. Toxicol. **35:** 319–323.
44. SHUKLA, A., K.N. AGARWAL & G.S. SHUKLA. 1989. Effect of latent iron deficiency on metal levels of rat brain regions. Biol. Trace Elem. Res. **22:** 141–152.
45. ERIKSON, K.M. et al. 2002. Manganese accumulates in iron-deficient rat brain regions in a heterogeneous fashion and is associated with neurochemical alterations. Biol. Trace Elem. Res. **87:** 143–156.
46. ROELS, H. et al. 1997. Influence of the route of administration and the chemical form ($MnCl_2$, MnO_2) on the absorption and cerebral distribution of manganese in rats. Arch. Toxicol. **71:** 223–230.
47. DORMAN, D.C. et al. 2001. Influence of particle solubility on the delivery of inhaled manganese to the rat brain: manganese sulfate and manganese tetroxide pharmacokinetics following repeated (14-day) exposure. Toxicol. Appl. Pharmacol. **170:** 79–87.
48. AISEN, P., R. AASA & A.G. REDFIELD. 1969. The chromium, manganese, and cobalt complexes of transferrin. J. Biol. Chem. **244:** 4628–4633.
49. CRITCHFIELD, J.W. & C.L. KEEN. 1992. Manganese +2 exhibits dynamic binding to multiple ligands in human plasma. Metabolism **41:** 1087–1092.
50. UEDA, F. et al. 1993. Rate of ^{59}Fe uptake into brain and cerebrospinal fluid and the influence thereon of antibodies against the transferrin receptor. J. Neurochem. **60:** 106–113.
51. ASCHNER, M. & J.L. ASCHNER. 1990. Manganese transport across the blood-brain barrier: relationship to iron homeostasis. Brain Res. Bull. **24:** 857–860.
52. ASCHNER, M., K.M. ERIKSON & D.C. DORMAN. 2003. Manganese dosimetry: species differences and implications for neurotoxicity. Crit. Rev. Toxicol. In press.
53. MIZOGUCHI, N. et al. 2001. Manganese elevations in blood of children with congenital portosystemic shunts. Eur. J. Pediatr. **160:** 247–250.
54. KEEN, C.L. & S. ZIDENBERG-CHERR. 1994. Manganese toxicity in humans and experimental animals. In Manganese in Health and Disease, pp. 193–205. CRC Press. Boca Raton, FL.
55. REHNBERG, G.L. et al. 1980. Chronic manganese oxide administration to preweanling rats: manganese accumulation and distribution. J. Toxicol. Environ. Health **6:** 217–226.
56. MALECKI, E.A. et al. 1999. Iron and manganese homeostasis in chronic liver disease: relationship to pallidal T1-weighted magnetic resonance signal hyperintensity. Neurotoxicology **20:** 647–652.
57. HERYNEK, V. et al. 2001. Chronic liver disease: relaxometry in the brain after liver transplantation. MAGMA **12:** 10–15.

58. MONTES, S. et al. 2001. Striatal manganese accumulation induces changes in dopamine metabolism in the cirrhotic rat. Brain Res. **891:** 123–129.
59. MURPHY, V.A. et al. 1991. Saturable transport of manganese(II) across the rat blood-brain barrier. J. Neurochem. **57:** 948–954.
60. RABIN, O. et al. 1993. Rapid brain uptake of manganese(II) across the blood-brain barrier. J. Neurochem. **61:** 509–517.
61. BRENNEMAN, K.A. et al. 2000. Direct olfactory transport of inhaled manganese ((54)MnCl(2)) to the rat brain: toxicokinetic investigations in a unilateral nasal occlusion model. Toxicol. Appl. Pharmacol. **169:** 238–248.
62. WADHWANI, K.C. et al. 1992. Saturable transport of manganese(II) across blood-nerve barrier of rat peripheral nerve. Am. J. Physiol. **262:** R284–R288.
63. TAKEDA, A., J. SAWASHITA & S. OKADA. 1998. Manganese concentration in rat brain: manganese transport from the peripheral tissues. Neurosci. Lett. **242:** 45–48.
64. MALECKI, E.A. et al. 1999. Existing and emerging mechanisms for transport of iron and manganese to the brain. J. Neurosci. Res. **56:** 113–122.
65. SUAREZ, N. & H. ERIKSSON. 1993. Receptor-mediated endocytosis of a manganese complex of transferrin into neuroblastoma (SHSY5Y) cells in culture. J. Neurochem. **61:** 127–131.
66. ASCHNER, M. & M. GANNON. 1994. Manganese (Mn) transport across the rat blood-brain barrier: saturable and transferrin-dependent transport mechanisms. Brain Res. Bull. **33:** 345–349.
67. CHUA, A.C. & E.H. MORGAN. 1997. Manganese metabolism is impaired in the Belgrade laboratory rat. J. Comp. Physiol. **B167:** 361–369.
68. MATHISON, S., R. NAGILLA & U.B. KOMPELLA. 1998. Nasal route for direct delivery of solutes to the central nervous system: fact or fiction? J. Drug Target **5:** 415–441.
69. GIANUTSOS, G., G.R. MORROW & J.B. MORRIS. 1997. Accumulation of manganese in rat brain following intranasal administration. Fundam. Appl. Toxicol. **37:** 102–105.
70. HENRIKSSON, J., J. TALLKVIST & H. TJALVE. 1999. Transport of manganese via the olfactory pathway in rats: dosage dependency of the uptake and subcellular distribution of the metal in the olfactory epithelium and the brain. Toxicol. Appl. Pharmacol. **156:** 119–128.
71. LIN, C.P. et al. 2001. Validation of diffusion tensor magnetic resonance axonal fiber imaging with registered manganese-enhanced optic tracts. Neuroimage **14:** 1035–1047.
72. SALEEM, K.S. et al. 2002. Magnetic resonance imaging of neuronal connections in the macaque monkey. Neuron **34:** 685–700.
73. TJALVE, H. & J. HENRIKSSON. 1999. Uptake of metals in the brain via olfactory pathways. Neurotoxicology **20:** 181–195.
74. TJALVE, H., C. MEJARE & K. BORG-NECZAK. 1995. Uptake and transport of manganese in primary and secondary olfactory neurones in pike. Pharmacol. Toxicol. **77:** 23–31.
75. DORMAN, D.C. et al. 2002. Olfactory transport: a direct route of delivery of inhaled manganese phosphate to the rat brain. J. Toxicol. Environ. Health **A65:** 1493–1511.
76. TJALVE, H. et al. 1996. Uptake of manganese and cadmium from the nasal mucosa into the central nervous system via olfactory pathways in rats. Pharmacol. Toxicol. **79:** 347–356.
77. DORMAN, D.C., J.G. OWENS & K.T. MORGAN. 1997. Olfactory Neurotoxicology. *In* Comprehensive Toxicology. Volume 11: Nervous System and Behavioral Toxicology, pp. 281–294. Elsevier. Cambridge, UK.
78. CALNE, D.B. et al. 1994. Manganism and idiopathic parkinsonism: similarities and differences. Neurology **44:** 1583–1586.
79. ERIKSSON, H. et al. 1992. Manganese induced brain lesions in *Macaca fascicularis* as revealed by positron emission tomography and magnetic resonance imaging. Arch. Toxicol. **66:** 403–407.
80. NAGATOMO, S. et al. 1999. Manganese intoxication during total parenteral nutrition: report of two cases and review of the literature. J. Neurol. Sci. **162:** 102–105.
81. BRENNEMAN, K.A. et al. 1999. Manganese-induced developmental neurotoxicity in the CD rat: is oxidative damage a mechanism of action? Neurotoxicology **20:** 477–487.
82. ROELS, H. et al. 1987. Relationship between eternal and internal parameters of exposure to manganese in workers from a manganese oxide and salt producing plant. Am. J. Ind. Med. **11:** 297–305.

83. GURGUEIRA, S.A. *et al*. 2002. Rapid increases in the steady-state concentration of reactive oxygen species in the lungs and heart after particulate air pollution inhalation. Environ. Health Perspect. **110:** 749–755.
84. MERGLER, D. *et al*. 1994. Nervous system dysfunction among workers with long-term exposure to manganese. Environ. Res. **64:** 151–180.
85. PAL, P.K., A. SAMI & D.B. CALNE. 1999. Manganese neurotoxicity: a review of clinical features, imaging, and pathology. Neurotoxicology **20:** 227–238.
86. KAWAMURA, R. *et al*. 1941. Intoxication by manganese in well water. Arch. Exp. Med. **18:** 145–169.
87. KONDAKIS, X.G. *et al*. 1989. Possible health effects of high manganese concentrations in drinking water. Arch. Environ. Health **44:** 175–178.
88. VIEREGGE, P. *et al*. 1995. Long term exposure to manganese in rural well water has no neurological effects. Can. J. Neurol. Sci. **22:** 286–289.
89. BEUTER, A. *et al*. 1994. Diadochokinesimetry: a study of patients with Parkinson's disease and manganese exposed workers. Neurotoxicology **15:** 655–664.
90. CARLSSON, M. & A. CARLSSON. 1990. Interactions between glutamatergic and monoaminergic systems within the basal ganglia—implications for schizophrenia and Parkinson's disease. Trends Neurosci. **13:** 272–276.
91. VERITY, M.A. 1999. Manganese neurotoxicity: a mechanistic hypothesis. Neurotoxicology **20:** 489–497.
92. DORMAN, D.C. *et al*. 2000. Neurotoxicity of manganese chloride in neonatal and adult CD rats following subchronic (21-day) high-dose oral exposure. J. Appl. Toxicol. **20:** 179–187.
93. NEWLAND, M.C. 1999. Animal models of manganese's neurotoxicity. Neurotoxicology **20:** 415–432.
94. PAPPAS, B.A. *et al*. 1997. Perinatal manganese exposure: behavioral, neurochemical, and histopathological effects in the rat. Neurotoxicol. Teratol. **19:** 17–25.
95. ALBERS, D.S. & M.F. BEAL. 2000. Mitochondrial dysfunction and oxidative stress in aging and neurodegenerative disease. J. Neura Transm. Suppl. **59:** 133–154.
96. BECKMAN, K.B. & B.N. AMES. 1998. Mitochondrial aging: open questions. Ann. N.Y. Acad. Sci. **854:** 118–127.
97. MELOV, S. 2002. Therapeutics against mitochondrial oxidative stress in animal models of aging. Ann. N.Y. Acad. Sci. **959:** 330–340.
98. WITHOLT, R., R.H. GWIAZDA & D.R. SMITH. 2000. The neurobehavioral effects of subchronic manganese exposure in the presence and absence of pre-parkinsonism. Neurotoxicol. Teratol. **22:** 851–861.
99. ANANTHARAM, V. *et al*. 2002. Caspase-3-dependent proteolytic cleavage of protein kinase Cδ is essential for oxidative stress–mediated dopaminergic cell death after exposure to methylcyclopentadienyl manganese tricarbonyl. J. Neurosci. **22:** 1738–1751.
100. KITAZAWA, M. *et al*. 2002. Oxidative stress and mitochondrial-mediated apoptosis in dopaminergic cells exposed to methylcyclopentadienyl manganese tricarbonyl. J. Pharmacol. Exp. Ther. **302:** 26–35.
101. MELOV, S. 2000. Mitochondrial oxidative stress: physiologic consequences and potential for a role in aging. Ann. N.Y. Acad. Sci. **908:** 219–225.
102. CALABRESE, V. *et al*. 2001 Mitochondrial involvement in brain function and dysfunction: relevance to aging, neurodegenerative disorders and longevity. Neurochem. Res. **26:** 739–764.
103. BLASS, J.P. 2001. Brain metabolism and brain disease: is metabolic deficiency the proximate cause of Alzheimer dementia? J. Neurosci. Res. **66:** 851–856.
104. ALBIN, R.L. 2000. Basal ganglia neurotoxins. Neurol. Clin. **18:** 665–680.
105. ASCHNER, M., U. SONNEWALD & K.H. TAN. 2002. Astrocyte modulation of neurotoxic injury. Brain Pathol. **12:** 475–481.
106. GWIAZDA, R.H. *et al*. 2002. Low cumulative manganese exposure affects striatal GABA, but not dopamine. Neurotoxicology **23:** 69–76.
107. ERIKSON, K.M. & M. ASCHNER. 2003. Manganese neurotoxicity and glutamate-GABA interaction. Neurochem. Int. **43:** 475–480.
108. LIPE, G.W. *et al*. 1999. Effect of manganese on the concentration of amino acids in different regions of the rat brain. J. Environ. Sci. Health **B34:** 119–132.

109. MAYNARD, L.S. & G.C. COTZIAS. 1954. The partition of manganese among organs and intracellular organelles of the rat. J. Biol. Chem. **214:** 489–495.
110. GAVIN, C.E., K.K. GUNTER & T.E. GUNTER. 1999. Manganese and calcium transport in mitochondria: implications for manganese toxicity. Neurotoxicology **20:** 445–453.
111. MALECKI, E.A. 2001. Manganese toxicity is associated with mitochondrial dysfunction and DNA fragmentation in rat primary striatal neurons. Brain Res. Bull. **55:** 225–228.
112. ROTH, J.A. *et al.* 2000. Manganese-induced rat pheochromocytoma (PC12) cell death is independent of caspase activation. J. Neurosci. Res. **61:** 162–171.
113. ZWINGMANN, C., D. LEIBFRITZ & A.S. HAZELL. 2003. Energy metabolism in astrocytes and neurons treated with manganese: relation among cell-specific energy failure, glucose metabolism, and intercellular trafficking using multinuclear NMR-spectroscopic analysis. J. Cereb. Blood Flow Metab. **23:** 756–771.
114. HAZELL, A.S. 2002. Astrocytes and manganese neurotoxicity. Neurochem. Int. **41:** 271–277.
115. TIFFANY-CASTIGLION, E. & Y. QIAN. 2001. Astroglia as metal depots: molecular mechanisms for metal accumulation, storage, and release. Neurotoxicology **22:** 577–592.
116. ARCHIBALD, F.S. & C. TYREE. 1987. Manganese poisoning and the attack of trivalent manganese upon catecholamines. Arch. Biochem. Biophys. **256:** 638–650.
117. ALI, S.F., *et al.* 1995. Manganese-induced reactive oxygen species: comparison between Mn^{+2} and Mn^{+3}. Neurodegeneration **4:** 329–334.
118. CHEN, J.Y. *et al.* 2001. Differential cytotoxicity of Mn(II) and Mn(III): special reference to mitochondrial [Fe-S] containing enzymes. Toxicol. Appl. Pharmacol. **175:** 160–168.
119. HAMAI, D., A. CAMPBELL & S.C. BONDY. 2001. Modulation of oxidative events by multivalent manganese complexes in brain tissue. Free Radical Biol. Med. **31:** 763–768.

Oxidative Basis of Manganese Neurotoxicity

DIEM HaMAI AND STEPHEN C. BONDY

Department of Community and Environmental Medicine, Center for Occupational and Environmental Health, University of California, Irvine, California, USA

ABSTRACT: Exposure to excessive levels of manganese, an essential trace element, can evoke severe psychiatric and extrapyramidal motor dysfunction closely resembling Parkinson's disease. The clinical manifestations of manganese toxicity arise from focal injury to the basal ganglia. This region, characterized by intense consumption of oxygen and significant dopamine content, can incur mitochondrial dysfunction, depletion of levels of peroxidase and catalase, and catecholamine biochemical imbalances following manganese exposure. The site specificity of the pathology and the nature of the cellular damage caused by manganese have been attributed to its capacity to produce cytotoxic levels of free radicals. However, support for such a pro-oxidant role for manganese has been largely limited to inferences drawn from histopathological observations. More recently, research efforts into the molecular details of manganese toxicity have provided evidence of an etiological relationship between oxidative stress and manganese-related neurodegeneration. This review focuses on studies that evaluate the redox chemistry of manganese during the neurodegenerative process and its molecular consequences.

KEYWORDS: manganese; neurodegeneration; physical chemistry; oxidative stress

INTRODUCTION

Trace amounts of the essential element, manganese (Mn), are sufficient to satisfy the nutritional requirements of the body, while exposure to elevated levels of Mn can lead to symptoms of irreversible neurological dysfunction similar to Parkinson's disease (PD).

The regional specificity of its neuropathology and the oxidative nature of the cellular damage induced by Mn often have been attributed to the capacity of the essential element to stimulate the generation of free radicals to cytotoxic levels. However, support for such a pro-oxidant role for Mn has been largely drawn from histological inferences. This review focuses on recent studies of the redox chemistry of Mn during the neurodegenerative process and its molecular consequences to evaluate the evidence for an etiological relationship between oxidative stress and Mn-related toxicity.

Address for correspondence: D. HaMai, Department of Community and Environmental Medicine, University of California, Irvine, Irvine, CA 92697-1825. Voice: 949-824-8642; fax: 949-824-2793. diemh@uci.edu

Ann. N.Y. Acad. Sci. 1012: 129–141 (2004). © 2004 New York Academy of Sciences.
doi: 10.1196/annals.1306.010

CLINICAL MANIFESTATIONS OF Mn INTOXICATION

Symptoms of Mn Neurotoxicity

Intoxication by Mn, or manganism, primarily occurs as a consequence of chronic inhalation of high concentrations of airborne Mn-containing particles,[1,2] However, several outbreaks of manganism have occurred after the consumption of water with high Mn levels.[3,4] Exposure to dialysis fluids contaminated with Mn[5] and total parenteral nutrition coupled with reduced biliary Mn secretions secondary to cholestasis have also resulted in intoxication.[6]

The symptomatic onset of Mn toxicity may be delayed up to 1 to 2 years after exposure and thereafter progresses slowly.[7,8] An akinetic-rigid parkinsonian syndrome is prominent. However, a prodrome consisting of psychiatric symptoms may present for several months prior to the movement disorder. "Manganese madness" includes symptoms of irritability, emotional lability, illusion, and hallucinations. The movement disorder is characterized by bradykinesia and rigidity with little resting tremor. Gait disturbance can be followed by dysarthria and deteriorating handwriting, as well as axial and extremity dystonia.[1,7] Dementia and signs of cerebellar dysfunction may also be present as in the case of classic PD. Dopaminergic pharmacotherapy is largely ineffective.[9] In less severe cases, the symptoms of Mn toxicity may manifest as neurological dysfunction that may be too subtle for unequivocal diagnosis of parkinsonism. Subclinical deficits in performance on tests, such as short-term memory capacity, eye-hand coordination, hand steadiness, and reaction time, have been reported following inhalation exposure to airborne levels of Mn between 0.2 and 5 mg/m^3.[10–12]

In a recent population-based case-control study of workers older than 50 years of age with occupational exposure to metals, the development of parkinsonism was more significantly associated with exposure to Mn lasting more than 20 years than any other single metal in the study (odds ratio = 10.6; 95% confidence interval: 1.06–105.8).[13] It has been estimated that 25–35% of idiopathic PD is diagnosed incorrectly.[14,15] This raises the possibility that cases of Mn-induced parkinsonism may be overlooked and misdiagnosed as PD.

Histopathology of Mn Neurotoxicity

The clinical manifestation of parkinsonism indicates injury to the extrapyramidal system, with specific insult to the basal ganglia. These nuclei, including the striatum and the globus pallidus (GP), as well as two associated nuclei, the substantia nigra (SN) and the subthalamic nucleus (STN), receive afferents from various regions of the central nervous system, such as the cerbral cortex, brain stem, thalamus, and subthalamic nuclei. The excitatory input is processed in the caudate and putamen regions of the striatum, which is then passed onto the internal segment of the GP. Once the information is modified and integrated, a major portion of the output is projected to the motor regions of the cortex via the thalamic motor nuclei. This latter group of brain nuclei function in concert to suppress involuntary movements. Loss of the dopaminergic neurons in the nigrostriatal pathway of the basal ganglia, which are inhibitory, leads to heightened activity of neurons in the GP. Since the efferents from the GP are also inhibitory, the sum of increased suppression of motor function produces the symptoms characteristic of Mn-related parkinsonism.

More specifically, rigidity and bradykinesia arise from the degeneration of neurons in the substantia nigra pars compacta, which project to the striatum. This is associated with depletion of dopamine, a blockade of dopamine action in the striatum, and loss of striatal dopaminoceptive neurons.[16] The dystonia is a consequence of neuronal degeneration, a decrease in myelinated fibers, and reactive gliosis in the pallidum. Notably, the internal segment of the GP, which extends to the striatum, putamen, and substantia nigra pars reticulata, sustains the most overt injury.[17–19] These lesions are focal to both pre- and postsynaptic terminals of the dopaminergic nigrostriatal pathway. The subcellular localization of Mn occurs in the mitochondria since Mn has a high affinity for the inner mitochondrial membrane.[20] There is extensive congruence between the patterns of the brain lesions sustained after exposure to Mn and the sites of its accumulation.

Oxidative Features in Mn Neuropathology

Salient features of the brain regions susceptible to Mn-provoked injury include their intense oxidative metabolism, major dopamine content, and high content of nonheme iron.[21] This raises the possibility that the mechanisms of Mn neurotoxicity relate to its potential for oxidative injury and promotion of dopamine auto-oxidation. These striatal regions, in which extensive oxidative metabolism occurs, are rich in mitochondria. Since the electron transport chain is the most significant physiological source of superoxide,[22] these regions are especially vulnerable to the injury that can occur from unchecked free radical production during oxidative phosphorylation. Injury to the dopaminergic system also suggests oxidative stress as a source of cell injury since the chemical structure of catecholamines predisposes them to oxidation, and their well-characterized metabolic routes can yield quinones and free radicals.[23]

MECHANISMS OF Mn-RELATED OXIDATIVE STRESS

Support for the involvement of oxidative stress remains largely limited to inferences from the known chemical properties of Mn and histopathological findings of dopaminergic neurotoxicity. More recently, evidence for the role of oxidative events in Mn toxicity at the molecular level has emerged. Mn treatment of primary cortical astrocytes leads to decrease of glial fibrillary acidic protein synthesis, energy production, glutamate transporter activity, and activity of the antioxidants, superoxide dismutase (SOD) and glutathione peroxidase.[24] DNA binding by nuclear factor kappa B (NF-κB) and activator protein (AP1), known to be activated by oxidative stress, is induced by treatment of murine PC-12 cells with Mn, suggesting that transcriptional activity is upregulated.[25]

Several explanations have been proposed to account for the targeting of the basal ganglia by Mn with the neurochemistry of the brain region. These include the specific vulnerability of the catechol dopamine to Mn, the impairment of cellular antioxidant defenses by the accumulation of the metal, and the disruption of mitochondrial oxidative energy metabolism. This has led to the conclusion that excessive levels of brain Mn induce oxidative stress leading to neurodegeneration. Detailed consideration of the redox status of Mn could lead to a clearer mechanistic explanation of these processes.

Dopaminergic Susceptibility

The dopaminergic neurodegeneration that underlies parkinsonism is certainly the most studied aspect of Mn neurotoxicity. It has been hypothesized that Mn interacts with catechols specific to dopaminergic neurons so as to rapidly deplete them and render such cells no longer viable.[23,26,27] Another study using a transfected dopaminergic nonneuronal cell system overexpressing aromatic amino acid decarboxylase compared the cytotoxicity of L-dopa, dopamine, and DOPAC in the presence of Mn. Results revealed no significant difference between the cytotoxicity of these catechols. Instead, the intracellular levels of total catechols determined the susceptibility of the transfected cells to Mn treatment.[28] This suggestion of a synergistic relationship was supported by the observation that Mn enhanced L-dopa-induced apoptosis in the catecholaminergic PC-12 cells.[29]

In another study, the cell viability and dopamine levels of PC-12 cells were assessed in the presence of different chemical species of Mn. The LC_{50} values of Mn^{2+}-chloride, Mn^{3+}-phosphate, and Mn^{3+}-transferrin (Mn^{3+}-tf) were determined as 65, 25, and 3 µM, respectively.[30] Here, 1.25 µM Mn^{3+}-tf rapidly depleted dopamine levels to 50% relative to controls within 2 h, whereas 10 µM Mn^{2+} and Mn^{3+} in the absence of transferrin only modestly decreased the presence of the catechol even after 42 h of exposure. The finding that trivalent Mn efficiently oxidizes catechols *in vitro* illustrates the differential susceptibility of catecholamines to oxidation and nucleophilic additions by different forms of Mn. This metabolic route can yield quinones and free radicals that lead to an oxidative biochemical environment.[31]

Mitochondrial Dysfunction

The mitochondrion is the primary site of Mn within the cell. Due to transport mechanisms that favor the influx of Mn over its efflux, Mn accumulates in the organelle and can thus readily disturb the process of oxidative phosphorylation. The electron transport chain is the most significant physiological source of superoxide,[22] and Mn serves as an essential cofactor to several mitochondrial enzymes. Hence, this region, which has an intrinsic tendency to accumulate Mn, is especially vulnerable to elevated influxes of Mn and the unchecked free radical production that may occur.

In subcellular preparations, Mn^{3+} was more effective than Mn^{2+} in inhibiting the enzyme kinetics activity of complex-1 in mitochondrial fractions and its consumption of the substrate NADH.[32] However, in whole cells, Mn^{2+} exerted a more profound effect on complex-1 and decreased its activity more than Mn^{3+}. This suggests that preferential uptake of Mn^{2+} by cells is critical to the disruption of mitochondrial function, whereas the direct suppression of complex-1 activity by Mn^{3+} can take place in isolated systems. Further analysis comparing the binding affinity of the two valence states of Mn to the iron-sulfur clusters of the active centers of mitochondrial enzymes may help to better delineate their mechanisms of toxicity.

The heightened vulnerability of mitochondria to Mn was further investigated in a recent study comparing the energy metabolism of astrocytes and neurons exposed to 100 µmol/L $MnCl_2$ for 5 days.[33] Analysis of cocultures of the two different cell types showed that Mn selectively impaired neuronal glucose oxidation and sharply reduced mitochondrial energy production. In contrast, both oxidative and anaerobic glucose metabolism of the astrocytes were increased in response to the treatment.

Weakened Antioxidant Cellular Defenses

Another mechanism that may explain the oxidative toxicity of Mn is the exhaustion of cellular antioxidant defenses. A decrease in enzymes that serve as active oxygen scavengers, such as peroxidase and catalase, in the SN following Mn administration[34] suggests their depletion by free radicals. Following chronic Mn treatment, cellular DNA synthesis, glial fibrillary acidic protein synthesis, energy production, glutamate transporter activity, and the enzymatic levels of the antioxidants, SOD and glutathione peroxidase, are all concurrently decreased in primary cortical astrocytes.[24]

Oxidative Potential

Despite extensive research efforts, the characterization of Mn as a pro-oxidant has been highly contentious.[26,35,36] The few specific determinations of the oxidative capacity of Mn and its ability to alter the kinetics of cellular free radical production have yielded ambiguous and even conflicting results. Pro-oxidant properties have been attributed to both the divalent and trivalent states of the metal[37–39] or to only the latter.[40,41] However, other reports contend that divalent Mn ion scavenges the reactive oxygen species (ROS), superoxide and hydroxyl radicals, even when SOD activity is inhibited[40] and attenuates oxygen toxicity in *Lactobacillus planetarium* and related bacteria deficient in the SOD enzyme.[27] Additional antioxidant effects, including catalase-like disproportionation of H_2O_2,[42] inhibition of iron-induced lipid peroxidation, and copper-dependent low-density lipoprotein conjugation,[43] have been found using the same divalent salt, $MnCl_2$. The report that the divalent form of Mn does not react with H_2O_2 to produce the hydroxyl radical at a measurable rate under biological conditions[44] challenges the finding that Mn can generate both the hydroxyl and superoxide radicals from H_2O_2 by way of the Fenton reaction, following elevation of physiological pH.[45] Conflicting findings such as these have complicated the characterization of Mn as either a pro-oxidant or an antioxidant.

Chemical Speciation during Oxidative Events

The disparate characterizations of Mn described earlier can be reconciled by the finding that the pro-oxidant activity associated with divalent Mn depends on the trace presence of trivalent metal ion, most probably manganic (Mn^{3+}).[46] Although both valences of Mn have been associated with accelerated generation of ROS, the findings reveal a low potential for Mn^{2+} to engage in radical generating reactions in biological environments. This suggests a resistance to oxidation by the divalent ion that contrasts with the readiness of the lower valence states of iron and copper to become oxidized to higher valence levels.[47] The oxidation rate of manganous ion has been calculated to be slower than that of ferrous ion by approximately 6 orders of magnitude.[48]

The significance of a trivalent metal in the apparent ability of Mn^{2+} to promote the formation of ROS is underscored by the complete inhibition of the pro-oxidant activity in the presence of 1/500th the concentration of desferrioxamine (DFO), a highly selective chelator of trivalent metal (FIG. 1). The stability constants of complexes between DFO and divalent ions such as Cu^{2+} and Zn^{2+} are at least 11 orders

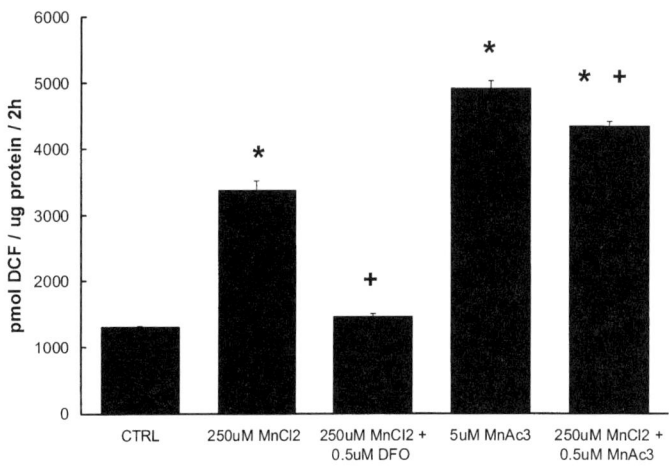

FIGURE 1. Apparent pro-oxidant activity of Mn^{2+} dependent upon Mn^{3+}. Mn^{2+}-based ROS formation is quenched by the trivalent metal chelator DFO at a 500:1 molar basis. The addition of a trace amount of Mn^{3+} restores the apparent pro-oxidant activity of Mn^{2+}. Values are the mean ± SE of three independent experiments realized in quadruplicate. *Value of manganese-treated fractions differs ($p < .01$) from that of the control; +value of fractions treated additionally with DFO differs ($p < .01$) from the corresponding value of fractions treated with one of the manganese salts alone.

of magnitude lower than those of trivalent ones.[44] The trace presence of a trivalent metal as a promoting factor in Mn^{2+}-related ROS generation could be inferred.

Further investigation revealed that Mn^{2+} may act in a pro- or antioxidant manner in the presence of traces of Mn^{3+} or Fe^{3+} ion, respectively. ROS generation attributed to divalent Mn was increased with nanomolar increases of Mn^{3+} (FIG. 1). This suggested that catalytic amounts of the more reactive trivalent ion promote Fenton-type redox cycling, whereas ferric iron attenuates this process. This implies a synergistic interaction between the divalent and trivalent states of Mn in the catalysis of ROS-generating reactions. The significance of Mn^{3+} in catalytic promotion of oxidative events is supported by the observation that Mn^{3+}, but not Mn^{2+}, accelerates the oxidation of ferrous iron. Low micromolar concentrations of manganic salts were sufficient to trigger an immediate oxidation of ferrous ion, whereas divalent Mn at concentrations of 100-fold higher did not promote the conversion of ferrous to ferric.[49]

These studies illustrate the significance of valence and ion speciation in metal-mediated oxidative events and establish the capacity of Mn to alter the cellular redox potential. Evaluations of other biochemical parameters of oxidative stress in neural systems, including dopaminergic neuronal viability, catechol levels, and mitochondrial enzyme function, must be considered within this context of the chemical range and reactivity of Mn.

REDOX CHEMISTRY OF Mn IN BIOLOGICAL SYSTEMS

Electrochemical Properties

Prior to interpretations of the physiological and biochemical changes, it is critical to gain understanding of the general coordination properties and redox tendencies of Mn. Detailed knowledge of this chemistry can serve as the framework in which theories on the etiological relationship between oxidative stress and Mn neurotoxicity may be more thoroughly evaluated.

The capacity of Mn to act as both a nutrient and a toxin in biological systems results from its chemical reactivity and its ability to assume a variety of stable, positively charged forms. The element engages in a wide range of chemical roles in biological systems, including acting as a cofactor defining protein structure, a simple Lewis acid influencing reaction rates, or a redox-active center of an enzyme facilitating the reduction and/or oxidation of substrates.[50] The structural properties of Mn—a small atomic radius, low ionization energy, low electronegativity, and an incomplete inner valence subshell—result in valence electrons that are held loosely

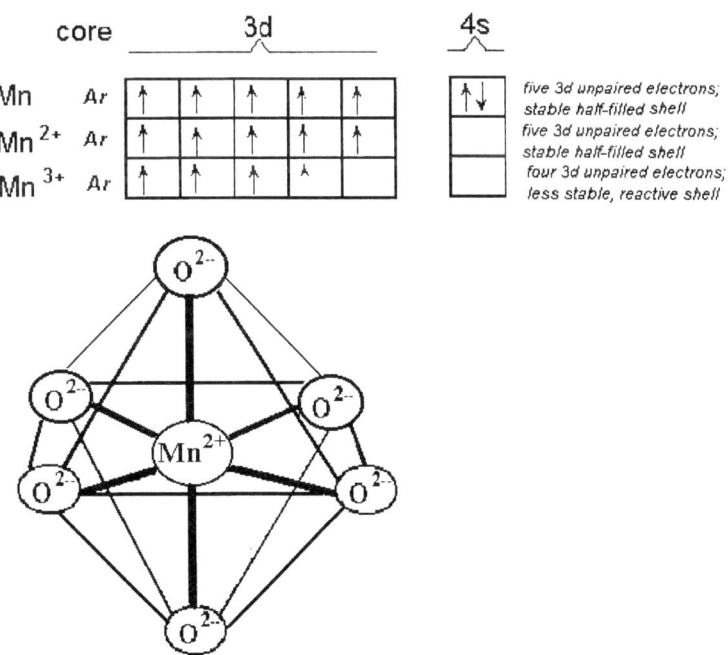

FIGURE 2. Electron configuration and coordination chemistry of Mn. For a transition metal, the number of valence electrons in the five orbitals of the "inner" *3d* subshell determines its chemical reactivity and the number of ligands to which it can bind for additional stabilization. The half-filled *3d* subshell of the Mn^{2+} ion imparts a stability that accounts for its poor reducing ability. The five unpaired *3d* electrons of the Mn^{2+} ion result in a preference for binding to six ligands. This yields an octahedral coordination complex, which offers maximal stability to most transition metals.

by the nucleus. The electrons are highly available for bonding and are readily lost with very little expenditure of energy.

Although Mn is classified as a hard Lewis acid and readily accepts electron pairs in covalent bonds, it can also serve as a good reducing agent, which gives up electrons. Their unfilled $3d$ orbital of transition metals gives them the capacity to act as an oxidizing agent and accept electrons in return. The presence of the unpaired electrons in the orbitals of the $3d$ subshell effectively imparts a chemical reactivity to transition metals that resembles those of radicals and enables them to readily accept and donate single electrons in redox reactions (FIG. 2).

The mechanisms underlying the reactivity between the Mn^{2+} complexes and HO_2/O_2^- are analogous to their iron counterparts.[51–53] The net reaction yields the removal of superoxide and the recycling of Mn^{2+}. The reaction of Mn^{2+} with superoxide radical occurs at a rate of 1.8×10^7 $M^{-1}s^{-1}$.[54] In contrast, the reaction rate between Mn^{2+} and the hydroxyl radical ($OH^\bullet$) has been found to be slower with a rate constant of 3.5×10^7 $M^{-1}s^{-1}$.[55]

The resistance of Mn^{2+} to oxidation differs from the readiness of the lower valence states of iron and copper to change to higher valence levels. For these latter transition metals, the lower valence state exhibits greater redox potential, higher reactivity, and greater readiness to catalyze free-radical-generating Fenton reactions. The higher valence state Mn^{3+} shows significant redox potential, while the lower valence state Mn^{2+} is a modest pro-oxidant only in the trace presence of another trivalent metal.

In distinction from other transition metals, Mn in aqueous, neutral pH solutions is most stable with the oxidation number of +2. The aqueous ion can exist at relatively high concentrations under physiological conditions.[56] Mn^{2+} is an S-state ion in the usual high-spin configuration with an approximately spherical distribution of the five unpaired $3d$ electrons, one positioned in each of the d orbitals, about the nucleus. The half-filled $3d$ shell of Mn^{2+} imparts the ion with stability analogous to that of the higher oxidation states of iron, Fe^{3+}, and those of closed-shell ions, such as those in the alkaline earth family (e.g., Mg, Ca).[57] This coordination symmetry maximizes the electron exchange energy and yields no ligand field stabilization energy. For these reasons, Mn^{2+} does not gain the thermodynamic stability that other elements of the first row of the transition series achieve upon forming complexes with available ligands. The equilibrium constants for such complexes are relatively low compared to the divalent cations of Fe, Co, Ni, or Cu.[56] The stability of this electron configuration and the reluctance to lose a d electron accounts for the poor reducing ability of Mn^{2+}.[50]

In contrast, the four Mn^{3+} unpaired electrons in the usual high-spin configuration are similar to those of the lower oxidation state of iron, Fe^{2+}, and result in Mn^{3+} having marked instability and high reduction and oxidation potential. In view of its high-spin d^4 electrons, Mn^{3+} is inclined either to lose one electron in the antibonding e_g set of orbitals to maximize ligand field stabilization or to gain an electron to maximize electron exchange energy.[50]

In spite of its electrochemical stability, the redox potential and the overall reactivity of the aqueous Mn^{2+} ion can be substantially altered by the ligand to which it binds. Because the electropositivity of the metal center is modified by the donor capacity of the ligands, the ability of the ion to act as a reductant or oxidant is strongly affected by the nature of the ligand.[58] Numerous enzymes and biological transformations utilize this aspect of Mn coordination chemistry in the metabolism of dioxygen and its reduced forms. The reduction potential of divalent Mn may be significantly enhanced and its oxidation may be catalyzed by its contact with surfaces of metal colloids, or metal-containing precipitates, such as oxy-hydroxides or other biological aggregates:[48]

$$Mn^{2+} + O_2^- \leftrightarrow [Mn^{2+} - O_2 \leftrightarrow Mn^{3+} - O_2^{\cdot -}] \leftrightarrow Mn^{3+} + O_2^{\cdot}.$$

Intermediate Complexes

The much higher transition probabilities of Mn^{3+} compared to Mn^{2+} result in the propensity of Mn^{3+} to form colloidal bodies in aqueous solutions.[57] A localized increase in the Mn^{2+} concentration on such an insoluble surface may induce a shift in the electron density at the interface between the two different metal ionic species. Mn^{3+}, situated at the surfaces of biological aggregates, may facilitate a fraction of Mn^{2+} to undergo oxidation to Mn^{3+} (FIG. 3). Thus, on such surfaces, the divalent ion

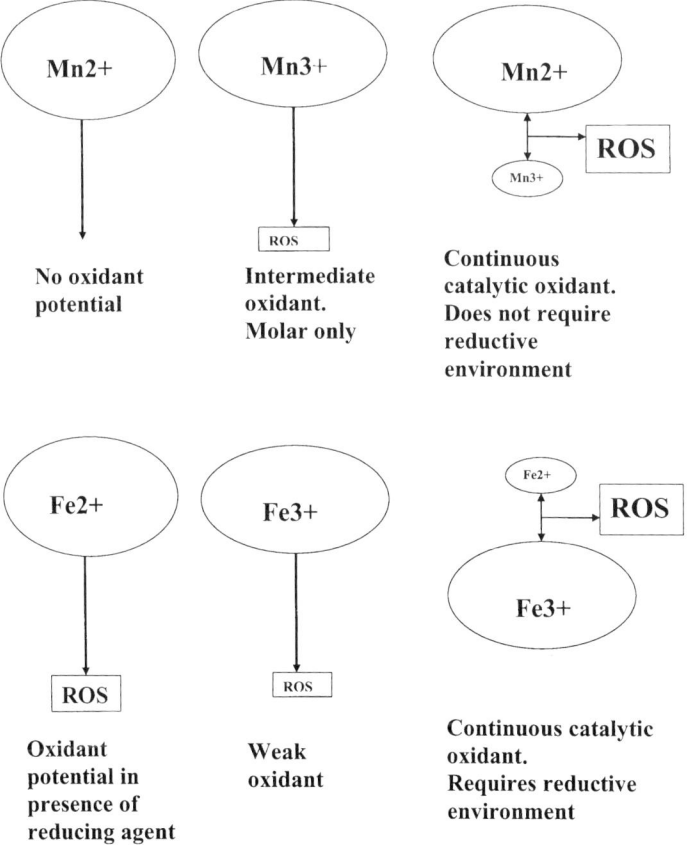

FIGURE 3. A comparison of oxidant properties of Mn and Fe.

may be able to overcome its intrinsic reluctance to oxidize or reduce. Divalent Mn has been observed to accelerate oxidation reactions such as that of reduced nicotinamide adenine dinucleotide (NADH).[27] In this case, the oxidation of Mn^{2+} by electron-rich oxygen-derived species such as O_2^- can produce Mn^{3+} species or Mn-O intermediate complexes that are substantially more oxidizing than the original reactants.

Deleterious Consequences of Mn Chemistry

Excessive stimulation or sustained production of free radicals by redox-active metals including Mn may overwhelm the antioxidant capacities of the cell and result in the direct damage of any nearby cellular constituent. The damage may include oxidative changes of both the metal-complexing ligand itself and molecules within close proximity that are susceptible to the reactive intermediates of the original free-radical-generating reaction. In aqueous solutions, redox-active metals may also oxidize the amino acid residues in proteins and peptides, resulting in their deamination and decarboxylation.[45] The formation of radicals from the oxidation process of proteins may eventually lead to the fragmentation and reconfiguration of proteins through cross-linking. Metal ions may also oxidize the polyunsaturated fatty acids of cell membranes to yield various products, including epoxy-fatty acids, alkanes, alkenes, and aldehydes.[59] Oxidative damage to mitochondrial and nuclear DNA can occur in the presence of copper, iron, and chromium. This arises from the direct targeting of the pyrimidines and purines in DNA, as well as from the cross-linking between DNA and proteins.[60] Oxygen radicals and other intermediates created by transition metals may thus chemically modify many cellular macromolecules.

Redox-active metals such as Mn may also inflict metabolic damage indirectly by acting through physiological systems that rely on free radicals as their basis for regulation and signaling. Nitric oxide and other ROS serve as regulatory mediators of a wide range of cellular functions, including cell adhesion and recruitment, programmed cell death, and redox-sensitive signaling pathways. Thus, oxidative stress may also affect metabolism by disruption of these signaling systems.[44] With regard to the brain, the acceleration of free radical production mediated by Mn could lead to enhanced turnover of neurotransmitters, particularly the catecholamine dopamine. Thus, excessive levels of redox-active metals such as Mn may result in the promotion of oxidative events in the brain.

CONCLUSIONS

Decades of research have been vested in efforts to understand the physiological and biochemical changes related to environmental exposures to Mn. However, only a few studies have yet investigated the molecular details of the mechanisms underlying Mn-induced neurodegeneration. Support for a causal relationship between oxidative stress and the toxicity of Mn has been based on the site-specificity of Mn accumulation and pathology and on the oxidative features of the cellular damage in this area. Past efforts to investigate the role of Mn in oxidative stress have largely ignored the relationship between valence and the capacity to promote oxidative cellular conditions. This oversight incorrectly assumes that the redox tendencies of Mn are irrelevant to the subsequent biological response.

The capacity of Mn to engage in single electron transfers and thus generate radicals is critical to its many biological functions. Yet, this same ability of Mn to react with oxygen and other substrates, and to transform them to highly oxidizing species under physiological conditions, may function as the molecular mechanism underlying Mn-induced neurotoxicity. Determination of the capacity of Mn to alter cellular redox state and investigations into the subsequent molecular changes occurring downstream could offer insights into the relationship between Mn-related oxidative damage and the regional selectivity of Mn-induced neurotoxicity.

Recent studies evaluating the significance of valence reveal that interactions between metal ions of different valences, even present as trace contaminants, may extensively modify the capacity of Mn to generate free radicals. Pronounced differences exist between its two biologically relevant valences in their capacity to promote free radical generation, accelerate iron oxidation, enhance dopamine autooxidation, and decrease mitochondrial complex-1 activity. These recent findings elucidate the atypical chemical processes by which Mn can act in a pro- or antioxidant manner and strengthen the basis for oxidative stress in Mn-induced neurodegeneration. Such experimental insights offer coherence to the many and often disparate descriptions of the mechanisms underlying the neurotoxicity of Mn.

ACKNOWLEDGMENTS

This work was supported by NIH Grant Nos. ES 7992 and AG 16794.

REFERENCES

1. MENA, I. et al. 1967. Chronic manganese poisoning: clinical picture and manganese turnover. Neurology **17:** 128–136.
2. TAKEDA, A. 2003. Manganese action in brain function. Brain Res. Rev. **41:** 79–87.
3. KAWAMURA, R. 1941. Intoxication by manganese in well water. Kisasato Arch. Exp. Med. **18:** 145–169.
4. KONDAKIS, X.G. et al. 1989. Possible health effects of high manganese concentration in drinking water. Arch. Environ. Health **44:** 175–178.
5. TAYLOR, P.A. & J.D.E. PRICE. 1982. Acute manganese intoxication and pancreatitis in a patient treated with a contaminated dialysate. Can. Med. Assoc. J. **126:** 503.
6. MEHTA, R. & J.J. REILLY. 1990. Manganese levels in a jaundiced long-term total parenteral nutrition patient: potentiation of haloperidol toxicity? J. Parenter. Enteral Nutr. **14:** 428.
7. CALNE, D.B. et al. 1994. Manganism and idiopathic parkinsonism: similarities and differences. Neurology **44:** 1583–1586.
8. HUANG, C-C. et al. 1993. Progression after chronic manganese exposure. Neurology **43:** 1479–1483.
9. LU, C.S. et al. 1994. Levodopa failure in chronic manganism. Neurology **44:** 1600–1602.
10. ROELS, H. et al. 1987. Epidemiological survey among workers exposed to manganese: effects on lung, central nervous system, and some biological indices. Am. J. Ind. Med. **11:** 307.
11. MERGLER, D. et al. 1994. Nervous system dysfunction among workers with long-term exposure to manganese. Environ. Res. **64:** 151–180.
12. CRUMP, K.S. & P. ROUSSEAU. 1999. Results from eleven years of neurological health surveillance at a manganese oxide and salt producing plant. Neurotoxicology **20:** 273–286.

13. GORELL, J.M. *et al.* 1999. Occupational exposure to manganese, copper, lead, iron, mercury, and zinc and the risk of Parkinson's disease. Neurotoxicology **20:** 239–248.
14. HUGHES, A.J. *et al.* 1992. Accuracy of clinical diagnosis of idiopathic Parkinson's disease: a clinico-pathological study of 100 cases. J. Neurol. Neurosurg. Psychiatry **55:** 181–184.
15. RAJPUT, A.H. *et al.* 1991. Accuracy of clinical diagnosis in parkinsonism—a prospective study. Can. J. Neurol. Sci. **18:** 275–278.
16. ALBIN, R.L. 2000. Basal ganglia neurotoxins. Neurol. Clinics **18:** 1–16.
17. BANTA, R.G. & W.R. MARKESBERY. 1977. Elevated manganese levels associated with dementia and extrapyramidal signs. Neurology **27:** 213–216.
18. YAMADA, M. *et al.* 1986. Chronic manganese poisoning: a neuropathological study with determination of manganese distribution in the brain. Acta Neuropathol. (Berlin) **70:** 273.
19. BARBEAU, A. 1984. Manganese and extrapyramidal disorders. Neurotoxicology **5:** 13–36.
20. LAI, J.C.K. *et al.* 1997. Relationships between manganese and other minerals. *In* Mineral and Metal Toxicology, pp. 297–303. CRC Press. Boca Raton, FL.
21. HILL, J.M. & R.C. SWITZER. 1984. The regional distribution and cellular localization of iron in the rat brain. Neuroscience **11:** 595–603.
22. CHANCE, B. *et al.* 1979. Hydroperoxide metabolism in mammalian organs. Physiol. Rev. **59:** 527–605.
23. GRAHAM, D.G. *et al.* 1978. Auto-oxidation versus covalent binding of quinones as the mechanism of toxicity of dopamine, 6-hydroxydopamine, and related compounds toward C1300 neuroblastoma cells *in vitro*. Mol. Pharmacol. **14:** 644–653.
24. CHEN, C-J. & S-L. LIAO. 2002. Oxidative stress involved in astrocytic alterations induced by manganese. Exp. Neurol. **175:** 216–225.
25. RAMESH, G.T. *et al.* 2002. Activation of early signaling transcription factor NFκB following low-level manganese exposure. Toxicol. Lett. **136:** 151–158.
26. DONALDSON, J. *et al.* 1982. Manganese neurotoxicity: a model for free radical mediated neurodegeneration. Can. J. Physiol. Pharmacol. **60:** 1398–1405.
27. ARCHIBALD, F.S. & I. FRIDOVICH. 1982. The scavenging of superoxide radical by manganous complexes *in vitro*. Arch. Biochem. Biophys. **214:** 452–463.
28. MONTINE, T.J. *et al.* 1994. Fibroblasts that express aromatic amino acid decarboxylase have increased sensitivity to the synergistic cytotoxicity of L-dopa and manganese. Toxicol. Appl. Pharmacol. **128:** 116–122.
29. MIGHELI, R. *et al.* 1999. Enhancing effect of manganese on L-DOPA-induced apoptosis in PC12 cells: role of oxidative stress. J. Neurochem. **73:** 1155–1163.
30. BARBER, D.S. *et al.* 2001. Dopamine depletion and cytotoxicity in PC-12 cells treated with Mn(II), Mn(III), and Mn-transferrin. Presented at the Annual Meeting of the Society of Toxicology, San Francisco.
31. GRAHAM, D.G. 1984. Catecholamine toxicity: a proposal for the molecular pathogenesis of manganese neurotoxicity and Parkinson's disease. Neurotoxicology **5:** 83–95.
32. CHEN, J.Y. *et al.* 2001. Differential toxicity of Mn(II) and Mn(III) on mitochondrial complex I activity *in vitro*. Presented at the Annual Meeting of the Society of Toxicology, San Francisco.
33. ZWINGMAN, C. *et al.* 2003. Energy metabolism in astrocytes and neurons treated with manganese: relation among cell-specific energy failure, glucose metabolism, and intercellular trafficking using multinuclear NMR-spectroscopic analysis. J. Cereb. Blood Flow Metab. **23:** 756–771.
34. COHEN, G. & R.E. HEIKKILA. 1974. The generation of hydrogen peroxide, superoxide radical, and hydroxyl radical by 6-hydroxydopamine, dialuric acid, and related cytotoxic agents. J. Biol. Chem. **249:** 2447–2452.
35. SHI, X.L. & N.S. DALUL. 1990. The glutathionyl radical formation in the reaction between manganese and glutathione and its neurotoxic implications. Med. Hypotheses **33:** 83–87.
36. ASCHNER, M. 1997. Manganese neurotoxicity and oxidative damage. *In* Metals and Oxidative Damage in Neurological Disease, pp. 79–93. Plenum. New York.
37. DONALDSON, J. *et al.* 1984. Enhanced autoxidation of dopamine as a possible basis of manganese neurotoxicity. Neurotoxicology **5:** 53–64.

38. SOLIMAN, E.F. et al. 1995. Manganese-induced oxidative stress as measured by a fluorescent probe: an in vitro study. J. Neurosci. Res. Commun. **17:** 185–193.
39. ALI, S.F. et al. 1995. Manganese-induced reactive oxygen species: comparison between Mn^{2+} and Mn^{3+}. Neurodegeneration **4:** 329–334.
40. HUSSAIN, S. & S.F. ALI. 1999. Manganese scavenges superoxide and hydroxyl radicals: an in vitro study in rats. Neurosci. Lett. **261:** 21–24.
41. SZIRAKI, I. et al. 1995. Novel protective effect of manganese against ferrous citrate–induced lipid peroxidation and nigrostriatal neurodegeneration in vivo. Brain Res. **698:** 285–287.
42. STADTMAN, E.R. et al. 1990. Manganese-dependent disproportionation of hydrogen peroxide in bicarbonate buffer. Proc. Natl. Acad. Sci. USA **87:** 384–388.
43. SZIRAKI, I. et al. 1999. Implications for atypical antioxidative properties of manganese in iron-induced brain lipid peroxidation and copper-dependent low-density lipoprotein conjugation. Neurotoxicology **20:** 455–466.
44. HALLIWELL, B. & J.M.C. GUTTERIDGE. 1999. In Aging, Nutrition, Disease, and Therapy, pp. 843–850. Oxford University Press. London/New York.
45. YIM, M.B. et al. 1993. Enzyme function of copper, zinc superoxide dismutase as a free radical generator. J. Biol. Chem. **268:** 4099–4105.
46. HAMAI, D. et al. 2001. Modulation of oxidative events by multivalent manganese complexes in brain tissue. Free Radical Biol. Med. **31:** 763–768.
47. HAMAI, D. et al. 2001. The chemistry of transition metals in relation to their potential role in neurodegenerative processes. Curr. Top. Med. Chem. **1:** 541–551.
48. MORGAN, J.J. 2000. Manganese in natural waters and earth's crust: its availability to organisms. In Manganese and Its Role in Biological Systems, pp. 2–30. Dekker. New York.
49. HAMAI, D. & S.C. BONDY. 2003. Pro- or anti-oxidant manganese: a suggested mechanism for resolution. Neurochem. Int. In press.
50. LARSON, E.J. & V.L. PECORARO. 1992. Introduction to manganese enzymes. In Manganese Redox Enzymes, pp. 1–28. VCH Pub. New York.
51. BARAL, S. et al. 1986. Chemistry of colloidal manganese oxides: formation in the reaction of hydroxyl radical with Mn^{2+} ions. J. Phys. Chem. **90:** 6025–6028.
52. CABELLI, D.E. & B.H.J. BIELSKI. 1984. Pulse radiolysis: study of the kinetics and mechanisms of the reactions between Mn(II) complexes and HO_2/O_2^- radicals. J. Phys. Chem. **88:** 3111–3115.
53. LATI, J. & D. MEYERSTEIN. 1978. Oxidation of first-row bivalent transition metal containing ethylenediaminetetra-acetate and nitrilotriacetate ligands by free radicals. J. Chem. Soc. Dalton Trans., pp. 1105–1118.
54. BIELSKI, B.H.J. et al. 1985. Reactivity of HO_2/O_2^- radicals in aqueous solutions. J. Phys. Chem. Ref. Data **14:** 1041–1100.
55. BUXTON, G.V. et al. 1988. Critical review of rate constants from reactions of hydrated electrons, hydrogen atoms, and hydroxyl radicals (OH/O^-) in aqueous solutions. J. Phys. Chem. Ref. Data **17:** 513–886.
56. COTTON, F.A. & R.G. WILKINSON. 1988. Advance Inorganic Chemistry. Interscience. New York.
57. REED, G.H. 1986. Manganese: an overview of chemical properties. In Manganese in Metabolism and Enzyme Function, pp. 313–323. Academic Press. New York.
58. ARNDT, D. 1981. Oxidation with manganese (III) and manganese (IV) complex salts. In Manganese Compounds as Oxidizing Agents in Organic Chemistry, pp. 1–25. Open Court. La Salle, IL.
59. SUNDERMAN, F.W., JR. 1986. Metals and lipid peroxidation. Acta Pharmacol. Toxicol. **59**(suppl. VII): 248–255.
60. KAZPRZAK, K.S. & G.S. BUZARD. 2000. The role of metals in oxidative damage and redox signaling derangement. In Molecular Biology and Toxicology of Metals, pp. 477–527. Taylor & Francis. London.

Divalent Metal Transporter 1 in Lead and Cadmium Transport

JOSEPH P. BRESSLER,[a,b] LUISA OLIVI,[b] JAE HOON CHEONG,[a,b,c] YONGBAE KIM,[a,d] AND DESMOND BANNON[a,b,e]

[a]*Department of Environmental Health Sciences, Bloomberg School of Public Health, Johns Hopkins University, Baltimore, Maryland, USA*

[b]*Kennedy-Krieger Institute, Baltimore, Maryland, USA*

[c]*School of Pharmacy, Sahmyook University, Seoul, Korea*

[d]*Department of Preventive Medicine, Soonchunhyan University, Chunan City, Korea*

ABSTRACT: The effect of exposure to cadmium (Cd) and lead (Pb) on human health has been recognized for many years and recent information suggests that minimal exposure levels are themselves too high. Common scenarios for Pb exposure include occupational, residential, and/or behavioral (hand-to-mouth activity) settings. The main source of Cd exposure for nonsmokers is dietary, through plants or animals that accumulate the metal. Specific cellular importers for Pb and Cd are unlikely as these metals are nonessential and toxic. Accordingly, in the intestine, the operational mechanism is assumed to be inadvertent uptake through pathways intended for essential nutrients such as iron. Results from experimental and epidemiological studies indicated that diets low in iron (Fe) result in increased absorption of Pb and Cd, suggesting common molecular mechanisms of Cd and Pb transport. Indeed, recent mechanistic studies found that the intestinal transporter for nonheme iron, divalent metal transporter 1 (DMT1), mediates the transport of Pb and Cd. DMT1 is regulated, in part, by dietary iron, and chemical species of Cd and Pb that are transported by DMT1 would be made available through digestion and are also found in plasma. Accordingly, the involvement of DMT1 in metal uptake offers a mechanistic explanation for why an iron-deficient diet is a risk factor for Pb and Cd poisoning. It also suggests that diets rich in iron-containing food could be protective against heavy metal poisoning.

KEYWORDS: divalent metal transporter 1 (DMT1); transport; cadmium (Cd); lead (Pb); diet

CADMIUM AND LEAD IN THE GENERAL ENVIRONMENT

The anthropogenic release of nonessential and toxic heavy metals to the environment and their health effects have been recognized for many years. Nowadays, low-level exposure to lead (Pb) in children is associated with impaired cognitive

---

Address for correspondence: Joseph Bressler, Department of Neurology, Kennedy-Krieger Institute, 707 North Broadway, Baltimore, MD 21205. Voice: 443-923-2677; fax: 443-923-2695.
bressler@kennedykrieger.org

[e]Present address: Desmond Bannon, U.S. Army, Aberdeen Proving Ground, MD 21010.

development[1,2] and delayed puberty in girls.[3] In adults, Pb exposure is associated with hypertension,[4,5] and recent studies have also shown adverse effects on cognition.[6] Approximately 5% of the children living in the United States still have blood Pb levels greater than 10 µg/dL[7] and are currently considered at risk for chronic Pb poisoning by the Centers for Disease Control and the Environmental Protection Agency. Furthermore, a recent study found an inverse association between IQ scores in children three and five years of age with blood Pb levels below 10 µg/dL, suggesting more children are at risk than was originally thought.[8] Exposure to low levels of cadmium (Cd) is associated with renal damage and was once thought to be mostly an occupational hazard, but attention has now focused on the general population. Recent studies have suggested that estimates of levels of Cd in the kidney that produce renal damage in the kidney were too high,[9] which suggests that the maximal tolerated Cd intake should be lowered.[10] Since the levels of Cd in the environment continue to rise,[11] large populations in Europe and elsewhere potentially have exposure levels that are now considered damaging.[11]

The government's regulatory response to these hazards often lags too far behind the scientific evidence. Although Pb was removed from paint and gasoline in the 1970s and 1980s, a NHANES study conducted in 1991–94 found that children living in housing developments that were constructed after 1973 have a mean blood Pb level of 2 µg/dL,[7] which suggests that exposure is from sources other than paint. An additional response would be to identify factors that increase the individual's risk, such as those in the diet, and remove them. This is because the most common exposure route for Pb is through the digestive system, whereas exposure to Cd, at least in nonsmokers, is through diets rich in shellfish, rice, and leafy vegetables, which contain relatively high levels of Cd.[10]

INTESTINAL TRANSPORT OF Fe, Cd, AND Pb

Absorption of Cd and Pb in the diet occurs through a process that appears similar to the absorption of essential metals such as iron,[12] calcium,[13] and zinc.[14] The absorption of these metals occurs predominantly in the upper intestine and can be divided into three steps: (1) a membrane transporter facilitates the transport of the metal at the brush border membrane, (2) serosal transfer to the basolateral side, and (3) basolateral transfer of the metal to the plasma. Early animal studies showed that maximal absorption of Pb[15,16] and Cd[17,18] also occurs in the upper intestine by a carrier-mediated process.[19,20]

From a human perspective, epidemiological studies have provided very strong evidence showing an association between low iron stores and blood Cd levels. In pregnant women in Sweden, higher blood Cd levels were also associated with reduced iron stores.[21,22] In human subjects, the average absorption of Cd was four times greater in people with low body iron stores compared to subjects with normal iron stores.[23] Additionally, epidemiological studies show an association, though not unequivocal, between low levels of iron stores and elevated blood Pb. For example, the blood Pb levels of iron-deficient children living in contaminated areas were almost 3 µg/dL higher than blood Pb levels of iron-replete children.[24] A study of mother-infant pairs found a negative association between the mothers' iron status and the blood Pb level in the newborn.[25] A possible explanation for this association is that the mechanism for iron transport in the intestine also mediates the transport of Pb and Cd. There is evidence to support this explanation in earlier studies in animal models showing that iron inhibits the gastrointestinal transport of Pb[26,27] and Cd[28,29] in rat intestine.

IRON TRANSPORT BY DIVALENT METAL TRANSPORTER 1 (DMT1)

The transporter for nonheme iron at the luminal intestinal surface is DMT1, which is an H^+-coupled and electrogenic membrane transporter that belongs to a large family of integral membrane proteins highly conserved throughout evolution.[30,31] In mammals, alternative splicing on the 3′-terminal exon of the DMT1 gene results in two distinct mRNA species. The mRNAs are distinguished by different C-terminal amino acid sequences and by the presence of an iron response element (IRE) on the 3′ untranslated region on one species. Recently, the expression of a previously unrecognized upstream 5′ exon (exon 1A) of the human and murine DMT1 gene has been shown that is tissue-specific and particularly prevalent in the duodenum and kidney.[32] Evidence supporting the involvement of DMT1 in intestinal absorption of iron is quite substantial (see refs. 33 and 34). In one of the studies that discovered DMT1, results from a surrogate assay that measured evoked currents associated with metal transport in *Xenopus* oocytes suggested that DMT1 transports a broad range of metals, including zinc, copper, manganese, nickel, and cobalt, as well as the toxic metals, Cd and Pb.[30]

Model systems that measured metal transport with radioisotopes or fluorescent dyes have confirmed that DMT1 mediates the transport of essential metals including copper,[35] manganese,[36] and cobalt.[37] In contrast, several studies measured the uptake of zinc directly[38–40] rather than the surrogate assay described above and failed to demonstrate DMT1-mediated uptake of zinc. DMT1, however, has been firmly established as a transporter for Cd. In Caco-2 cells, knocking down DMT1 expression with a ribozyme construct decreased transport of Cd;[41] Cd uptake was greater at an acidic pH than at neutral pH and was inhibited by Fe;[42] and increasing expression of DMT1 resulted in increased transport of Cd.[37,43,44] Activation of protein kinase C results in increased expression of DMT1 and increased uptake of Cd and Fe in kidney epithelial cells.[45] Most importantly, iron-deficient rats displayed dramatically higher levels of intestinal DMT1 and greater transport of Cd.[46,47]

DMT1 also appears to be a transporter for Pb. Overexpression of DMT1 in yeast and human fibroblasts resulted in increased transport of Pb, which was greater at an acidic pH and inhibited by Fe.[48] DMT1 also appears to be a transporter for Pb in glial cells. When DMT1 levels are raised, transport of Pb was increased at low pH and inhibited by Fe (accepted to *Toxicological Sciences*). In the Caco-2 cell model, however, knocking down DMT1 did not affect uptake of Pb.[41] One possible explanation is that there is more than one transporter for Pb in Caco-2 cells. Because Pb is not an essential metal, it is unlikely that a specific transporter has evolved for its transport. Caco-2 cells are grown with adequate concentrations of Fe so that the basal levels of DMT1 are low, which allows other transporters to transport Pb. Accordingly, knocking down DMT1 under these conditions would not affect uptake of Pb. If, however, cells were grown under iron-poor conditions, we would expect to observe DMT1-mediated transport of Pb because levels of DMT1 would rise.

BIOAVAILABILITY OF Pb AND Cd IN THE ENVIRONMENT

Whether environmental Cd and Pb are transported by DMT1 will depend on their bioavailability. Iron is easily oxidized and dietary iron is predominantly in the ferric form as a high-molecular-weight species and is reduced to Fe^{++} (which is the preferred substrate of DMT1) by enzymes, the acidity of gastric juices, and the presence of ascorbic acid. Additionally, a duodenal reductase (Dcytb), found in the apical duodenum of enterocytes, appears to be a variant of cytochrome b and has been shown important for the absorption of iron.[49] Toxic metals such as Cd and Pb are divalent and would not require a reductase for transport by DMT1. Rather, the bioavailability of Cd, for example, depends upon its interaction with macromolecules in food. Most Cd in food is bound to metallothionein (MT) (see ref. 50) and phytochelatin.[51] The Cd/MT conjugate is likely partially degraded by gastric juices,[52] suggesting that Cd is released from these macromolecules in the stomach and it is thus available for transport by DMT1. A major source of Pb is in paint in the form of lead carbonate, which is partly soluble in acidic pH and thus available for DMT1-mediated transport.[53] In addition to paint, other sources of Pb, for example, in drinking water or in soil, pose a health hazard because blood Pb levels in the low-microgram range are observed in children living in housing built after 1973.[7] Pb in soil is particularly important because children accidentally ingest 50–200 mg of soil per day by hand-to-mouth behavior.[54,55] A recent study suggests that Pb complexes formed in the small intestine can release free Pb, which then can contribute to transport across the intestinal epithelium.[56]

An important consideration in explaining the regulation of absorption of iron is the coordination of the three steps of absorption: (1) transport by DMT1 or other transporters; (2) transcellular transport to the basal surface that requires hephaestin and its ferroxidase activity;[57,58] (3) basolateral export into the blood to bind to transferrin, which is mediated by the membrane iron exporter Ireg1[59] (also referred to as ferroportin and MTP1[60]). Levels of Dcytb, DMT1, Ireg1, and hephaestin increase under iron-poor conditions, allowing for more iron absorption.[61] The contribution of Dcytb and Ireg1 to the higher levels of Pb and Cd absorption under iron-poor conditions is unclear. Dcytb is probably not involved because Pb and Cd are divalent and would not require reductase activity. The specificity of Ireg1 for different metals has not been examined.

Besides Ireg1, another possible candidate for transporting Cd and Pb at the basal surface is Ca-ATPase, which mediates the export of calcium at the basal surface.[62] Ca-ATPase has been shown to catalyze the efflux of Pb in erythrocytes.[63] Although the efflux of Cd by Ca-ATPase has not been reported, Cd is a competitive and potent inhibitor of calcium for intestinal Ca-ATPase.[64] Calbindin, a member of the calmodulin family, which binds Pb and Cd,[65,66] might also be involved in the absorption of Cd and Pb. Calbindin mediates the trafficking of calcium from the luminal surface to the basal surface in the intestine. The expression of calbindin is regulated by vitamin D[67,68] and might explain the effects of vitamin D on the absorption of Pb[69] and Cd.[70] The intestinal calcium transporter on the luminal surface is also regulated by vitamin D, but it does not appear to transport Pb and Cd.[71]

Although DMT1, Dcytb, and Ireg1 each have IREs, DMT1 and Dcytb expression (but not Ireg1 expression) is affected by dietary iron levels. For example, levels of DMT1 and Dcytb mRNA rapidly decrease in iron-deficient rats after an intragastric

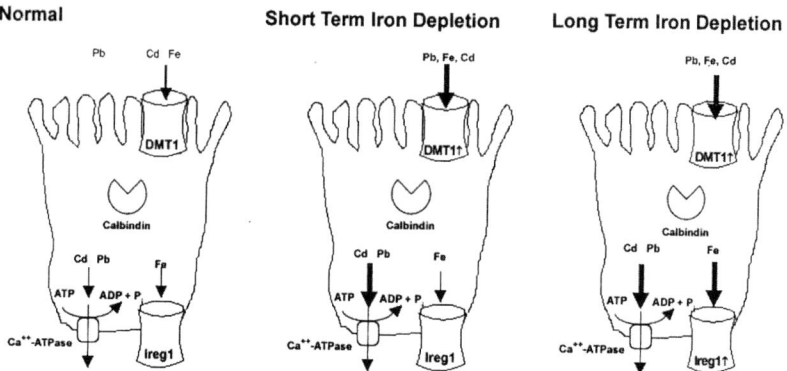

FIGURE 1. Individuals on an iron-replete diet will express low levels of proteins involved in iron transport in the intestine such as DMT1 and Ireg1. A short-term iron-poor diet will result in increased levels of DMT1, but not Ireg1, which is regulated by iron stores. Higher levels of DMT1 could result in increased absorption of Cd and Pb because of calbindin and Ca-ATPase. Longer time on an iron-deplete diet will result in increased levels of Ireg1.

bolus of iron, with no change in levels of Ireg1 mRNA.[72] Rather, Ireg1 expression appears regulated by the body's iron stores.[73] Hence, individuals on an iron-deplete diet for only a few days will display higher levels of DMT1, but might not absorb more iron because levels of Ireg1 have not increased. These individuals, however, could absorb more Cd and Pb because of the participation of calbindin and Ca-ATPase (FIG. 1). Accordingly, individuals not receiving adequate levels of iron for a few days would be at greater risk to Pb and Cd poisoning and yet not have lower iron stores or display clinical signs of iron deficiency.

DMT1 AND CELL SURFACE TRANSPORT OF METALS

Whether DMT1 is a cell surface transporter for iron in types of cells other than the intestine is unclear. Most cells acquire iron through the transferrin cycle. Several studies have provided evidence indicating that DMT1 mediates the transfer of endosomal free Fe^{2+} into the cytoplasm in the transferrin cycle.[74] It is possible, however, that DMT1 is also a cell surface transporter for iron as well as for other metals. Although the transferrin cycle is the major mechanism by which cells obtain iron, there is evidence for transferrin-independent pathways. For example, in the hpx/hpx mouse that lacks transferrin,[75] or in individuals with the very rare disorder, atransferrinemia, dietary iron is not used for erythrocyte production, but iron is deposited in liver.[50,76] Similarly, when transferrin is saturated, the nontransferrin pool of iron in plasma increases, resulting in buildup of iron in the liver.[77] Transferrin-independent iron uptake was demonstrated in several cell lines, and astrocytes do not express transferrin receptors *in vivo*.[78] Although DMT1 may be a cell surface iron transporter, it requires H^+, and extracellular fluid is at neutral or near-neutral pH. Nonetheless, DMT1 has been shown to transport metals at pH 7.4. When the non-IRE species of DMT1 was elevated by transfecting fibroblasts with a DMT1 expression vector, cobalt transport[79] was increased at pH values below 8.0 (although optimal transport occurred with an acidic extracellular fluid), but a similar study did not find increased transport of manganese.[36] DMT1 was expressed at the cell surface in these transfected fibroblasts. Knocking down expression of the IRE form of DMT1 diminished the transport of iron in Caco-2 cells at pH 7.4.[41] The concentration of H^+ at pH 7.4 is 40 nM, which could be sufficient, though not optimal, for DMT1-mediated metal transport.[34] Also important, the pH of the extracellular fluid might not reflect the pH of the microenvironment of DMT1, which can be modified by transporters such as H^+-ATPase and the Na-H^+ exchanger.

Another argument against DMT1 as a cell surface iron transporter is that the preferred form of iron, the ferrous form, potentially will increase oxidative stress by catalyzing the Fenton reaction and producing the toxic hydroxyl radical. Iron chelated to natural ligands, including citrate, would be a powerful catalyst in extracellular fluid.[80] Interestingly, DMT1 might function in a mechanism by which the cell detoxifies the extracellular fluid by transporting iron into the cell. A recent study has shown increased expression of the non-IRE form of DMT1 and a decrease in iron in the extracellular fluid in bronchial epithelial cells exposed to iron citrate.[81]

In light of the possibility that DMT1 is a cell surface iron transporter, DMT1 could be a cell surface transporter for Cd in the liver. Free Cd absorbed by the intestine will bind to proteins and small molecules in the plasma, but some Cd will remain free depending on the binding constants. Cd bound to protein will be taken up into hepatocytes by endocytosis, but DMT1 might transport free cadmium. Cd/MT-conjugate, not-free Cd is taken up by the kidney, which likely does not involve DMT1.[82] The S1 and S2 segments of the proximal tubule in renal cortex have been shown to take up much of the Cd/MT conjugate[83] and, interestingly, the main site of renal toxicity is the S1 segment. DMT1 might be involved in Cd toxicity: for example, Cd-mediated nephrotoxicity might result from Cd/MT undergoing endocytosis and sorting to endosomes.[84,85] DMT1 colocalized with LAMP-1, a marker for late endosomes/lysosomes, in the S-1 segment.[86] It is possible that the Cd/MT conjugate undergoes degradation in endosomes and the free Cd is released into the cytoplasm by DMT1.

In children, the major organ affected by Pb is the brain. Pb in the blood is almost 95% in erythrocytes and likely bound to protein and other constituents in plasma. There is essentially no endocytosis in the blood-brain barrier, which suggests that both free Pb and Pb bound to small molecular ligands are transported into the brain. DMT1 is a possible candidate for transporting free Pb because it is expressed by end-feet of astrocytes around blood vessels and by brain capillary endothelial cells.[87] The end-feet are separated from endothelial cells by basement membrane and are thought to participate in the transport of nutrients from the blood vessels. Taken together with the earlier discussed finding that elevated expression of DMT1 results in increased transport of Pb in a glial cell line, we suggest that DMT1 mediates the transport of Pb in astrocytes. The involvement of DMT1 in Pb transport in astrocytes explains an earlier study that showed that astrocytes accumulated Pb in the brain when rats are fed Pb in their drinking water.[88]

CONCLUSIONS

Evidence has been presented demonstrating that DMT1 can act as a transporter for Cd and Pb. Furthermore, the involvement of DMT1 in the transport of Cd in models of intestinal absorption *in vivo* and *in vitro* has been shown. Additional studies are needed to examine the involvement of DMT1 for the intestinal transport of Pb. DMT1 is also a candidate transporter for Cd in the liver and Pb in astrocytes. We have not discussed other transporters for Pb and Cd, including calcium channels and anion transporters, nor have we discussed the possibility that Cd and/or Pb would be bound to amino acids or organic anions and transported by amino acid or organic anion transporters, respectively. It is unlikely that a transporter has evolved for nonessential metals. The identification of DMT1 as a transporter for Cd and Pb helps explain the relation between low iron status and higher levels of Cd and Pb in the blood. It also provides a rationale for administering iron supplements to children with elevated blood levels of these metals. Careful use of iron supplements to populations at risk for Pb and Cd poisoning might also be considered, together with a concerted effort to remove these toxic metals from our environment.

REFERENCES

1. BELLINGER, D., A. LEVITON, C. WATERNAUX et al. 1987. Longitudinal analyses of prenatal and postnatal lead exposure and early cognitive development. N. Engl. J. Med. **316:** 1037–1043.
2. MCMICHAEL, A.J., P.A. BAGHURST, N.R. WIGG et al. 1988. Port Pirie Cohort Study: environmental exposure to lead and children's abilities at the age of four years. N. Engl. J. Med. **319:** 468–475.
3. SELEVAN, S.G., D.C. RICE, K.A. HOGAN et al. 2003. Blood lead concentration and delayed puberty in girls. N. Engl. J. Med. **348:** 1527–1536.
4. HERTZ-PICCIOTTO, I. & J. CROFT. 1993. Review of the relation between blood lead and blood pressure. Epidemiol. Rev. **15:** 352–373.
5. CHENG, Y., J. SCHWARTZ, D. SPARROW et al. 2001. Bone lead and blood lead levels in relation to baseline blood pressure and the prospective development of hypertension: the Normative Aging Study. Am. J. Epidemiol. **153:** 164–171.
6. STEWART, W.F., B.S. SCHWARTZ, D. SIMON et al. 1999. Neurobehavioral function and tibial and chelatable lead levels in 543 former organolead workers. Neurology **52:** 1610–1617.
7. PIRKLE, J.L., R.B. KAUFMANN, D.J. BRODY et al. 1998. Exposure of the U.S. population to lead, 1991–1994. Environ. Health Perspect. **106:** 745–750.
8. CANFIELD, R.L., C.R. HENDERSON, JR., D.A. CORY-SLECHTA et al. 2003. Intellectual impairment in children with blood lead concentrations below 10 micrograms per deciliter. N. Engl. J. Med. **348:** 1517–1526.
9. JARUP, L., L. HELLSTROM, T. ALFVEN et al. 2000. Low level exposure to cadmium and early kidney damage: the OSCAR study. Occup. Environ. Med. **57:** 668–672.
10. SATARUG, S., M.R. HASWELL-ELKINS & M.R. MOORE. 2000. Safe levels of cadmium intake to prevent renal toxicity in human subjects. Br. J. Nutr. **84:** 791–802.
11. JARUP, L., M. BERGLUND, C.G. ELINDER et al. 1998. Health effects of cadmium exposure—a review of the literature and a risk estimate. Scand. J. Work Environ. Health **24:** 1–51.
12. MUIR, W.A., U. HOPFER & M. KING. 1984. Iron transport across brush-border membranes from normal and iron-deficient mouse upper small intestine. J. Biol. Chem. **259:** 4896–4903.
13. BRONNER, F. 1998. Calcium absorption—a paradigm for mineral absorption. J. Nutr. **128:** 917–920.

14. HOADLEY, J.E., A.S. LEINART & R.J. COUSINS. 1987. Kinetic analysis of zinc uptake and serosal transfer by vascularly perfused rat intestine. Am. J. Physiol. **252**: G825–G831.
15. CONRAD, M.E. & J.C. BARTON. 1978. Factors affecting the absorption and excretion of lead in the rat. Gastroenterology **74**: 731–740.
16. BARTON, J.C. 1984. Active transport of lead-210 by everted segments of rat duodenum. Am. J. Physiol. **247**: G193–G198.
17. SORENSEN, J.A., J.B. NIELSEN & O. ANDERSEN. 1993. Identification of the gastrointestinal absorption site for cadmium chloride *in vivo*. Pharmacol. Toxicol. **73**: 169–173.
18. ANDERSEN, O., J.B. NIELSEN, J.A. SORENSEN *et al.* 1994. Experimental localization of intestinal uptake sites for metals (Cd, Hg, Zn, Se) *in vivo* in mice. Environ. Health Perspect. **102**(suppl. 3): 199–206.
19. FOULKES, E.C. 1979. Some determinants of intestinal cadmium transport in the rat. J. Environ. Pathol. Toxicol. **3**: 471–481.
20. AUNGST, B.J. & H.L. FUNG. 1981. Kinetic characterization of *in vitro* lead transport across the rat small intestine: mechanism of intestinal lead transport. Toxicol. Appl. Pharmacol. **61**: 39–47.
21. BERGLUND, M., A. AKESSON, B. NERMELL *et al.* 1994. Intestinal absorption of dietary cadmium in women depends on body iron stores and fiber intake. Environ. Health Perspect. **102**: 1058–1066.
22. AKESSON, A., M. BERGLUND, A. SCHUTZ *et al.* 2002. Cadmium exposure in pregnancy and lactation in relation to iron status. Am. J. Public Health **92**: 284–287.
23. FLANAGAN, P.R., J.S. MCLELLAN, J. HAIST *et al.* 1978. Increased dietary cadmium absorption in mice and human subjects with iron deficiency. Gastroenterology **74**: 841–846.
24. BRADMAN, A., B. ESKENAZI, P. SUTTON *et al.* 2001. Iron deficiency associated with higher blood lead in children living in contaminated environments. Environ. Health Perspect. **109**: 1079–1084.
25. SCHELL, L.M., M. DENHAM, A.D. STARK *et al.* 2003. Maternal blood lead concentration, diet during pregnancy, and anthropometry predict neonatal blood lead in a socio-economically disadvantaged population. Environ. Health Perspect. **111**: 195–200.
26. BARTON, J.C., M.E. CONRAD, S. NUBY *et al.* 1978. Effects of iron on the absorption and retention of lead. J. Lab. Clin. Med. **92**: 536–547.
27. FLANAGAN, P.R., J. HAIST & L.S. VALBERG. 1980. Comparative effects of iron deficiency induced by bleeding and a low-iron diet on the intestinal absorptive interactions of iron, cobalt, manganese, zinc, lead and cadmium. J. Nutr. **110**: 1754–1763.
28. VALBERG, L.S., J. SORBIE & D.L. HAMILTON. 1976. Gastrointestinal metabolism of cadmium in experimental iron deficiency. Am. J. Physiol. **231**: 462–467.
29. RAGAN, H.A. 1977. Effects of iron deficiency on the absorption and distribution of lead and cadmium in rats. J. Lab. Clin. Med. **90**: 700–706.
30. GUNSHIN, H., B. MACKENZIE, U.V. BERGER *et al.* 1997. Cloning and characterization of a mammalian proton-coupled metal-ion transporter. Nature **388**: 482–488.
31. FLEMING, M.D., M.A. ROMANO, M.A. SU *et al.* 1998. Nramp2 is mutated in the anemic belgrade (b) rat: evidence of a role for Nramp2 in endosomal iron transport. Proc. Natl. Acad. Sci. USA **95**: 1148–1153.
32. HUBERT, N. & M.W. HENTZE. 2002. Previously uncharacterized isoforms of divalent metal transporter (DMT)–1: implications for regulation and cellular function. Proc. Natl. Acad. Sci. USA **99**: 12345–12350.
33. ROLFS, A. & M.A. HEDIGER. 2001. Intestinal metal ion absorption: an update. Curr. Opin. Gastroenterol. **17**: 177–183.
34. GARRICK, M.D., K.G. DOLAN, C. HORBINSKI *et al.* 2003. DMT1: a mammalian transporter for multiple metals. Biometals **16**: 41–54.
35. ARREDONDO, M., P. MUNOZ *et al.* 2003. DMT1, a physiologically relevant apical Cu^{1+} transporter of intestinal cells. Am. J. Physiol. Cell. Physiol. **284**: C1525–C1530.
36. FORBES, J.R. & P. GROS. 2003. Iron, manganese, and cobalt transport by Nramp1 (Slc11a1) and Nramp2 (Slc11a2) expressed at the plasma membrane. Blood. In press.
37. PICARD, V., G. GOVONI, N. JABADO *et al.* 2000. Nramp 2 (DCT1/DMT1) expressed at the plasma membrane transports iron and other divalent cations into a calcein-accessible cytoplasmic pool. J. Biol. Chem. **275**: 35738–35745.

38. CONRAD, M.E., J.N. UMBREIT, E.G. MOORE et al. 2000. Separate pathways for cellular uptake of ferric and ferrous iron. Am. J. Physiol. Gastrointest. Liver Physiol. **279**: G767–G774.
39. TANDY, S., M. WILLIAMS, A. LEGGETT et al. 2000. Nramp2 expression is associated with pH-dependent iron uptake across the apical membrane of human intestinal Caco-2 cells. J. Biol. Chem. **275**: 1023–1029.
40. SACHER, A., A. COHEN & N. NELSON. 2001. Properties of the mammalian and yeast metal-ion transporters DCT1 and Smf1p expressed in *Xenopus laevis* oocytes. J. Exp. Biol. **204**: 1053–1061.
41. BANNON, D.I., R. ABOUNADER, P.S.J. LEES & J.P. BRESSLER. 2003. Effect of DMT1 knockdown on iron, cadmium, and lead uptake in Caco-2 cells. Am. J. Physiol. **28**: C44–C50.
42. ELISMA, F. & C. JUMARIE. 2001. Evidence for cadmium uptake through Nramp2: metal speciation studies with Caco-2 cells. Biochem. Biophys. Res. Commun. **285**: 662–668.
43. TALLKVIST, J., C.L. BOWLUS & B. LONNERDAL. 2001. DMT1 gene expression and cadmium absorption in human absorptive enterocytes. Toxicol. Lett. **122**: 171–177.
44. OKUBO, M., K. YAMADA, M. HOSOYAMADA et al. 2003. Cadmium transport by human Nramp 2 expressed in *Xenopus laevis* oocytes. Toxicol. Appl. Pharmacol. **187**: 162–167.
45. OLIVI, L., J. SISK & J. BRESSLER. 2001. Involvement of DMT1 in uptake of Cd in MDCK cells: role of protein kinase C. Am. J. Physiol. Cell. Physiol. **281**: C793–C800.
46. LEAZER, T.M., Y. LIU & C.D. KLAASSEN. 2002. Cadmium absorption and its relationship to divalent metal transporter-1 in the pregnant rat. Toxicol. Appl. Pharmacol. **185**: 18–24.
47. PARK, J.D., N.J. CHERRINGTON & C.D. KLAASSEN. 2002. Intestinal absorption of cadmium is associated with divalent metal transporter 1 in rats. Toxicol. Sci. **68**: 288–294.
48. BANNON, D., M.E. PORTNOY, L. OLIVI et al. 2002. Uptake of lead and iron by divalent metal transporter 1 in yeast and mammalian cells. Biochem. Biophys. Res. Commun. **295**: 978–984.
49. MIRET, S., R.J. SIMPSON & A.T. MCKIE. 2003. Physiology and molecular biology of dietary iron absorption. Annu. Rev. Nutr. In press.
50. PETERING, D.H. & B.A. FOWLER. 1986. Roles of metallothionein and related proteins in metal metabolism and toxicity: problems and perspectives. Environ. Health Perspect. **65**: 217–224.
51. COBBETT, C.S. 2000. Phytochelatins and their roles in heavy metal detoxification. Plant Physiol. **123**: 825–832.
52. KLEIN, D., H. GREIM & K.H. SUMMER. 1986. Stability of metallothionein in gastric juice. Toxicology **41**: 121–129.
53. HARRISON, R.M. & D.P. LAXEN. 1980. Physicochemical speciation of lead in drinking water. Nature **286**: 791–793.
54. CALABRESE, E.J., E.J. STANEK, R.C. JAMES et al. 1997. Soil ingestion: a concern for acute toxicity in children. Environ. Health Perspect. **105**: 1354–1358.
55. STANEK, E.J., III & E.J. CALABRESE. 2000. Daily soil ingestion estimates for children at a Superfund site. Risk Anal. **20**: 627–635.
56. OOMEN, A.G., J. TOLLS, A.J. SIPS et al. 2003. *In vitro* intestinal lead uptake and transport in relation to speciation. Arch. Environ. Contam. Toxicol. **44**: 116–124.
57. VULPE, C.D., Y.M. KUO, T.L. MURPHY et al. 1999. Hephaestin, a ceruloplasmin homologue implicated in intestinal iron transport, is defective in the sla mouse. Nat. Genet. **21**: 195–199.
58. ANDERSON, G.J., D.M. FRAZER, A.T. MCKIE et al. 2002. The ceruloplasmin homolog hephaestin and the control of intestinal iron absorption. Blood Cells Mol. Dis. **29**: 367–375.
59. MCKIE, A.T., P. MARCIANI, A. ROLFS et al. 2000. A novel duodenal iron-regulated transporter, IREG1, implicated in the basolateral transfer of iron to the circulation. Mol. Cell **5**: 299–309.
60. ABBOUD, S. & D.J. HAILE. 2000. A novel mammalian iron-regulated protein involved in intracellular iron metabolism. J. Biol. Chem. **275**: 19906–19912.
61. MCKIE, A.T., D. BARROW, G.O. LATUNDE-DADA et al. 2001. An iron-regulated ferric reductase associated with the absorption of dietary iron. Science **291**: 1755–1759.

62. BRONNER, F. 2003. Mechanisms of intestinal calcium absorption. J. Cell. Biochem. **88**: 387–393.
63. SIMONS, T.J.B. 1988. Active transport of lead by the calcium pump in human red cell ghosts. J. Physiol. Lond. **405**: 383–389.
64. VERBOST, P.M., M.H. SENDEN & C.H. VAN OS. 1987. Nanomolar concentrations of Cd^{2+} inhibit Ca^{2+} transport systems in plasma membranes and intracellular Ca^{2+} stores in intestinal epithelium. Biochim. Biophys. Acta **902**: 247–252.
65. HABERMANN, E., K. CROWELL & P. JANICKI. 1983. Lead and other metals can substitute for Ca^{2+} in calmodulin. Arch. Toxicol. **54**: 61–70.
66. HABERMANN, E. & G. RICHARDT. 1986. Intracellular calcium binding proteins as targets for heavy metal ions. Trends Pharmacol. Sci. **11**: 298–300.
67. BUCKLEY, M. & F. BRONNER. 1980. Calcium-binding protein biosynthesis in the rat: regulation by calcium and 1,25-dihydroxyvitamin D3. Arch. Biochem. Biophys. **202**: 235–241.
68. VAN CROMPHAUT, S.J., M. DEWERCHIN, J.G. HOENDEROP et al. 2001. Duodenal calcium absorption in vitamin D receptor–knockout mice: functional and molecular aspects. Proc. Natl. Acad. Sci. USA **98**: 13324–13329.
69. FULLMER, C.S. 1990. Intestinal lead and calcium absorption: effect of 1,25-dihydroxycholecalciferol and lead status. Proc. Soc. Exp. Biol. Med. **194**: 258–264.
70. PIGMAN, E.A., J. BLANCHARD & H.E. LAIRD II. 1997. A study of cadmium transport pathways using the Caco-2 cell model. Toxicol. Appl. Pharmacol. **142**: 243–247.
71. WOOD, R.J., L. TCHACK & S. TAPARIA. 2001. 1,25-Dihydroxyvitamin D3 increases the expression of the CaT1 epithelial calcium channel in the Caco-2 human intestinal cell line. BMC Physiol. **1**: 11.
72. FRAZER, D.M., S.J. WILKINS, E.M. BECKER et al. 2003. A rapid decrease in the expression of DMT1 and Dcytb, but not Ireg1 or hephaestin explains the mucosal block phenomenon of iron absorption. Gut **52**: 340–346.
73. FRAZER, D.M., S.J. WILKINS, E.M. BECKER et al. 2002. Hepcidin expression inversely correlates with the expression of duodenal iron transporters and iron absorption in rats. Gastroenterology **123**: 835–844.
74. AISEN, P., C. ENNS & M. WESSLING-RESNICK. 2001. Chemistry and biology of eukaryotic iron metabolism. Int. J. Biochem. Cell. Biol. **33**: 940–959.
75. BERNSTEIN, S.E. 1987. Hereditary hypotransferrinemia with hemosiderosis, a murine disorder resembling human atransferrinemia. J. Lab. Clin. Med. **110**: 690–705.
76. HAMILL, R.L., J.C. WOODS & B.A. COOK. 1991. Congenital atransferrinemia: a case report and review of the literature. Am. J. Clin. Pathol. **96**: 215–218.
77. CRAVEN, C.M., J. ALEXANDER, M. ELDRIDGE et al. 1987. Tissue distribution and clearance kinetics of non-transferrin-bound iron in the hypotransferrinemic mouse: a rodent model for hemochromatosis. Proc. Natl. Acad. Sci. USA **84**: 3457–3461.
78. MOOS, T., P.S. OATES & E.H. MORGAN. 1998. Expression of the neuronal transferrin receptor is age dependent and susceptible to iron deficiency. J. Comp. Neurol. **398**: 420–430.
79. PICARD, V., G. GOVONI, N. JABADO et al. 2000. NRAMP 2 (DCT1/DMT1) expressed at the plasma membrane transports iron and other divalent cations into a calcein-accessible cytoplasmic pool. J. Biol. Chem. **275**: 35738–35745.
80. PIERRE, J.L. & M. FONTECAVE. 1999. Iron and activated oxygen species in biology: the basic chemistry. Biometals **12**: 195–199.
81. WANG, X., A.J. GHIO, F. YANG et al. 2002. Iron uptake and Nramp2/DMT1/DCT1 in human bronchial epithelial cells. Am. J. Physiol. Lung Cell. Mol. Physiol. **282**: L987–L995.
82. ZALUPS, R.K. & S. AHMAD. 2003. Molecular handling of cadmium in transporting epithelia. Toxicol. Appl. Pharmacol. **186**: 163–188.
83. DORIAN, C., V.H. GATTONE II & C.D. KLAASSEN. 1992. Accumulation and degradation of the protein moiety of cadmium-metallothionein (CdMT) in the mouse kidney. Toxicol. Appl. Pharmacol. **117**: 242–248.
84. ERFURT, C., E. ROUSSA & F. THEVENOD. 2003. Apoptosis by cadmium or cadmium-metallothionein (CdMT) in proximal tubule cells: different uptake routes and permissive role of endo-/lysosomal CdMT uptake. Am. J. Physiol. Cell. Physiol. In press.

85. THEVENOD, F. 2003. Nephrotoxicity and the proximal tubule: insights from cadmium. Nephron Physiol. **93:** p87–p93.
86. FERGUSON, C.J., M. WAREING, D.T. WARD *et al.* 2001. Cellular localization of divalent metal transporter DMT-1 in rat kidney. Am. J. Physiol. Renal Physiol. **280:** F803–F814.
87. WANG, X.S., W.Y. ONG & J.R. CONNOR. 2001. A light and electron microscopic study of the iron transporter protein DMT-1 in the monkey cerebral neocortex and hippocampus. J. Neurocytol. **30:** 353–360.
88. THOMAS, J.A., F.D. DALLENBACH & M. THOMAS. 1973. The distribution of radioactive lead (210Pb) in the cerebellum of developing rats. J. Pathol. **109:** 45–50.

Redox-Active Metals, Oxidative Stress, and Alzheimer's Disease Pathology

XUDONG HUANG,[a,b,c] ROBERT D. MOIR,[a,b,d] RUDOLPH E. TANZI,[a,b,d] ASHLEY I. BUSH,[a,b,c,e] AND JACK T. ROGERS[a,b,c]

[a]*Laboratory for Oxidation Biology,* [b]*Genetics and Aging Research Unit,* [c]*Department of Psychiatry,* [d]*Department of Neurology, Massachusetts General Hospital and Harvard Medical School, Charlestown, Massachusetts 02129, USA*

[e]*Mental Health Research Institute of Victoria and Department of Pathology, University of Melbourne, Parkville, Victoria 3052, Australia*

ABSTRACT: Considerable evidence is mounting that dyshomeostasis of the redox-active biometals, Cu and Fe, and oxidative stress contribute to the neuropathology of Alzheimer's disease (AD). Present data suggest that metals can interact directly with Aβ peptide, the principal component of β-amyloid that is one of the primary lesions in AD. The binding of metals to Aβ modulates several physiochemical properties of Aβ that are thought to be central to the pathogenicity of the peptide. First, we and others have shown that metals can promote the *in vitro* aggregation into tinctorial Aβ amyloid. Studies have confirmed that insoluble amyloid plaques in postmortem AD brain are abnormally enriched in Cu, Fe, and Zn. Conversely, metal chelators dissolve these proteinaceous deposits from postmortem AD brain tissue and attenuate cerebral Aβ amyloid burden in APP transgenic mouse models of AD. Second, we have demonstrated that redox-active Cu(II) and, to a lesser extent, Fe(III) are reduced in the presence of Aβ with concomitant production of reactive oxygen species (ROS), hydrogen peroxide (H_2O_2) and hydroxyl radical (OH•). These Aβ/metal redox reactions, which are silenced by redox-inert Zn(II), but exacerbated by biological reducing agents, may lead directly to the widespread oxidation damages observed in AD brains. Moreover, studies have also shown that H_2O_2 mediates Aβ cellular toxicity and increases the production of both Aβ and amyloid precursor protein (APP). Third, the 5′ untranslated region (5′UTR) of APP mRNA has a functional iron-response element (IRE), which is consistent with biochemical evidence that APP is a redox-active metalloprotein. Hence, the redox interactions between Aβ, APP, and metals may be at the heart of a pathological positive feedback system wherein Aβ amyloidosis and oxidative stress promote each other. The emergence of redox-active metals as key players in AD pathogenesis strongly argues that amyloid-specific metal-complexing agents and antioxidants be investigated as possible disease-modifying agents for treating this horrible disease.

KEYWORDS: metals; oxidative stress; Alzheimer's disease; amyloid precursor protein; Aβ amyloidosis; metal chelator; antioxidant

Address for correspondence: Xudong Huang, Ph.D., Laboratory for Oxidation Biology, Massachusetts General Hospital and Harvard Medical School, Charlestown, MA 02129. Voice: 617-724-9778; fax: 617-724-1823.
huangx@helix.mgh.harvard.edu

INTRODUCTION

Alzheimer's disease (AD) affects 4.5 million Americans and at least $100 billion is spent a year on direct care alone. The problem is worsening as life expectancy continues to increase. By 2050, the number of AD patients is projected to approach 13.2 million in the United States if no cure or preventive measure for AD is found for the disease. Genetic, biochemical, and neuropathological data strongly point to Aβ amyloid formation as a central event in AD pathogenesis.[1] Considerable effort has been expended on identifying the pathways of Aβ metabolism. However, the neurochemical factors that promote the age-dependent Aβ amyloidosis have received far less attention. Only recently has the importance of Aβ's interactions with metals in AD pathogenesis begun to be recognized. In this article, we will attempt to review the accumulating data that support the significant roles of cerebral biometal dysregulation and oxidative stress in Alzheimer's Aβ amyloidogenesis.[2,3] Thus, amyloid-targeted metal chelation and antioxidant therapies as valid therapeutic strategies for interdicting AD pathogenesis will be justified.

ALZHEIMER'S DISEASE IS A POLYGENIC AND MULTIFACTORIAL COMPLEX DISEASE

AD is the most common senile dementing disorder. It is manifested by a gradual onset with a progressive and irreversible cognitive decline. Memory impairment appears in the earliest stage of the disease, although patients' motor and sensory functions are usually not affected until later stages.[4] AD is a genetically complex disease and the majority of AD cases are sporadic, while 5–10% of cases are early-onset familial AD (FAD) with an autosomal dominant inheritance pattern. The neuropathology of AD is characterized by the accumulation of insoluble Aβ-amyloid protein, neurofibrillary tangles (NFTs, the misfolded microtubule-associated tau protein), and neuropil threads in postmortem AD brains.[5,6]

Aβ (39–43 amino acid residues, ≈4 kDa) is the main constituent of both senile plaques and cerebrovascular amyloid deposits.[5,6] This Aβ peptide is generated from a much larger amyloid precursor protein (APP).[7–9] APP is a member of a ubiquitously expressed, type-1 integral membrane glycoprotein family.[10] Multiple APP isoforms are generated by alternatively splicing APP mRNA, but the function(s) of APP remains unclear. Normal catabolism of APP is known to involve synergistic cleavage by β-secretase and γ-secretase and production of a pool of Aβ peptides with carboxyl-terminal heterogeneity:[1] Aβ1–40 (40 amino acid residues) is the major soluble Aβ species, which is found in the CSF at low nanomolar concentrations;[11] Aβ1–42 (42 residues) is a minor Aβ species, but more fibrillogenic than Aβ1–40, and is heavily enriched in interstitial plaque amyloid.[12] It is generally agreed that Aβ peptide neurotoxicity is dependent upon its conformational state.[13,14]

Although the etiopathology of AD is unclear, identification of novel AD genes remains to be both an exciting and challenging task.[15] Mutations in the genes for APP (on chromosome 21), presenilin-1[16] (chromosome 14), and presenilin-2[17] (chromosome 1) have been associated with early-onset FAD.[18,19] The presenilin proteins are associated with proteolytic activity (γ-secretase activity) that cleaves APP to generate Aβ. The ε4 allele of the apolipoprotein E (ApoE) gene (on chromosome 19) and a mutation in the gene for α_2-macroglobulin (α_2M) (on chromosome 12) have been identified as risk factors for late-onset AD.[20,21] Interestingly, α_2M is a zinc-binding protein and is a major ligand for low-density lipoprotein receptor–related protein (LRP) found accumulated in senile plaques.[22,23] Both ApoE and α_2M show specific high-affinity binding for Aβ and they are believed to act as molecular chaperones for Aβ clearance.[15] Additionally, interleukin-1 (IL-1) gene polymorphisms are found to be associated with AD,[24] while the hemochromatosis gene, HFE, may affect the age of onset of sporadic AD.[25] More recently, insulin-degrading enzyme (IDE) gene (on chromosome 10) was shown to be linked to AD, and hypofunction of IDE may be associated with Aβ amyloid pathology of some forms of AD.[26,27] Indeed, IDE has been shown to regulate Aβ catabolism in which Aβ serves as its substrate.[28] These findings further confirm the proximity of Aβ peptide to the pathogenesis of the disorder. However, these studies also show that Aβ metabolism is sensitive to a range of influences, and several mechanisms can cause a shift towards pathogenic pathways that lead to AD.

Elucidating the neurochemical factors that lead to Aβ amyloid deposition is likely to yield important insights into the hierarchy of pathophysiological events in AD. A number of studies have characterized the physiochemical properties of the different Aβ peptides in a range of neurochemical environments. The *in vitro* solubility of synthetic Aβ1–42, in neutral aqueous solutions, is less than Aβ1–39 and Aβ1–40.[29] Soluble Aβ1–39/40 can be destabilized by seeding with Aβ1–42 fibrils.[30] In addition, Aβ1–42 is enriched in amyloid plaque cores from AD patients. Indeed, increasing evidence suggests that heightened levels of Aβ1–42 accelerate amyloid deposition in FAD. The FAD-linked APP670/671 Swedish mutation has been shown to increase the secretion of Aβ species severalfold, while the APP717 London mutation (downstream from the carboxyl terminus of Aβ) increases the proportion of Aβ1–42 produced.[31] Increased soluble Aβ1–42 has also been found in the brains of individuals affected by Down syndrome (DS), a condition complicated by premature AD.[32] The emerging consensus from these and other findings is that FAD-linked presenilin mutations share a common pathological mechanism: the mutations all appear to increase Aβ1–42 production.[33,34]

The presence or overproduction of Aβ1–42 alone appears to be insufficient to initiate Aβ amyloid deposition. Aβ levels in the CSF are not elevated in late-onset AD.[35,36] In fact, there is evidence that Aβ1–42 levels may be decreased in the CSF of AD subjects.[37] In addition, elevation of cortical Aβ concentrations as an initiator of amyloid deposition is inconsistent with the focal nature of these deposits (concentrated around synapses and the cerebrovascular lamina media). Overexpression of APP and consequential Aβ overproduction in transgenic mice rarely results in mice bearing full-blown AD-like neuropathology.[38,39] Finally, overexpression of Aβ1–42 from birth, which occurs in genetic forms of AD, namely, FAD and DS,[40] does not induce amyloid deposition in childhood. In these cases, Aβ deposition still occurs in an age-dependent, albeit accelerated manner. From these sets of observations, it seems highly unlikely that Aβ overproduction alone initiates Aβ amyloid deposition. It appears more likely that additional neurochemical factors are required for Aβ amyloidosis.

There are now compelling data that link biometals with Aβ metabolism and AD pathology. How cerebral homeostatic failure of biometals and Aβ combine to cause neuronal demise is presently the subject of intense study. Evidence is mounting that redox-active metals and Aβ may interact to elevate oxidative stress. In the next section, we will discuss data that implicate the significant roles of redox-active metal dysregulation and oxidative stress in Alzheimer's Aβ amyloid pathology.

REDOX-ACTIVE METALS AND OXIDATIVE STRESS ARE INTIMATELY INVOLVED IN ALZHEIMER'S PATHOLOGY

Our interest in the possible role of biometals in AD pathology was originally stimulated by a series of experiments that demonstrated strong (low micromolar and submicromolar metal concentrations) and specific binding between Aβ, APP, and the Zn(II), Cu(II), and (to a lesser extent) Fe(III).[41–43] Notably, Aβ1–42 has much stronger affinity for metals than Aβ1–40. Indeed, the binding constant reported for Aβ1–42 and Cu(II) is the second highest for any biomolecule[44] and is exceeded only by Cu/Zn–superoxide dismutase (Cu/Zn-SOD). Additionally, it has been proposed that ApoE binds Aβ and mediates its clearance via LRP, and the clearing efficiency is ApoE isoform–dependent. ApoE4 is the least effective chaperone among ApoE isoforms, retarding Aβ clearance and allowing Aβ aggregation to occur more readily.[15] Interestingly, we recently discovered that metals readily modulate the Aβ/ApoE binding, and the modulation is different for ApoE4 compared to other ApoE isoforms.[45] These initial studies from our own group and other laboratories have accumulated considerable evidence linking the clearance and aggregation of Aβ to its direct interactions with biometals.

The brain is a specialized organ that normally concentrates Cu, Fe, and Zn in the neocortex.[46] There is a large body of evidence indicating that the homeostasis of Cu, Fe, and Zn, and their respective binding proteins, is significantly altered in the AD brain.[3] For example, a recent well-controlled study using microparticle-induced X-ray emission (micro-PIXE) analysis of the cortical and accessory basal nuclei of the amygdala indicated that these metals accumulate in the neuropil of the AD brain, where their concentrations are 3- to 5-fold increased compared to age-matched controls.[46] Interestingly, the concentration of these metal ions, in particular, the redox-active Cu and Fe that are implicated in free radical reactions,[47] are normally the most concentrated in those brain regions most affected by AD pathology. Cu(II) has been shown to be released in a Ca(II)-dependent manner from vesicles of peptidergic neurons and calculated to reach concentrations as high as 15 µM.[48] Evidence for abnormal Cu homeostasis in AD includes a 2.2-fold increase in the concentration of CSF Cu,[49] with an accompanying increase in cerebral and CSF ceruloplasmin in AD patients.[50,51] There is also considerable data on supporting abnormal tissue levels of Fe and Fe-binding proteins in AD.[52,53] Further, the interaction between iron regulatory proteins (IRPs) and their iron-response element (IRE) in AD brains has been altered.[54] Most significantly, redox-active Fe is found within amyloid deposits of both human brain[55] and the neocortex of APP transgenic mice models of AD.[56] From the available experimental data, it seems that a reasonable strategy for therapeutic interdiction in AD may be the use of small molecules to specifically deplete these deposits of excess Cu, Zn, and Fe. Consistent with this posit, we have recently shown that metal chelators markedly enhance the resolubilization of Aβ deposits from postmortem AD brain samples,[57] supporting the possibility that Cu, Fe, and Zn ions play a significant role in assembling these deposits.

A large body of data indicate that oxidative stress is also a salient pathological feature of AD. Metabolic signs of oxidative stress in the neocortex of AD patients; free radical–mediated damage of brain proteins, lipids, and DNA; systemic signs of oxidative stress; and the response of antioxidant systems have all been observed in AD tissues.[58–60] Histological studies have also shown many oxidative stress markers localized with amyloid plaques, neuritic plaques, and NFTs, including 4-hydroxynonenal,[61] pyrraline and pentosidine,[62] 3-nitrotyrosine,[63] AGE-modified tau,[62] redox-active Fe,[55] and neurofilament-related protein carbonyls,[64] as well as elevations in Cu/Zn-SOD, Mn-SOD, catalase,[65,66] ferritin,[67] and HO-1.[68] A strong spatial correlation has also been reported between brain regions most loaded with Aβ amyloid and markers of lipid peroxidation and elevated antioxidant enzyme activity.[69]

The biochemical relationship between Aβ amyloid deposition in AD and oxidative stress is particularly complex and intriguing. However, amyloid-bearing transgenic animals develop similar oxidative damage in the neocortex and have similarly elevated redox-active Fe within amyloid plaques in comparison to that seen in the AD-affected brain.[56] This finding is consistent with direct involvement of Aβ in oxidative stress observed in AD brain.

There is also *in vitro* evidence to suggest that Aβ deposits may be a source of oxidative damages in AD brain. Synthetic Aβ peptides induce lipid peroxidation of synaptosomes[70] and exert cellular toxicity and vascular atrophy through mechanisms involving the generation of cellular H_2O_2[71] and O_2^-, and both effects can be attenuated by Cu/Zn-SOD[72] and a synthetic O_2^-/H_2O_2 scavenging compound.[73] Aβ1–40 has also been reported to generate the OH• by mechanisms that are as yet unclear.[74] Vitamin E and the spin-trap compound PBN have been shown to protect against Aβ-mediated neurotoxicity *in vitro*.[75,76]

We have shown that Aβ is able to reduce Cu(II) and, to a lesser extent, Fe(III) to Cu(I) and Fe(II) with concurrent generation of reactive oxygen species (ROS), H_2O_2 and OH•. This mechanistic scheme is outlined in the following reactions:[77]

$$(A\beta)_2 + M^{(n+1)+} \rightarrow A\beta:A\beta^{+\bullet} + M^{n+}.$$

Reduced Fe(II)/Cu(I) reacts with molecular oxygen (O_2) to generate the superoxide anion (O_2^-):

$$M^{n+} + O_2 \rightarrow M^{(n+1)+} + O_2^-.$$

The O_2^- generated undergoes dismutation to H_2O_2 and O_2 either catalyzed by SOD or spontaneously:

$$O_2^- + O_2^- + 2H^+ \rightarrow H_2O_2 + O_2.$$

The reaction of reduced metals with H_2O_2 generates the highly reactive OH• by the Fenton reaction. Cu(I) catalyzes this reaction at a rate constant magnitude higher than that for Fe(II).[3]

$$M^{n+} + H_2O_2 \rightarrow M^{(n+1)+} + OH^\bullet + OH^-.$$

Additionally, the Haber-Weiss reaction can form OH• in a reaction catalyzed by $M^{(n+1)+}/M^{n+}$:

$$O_2^- + H_2O_2 \rightarrow OH^\bullet + OH^- + O_2.$$

Radicalized Aβ may eventually react with O_2 and be transformed into oligomerized and/or fragmented insoluble aggregates. Generation of both reduced metals and OH$^•$ was greatest for Aβ1–42 ≫ Aβ1–40 > rat Aβ1–40. This relationship reflects both the relative importance of these Aβ isoforms in amyloid pathology as well as toxicity of these peptides observed in cell culture.[78] Indeed, we have data (unpublished) showing that redox-active Cu and Fe ions may be involved in formation of Aβ oligomers that are both neurotoxic and synaptotoxic.[79–81] Our observations for cell-free metal-catalyzed Aβ redox chemistry have been confirmed by another lab.[82] The abnormal metal-mediated redox activity of Aβ can be silenced by redox-inert Zn(II),[83] but enhanced by biological reducing agents.[84,85]

AMYLOID-TARGETING METAL CHELATORS AND ANTIOXIDANTS ARE PROMISING THERAPEUTICS FOR ALZHEIMER'S DISEASE

Antioxidants and specific metal-complexing agents appear promising as therapeutics for AD treatment. Indeed, antioxidant compounds such as α-tocopherol (vitamin E) have been reported to mildly slow AD progression in patients with moderately severe impairment.[86] However, the vitamin E does not cross the blood-brain barrier (BBB) readily, although very low levels actually enter the brain.[87] Hence, the molecular mechanism responsible for beneficial effects of vitamin E on AD patients is unclear. Thus, the pharmacological action mode of vitamin E will need to be fully elucidated before more intelligent drug design for its derivatives can be applied to increase the moderate therapeutic efficacy of vitamin E. An early report shows that sustained intramuscular administration of the metal chelator, desferrioxamine (DFO), may also slow the clinical progression of AD dementia.[88] DFO also attenuates APP expression through the 5′ untranslated region (5′UTR) of its mRNA,[89] which appears to be a druggable target for AD.[90] However, DFO's inherent toxicity and low BBB permeability[91] limit its therapeutic use. It is noted that APP 5′UTR contains a fully functional IRE,[89] and APP itself is a redox-active metalloprotein that is able to reduce bound Cu(II) to Cu(I) at its copper-binding site.[92] This may explain why APP expression can be affected by metal chelators.

At present, the compound most likely to be of immediate clinical value is a metal chelator, clioquinol (CQ). CQ is a well-characterized drug with few side effects and it has been used clinically since the 1920s. Studies with CQ in APP transgenic mice have demonstrated excellent efficacy in reduction of cerebral Aβ amyloid deposits.[93] Moreover, a mild cognitive improvement has been reported for a small-scale open study of 30 AD patients following a 3-week treatment regime with CQ.[94] Most encouragingly, a recent pilot phase II trial of CQ with double-blind placebo control shows that the CQ treatment significantly arrests cognitive decline and lowers plasma Aβ1–42 levels in AD subjects.[95]

Although metal chelation may be a promising therapeutic strategy for AD, a significant problem faces the wide use of chelation therapy. The present metal chelators usually have poor target specificity. Long-term use of these agents is likely to perturb the homeostasis of many biometals and normal physiological functions of essential metal-requiring biomolecules. The very mild and transient benefits reported for vitamin E and related compounds suggest that general purpose antioxidants will also be inadequate for AD treatment as they are usually not disease-specific. What may be required is a new class of metal-complexing and antioxidant agents that specifically target amyloid. We have recently designed and synthesized novel bifunctional agents that contain both amyloid-binding and metal-chelating or ROS-scavenging moieties. Preliminary studies indicate that they are able to attenuate (i) APP expression probably through the 5′UTR region of APP mRNA and (ii) metal-induced Aβ aggregation and oligomerization (Huang et al., unpublished data). Studies are currently under way to validate these putative lead compounds, which may have improved target specificity and clinical safety, as amyloid-specific metal chelators and antioxidants for AD treatment.

CONCLUDING REMARKS

A growing body of experimental evidence and clinical data suggest that a large portion of AD pathology is mediated by oxidative stress that may result from the direct interplay of brain redox-active metals and Aβ amyloid. Many mechanistic aspects of this model are unclear and further investigation is certainly required. However, an objective assessment of available experimental and clinical data clearly indicates that Aβ/metal redox interaction is among the most promising therapeutic targets for current AD drug development. It is to be hoped that the next few years will see CNS drug development programs take advantage of the explosion in findings on the roles of redox-active metals in pathogenesis of AD and other neurological disorders.

ACKNOWLEDGMENTS

X. Huang is supported by grants from the NIMH/NIH (No. 5-K01-MH02001) and AFAR. X. Huang and R. D. Moir contributed equally to this paper.

REFERENCES

1. SELKOE, D.J. 2001. Alzheimer's disease: genes, proteins, and therapy. Physiol. Rev. **81:** 741–766.
2. MARKESBERY, W.R. 1997. Oxidative stress hypothesis in Alzheimer's disease. Free Radical Biol. Med. **23:** 134–147.
3. ATWOOD, C.S. et al. 1999. Role of free radicals and metal ions in the pathogenesis of Alzheimer's disease. Metal Ions Biol. Syst. **36:** 309–364.
4. CUMMINGS, J.L. et al. 1998. Alzheimer's disease: etiologies, pathophysiology, cognitive reserve, and treatment opportunities. Neurology **51S:** S2–S17.
5. GLENNER, G.G. & C.W. WONG. 1984. Alzheimer's disease: initial report of the purification and characterization of a novel cerebrovascular amyloid protein. Biochem. Biophys. Res. Commun. **120:** 885–890.
6. MASTERS, C.L. et al. 1985. Amyloid plaque core protein in Alzheimer disease and Down syndrome. Proc. Natl. Acad. Sci. USA **82:** 4245–4249.
7. KANG, J. et al. 1987. The precursor of Alzheimer's disease amyloid A4 protein resembles a cell-surface receptor. Nature **325:** 733–736.
8. ROBAKIS, N.K. et al. 1987. Chromosome 21q21 sublocalisation of gene encoding β-amyloid peptide in cerebral vessels and neuritic (senile) plaques of people with Alzheimer disease and Down syndrome [letter]. Lancet **1:** 384–385.

9. TANZI, R.E. et al. 1987. Amyloid β protein gene: cDNA, mRNA distribution, and genetic linkage near the Alzheimer locus. Science **235**: 880–884.
10. TANZI, R.E. et al. 1988. Protease inhibitor domain encoded by an amyloid protein precursor mRNA associated with Alzheimer's disease. Nature **331**: 528–530.
11. VIGO-PELFREY, C. et al. 1993. Characterization of β-amyloid peptide from human cerebrospinal fluid. J. Neurochem. **61**: 1965–1968.
12. PRELLI, F. et al. 1988. Different processing of Alzheimer's β-protein precursor in the vessel wall of patients with hereditary cerebral hemorrhage with amyloidosis–Dutch type. Biochem. Biophys. Res. Commun. **151**: 1150–1155.
13. PIKE, C.J. et al. 1991. In vitro aging of β-amyloid protein causes peptide aggregation and neurotoxicity. Brain Res. **563**: 311–314.
14. LORENZO, A. & B.A. YANKNER. 1994. Beta-amyloid neurotoxicity requires fibril formation and is inhibited by Congo red. Proc. Natl. Acad. Sci. USA **91**: 12243–12247.
15. TANZI, R.E. & L. BERTRAM. 2001. New frontiers in Alzheimer's disease genetics. Neuron **32**: 181–184.
16. SHERRINGTON, R. et al. 1995. Cloning of a gene bearing missense mutations in early-onset familial Alzheimer's disease. Nature **375**: 754–760.
17. LEVY-LAHAD, E. et al. 1995. Candidate gene for the chromosome 1 familial Alzheimer's disease locus. Science **269**: 973–977.
18. CHARTIER-HARLIN, M.C. et al. 1991. Early-onset Alzheimer's disease caused by mutations at codon 717 of the β-amyloid precursor protein gene. Nature **353**: 844–846.
19. MURRELL, J. et al. 1991. A mutation in the amyloid precursor protein associated with hereditary Alzheimer's disease. Science **254**: 97–99.
20. SAUNDERS, A.M, et al. 1993. Association of apolipoprotein E allele epsilon4 with late-onset familial and sporadic Alzheimer's disease. Neurology **43**: 1467–1472.
21. BLACKER, D. et al. 1998. Alpha-2 macroglobulin is genetically associated with Alzheimer disease. Nat. Genet. **19**: 357–360.
22. REBECK, G.W. et al. 1995. Multiple, diverse senile plaque-associated proteins are ligands of an apolipoprotein E receptor, the alpha 2-macroglobulin receptor/low-density lipoprotein receptor–related protein. Ann. Neurol. **37**: 211–217.
23. DU, Y. et al. 1997. α2-Macroglobulin as Aβ-amyloid peptide-binding plasma protein. J. Neurochem. **69**: 299–305.
24. NICOLL, J.A. et al. 2000. Association of interleukin-1 gene polymorphisms with Alzheimer's disease. Ann. Neurol. **47**: 365–368.
25. SAMPIETRO, M. et al. 2001. The hemochromatosis gene affects the age of onset of sporadic Alzheimer's disease. Neurobiol. Aging **22**: 563–568.
26. BERTRAM, L. et al. 2000. Evidence for genetic linkage of Alzheimer's disease to chromosome 10q. Science **290**: 2302–2303.
27. FARRIS, W. et al. 2003. Insulin-degrading enzyme regulates the levels of insulin, amyloid beta-protein, and the beta-amyloid precursor protein intracellular domain *in vivo*. Proc. Natl. Acad. Sci. USA **100**: 4162–4167.
28. QIU, W.Q. et al. 1998. Insulin-degrading enzyme regulates extracellular levels of amyloid beta-protein by degradation. J. Biol. Chem. **273**: 32730–32738.
29. HILBICH, C. et al. 1991. Aggregation and secondary structure of synthetic amyloid β A4 peptides of Alzheimer's disease. J. Mol. Biol. **218**: 149–163.
30. JARRETT, J.T., E.P. BERGER & P.T. LANSBURY, JR. 1993. The carboxy terminus of the β amyloid protein is critical for the seeding of amyloid formation: implications for the pathogenesis of Alzheimer's disease. Biochemistry **32**: 4693–4697.
31. SUZUKI, A. et al. 1994. High expression on Kunitz-type protease inhibitor-containing substances in the cerebral vessels of patients with Down syndrome. Tohoku J. Exp. Med. **174**: 181–187.
32. TELLER, J.K. et al. 1996. Presence of soluble amyloid β-peptide precedes amyloid plaque formation in Down's syndrome [see comments]. Nat. Med. **2**: 93–95.
33. CITRON, M. et al. 1997. Mutant presenilins of Alzheimer's disease increase production of 42-residue amyloid β-protein in both transfected cells and transgenic mice [see comments]. Nat. Med. **3**: 67–72.

34. XIA, W. *et al.* 1997. Enhanced production and oligomerization of the 42-residue amyloid β-protein by Chinese hamster ovary cells stably expressing mutant presenilins. J. Biol. Chem. **272:** 7977–7982.
35. SHOJI, M. *et al.* 1992. Production of the Alzheimer amyloid β protein by normal proteolytic processing. Science **258:** 126–129.
36. NAKAMURA, T. *et al.* 1994. Amyloid β protein levels in cerebrospinal fluid are elevated in early-onset Alzheimer's disease. Ann. Neurol. **36:** 903–911.
37. MOTTER, R. *et al.* 1995. Reduction of β-amyloid peptide 42 in the cerebrospinal fluid of patients with Alzheimer's disease. Ann. Neurol. **38:** 643–648.
38. HSIAO, K.K. *et al.* 1995. Age-related CNS disorder and early death in transgenic FVB/N mice overexpressing Alzheimer amyloid precursor proteins. Neuron **15:** 1203–1218.
39. HSIAO, K. *et al.* 1996. Correlative memory deficits, Aβ elevation, and amyloid plaques in transgenic mice. Science **274:** 99–102.
40. WISNIEWSKI, K.E., H.M. WISNIEWSKI & G.Y. WEN. 1985. Occurrence of neuropathological changes and dementia of Alzheimer's disease in Down's syndrome. Ann. Neurol. **17:** 278–282.
41. BUSH, A.I. *et al.* 1994. Rapid induction of Alzheimer Aβ amyloid formation by zinc. Science **265:** 1464–1467.
42. HUANG, X. *et al.* 1997. Zinc-induced Alzheimer's Aβ1–40 aggregation is mediated by conformational factors. J. Biol. Chem. **272:** 26464–26470.
43. ATWOOD, C.S. *et al.* 1998. Dramatic aggregation of Alzheimer Aβ by Cu(II) is induced by conditions representing physiological acidosis. J. Biol. Chem. **273:** 12817–12826.
44. ATWOOD, C.S. *et al.* 2000. Characterization of copper interactions with Alzheimer amyloid beta peptides: identification of an attomolar-affinity copper binding site on amyloid beta1–42. J. Neurochem. **75:** 1219–1233.
45. MOIR, R.D. *et al.* 1999. Differential effects of apolipoprotein E isoforms on metal-induced aggregation of A beta using physiological concentrations. Biochemistry **38:** 4595–4603.
46. LOVELL, M.A. *et al.* 1998. Copper, iron, and zinc in Alzheimer's disease senile plaques. J. Neurol. Sci. **158:** 47–52.
47. HALLIWELL, B. & J.M.C. GUTTERIDGE. 1984. Oxygen toxicity, oxygen radicals, transition metals, and disease. Biochem. J. **219:** 1–14.
48. HARTTER, D.E. & A. BARNEA. 1988. Brain tissue accumulates 67copper by two ligand-dependent saturable processes: a high affinity, low capacity and a low affinity, high capacity process. J. Biol. Chem. **263:** 799–805.
49. BASUN, H. *et al.* 1991. Metals and trace elements in plasma and cerebrospinal fluid in normal aging and Alzheimer's disease. J. Neural Transm. Parkinson's Dis. Dement. Sect. **3:** 231–258.
50. CONNOR, J.R. *et al.* 1993. Ceruloplasmin levels in the human superior temporal gyrus in aging and Alzheimer's disease. Neurosci. Lett. **159:** 88–90.
51. LOEFFLER, D.A. *et al.* 1994. Ceruloplasmin is increased in cerebrospinal fluid in Alzheimer's disease, but not Parkinson's disease. Alzheimer Dis. Assoc. Disord. **8:** 190–197.
52. ROBINSON, S.R. *et al.* 1995. Most amyloid plaques contain ferritin-rich cells. Alzheimer's Res. **1:** 191–196.
53. CONNOR, J.R. *et al.* 1995. A quantitative analysis of isoferritins in select regions of aged, parkinsonian, and Alzheimer's diseased brains. J. Neurochem. **65:** 717–724.
54. PINERO, D.J., J. HU & J.R. CONNOR. 2000. Alterations in the interaction between iron regulatory proteins and their iron responsive element in normal and Alzheimer's diseased brains. Cell. Mol. Biol. (Noisy-Le-Grand) **46:** 761–776.
55. SMITH, M.A. *et al.* 1997. Iron accumulation in Alzheimer's disease is a source of redox-generated free radicals. Proc. Natl. Acad. Sci. USA **94:** 9866–9868.
56. SMITH, M.A. *et al.* 1998. Amyloid-beta deposition in Alzheimer transgenic mice is associated with oxidative stress. J. Neurochem. **70:** 2212–2215.
57. CHERNY, R.A. *et al.* 1999. Aqueous dissolution of Alzheimer's disease Aβ amyloid deposits by biometal depletion. J. Biol. Chem. **274:** 23223–23228.

58. CEBALLOS-PICOT, I. *et al.* 1996. Peripheral antioxidant enzyme activities and selenium in elderly subjects and in dementia of Alzheimer's type—place of the extracellular glutathione peroxidase. Free Radical Biol. Med. **20:** 579–587.
59. HENSLEY, K. *et al.* 1995. Brain regional correspondence between Alzheimer's disease histopathology and biomarkers of protein oxidation. J. Neurochem. **65:** 2146–2156.
60. MECOCCI, P., U. MACGARVEY & M.F. BEAL. 1994. Oxidative damage to mitochondrial DNA is increased in Alzheimer's disease. Ann. Neurol. **36:** 747–751.
61. SAYRE, L.M. *et al.* 1997. 4-Hydroxynonenal-derived advanced lipid peroxidation end products are increased in Alzheimer's disease. J. Neurochem. **68:** 2092–2097.
62. SMITH, M.A. *et al.* 1994. Advanced Maillard reaction end products are associated with Alzheimer disease pathology. Proc. Natl. Acad. Sci. USA **91:** 5710–5714.
63. SMITH, M.A. *et al.* 1997. Widespread peroxynitrite-mediated damage in Alzheimer's disease. J. Neurosci. **17:** 2653–2657.
64. SMITH, C.D. *et al.* 1991. Excess brain protein oxidation and enzyme dysfunction in normal aging and Alzheimer's disease. Proc. Natl. Acad. Sci. USA **88:** 10540–10543.
65. PAPPOLLA, M.A. *et al.* 1992. Immunohistochemical evidence of antioxidant stress in Alzheimer's disease. Am. J. Pathol. **140:** 621–628.
66. FURUTA, A. *et al.* 1995. Localization of superoxide dismutases in Alzheimer's disease and Down's syndrome neocortex and hippocampus. Am. J. Pathol. **146:** 357–367.
67. GRUNDKE-IQBAL, I. *et al.* 1990. Ferritin is a component of the neuritic (senile) plaque in Alzheimer dementia. Acta Neuropathol. (Berlin) **81:** 105–110.
68. SMITH, M.A. *et al.* 1994. Heme oxygenase-1 is associated with the neurofibrillary pathology of Alzheimer's disease. Am. J. Pathol. **145:** 42–47.
69. LOVELL, M.A. *et al.* 1995. Elevated thiobarbituric acid–reactive substances and antioxidant enzyme activity in the brain in Alzheimer's disease. Neurology **45:** 1594–1601.
70. BUTTERFIELD, D.A. *et al.* 1994. β-Amyloid peptide free radical fragments initiate synaptosomal lipoperoxidation in a sequence-specific fashion: implications to Alzheimer's disease. Biochem. Biophys. Res. Commun. **200:** 710–715.
71. BEHL, C. *et al.* 1994. Hydrogen peroxide mediates amyloid β protein toxicity. Cell **77:** 817–827.
72. THOMAS, T. *et al.* 1996. β-Amyloid-mediated vasoactivity and vascular endothelial damage [see comments]. Nature **380:** 168–171.
73. BRUCE, A.J., B. MALFROY & M. BAUDRY. 1996. β-Amyloid toxicity in organotypic hippocampal cultures: protection by Euk-8, a synthetic catalytic free radical scavenger. Proc. Natl. Acad. Sci. USA **93:** 2312–2316.
74. TOMIYAMA, T. *et al.* 1996. Inhibition of amyloid β protein aggregation and neurotoxicity by rifampicin: its possible function as a hydroxyl radical scavenger. J. Biol. Chem. **271:** 6839–6844.
75. YATIN, S.M., S. VARADARAJAN & D.A. BUTTERFIELD. 2000. Vitamin E prevents Alzheimer's amyloid beta-peptide (1–42)–induced neuronal protein oxidation and reactive oxygen species production. J. Alzheimer's Dis. **2:** 123–131.
76. HARRIS, M.E. *et al.* 1995. Direct evidence of oxidative injury produced by the Alzheimer's beta-amyloid peptide (1–40) in cultured hippocampal neurons. Exp. Neurol. **131:** 193–202.
77. HUANG, X. *et al.* 1999. The Aβ peptide of Alzheimer's disease directly produces hydrogen peroxide through metal ion reduction. Biochemistry **38:** 7609–7616.
78. HUANG, X. *et al.* 1999. Cu(II) potentiation of Alzheimer Aβ neurotoxicity: correlation with cell-free hydrogen peroxide production and metal reduction. J. Biol. Chem. **274:** 37111–37116.
79. LAMBERT, M.P. *et al.* 1998. Diffusible, nonfibrillar ligands derived from Aβ1–42 are potent central nervous system neurotoxins. Proc. Natl. Acad. Sci. USA **95:** 6448–6453.
80. WALSH, D.M. *et al.* 2002. Naturally secreted oligomers of amyloid beta protein potently inhibit hippocampal long-term potentiation *in vivo*. Nature **416:** 535–539.
81. KAYED, R. *et al.* 2003. Common structure of soluble amyloid oligomers implies common mechanism of pathogenesis. Science **300:** 486–489.
82. DIKALOV, S.I. *et al.* 1999. Amyloid β peptides do not form peptide-derived free radicals spontaneously, but can enhance metal-catalyzed oxidation of hydroxylamines to nitroxides. J. Biol. Chem. **274:** 9392–9399.

83. CUAJUNGCO, M.P. et al. 2000. Evidence that the β-amyloid plaques of Alzheimer's disease represent the redox-silencing and entombment of Aβ by zinc. J. Biol. Chem. **275:** 19439–19442.
84. WHITE, A.R. et al. 2001. Homocysteine potentiates copper- and amyloid beta peptide–mediated toxicity in primary neuronal cultures: possible risk factors in the Alzheimer's-type neurodegenerative pathways. J. Neurochem. **76:** 1509–1520.
85. OPAZO, C. et al. 2002. Metalloenzyme-like activity of Alzheimer's disease beta-amyloid: Cu-dependent catalytic conversion of dopamine, cholesterol, and biological reducing agents to neurotoxic $H(2)O(2)$. J. Biol. Chem. **277:** 40302–40308.
86. SANO, M. et al. 1997. A controlled trial of selegiline, alpha-tocopherol, or both as treatment for Alzheimer's disease: the Alzheimer's Disease Cooperative Study. N. Engl. J. Med. **336:** 1216–1222.
87. PAPPERT, E.J. et al. 1996. α-Tocopherol in the ventricular cerebrospinal fluid of Parkinson's disease patients: dose-response study and correlations with plasma levels. Neurology **47:** 1037–1042.
88. CRAPPER MCLACHLAN, D.R. et al. 1991. Intramuscular desferrioxamine in patients with Alzheimer's disease. Lancet **337:** 1304–1308.
89. ROGERS, J.T. et al. 2002. An iron-responsive element type II in the 5′-untranslated region of the Alzheimer's amyloid precursor protein transcript. J. Biol. Chem. **277:** 45518–45528.
90. ROGERS, J.T. et al. 2002. Alzheimer's disease drug discovery targeted to the APP mRNA 5′ untranslated region. J. Mol. Neurosci. **19:** 77–82.
91. PORTER, J.B. & E.R. HUEHNS. 1989. The toxic effects of desferrioxamine. Bailliere's Clin. Haematol. **2:** 459–474.
92. MULTHAUP, G. et al. 1996. The amyloid precursor protein of Alzheimer's disease in the reduction of copper(II) to copper(I). Science **271:** 1406–1409.
93. CHERNY, R.A. et al. 2001. Treatment with a copper-zinc chelator markedly and rapidly inhibits beta-amyloid accumulation in Alzheimer's disease transgenic mice. Neuron **30:** 665–676.
94. REGLAND, B. et al. 2001. Treatment of Alzheimer's disease with clioquinol. Dement. Geriatr. Cognit. Disord. **12:** 408–414.
95. RITCHIE, C.W. et al. 2003. Metal-protein attenuation with iodochlorhydroxyquin (clioquinol) targeting Aβ amyloid deposition and toxicity in Alzheimer's disease: a pilot phase 2 clinical trial. Arch. Neurol. **60:** 1685–1691.

Selective Cu^{2+}/Ascorbate-Dependent Oxidation of Alzheimer's Disease β-Amyloid Peptides

CHRISTIAN SCHÖNEICH

Department of Pharmaceutical Chemistry, University of Kansas, Lawrence, Kansas 66047, USA

ABSTRACT: This review summarizes tandem mass spectrometric investigations on the selectivity of metal-catalyzed oxidation of β-amyloid peptide (βAP) and related sequences. A remarkable feature of the Cu^{2+}/ascorbate-dependent oxidation of these peptides is the switch from predominantly His oxidation in the neurotoxic peptide βAP1–40 to predominantly Tyr oxidation in the nonneurotoxic reverse sequence βAP40–1. Within βAP1–40, His^{13} and His^{14} of the high-affinity Cu^{2+}-binding site are most sensitive to oxidation. Eventually, the oxidation of one or both of these His residues could result in a less redox-active βAP-Cu^{2+} complex, lowering the incidence of βAP-Cu^{2+}-dependent Fenton-type reactions for the benefit of surrounding biological tissue.

KEYWORDS: Alzheimer's disease; ascorbate; β-amyloid peptide (βAP); Cu^{2+}; oxidation; sequence

INTRODUCTION

The deposition of β-amyloid peptide (βAP) into senile plaque is an important factor in the pathogenesis of Alzheimer's disease.[1–5] While βAP1–40 represents the major circulating βAP sequence, the predominant sequence incorporated into plaque is βAP1–42.[6,7] An important hallmark of βAP is its pronounced tendency to aggregate, ultimately forming fibrils in which βAP adopts a β-sheet conformation.[8–13] For long, βAP aggregation has been considered the key parameter by which βAP formation influences the progression of Alzheimer's disease. However, increasing experimental evidence is mounting that aggregation may not be the only mechanism of βAP action.[14–23] βAP binds Cu^{2+} and Zn^{2+} with very high affinity and the incubation of full-length βAP with Cu^{2+} leads to the reduction of Cu^{2+} to Cu^+.[19–21,23] This electron transfer depends on the simultaneous presence of three His residues in the N-terminus, His^6, His^{13}, and His^{14}, and Met^{35} in the C-terminus of the peptide. Initially, it was reported that the aerobic incubation of βAP with Cu^{2+} resulted in the formation of H_2O_2.[20] However, recently, Opazo *et al.*[24] demonstrated that the Cu^{2+}-catalyzed generation of measurable quantities of H_2O_2 requires the additional presence of reductants such as, for example, ascorbate, dopamine, or cholesterol. The high affinity of βAP to Cu^{2+}, the facile conversion to Cu^+, and the ability of

---

Address for correspondence: Christian Schöneich, Department of Pharmaceutical Chemistry, University of Kansas, 2095 Constant Avenue, Lawrence, KS 66047. Voice: 785-864-4880; fax: 785-864-5736.
schoneic@ukans.edu

Ann. N.Y. Acad. Sci. 1012: 164–170 (2004). © 2004 New York Academy of Sciences.
doi: 10.1196/annals.1306.013

biological reductants to trigger the production of H_2O_2 suggest that the βAP-Cu^{2+} complex may be a source for oxidative stress *in vivo*, potentially through Fenton-type reactions. In fact, Alzheimer's disease brain is characterized by elevated levels of oxidative stress and high levels of redox-active transition metals such as Cu and Fe.[25–27] Moreover, recent studies show the induction of βAP plaque formation in cholesterol-fed rabbits through trace levels of Cu^{2+} in their drinking water, suggesting a potential correlation between Cu^{2+} uptake and Alzheimer's disease etiology.[28] *In vivo* imaging studies of McLellan *et al.* confirm that reactive oxygen species are generated especially in the immediate environment of dense core, but not diffuse plaques.[29] On the other hand, Cu^{2+} reduction is significantly more efficient for freshly solubilized βAP as compared with βAP incubated with buffer for 24 hours prior to the experimental measurement of Cu^{2+} reduction.[23] The latter data would suggest a higher efficiency of low-molecular-weight βAP aggregates to reduce Cu^{2+} as compared to higher-molecular-weight aggregates. However, this experiment was not performed in the additional presence of biological reductants, which are available in the *in vivo* imaging studies. The potential of βAP to induce oxidative stress *in vivo* was elegantly demonstrated through the expression of native βAP1–42 in transgenic *C. elegans*, leading to plaque formation and significantly elevated levels of protein carbonyls, a marker for protein oxidation.[30] Oxidative stress preceded the fibrillar deposition of βAP.[31]

The metal-catalyzed formation of H_2O_2 through βAP suggests that the peptide itself may be modified as a consequence of such a process. In fact, βAP purified from Alzheimer's disease brain shows a significant loss of His residues (ca. 1 mol/mol βAP)[32] and the formation of methionine sulfoxide.[33] In order to examine the reactivity of specific amino acids of βAP as well as the potential mechanism(s), we have subjected several βAP sequences *in vitro* to metal-catalyzed oxidation and analyzed the products by HPLC–tandem mass spectrometry.[34,35]

SELECTIVE Cu^{2+}-CATALYZED OXIDATION OF βAP

The primary sequences of several βAP congeners are shown below. Here, βAP1–40 represents a neurotoxic,[36] physiologically relevant sequence with a 2-amino-acid deletion at the C-terminus of βAP1–42, which is most abundant in plaques. βAP40–1 constitutes the synthetic reverse sequence, which is not neurotoxic.[36] The other sequences, βAP1–16, βAP1–28, and βAP10–20, are included for mechanistic purposes:

βAP1–40:
$D^1AEFRH^6DSGY^{10}EVH^{13}H^{14}QKLVFFAEDVGSNKGAIIGLM^{35}VGGVV^{40}$

βAP1–28: $D^1AEFRH^6DSGY^{10}EVH^{13}H^{14}QKLVFFAEDVGSNK^{28}$

βAP1–16: $D^1AEFRH^6DSGY^{10}EVH^{13}H^{14}QK^{16}$

βAP10–20: $Y^{10}EVH^{13}H^{14}QKLVFF^{20}$

βAP40–1:
$V^1VGGVM^6LGIIAGKNSGVDEAFFVLKQH^{27}H^{28}VEY^{31}GSDH^{35}RFEAD^{40}$

The exposure of all sequences to Cu^{2+} and ascorbate resulted in the highly selective oxidation of distinct amino acids, displayed in SCHEME 1.[34,35] The primary oxidation targets in βAP1–16, βAP1–28, and βAP1–40 were His^{13} and His^{14}, followed by His^6. Tandem mass spectrometric analysis confirmed the insertion of

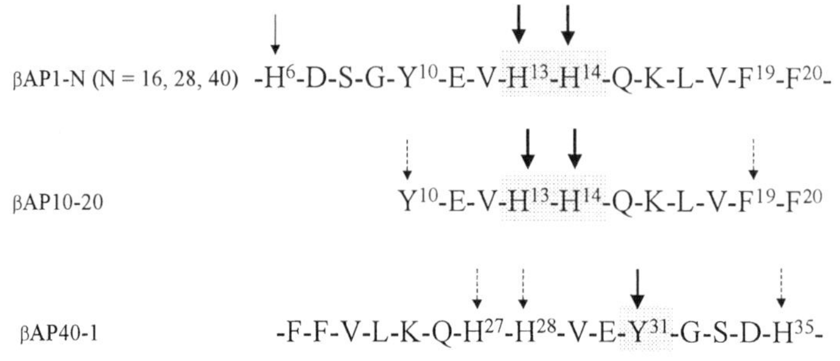

                     ↓   ↓ ↓
βAP1-N (N = 16, 28, 40) -H⁶-D-S-G-Y¹⁰-E-V-H¹³-H¹⁴-Q-K-L-V-F¹⁹-F²⁰-

                                ↓ ↓
βAP10-20   Y¹⁰-E-V-H¹³-H¹⁴-Q-K-L-V-F¹⁹-F²⁰

                                ↓ ↓   ↓   ↓
βAP40-1   -F-F-V-L-K-Q-H²⁷-H²⁸-V-E-Y³¹-G-S-D-H³⁵-

SCHEME 1. Selectivity of oxygen incorporation into selected βAP sequences. The bold arrows indicate high oxidation yields, underscored by the gray shadows, whereas the dashed arrows indicate low oxidation yields.

oxygen into these His residues, indicating the formation of 2-oxo-His. There was no evidence for any significant oxidation of Tyr^{10}. The preference for His oxidation still held for βAP10–20, although this peptide showed additional evidence for the hydroxylation of Tyr^{10} and Phe^{19}/Phe^{20}. In contrast, the oxidation of βAP40–1 showed a strong preference for hydroxylation of Tyr^{31} and only small yields of 2-oxo-His at all three His residues. Importantly, no mass spectrometric evidence for bityrosine formation was obtained for all investigated βAP sequences.

By analogy to previous mechanistic studies,[37–39] we propose that His oxidation and Tyr hydroxylation are carried out site-specifically through the reaction of hydroxyl radicals and/or their metal-bound equivalents. A putative mechanism for His oxidation is displayed in SCHEME 2. Theoretical calculations show that the addi-

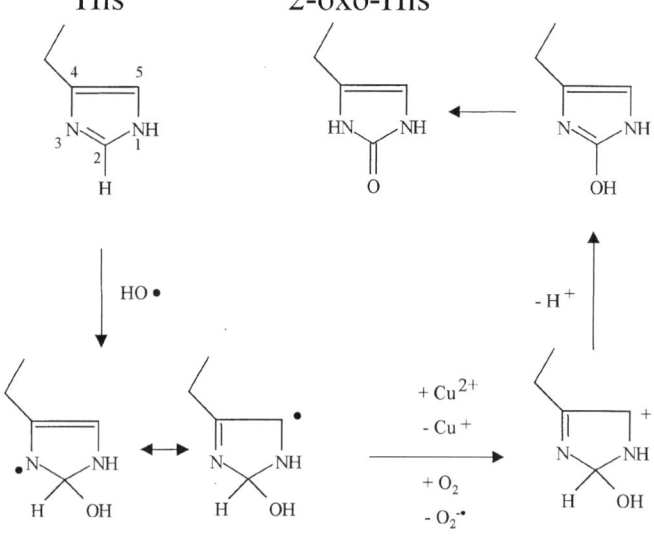

SCHEME 2. Mechanism of 2-oxo-His formation through hydroxyl radicals.

tion of hydroxyl radicals to the C-2 position of His represents the energetically most favorable pathway,[40] although experiments with free hydroxyl radicals and His have demonstrated significant reaction also with the C-5 position.[40,41] However, it is unlikely that hydroxyl radical addition to the C-5 position leads to stable products of 16 atomic mass units higher in molecular weight compared to native His (i.e., to a monooxygenation product)[38] unless potentially reversible water elimination[41] converts the initial C-5 addition product into a C-2 addition product.[38]

Spectroscopic analysis in solution indicates[21] that the Cu^{2+}-binding site of βAP comprises His[6], Tyr[10], His[13], and His[14]; in contrast, Raman spectroscopic data of senile plaque cores demonstrate that Cu^{2+} is coordinated to three His residues, but not Tyr.[33] Therefore, the absence of Cu^{2+}-catalyzed oxidation of Tyr[10] in solution is remarkable, considering that Tyr[10] is part of the metal-binding site. As a potential rationale, Cu^{2+} coordination of Tyr could sufficiently lower the electron density of the aromatic π system such that the reaction of hydroxyl radicals with the nearby His residues is kinetically preferred. Alternatively, the reduction of Cu^{2+} to Cu^+ could result in a transient disconnection of the Tyr residue from Cu^+. The latter rationale seems unlikely, though, on the basis of the results with βAP40–1. Here, Tyr hydroxylation constitutes the predominant reaction product. If carried out through hydroxyl radicals (or their metal-bound equivalents), this reaction must proceed site-specifically as otherwise we would have detected numerous additional oxidation products of the βAP peptides as a result of the competitive reaction of hydroxyl radicals with other amino acid targets (of course, under such conditions, a large fraction of free hydroxyl radicals would also directly react with ascorbate). One can write alternative mechanisms of Tyr hydroxylation, for example, two one-electron transfer processes followed by the addition of HO^- (see SCHEME 3).

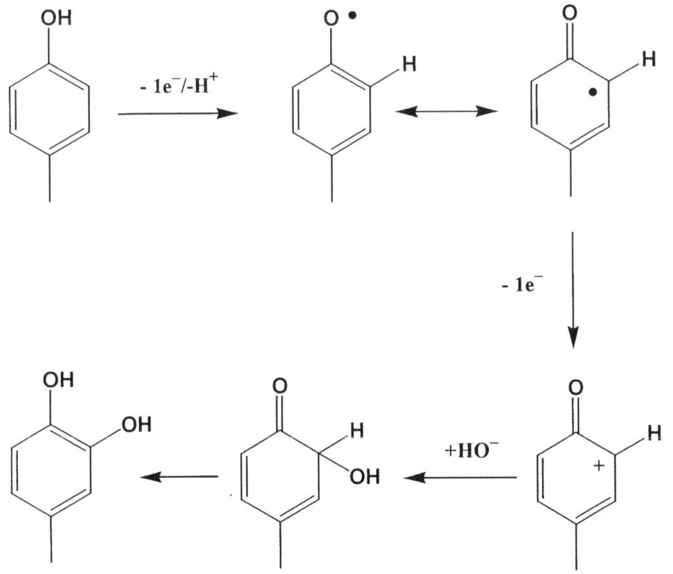

SCHEME 3. Mechanism of ascorbate-independent hydroxylation of Tyr.

Such mechanisms would only require the presence of Cu^{2+} and not ascorbate. However, over our experimental timescale, the oxidation of both His and Tyr in the respective βAP congeners required the presence of ascorbate, arguing against direct electron transfer from Tyr to Cu^{2+}. More evidence for the involvement of hydroxyl radicals and/or complexed hydroxyl radicals in the oxidation of βAP comes from the experimentally detected hydroxylation of the Phe residue(s) in βAP10–20. Although we have not compared the relative yields of ortho-, meta-, and para-hydroxylation of Phe^{19}/Phe^{20} with theoretically expected patterns based on the known chemistry of free hydroxyl radicals with Phe, we can probably exclude alternative mechanisms, which would require electron transfer to generate intermediate phenyl radical cations.

The physiological consequences of βAP His oxidation are presently unclear. One report indicates that βAP isolated from Alzheimer's disease brain had lost ca. 1 mol His/mol peptide,[32] but the potential oxidation product was not characterized. It can be expected that His oxidation in βAP leads to a change in metal affinity and redox characteristics. For example, Huang et al.[20] demonstrated an approximately three-fold lower ability of rat βAP1–40 to reduce Cu^{2+} to Cu^+ compared with human βAP1–40. The rat sequence shows three amino acid substitutions, Arg^5 → Gly, Tyr^{10} → Phe, and His^{13} → Arg. The fact that rat 1–40 still reduces Cu^{2+} suggests that Tyr^{10} is not required for Cu^{2+} reduction, consistent with the role of Met^{35} as electron donor.[23] Instead, one feature contributing to the less efficient reduction of Cu^{2+} may be the geometry of and affinity for Cu^{2+} binding due to the lack of His^{13} in rat βAP1–40. His^{13} is the most sensitive amino acid towards Cu^{2+}-catalyzed oxidation in human βAP1–40, and conversion of His^{13} into 2-oxo-His may transform βAP-Cu^{2+} into a less efficient electron acceptor for the reduction of the peptide-bound Cu^{2+}. In this sense, the oxidation of His^{13} could actually be protective for any surrounding biological tissue if the covalent modification of His^{13} leads to a less efficient generation of Cu^+ and associated Fenton-type processes.

ACKNOWLEDGMENTS

We acknowledge support of our research by the NIA (No. 2PO1AG12993).

REFERENCES

1. SELKOE, D.J. 1996. Amyloid β-protein and the genetics of Alzheimer's disease. J. Biol. Chem. **271**: 18295–18298.
2. WOLFE, M.S., J. DE LOS ANGELES, D.D. MILLER et al. 1999. Are presenilins intramembrane-cleaving proteases? Implications for the molecular mechanism of Alzheimer's disease. Biochemistry **38**: 11223–11230.
3. TANZI, R.E. 1999. A genetic dichotomy model for the inheritance of Alzheimer's disease and common age-related disorders. J. Clin. Invest. **104**: 1175–1179.
4. XIA, W., W.J. RAY, B.L. OSTASZEWSKI et al. 2000. Presenilin complexes with the C-terminal fragments of amyloid precursor protein at the sites of amyloid β-protein generation. Proc. Natl. Acad. Sci. USA **97**: 9299–9304.
5. MARTIN, G.M. 2000. Molecular mechanisms of late life dementias. Exp. Gerontol. **35**: 439–443.
6. HAASS, C., E.H. KOO, A. MELLON et al. 1992. Targeting of cell-surface β-amyloid precursor protein to lysosomes: alternative processing into amyloid-bearing fragments. Nature **357**: 500–503.

7. SEUBERT, P., C. VIGO-PELFREY, F. ESCH et al. 1992. Isolation and quantification of soluble Alzheimer's β-peptide from biological fluids. Nature **359:** 325–327.
8. BUSH, A.I., W.H. PETTINGELL, G. MULTHAUP et al. 1994. Rapid induction of Alzheimer Aβ amyloid formation by zinc. Science **265:** 1464–1467.
9. SHEN, C-L. & R.M. MURPHY. 1995. Solvent effects on self-assembly of β-amyloid peptide. Biophys. J. **69:** 640–651.
10. WALSH, D.M., A. LOMAKIN, G.B. BENEDEK et al. 1997. Amyloid β-protein fibrillogenesis. J. Biol. Chem. **272:** 22364–22372.
11. TJERNBERG, L.O., D.J.E. CALLAWAY, A. TJERNBERG et al. 1999. A molecular model of Alzheimer amyloid β-peptide fibril formation. J. Biol. Chem. **274:** 12619–12625.
12. SIPE, J.D. & A.S. COHEN. 2000. History of the amyloid fibril. J. Struct. Biol. **130:** 88–98.
13. KISILEVSKI, R. 2000. Amyloidogenesis—unquestioned answers and unanswered questions. J. Struct. Biol. **130:** 99–108.
14. HENSLEY, K., J.M. CARNEY, M.P. MATTSON et al. 1994. A model for β-amyloid aggregation and neurotoxicity based on free radical generation by the peptide: relevance to Alzheimer disease. Proc. Natl. Acad. Sci. USA **91:** 3270–3274.
15. HARRIS, M.E., K. HENSLEY, D.A. BUTTERFIELD et al. 1995. Direct evidence for oxidative injury produced by the Alzheimer's β-amyloid peptide (1–40) in cultured hippocampal neurons. Exp. Neurol. **131:** 193–202.
16. BUTTERFIELD, D.A. 1997. β-Amyloid-associated free radical oxidative stress and neurotoxicity: implications for Alzheimer's disease. Chem. Res. Toxicol. **10:** 495–506.
17. MATTSON, M.P., R.J. MARK, K. FURUKAWA & A.J. BRUCE. 1997. Disruption of brain cell ion homeostasis in Alzheimer's disease by oxy radicals, and signalling pathways that protect therefrom. Chem. Res. Toxicol. **10:** 507–517.
18. SAYRE, L.M., M.G. ZAGORSKI, W.K. SUREWICZ et al. 1997. Mechanisms of neurotoxicity associated with amyloid β deposition and the role of free radicals in the pathogenesis of Alzheimer's disease: a critical appraisal. Chem. Res. Toxicol. **10:** 518–526.
19. HUANG, X., M.P. CUAJUNGCO, C.S. ATWOOD et al. 1999. Cu(II) potentiation of Alzheimer Aβ neurotoxicity: correlation with cell-free hydrogen peroxide production and metal reduction. J. Biol. Chem. **274:** 37111–37116.
20. HUANG, X., C.S. ATWOOD, M.A. HARTSHORN et al. 1999. The Aβ peptide of Alzheimer's disease directly produces hydrogen peroxide through metal ion reduction. Biochemistry **38:** 7609–7616.
21. CURTAIN, C.C., F. ALI, I. VOLITAKIS et al. 2001. Alzheimer's disease amyloid-β binds copper and zinc to generate an allosterically ordered membrane-penetrating structure containing superoxide-dismutase-like subunits. J. Biol. Chem. **276:** 20466–20473.
22. VARADARAJAN, S., S. YATIN, M. AKSENOVA & D.A. BUTTERFIELD. 2000. Alzheimer's amyloid β-peptide–associated free radical oxidative stress and neurotoxicity. J. Struct. Biol. **130:** 184–208.
23. VARADARAJAN, S., J. KANSKI, M. AKSENOVA et al. 2001. Different mechanisms of oxidative stress and neurotoxicity for Alzheimer's Aβ(1–42) and Aβ(25–35). J. Am. Chem. Soc. **123:** 5625–5631.
24. OPAZO, C., X. HUANG, R.A. CHERNY et al. 2002. Metalloenzyme-like activity of Alzheimer's disease β-amyloid. J. Biol. Chem. **277:** 40302–40308.
25. SMITH, M.A., P.L.R. HARRIS, L.M. SAYRE & G. PERRY. 1997. Iron accumulation in Alzheimer disease is a source of redox-generated free radicals. Proc. Natl. Acad. Sci. USA **94:** 9866–9868.
26. SMITH, M.A., K. HIRAI, K. HSIAO et al. 1998. Amyloid-β deposition in Alzheimer transgenic mice is associated with oxidative stress. J. Neurochem. **70:** 2212–2215.
27. SAYRE, L.M., G. PERRY, P.L.R. HARRIS et al. 2000. In situ oxidative catalysis by neurofibrillary tangles and senile plaques in Alzheimer's disease: a central role for bound transition metals. J. Neurochem. **74:** 270–279.
28. SPARKS, D.L. & B.G. SCHREURS. 2003. Trace amounts of copper in water reduce β-amyloid plaques and learning deficits in a rabbit model of Alzheimer's disease. Proc. Natl. Acad. Sci. USA **100:** 11065–11069.
29. MCLELLAN, M.E., S.T. KAJDASZ, B.T. HYMAN & B.J. BACSKAI. 2003. In vivo imaging of reactive oxygen species specifically associated with thioflavine S–positive amyloid plaques by multiphoton microscopy. J. Neurosci. **23:** 2212–2217.

30. YATIN, S.M., S. VARADARAJAN, C.D. LINK & D.A. BUTTERFIELD. 1999. *In vitro* and *in vivo* oxidative stress associated with Alzheimer's amyloid β-peptide (1–42). Neurobiol. Aging **20:** 325–330.
31. DRAKE, J., C.D. LINK & A.D. BUTTERFIELD. 2003. Oxidative stress precedes fibrillar deposition of Alzheimer's disease amyloid β-peptide (1–42) in a transgenic *Caenorhabditis elegans* model. Neurobiol. Aging **24:** 415–420.
32. ATWOOD, C.S., X. HUANG, A. KHATRI *et al.* 2000. Copper catalyzed oxidation of Alzheimer Aβ. Cell. Mol. Biol. **46:** 777–783.
33. DONG, J., C.S. ATWOOD, V.E. ANDERSON *et al.* 2003. Metal-binding and oxidation of amyloid-β within isolated senile plaque cores: Raman microscopic evidence. Biochemistry **42:** 2768–2773.
34. SCHÖNEICH, CH. & T.D. WILLIAMS. 2002. Cu(II)-catalyzed oxidation of β-amyloid peptide targets His[13] and His[14] over His[6]: detection of 2-oxo-histidine by HPLC-MS/MS. Chem. Res. Toxicol. **15:** 717–722.
35. SCHÖNEICH, CH. & T.D. WILLIAMS. 2003. Cu(II)-catalyzed oxidation of Alzheimer's disease β-amyloid peptide and related sequences: remarkably different selectivities of neurotoxic βAP1–40 and non-toxic βAP40–1. Cell. Mol. Biol. **49:** 753–761.
36. AKSENOV, M.Y., M.V. AKSENOVA, M.E. HARRIS *et al.* 1995. Enhancement of β-amyloid peptide Ab(1–40)–mediated neurotoxicity by glutamine synthetase. J. Neurochem. **65:** 1899–1902.
37. ZHAO, F., E. GHEZZO-SCHÖNEICH, G.I. ACED *et al.* 1997. Metal-catalyzed oxidation of histidine in human growth hormone. J. Biol. Chem. **272:** 9019–9029.
38. SCHÖNEICH, CH. 2000. Mechanisms of metal-catalyzed oxidation of histidine to 2-oxo-histidine in peptides and proteins. J. Pharm. Biomed. Anal. **21:** 1093–1097.
39. MVULA, E., M.N. SCHUCHMANN & C. VON SONNTAG. 2001. Reactions of phenol-OH-adduct radicals: phenoxyl radical formation by water elimination vs. oxidation by dioxygen. J. Chem. Soc. Perkin Trans. **2:** 264–268.
40. LASSMANN, G., L.A. ERIKSSON, F. HIMO *et al.* 1999. Electronic structure of a transient histidine radical in liquid aqueous solution: EPR continuous-flow studies and density functional calculations. J. Phys. Chem. **A103:** 1283–1290.
41. SAMUNI, A. & P. NETA. 1973. Electron spin resonance study of the reaction of hydroxyl radicals with pyrrole, imidazole, and related compounds. J. Phys. Chem. **77:** 1629–1635.

Redox Metals in Alzheimer's Disease

BOZHO M. TODORICH AND JAMES R. CONNOR

George M. Leader Family Laboratory for Alzheimer's Disease Research, Department of Neural and Behavioral Sciences, Pennsylvania State University School of Medicine, Hershey, Pennsylvania, USA

> ABSTRACT: Redox metals in the brain play many important roles in maintenance of cellular function. The maintenance of their homeostasis is of paramount importance to a number of diseases such as Alzheimer's disease and multiple sclerosis. Iron, copper, and zinc are metals of special interest in the pathogenesis of these disorders. This review will focus primarily on iron.
>
> KEYWORDS: Alzheimer's disease (AD); redox metal; iron; copper; zinc; brain

INTRODUCTION

Redox metals in the brain play many important roles in maintenance of cellular function. The maintenance of their homeostasis is of paramount importance to a number of diseases such as Alzheimer's disease (AD) and multiple sclerosis. Iron (Fe), copper (Cu), and zinc (Zn) are metals of special interest in the pathogenesis of these disorders. This review will focus primarily on iron.

IRON

Iron is the most abundant transition metal in the body and in the brain. It exists in two redox-active forms: ferrous (Fe^{+2}) and ferric (Fe^{+3}) forms.[1,2] This strong redox potential affords it versatile functions as a cofactor and biocatalyst in many vital as well as potentially damaging reactions in the cell.[3] Iron is an integral part of the metabolic activity of cells. Iron is an essential cofactor in cytochrome oxidase, glucose-6-phosphate dehydrogenase, NADH dehydrogenase, aldehyde dehydrogenase, succinate dehydrogenase, and aconitase enzymes, all of which are enzymes involved in production of ATP.[1,2] Iron plays a crucial role in the function of hemoglobin and the cytochromes a, b, and c, and in the biosynthesis of cholesterol and lipids, which are necessary precursors in the formation of neuronal membranes and myelin.[1,2] Because of the relatively high oxygen utilization in the brain and the relatively high content of myelin, there is an especially important link between iron and brain function.[4] Iron is unevenly distributed in the brain and is enriched in areas such as the substantia nigra, a major site of dopaminergic neurons in the CNS. Iron is also

---

Address for correspondence: James R. Connor, George M. Leader Family Laboratory for Alzheimer's Disease Research, Department of Neural and Behavioral Sciences, Pennsylvania State University School of Medicine, Hershey, PA 17033. Voice: 717-531-6408; fax: 717-531-5184.
jrc3@psu.edu

involved in the production of GABA and norepinephrine.[1,5] Thus, not only is the concentration of iron in a given brain region or even cell critical to normal brain function, but the timely delivery of the iron is of equal importance.

If the amounts or timings of iron availability are inappropriate, iron can participate in numerous toxic reactions in the cell via the Fenton reaction that produces free radicals.[3] Free radicals damage proteins, nucleic acids, and lipids, causing irreversible damage to cells and inducing cell death by apoptosis or necrosis.[1,6] The severity of iron-related damage can be illustrated by numerous conditions associated with the iron dyshomeostasis: Alzheimer's disease (AD), multiple sclerosis (MS), Parkinson's disease (PD), tardive dyskinesia, Pick's disease, Huntington's disease, Hallervorden-Spatz disease, Friedreich's ataxia, and aceruloplasminemia have all been associated with abnormal iron accumulation in one or more brain regions.[1,7] There are exquisite mechanisms in place for the regulation of iron both into the brain and at the cellular level. Therefore, the question that will be addressed in this review is how loss of iron homeostasis can occur in the different diseases.

Several proteins have been identified in the brain that regulate iron homeostasis.[8] Ferritin sequesters iron in the cytosol and is primarily responsible for regulating the amount of labile iron in the cell.[9] Transferrin (Tf) is responsible for cellular acquisition of iron from the plasma via Tf-receptor-mediated endocytosis.[2,10] Tf receptors are prominently expressed on neurons and on the endothelial cells of the brain microvasculature. Iron regulatory proteins are sensors of the labile iron pool and regulate levels of expression of ferritin and Tf receptor in cells by binding to a portion of mRNA of these proteins termed the iron-responsive element (IRE).[1,2,11] Consequently, posttranscriptional regulation of the iron-acquisition proteins affords maintenance of intracellular Fe at relatively stable concentrations. Alterations in the makeup of the IRPs may contribute to iron overload in AD.[12] Relatively recently, a transmembrane iron transporter, DMT1 (divalent metal transporter 1), was identified, which is the first iron transporter in mammalian cells. One of its functions is the transfer of iron across the apical surface of the gut epithelial lining cells.[13] DMT1 is also found in the brain and is critical for iron transport into the brain.[14,15]

COPPER AND ZINC

Similar to iron, copper (Cu) has a dual redox potential, existing in both cuprous (Cu^{+1}) and cupric (Cu^{+2}) forms. It functions as a cofactor in several enzymes, such as cytochrome c oxidase and superoxide dismutase (SOD).[16] Zinc (Zn), although redox-inactive, maintains important functions in neurotransmission.[17] The largest labile Zn pool in the brain is in the synapses of corticofugal glutaminergic fibers in the hippocampus.[17] Cu and Zn are also involved in inflammatory functions, such as synthesis and secretion of the interleukin-2 that may be important in neurodegenerative diseases.[16] These metals are discussed in greater detail elsewhere in this volume.

METALS AND AD

The loss of homeostasis of copper and iron in the brain is devastating to neurological function because both deficiency and overaccumulation result in pathologic consequences. For example, deficiency of iron in the brain can result in the metabolic impairment of the neurons and glia that negatively impacts motor and cognitive functions.[18] On the other hand, iron overaccumulation is involved in many diseases including AD.[1,2] In AD patients, overaccumulation of iron in the hippocampus, cerebral cortex, and basal nucleus of Meynert colocalizes with the lesions, neurofibrillary tangles, and neuritic plaques.[19] These affected brain areas are of special interest for researchers because they represent centers of memory and thought processing, all lost in the clinical picture of AD.[2] Abnormal iron deposition in neuritic plaques can associate with Aβ and may promote increased formation of oxygen radicals.[20]

The ability of iron to interact with oxygen is critical for energy production in the cell, but the same property makes it capable of producing damaging free radicals via the Fenton reaction:

$$M^{+n} \text{ (Fe or Cu)} + H_2O \rightarrow M^{+(n-1)} + \cdot OH^- \text{ (hydroxyl radical)}.$$

Reactive oxygen species (ROS) resulting from this reaction, such as the hydroxyl radicals, nonselectively damage intracellular structures via lipid peroxidation, protein cross-linking and carboxylation, and DNA mutation via base modification, cross-linking, H-bond interference, and strand breakage.[21] Free radical damage observed in the areas of AD neurodegeneration and the mechanisms by which free radicals can lead to AD lesions are being intensely investigated. β-Amyloid itself is reportedly a substrate for hydroxyl radicals.[22] There is evidence that β-amyloid extracted from postmortem AD brains has undergone oxidative stress such as carbonyl adduct formation, histidine loss, and dityrosine cross-linking, making this protein less water-soluble and less susceptible to degradation by the proteases.[22] Deposition of β-amyloid fragment and AβPP cleavage and synthesis are promoted by the presence of iron.[23]

Several studies utilizing magnetic resonance imaging (MRI) have found a positive correlation between aging and iron accumulation in the brain, specifically, in the striatum, caudate nucleus, putamen, and globus pallidus.[24,25] As a consequence of iron accumulation, especially if ferritin does not increase with the increasing iron as has been reported in AD and PD,[26] the brain becomes increasingly vulnerable to the iron-catalyzed oxidative stress. Hence, the age-related accumulation of iron may underlie or at least contribute to the commonly accepted theory that age is the biggest risk factor for most neurodegenerative diseases.[27]

HEREDITARY HEMOCHROMATOSIS AND AD

A genetic predisposition to AD has been identified in reference to the familial mutations of the genes encoding amyloid precursor protein, presenilin 1 and 2, and apolipoprotein E. These account for <2% of all cases of familial AD (FAD),[28] but do not address the numerous cases of sporadic AD. Because of the compelling evidence that an imbalance of iron contributes to the pathogenesis of AD in the context of lesion formation, neurofibrillary tangle formation, and oxidative stress, the possibility that a genetic mutation impacts iron status is being considered by a number of research groups.[19] The focus has been on the Hfe mutation that is associated with the iron overload disease known as hemochromatosis.[29]

Hemochromatosis is an autosomal-recessive disease that involves deposition of iron into various organs, such as liver, intestine, pancreas, and the heart.[29] The gene involved in hemochromatosis, Hfe, occupies a locus at chromosome 6p21.[19,30] The function of the protein has not been clearly identified, but is thought to couple with the Tf receptor on the cell surface and decrease the affinity of the receptor for iron-loaded Tf in the extracellular fluid (ECF).[19,30] A mutation of the Hfe renders it incapable of coupling with the Tf receptor and results in an increase of iron uptake into the cells.[19]

Patients with clinical diagnosis of hemochromatosis are usually homozygous for one of the Hfe mutations, while some clinical symptoms may occur conditionally in the heterozygous individuals who are thought to subclinically load iron.[29] Traditionally, it has been thought that the brain was protected from the hemochromatosis-induced iron overload because of the blood-brain barrier (BBB).[19] However, critical review of the literature indicates that there is little support for this concept and, in fact, two papers describing iron content of brains in hemochromatosis find increased stainable iron in numerous brain regions that reside behind the BBB.[31,32] In addition, the Hfe protein is expressed on ependymal cells of the brain-ventricle interface and the capillary endothelium of the brain, which comprises the BBB along with Tf receptors and is thus in position to influence brain iron uptake.[19,33,34] Hfe is induced in cells associated with neuritic plaques and in neurons in AD brains.[16] Therefore, expression of the mutant form of Hfe could promote an increase in iron accumulation in cells in AD, increasing the potential for oxidative stress and predisposing the carrier to development of AD pathologies. Why Hfe is induced in AD brains in these areas is not known and is currently under investigation.

Two mutations in the Hfe gene have been identified, C282Y and H63D, with a total combined frequency of 20–40% occurrence in people of European ancestry.[29] This reported frequency of the Hfe mutation places it as one of the most frequently occurring genetic mutations. Four studies have recently suggested a link between the Hfe mutations and the risk for AD. For instance, a study conducted by Sampietro *et al.* at the University of Milan found that presence of the Hfe H63D mutation (global allelic frequency = 0.08) represents a risk factor for an earlier onset of AD.[28] Patients carrying this mutant allele had a mean age onset of AD symptoms at 71.7 ± 6.0 years, while the mutation-lacking controls had a mean age onset of 76.6 ± 5.8 years.[28]

The presence of the E4 allele of apolipoprotein E (APOE) is an apparent risk factor for the development of FAD. A paper by Moalem *et al.* found that presence of the Hfe mutations, specifically C282Y, appears to be predisposing for AD in APOE-negative males, while tending to be relatively protective for APOE-negative females.[28,30] Such a paradox perhaps illustrates the complex biochemical and hormonal interactions and other epigenetic factors that evolve differently in males and females. A study using a Spanish population found that the H63D mutation of the Hfe gene failed to induce early AD onset alone, but was able to do so in combination with the APOE mutation.[35] The most recent paper to examine a potential contribution of the Hfe mutation to AD found that Hfe mutations are associated with increased oxidative stress and Braak stage.[36]

Additionally, Tf subtype C2 was also found to have increased frequency in the AD patients when compared to age-matched controls.[37] Although in need of further investigation, perhaps defective iron binding to Tf C2 results in overaccumulation of Fe in the AD lesions exacerbating oxidative stress.[28] Interestingly, AD subjects homozygous for the APOE4 allele are twice as likely to have a Tf C2 allele than those who had no APOE4 or a single copy of APOE4.[37] No studies to date have attempted to address the link between Tf C2 and Hfe mutation frequency and brain iron overload in AD patients.

COPPER, ZINC, AND AD PATHOGENESIS

Copper induces β-amyloid aggregation under slightly acidic conditions (pH 6.6–7.0).[21] Copper, like iron, can also participate in the Fenton reaction to generate ROS.[21,38] Recent studies have found conflicting results regarding the amount of copper in the brain and the production of plaques.[39–41] Some studies link impaired functioning of Zn/Cu-SOD to either abnormally high or low levels of copper.[42,43] SOD is involved in production of H_2O_2 from the highly reactive peroxide radical. Consequentially, alterations in copper metabolism are associated with an increase in oxidative stress responses in the liver and brain.[42,43] Copper imbalance could decrease the activity of Cu/Zn-SOD contributing to the oxidative stress that is part of the AD pathogenesis.[42,43] An interesting parallel to iron overload in the brain in hemochromatosis is Wilson's disease. This rare disorder is characterized by accumulation of copper in the liver and subsequently to other organs, mainly the brain and kidneys.[44] An exhaustive literature search did not find any studies that examined or identified a connection between Wilson's disease and AD.

Zinc was shown to precipitate the β-amyloid plaques at the physiologic pH (7.4) by binding the histidine residues on the β-amyloid.[2] Plaque aggregation is increased with intense neuronal activity in association with release of zinc from the dopaminergic synaptic vesicles in the brain.[17,45] Recently, Lee *et al.* crossed a mouse model for AD with a mouse lacking a ZnT3 protein, a zinc transporter at the dopamine synapses. The progeny of this cross were AD mice that had diminished Zn levels at synapses and markedly lower load of insoluble β-amyloid. These data implicate zinc in AD pathogenesis.[45]

AD AND METAL CHELATOR THERAPY

Therapy to Prevent/Alleviate the Metal Imbalance in the Brain

Accumulation of β-amyloid represents the pathological hallmark of AD. Hence, it is understandable that research up to date has focused on preventing its production with agents such as secretease inhibitors or facilitating its elimination with agents like vaccine.[17] Considering that metals such as zinc, copper, and iron induce aggregation of metalloprotein β-amyloid and facilitate in free radical–mediated oxidative stress reactions, metal chelation has gained increasing popularity in AD research.[17]

Metal chelation is the strategy by which a small organic ligand is introduced to bind metal ions, making them nonbioavailable and facilitating their elimination.[46] Consequently, redox metals are not available to bind other larger ligands such as the β-amyloid, promoting its deposition, nor to promote oxidative stress reactions.[21,46] Many chelators for iron, copper, and zinc appear in the literature, but we will briefly review only the most prominent ones, including desferrioxamine and clioquinol.

Desferrioxamine (DFO) is best known as an iron chelator, even though its first application to AD was under the assumption that it chelated aluminum.[46] In this study, DFO, given orally, halved the rate of decline of AD patients.[47] Problems with DFO therapy include its high polarity, which causes difficulty in crossing the BBB.[46] Further, it is toxic and decomposes soon after administration.[21] DFO administration is expensive and inconvenient, requiring painful intramuscular injections twice daily.[21,46]

Clioquinol (CQ), formerly used as an antibiotic, is a hydrophobic agent with high affinity for Zn, Cu, and (to some degree) Fe.[17] In a mouse model of AD, 9-week oral treatment with CQ resulted in a 49% reduction in β-amyloid levels and significantly decreased concentrations of Cu, Zn, and Fe in the brain.[17] Because Zn inhibits H_2O_2 formation and β-amyloid aggregation, CQ likely exhibits the therapeutic effect of hindering oxidative stress by chelating Cu and Fe, and facilitating β-amyloid disaggregation by chelating Zn all in one.[17]

D-Penicillamine and trientine are well-known copper chelators for treatment of Wilson's disease.[44] However, Squitti *et al.* failed to show any therapeutic potential in AD, even though there was a marked decrease in oxidative stress biomarkers in AD patients in a clinical trial.

One major problem with chelator therapy is delivery to the brain. Chelators ingested or given intravenously have to cross the BBB. Therefore, chelators must be lipophilic and have a low molecular weight. Many chelators fall short of this ideal because the small lipophilic chelators are often too toxic to be considered. Areas of the brain exist that are not protected by the BBB, but these areas undergo minimal or no changes in AD patients. Despite these challenges, there is hope that new developments in pharmacotherapy will bolster properties of metal chelators and make them more effective in fighting AD.

REFERENCES

1. CONNOR, J.R. 1997. Evidence for iron mismanagement in the brian in neurological disorders. *In* Metals and Oxidative Damage in Neurological Disorders, pp. 23–39. Plenum. New York.
2. HENDERSON, R.J. & J.R. CONNOR. 2003. Iron's involvement in the molecular mechanisms of Alzheimer's disease pathogenesis. *In* Neuronal and Vascular Plasticity: Elucidating Basic Cellular Mechanisms for Future Therapeutic Discovery. Kluwer. Dordrecht/Detroit.
3. KAUR, D. & J.K. ANDERSEN. 2002. Ironing out Parkinson's disease: is therapeutic treatment with iron chelators a real possibility? Aging Cell **1**(1): 17–21.
4. BEARD, J.L. & J.R. CONNOR. 2003. Iron status and neural functioning. Annu. Rev. Nutr. **23**: 41–58.
5. BENKOVIC, S. & J.R. CONNOR. 1993. Ferritin, transferrin, and iron in normal and aged rat brains. J. Comp. Neurol. **338**: 97–113.
6. TOYOKUNI, S. 2002. Iron and carcinogenesis: from Fenton reaction to target genes. Redox Rep. **7**(4): 189–197.
7. WAGNER, K.R., F.R. SHARP, T.D. ARDIZZONE *et al.* 2003. Heme and iron metabolism: role in cerebral hemorrhage. J. Cereb. Blood Flow Metab. **23**(6): 629–652.
8. PINERO, D. & J.R. CONNOR. 2000. Iron in the brain: an important contributor in normal and diseased states. Neuroscientist **6**(6): 435–453.
9. KAKHLON, O. & Z.I. CABANTCHIK. 2002. The labile iron pool: characterization, measurement, and participation in cellular processes. Free Radical Biol. Med. **33**(8): 1037–1046.

10. THOMPSON, K., S. SHOHAM & J.R. CONNOR. 2001. Iron and neurodegenerative disorders. Brain Res. Bull. **55**(2): 155–164.
11. BOUTON, C. & J.C. DRAPIER. 2003. Iron regulatory proteins as NO signal transducers. Sci. STKE **182**: 17.
12. PINERO, D.J., J. HU & J.R. CONNOR. 2000. Alterations in the interaction between iron regulatory proteins and their iron responsive element in normal and Alzheimer's diseased brains. Cell. Mol. Biol. **46**(4): 761–776.
13. ANDREWS, N. 1999. The iron transporter DMT1. Int. J. Biochem. Cell Biol. **31**(10): 991–994.
14. BURDO, J.R., S.L. MENZIES, I.A. SIMPSON et al. 2001. Distribution of divalent metal transporter 1 and metal transport protein 1 in the normal and Belgrade rat. J. Neurosci. Res. **66**: 1198–1207.
15. BURDO, J.R., J. MARTIN, S.L. MENZIES et al. 1999. Cellular distribution of iron in the brain of the Belgrade rat. Neuroscience **93**(3): 1189–1196.
16. FAILLA, M. 2003. Trace elements and host defense: recent advances and continuing challenges. J. Nutr. **133**: 1443S–1447S.
17. BUSH, A.I. 2002. Metal complexing agents as therapies for Alzheimer's disease. Neurobiol. Aging **23**(6): 1031–1038.
18. BEARD, J.L. & J.R. CONNOR. 2003. Iron status and neural functioning. Annu. Rev. Nutr. **23**: 41–58.
19. CONNOR, J.R., E.A. MILWARD, S. MOALEM et al. 2001. Is hemochromatosis a risk factor for Alzheimer's disease? J. Alzheimer's Dis. **3**(5): 471–477.
20. SMITH, M.A., L.R. PEGGY, L.M. HARRIS et al. 1997. Iron accumulation in Alzheimer disease is a source of redox-generated free radicals. Proc. Natl. Acad. Sci. USA **94**: 9866–9868.
21. FINEFROCK, A.E., A.I. BUSH & P.M. DORAISWAMY. 2003. Current status of metals as therapeutic targets in Alzheimer's disease. J. Am. Geriatr. Soc. **51**(8): 1143–1148.
22. ATWOOD, C.S., X. HUANG & R.D. MOIR. 1999. Role of free radicals and metal ions in the pathogenesis of Alzheimer's disease. In Metal Ions in Biological Systems, pp. 309–364. Dekker. New York.
23. ROGERS, J.T., J.D. RANDALL, C.M. CAHILL et al. 2002. An iron-responsive element type II in the 5'-untranslated region of the Alzheimer's amyloid precursor protein transcript. J. Biol. Chem. **277**: 45518–45528.
24. MARTIN, W.R., F.Q. YE & P.S. ALLEN. 1998. Increasing striatal iron content associated with normal aging. Mov. Disord. **13**(2): 281–286.
25. BARTZOKIS, G., M. BECKSON, D.B. HANCE et al. 1997. NMR evaluation of age-related increase of brain iron in young adult and older normal males. Magn. Reson. Imaging **15**(1): 29–35.
26. CONNOR, J.R., B.S. SNYDER, P. AROSIO et al. 1995. A quantitative analysis of isoferritins in select regions of aged, parkinsonian, and Alzheimer's diseased brains. J. Neurochem. **65**: 717–724.
27. MATTSON, M., W.A. PEDERSEN, W. DUAN et al. 1999. Cellular and molecular mechanisms underlying perturbed energy metabolism and neuronal degeneration in Alzheimer's and Parkinson's disease. Ann. N.Y. Acad. Sci. **893**: 154–175.
28. SAMPIETRO, M., L. CAPUTO, A. CASATTA et al. 2001. The hemochromatosis gene affects the age of onset of sporadic Alzheimer's disease. Neurobiol. Aging **22**(4): 563–568.
29. HANSON, E.H., G. IMPERATORE & W. BURKE. 2001. HFE gene and hereditary hemochromatosis: a HuGE review. Am. J. Epidemiol. **154**: 193–206.
30. MOALEM, S., M.E. PERCY, D.F. ANDREWS et al. 2000. Are hereditary hemochromatosis mutations involved in Alzheimer disease? Am. J. Med. Genet. **93**: 58–66.
31. CHAMMERMEYER, J. 1947. Deposition of iron in the paraventricular areas of the human brain in hemochromatosis. J. Neuropathol. Exp. Neurol. **6**: 111–127.
32. SHELDON, J.H. 1935. Hemochromatosis, pp. 155–159. Oxford University Press. London/New York.
33. STREMMEL, W., H.D. RIEDEL, C. NIEDERAU & G. STROHMEYER. 1993. Pathogenesis of genetic hemochromatosis. Eur. J. Clin. Invest. **23**: 321.

34. BASTIN, J.M., M. JONES, C.A. O'CALLAGHAN et al. 1998. Kupffer cell staining by an HFE-specific monoclonal antibody: implications for hereditary haemochromatosis. Br. J. Haematol. **103**(4): 931–941.
35. COMBARROS, O., M. GARCIA-ROMAN, A. FONTALBA et al. 2003. Interaction of the H63D mutation in the hemochromatosis gene with the apolipoprotein E epsilon 4 allele modulates age at onset of Alzheimer's disease. Dement. Geriatr. Cognit. Disord. **15**(3): 151–154.
36. PULLIAM, J.F., C.D. JENNINGS, R.J. KRYSCIO et al. 2003. Association of HFE mutations with neurodegeneration and oxidative stress in Alzheimer's disease and correlation with APOE. Am. J. Med. Genet. **119B**(1): 48–53.
37. NAMEKATA, K., M. IMAGAWA, A. TERASHI et al. 1997. Association of transferrin C2 allele with late-onset Alzheimer's disease. Hum. Genet. **101**(2): 126–129.
38. HUANG, X., C.S. ATWOOD & M.A. HARTSHORN. 1999. The peptide of Alzheimer's disease directly produces hydrogen peroxide through metal ion reduction. Biochemistry **38**: 7609–7616.
39. PHINNEY, A.L., B. DRISALDI, S.D. SCHMIDT et al. 2003. *In vivo* reduction of amyloid-{beta} by a mutant copper transporter. Proc. Natl. Acad. Sci. USA. In press.
40. BAYER, T.A., S. SCHAFER, A. SIMONS et al. 2003. Dietary Cu stabilizes brain superoxide dismutase-1 activity and reduces amyloid Aβ production in APP23 transgenic mice. Proc. Natl. Acad. Sci. USA. In press.
41. SPARKS, D.L. & B.G. SCHREURS. 2003. Trace amounts of copper in water induce beta-amyloid plaques and learning deficits in a rabbit model of Alzheimer's disease. Proc. Natl. Acad. Sci. USA **100**(19): 11065–11069.
42. ROSSI, L., M.R. CIRIOLO, E. MARCHESE et al. 1994. Differential decrease of copper content and of copper binding to superoxide dismutase in liver, heart, and brain of copper-deficient rats. Biochem. Biophys. Res. Commun. **203**: 1028–1034.
43. ROSSI, L., R. SQUITTI, P. PASQUALETTI et al. 2002. Red blood cell copper, zinc superoxide dismutase activity is higher in Alzheimer's disease and is decreased by D-penicillamine. Neurosci. Lett. **329**(2): 137–140.
44. EL-YOUSSEF, M. 2003. Wilson disease. Mayo Clin. Proc. **78**(9): 1126–1136.
45. LEE, J.Y., T.B. COLE, R.D. PALMITER et al. 2002. Contribution by synaptic zinc to the gender-disparate plaque formation in human Swedish mutant APP transgenic mice. Proc. Natl. Acad. Sci. USA **99**(11): 7705–7710.
46. MALECKI, E.A. & J.R. CONNOR. 2002. The case for iron chelation and/or antioxidant therapy in Alzheimer's disease. Drug Dev. Res. **56**: 526–530.
47. CRAPPER MCLACHLAN, D.R., A.J. DALTON, T.P.A. KRUK et al. 1991. Intramuscular desferrioxamine in patients with Alzheimer's disease. Lancet **337**: 1304–1308.

Oxidative Stress and Redox-Active Iron in Alzheimer's Disease

KAZUHIRO HONDA, GEMMA CASADESUS, ROBERT B. PETERSEN, GEORGE PERRY, AND MARK A. SMITH

Institute of Pathology, Case Western Reserve University, Cleveland, Ohio, USA

ABSTRACT: Many lines of evidence indicate that oxidative stress is one of the earliest events in the genesis of Alzheimer's disease (AD). Iron is a transition metal capable of generating hydroxyl radicals, the most potent reactive oxygen species. Consequently, a disruption in the metabolism of iron has been postulated to have a role in the pathogenesis of AD. Indeed, both senile plaques and neurofibrillary tangles, the major pathological landmarks of AD, as well as neurons in the earliest stages of the disease, show elevated iron deposition. However, it is clear that the iron bound to lesion-associated proteins such as amyloid-β and tau plays only a minor, late role in the disease, with the RNA-associated iron found in the neuronal cytoplasm occurring early and being of paramount importance. In this regard, it is probably not surprising that there is significant oxidation of cytoplasmic RNA among the populations of neurons vulnerable to AD. In this review, we consider the role of iron-induced oxidative stress as a key event in AD pathophysiology.

KEYWORDS: Alzheimer's disease; amyloid-β; iron; mitochondria; redox; RNA

INTRODUCTION

Alzheimer's disease (AD) is characterized clinically as a progressive dementia and pathologically by the presence of senile plaques and neurofibrillary tangles in cortical regions of the brain. While the etiology of AD remains largely unclear, there is accumulating evidence that oxidative stress plays an important role in disease pathophysiology. Although brain represents only 2–3% of total body mass in humans, it requires 20% of basal oxygen consumption. During respiration, the mitochondria produce superoxide as a by-product of reducing oxygen molecules. Superoxide dismutase, a major player in the body's defense against oxidative stress, catalyzes the formation of hydrogen peroxide from superoxide. Iron is recognized as a major cause of oxidative stress in AD based on the considerable amount of iron deposition observed in the AD brain[1] and the fact that iron is a transition metal that is involved in the formation of hydroxyl radicals, the most potent reactive oxygen species, from hydrogen peroxide (Fenton reaction).

Address for correspondence: Mark A. Smith, Ph.D., Institute of Pathology, Case Western Reserve University, 2085 Adelbert Road, Cleveland, OH 44106. Voice: 216-368-3670; fax: 216-368-8964.

mark.smith@case.edu

ABNORMALITY OF IRON METABOLISM IN AD

Iron homeostasis in neurons is predominantly maintained by transferrin and ferritin. In the AD brain, several aspects of altered iron homeostasis have been reported. In addition to a high concentration of iron, senile plaques contain transferrin, which is usually found in oligodendrocytes.[2] An abnormal distribution of ferritin has also been reported in AD.[3] The expression level of these macromolecules is tightly controlled by two iron regulatory proteins (IRPs), IRP-1 and IRP-2. Interestingly, extracellular hydrogen peroxide causes the rapid induction of IRP-2, which is significantly altered in relation to both the pathological lesions[4] and redox-active iron distribution in AD.[5]

AMYLOID-β AND REDOX-ACTIVE IRON

Oxidative stress appears to be an early event in sporadic AD. In AD brain, there is a significant association of redox-active iron with both senile plaques and neurofibrillary tangles (NFTs).[5] A variety of studies have demonstrated that iron plays a pivotal role in the oxidative damage observed in proteins, sugars, lipids, and nucleic acids. The mechanism of neurotoxicity would likely be the generation of hydroxyl radicals through the Fenton reaction. Therefore, a critical observation is that the iron in senile plaques is redox-active and has the ability to generate reactive oxygen species.[6] In *in vitro* studies, amyloid-β (Aβ) binds iron[7] and the bound iron facilitates aggregation.[8,9] A recent study suggested that the toxicity of Aβ to cultured cells can be potentiated by iron binding and alleviated by an iron chelator.[10] Also, an *in vivo* model has been used to show that injection of Aβ with iron into rat brain causes more significant damage to neurons than Aβ treatment alone.[11] However, another interesting aspect of this report is that the presence of Aβ attenuates the neuronal loss that would normally be seen around an iron injection site. In fact, some reports show that Aβ deposition markedly reduces the extent of oxidative stress in sporadic[12] and also familial AD cases.[13] These results may suggest a complex behavior of Aβ that depends on the oxidation state of the environment. In addition to Aβ,[14] previous studies have demonstrated that tau protein[15] and apolipoprotein E[16] are also able to bind metals.

OXIDATION OF CYTOPLASMIC RNA

We recently demonstrated that oxidized nucleic acids were commonly observed in the cytoplasm of the neurons that are especially vulnerable to degeneration in AD.[17] Since there was a significant reduction in staining of oxidized nucleic acids by S1 nuclease, an enzyme that specifically digests single-stranded DNA or RNA, cytoplasmic RNA is considered to be the primary target of oxidative stress in the neuronal cell body. If this oxidation is mediated by hydroxyl radicals, taking into consideration the very short distance that hydroxyl radicals can diffuse, the site of hydroxyl radical generation must be very close to the RNA. Thus, it seems likely that redox metals, including iron, bound to RNA are the likely candidates for the oxidative modification of the nucleic acids. In support of this hypothesis, a very recent study demonstrated that messenger RNA is oxidized in AD brain.[18]

WHAT IS THE SOURCE OF INCREASED IRON?

Previously, we and others showed that heme oxygenase-1 (HO-1) is induced in AD.[19–21] HO-1 catalyzes the conversion of heme to biliverdin and iron. Biliverdin is subsequently reduced to the antioxidant, bilirubin. Since HO-1 is induced in proportion to the level of heme,[22] the induction of HO-1 suggests that there may be abnormal turnover of heme in AD.

It is well known that many heme-containing enzymes are found in mitochondria and, therefore, it is not too surprising that there are mitochondria abnormalities in AD brain. For example, there is a 3- to 4-fold increase in the mitochondrial protein cytochrome oxidase-1 and in mitochondrial DNA in the vulnerable neurons in AD without an increase in mitochondrial enzyme activity. Furthermore, ultrastructural observation suggested a high rate of mitochondrial turnover and redox activity in the residual body of lipofuscin.[23] These observations suggest that mitochondrial enzyme turnover occurs in the residual body of lipofuscin and that the lysosome is the likely source of heme. The increase in heme induces synthesis of more HO-1. This evidence supports the hypothesis that mitochondrial turnover promotes oxidative stress via increase of redox-active iron.

CONCLUSIONS

If an excess of iron in AD causes the neurodegeneration observed, the removal of iron may be beneficial as a treatment for AD.[24] Consistent with this hypothesis, administration of deferoxamine, an iron chelator, significantly decreases the progression of disease in AD patients.[25,26] Moreover, these observations provide a cautionary note with respect to the idea that simply getting rid of the presumptive pathological component, Aβ, would be beneficial in the treatment of AD.[27] As mentioned above, Aβ may play a role in sequestering reactive oxygen species. Ideally, future treatments related to preventing metal-associated damage will be to sequester the redox-active metals causing oxidative stress and neurodegeneration, while leaving that required for neuronal metabolism. This requires targeting of the metal chelators to the cytoplasm as well as the nucleus to prevent oxidation of nucleic acids.

ACKNOWLEDGMENTS

Work in the authors' laboratories is supported by funding from the National Institutes of Health (NS38648), the Alzheimer's Association, and a Philip Morris USA External Research Program Postdoctoral Fellowship to K. Honda.

REFERENCES

1. LOVELL, M.A. *et al.* 1998. Copper, iron, and zinc in Alzheimer's disease senile plaques. J. Neurol. Sci. **158:** 47–52.
2. CONNOR, J.R. *et al.* 1992. A histochemical study of iron, transferrin, and ferritin in Alzheimer's diseased brains. J. Neurosci. Res. **31:** 75–83.
3. ROBINSON, S.R. *et al.* 1995. Most amyloid plaques contain ferritin-rich cells. Alzheimer's Res. **1:** 191–196.

4. SMITH, M.A. *et al.* 1998. Abnormal localization of iron regulatory protein in Alzheimer's disease. Brain Res. **788:** 232–236.
5. SMITH, M.A. *et al.* 1997. Iron accumulation in Alzheimer disease is a source of redox-generated free radicals. Proc. Natl. Acad. Sci. USA **94:** 9866–9868.
6. BISHOP, G.M. & S.R. ROBINSON. 2001. Quantitative analysis of cell death and ferritin expression in response to cortical iron: implications for hypoxia-ischemia and stroke. Brain Res. **907:** 175–187.
7. GARZON-RODRIGUEZ, W. *et al.* 1999. Binding of Zn(II), Cu(II), and Fe(II) ions to Alzheimer's Aβ peptide studied by fluorescence. Bioorg. Med. Chem. Lett. **9:** 2243–2248.
8. MANTYH, P.W. *et al.* 1993 Aluminum, iron, and zinc ions promote aggregation of physiological concentrations of β-amyloid peptide. J. Neurochem. **61:** 1171–1174.
9. KURODA, Y. & M. KAWAHARA. 1994. Aggregation of amyloid β-protein and its neurotoxicity: enhancement by aluminum and other metals. Tohoku J. Exp. Med. **174:** 263–268.
10. ROTTKAMP, C.A. *et al.* 2001. Redox-active iron mediates amyloid-β toxicity. Free Radical Biol. Med. **30:** 447–450.
11. BISHOP, G.M. & S.R. ROBINSON. 2000. β-Amyloid helps to protect neurons from oxidative stress. Neurobiol. Aging **21**(suppl. 1S)**:** S226.
12. NUNOMURA, A. *et al.* 2001. Oxidative damage is the earliest event in Alzheimer disease. J. Neuropathol. Exp. Neurol. **60:** 759–767.
13. NUNOMURA, A. *et al.* 2000. Increasing amyloid Aβ 42 deposition is associated with decreasing neuronal RNA oxidation in Down's syndrome and familial Alzheimer's disease. Brain Pathol. **10:** 783.
14. CUAJUNGCO, M.P. *et al.* 2000. Evidence that the β-amyloid plaques of Alzheimer's disease represent the redox-silencing and entombment of Aβ by zinc. J. Biol. Chem. **275:** 19439–19442.
15. SAYRE, L.M. *et al.* 2000. *In situ* oxidative catalysis by neurofibrillary tangles and senile plaques in Alzheimer's disease: a central role for bound transition metals. J. Neurochem. **74:** 270–279.
16. MIYATA, M. & J.D. SMITH. 1996. Apolipoprotein E allele-specific antioxidant activity and effects on cytotoxicity by oxidative insults and beta-amyloid peptides. Nat. Genet. **14:** 55–61.
17. NUNOMURA, A. *et al.* 1999. RNA oxidation is a prominent feature of vulnerable neurons in Alzheimer disease. J. Neurosci. **19:** 1959–1964.
18. SHAN, X. *et al.* 2003. The identification and characterization of oxidized RNAs in Alzheimer's disease. J. Neurosci. **23:** 4913–4921.
19. SMITH, M.A. *et al.* 1994. Heme oxygenase-1 is associated with the neurofibrillary pathology of Alzheimer's disease. Am. J. Pathol. **145:** 42–47.
20. PREMKUMAR, D.R.D. *et al.* 1995. Induction of heme oxygenase-1 mRNA and protein in neocortex and cerebral vessels in Alzheimer's disease. J. Neurochem. **65:** 1399–1402.
21. SCHIPPER, H.M. *et al.* 1995. Expression of heme oxygenase-1 in the senescent and Alzheimer-diseased brain. Ann. Neurol. **37:** 758–768.
22. KEYSE, S.M. & R.M. TYRRELL. 1989. Heme oxygenase is the major 32-kDa stress protein induced in human skin fibroblasts by UVA radiation, hydrogen peroxide, and sodium arsenite. Proc. Natl. Acad. Sci. USA **86:** 99–103.
23. HIRAI, K. *et al.* 2001. Mitochondrial abnormalities in Alzheimer's disease. J. Neurosci. **21:** 3017–3023.
24. CRAPPER MCLACHLAN, D.R. *et al.* 1991. Intramuscular desferrioxamine in patients with Alzheimer's disease. Lancet **337:** 1304–1308.
25. MCLACHLAN, D.R. *et al.* 1993. Desferrioxamine and Alzheimer's disease: video home behavior assessment of clinical course and measures of brain aluminum. Ther. Drug Monit. **15:** 602–607.
26. PERRY, G. *et al.* 1998. Reactive oxygen species mediate cellular damage in Alzheimer disease. J. Alzheimer's Dis. **1:** 45–55.
27. PERRY, G. *et al.* 2000. Amyloid-β junkies. Lancet **355:** 757.

Magnetic Iron Compounds in Neurological Disorders

JON DOBSON

Centre for Science and Technology in Medicine, Keele University, Hartshill, Stoke-on-Trent, United Kingdom

ABSTRACT: Although iron plays an important role in many aspects of human neurophysiology, it also can be toxic under certain circumstances. Anomalous amounts of iron are known to be associated with most types of neurodegenerative disorders such as Alzheimer's, Parkinson's, and Huntington's diseases. To date, little is known about the specific iron compounds present in this tissue and there is recent evidence to suggest that some forms are magnetic. This raises important questions with regard to the role of magnetic iron compounds in disease initiation and progression and, indeed, the origin of these compounds. This paper reviews recent work on the identification and analysis of magnetic iron compounds associated with neurological disorders.

KEYWORDS: iron; neurodegenerative; magnetite; epilepsy; Alzheimer's disease; Parkinson's disease; Huntington's disease

INTRODUCTION AND BACKGROUND

Iron is an important component of physiological processes not only in humans, but in all living organisms. Its ability to change valence gives it a vital role in energetic processes that rely on electron transport and also enables it to transport molecular oxygen around the human body via hemoglobin. However, in addition to these necessary physiological roles, iron can also be toxic. It is therefore important for the body to store iron in an accessible, but nontoxic form when it is not being used.

The primary form of iron storage in virtually all organisms (including humans) is in the core of the protein ferritin. The ferritin protein is a hollow spheroid shell approximately 12 nm in diameter made up of 24 subunits. The central void in the shell is 8 nm in diameter and is normally occupied by the iron biomineral, ferrihydrite, a hydrated iron oxide ($5Fe_2O_3 \cdot 9H_2O$) that generally contains only Fe^{3+} (ferric iron). Iron is transported into the ferritin shell via 3-fold and 4-fold channels at the junctions of the protein's subunits. During this process, potentially toxic Fe^{2+} (ferrous iron) is oxidized for storage as ferrihydrite. Up to 4500 Fe atoms can be stored in the core and this is the form in which most of the iron in the body is stored.[1]

Address for correspondence: Jon Dobson, Centre for Science and Technology in Medicine, Keele University, Thornburrow Drive, Hartshill, Stoke-on-Trent ST4 7QB, United Kingdom. Voice: +44-(0)-1782-554-253; fax: +44-(0)-1782-717-079.
  jdobson@keele.ac.uk

In humans, iron has been implicated in neurodegenerative disorders such as Alzheimer's disease (AD), Parkinson's disease (PD), and Huntington's disease (HD) (e.g., refs. 2 and 3). In the case of AD, elevated levels of iron have been known to be associated with the pathology for 50 years,[4] although the role of iron in these diseases and the nature of the specific iron compounds present are still not clear. In addition, these elevated iron levels do not necessarily correlate with elevated levels of ferritin or the extracellular iron transport protein transferrin.[5] In fact, in several regions of the brains of AD and PD patients, a reduction in transferrin, indicating reduced mobility and sequestration of iron, has been reported.[5,6]

BIOGENIC MAGNETITE IN THE BRAIN

Although the association of anomalous concentrations of iron with neurodegenerative tissue has been known since 1953, methods for assaying iron in diseased tissue have not advanced greatly since that time. Most staining techniques used to identify iron in histological sections are not quantitative and only reveal the presence of Fe^{3+}; however, recent progress has resulted in the identification of Fe^{2+} associated with neurodegenerative tissue.[2,7] This work has demonstrated, using modified iron staining techniques, that redox-active iron is closely associated with AD plaques and neurofibrillary tangles. This work also has shown that lesion-associated iron is distinct from iron sequestered in ferritin and has provided indirect evidence of the presence of Fe^{2+} in AD tissue. Despite these advances, iron anomalies associated with neurodegenerative tissue are not well characterized, and the nature and amount of various iron compounds present in general are not known.

In 1992, Joseph Kirschvink's group at the California Institute of Technology published results of highly sensitive Superconducting Quantum Interference Device (SQUID) magnetometry studies indicating that the iron oxide biomineral, magnetite, is present in human brain tissue, including AD tissue.[8] Magnetite (Fe_3O_4) is a ferrimagnetic iron oxide with alternating lattices of Fe^{2+} and Fe^{3+}, which are antiferromagnetically coupled. This alternation of ferric and ferrous lattices and their corresponding differences in the number of unpaired electron spins give magnetite its strong magnetization, particularly when compared to ferrihydrite, which is antiferromagnetic with an equal number of unpaired spins coupled in an antiparallel fashion. In ferrihydrite, the net magnetization arises only from defects in the crystal structure or uncompensated ("frustrated") surface spins.

Although Kirschvink's group went to great lengths to control for contamination, the results of this study were controversial, particularly as it was conducted on cadaver tissue and there was potential for postmortem changes in iron chemistry upon cell death. In order to examine the possibilities of contamination and postmortem changes in brain chemistry, our group undertook a series of studies on tissue removed from the human hippocampus. As these experiments were performed on tissue resected during amygdalohippocampectomies (a surgical procedure in which the damaged hippocampus of focal epilepsy patients is removed) and frozen at 77 K, postmortem artifacts could be controlled. In addition, we conducted experiments to determine the potential magnetic iron contamination levels in the samples and all results were compared with the same measurements made on cadaver tissue.

The results of this work demonstrated clearly that biogenic magnetite is present in human brain tissue and confirmed Kirschvink's earlier results, although we reported more variation in the magnetite concentration and higher levels.[9,10] This is due to the fact that measurements were made of the remanent magnetization (the magnetization remaining after the removal of an applied field) at different temperatures in the two studies. In our studies conducted at 77 K, very fine superparamagnetic magnetite particles (generally less than 50 nm in diameter) would contribute to the magnetization signal; in contrast, at 273 K (the temperature at which Kirschvink's studies were conducted), there would be no contribution from these particles and the measured magnetization would be smaller.

Although Kirschvink's group examined AD tissue and magnetite was found, they did not report a significant increase in the amount of magnetite in AD tissue vs. controls. However, in that study, the sample number was small and the degree of AD pathogenesis in the analyzed tissue was not clear. Also, as the remanent magnetization measurements were made at 273 K, the contribution to the sample's magnetization from any fine particles would not be detected. This is particularly important in light of later studies by our group.

EPILEPSY

In addition to the correlation between iron accumulation and neurodegenerative conditions associated with aging, iron also is known to induce epileptic activity, primarily as a result of intracranial bleeding due to head trauma. The model of iron-induced epilepsy was first introduced by Willmore and others in 1978 and the electrophysiological responses have been characterized in many studies since (e.g., refs. 11–13). More recently, evidence has been presented that shows that iron overload diseases may even predispose to epilepsy.[14] Although the initial seizure response to trauma and excess iron is relatively swift (e.g., ref. 15), the long-term consequences of intracranial bleeding on the formation of various iron compounds in the brain are unclear.

In order to examine the role that iron plays in epilepsy, our group undertook an investigation of biogenic magnetite in patients suffering with Mesial Temporal Lobe Epilepsy (MTLE). Although SQUID magnetometry measurements of tissue samples revealed that magnetite is present in the epileptogenic focus, increased levels of magnetite associated with MTLE were not demonstrated.[16] The results did show, however, that there was preliminary evidence of a difference in the distribution of particles within the epileptogenic tissue when compared to normal tissue. This was evidenced by the fact that particles' magnetic interactions were not as strong in the MTLE patients as in control tissue, an indication that they are possibly more homogeneously dispersed. It is not clear from this work that magnetite plays a role in epilepsy or that it is responsible for brain electrical responses observed during weak-field magnetic stimulation, but our investigations are continuing.[17–20]

AGING

As iron is known to be associated with neurodegenerative diseases that are characteristic of aging, all of the data on biogenic magnetite concentrations in the

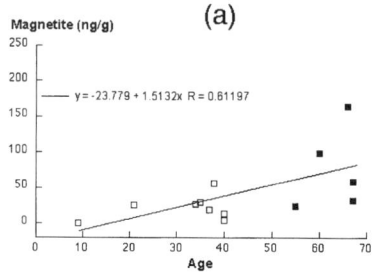 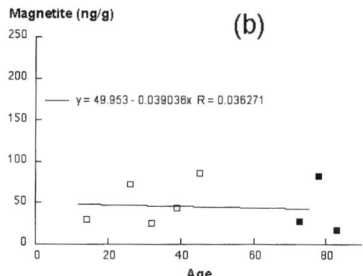

FIGURE 1. (**a**) Tissue concentration of magnetite as measured using IRM methods (ng of magnetite per g of tissue) vs. age for male subjects. *Solid squares* are nonepileptic subjects; *open squares* are epileptic subjects [$n = 13$]. (**b**) Tissue concentration of magnetite vs. age for female subjects. *Solid squares* are nonepileptic subjects; *open squares* are epileptic subjects [$n = 8$]. (Figure redrawn from ref. 21.)

human brain obtained by our group recently were analyzed as a function of age. All of the samples were taken from the hippocampus, a structure in the brain that is intimately associated with memory and learning and, therefore, is potentially most likely to play a role in age-related pathogenesis due to the presence of magnetite. These data included both MTLE patients and control tissue, and the subjects ranged in age from 9 to 78 years.[21]

When all samples were analyzed as a group, there was no significant correlation between age and biogenic magnetite concentration. However, when the samples were analyzed based on gender, there was a clear trend towards increasing magnetite level with age in the data for males, while female subjects demonstrated no such trend (FIG. 1).

It is interesting to note that, in rat models, increases in magnetite concentration in the brain were directly related to iron loading via diet and transfusion.[22] As women suffer periodic loss and reformation of iron-containing hemoglobin during their reproductive years (hence the high levels of iron deficiency in premenopausal women), the discrepancy in the data between men and women in this study could be a reflection of this process. However, as the trend in males was not highly significant, these results should be considered preliminary at this stage.

NEURODEGENERATIVE DISEASES

More recently, our group has developed new techniques for SQUID magnetometry studies of human brain tissue that have the advantage of allowing us to measure the total magnetite concentration rather than a temperature-dependent "apparent" concentration based on isothermal remanence as discussed previously. This is achieved by measuring the magnetization as a function of applied field (known as a hysteresis loop or M vs. H curve). As all materials respond to the application of a magnetic field, we have developed methods to model and subtract components of the signal due to diamagnetic tissue, paramagnetic signals such as heme iron, and the antiferromagnetic superparamagnet ferrihydrite (the magnetic core of the ferritin protein).

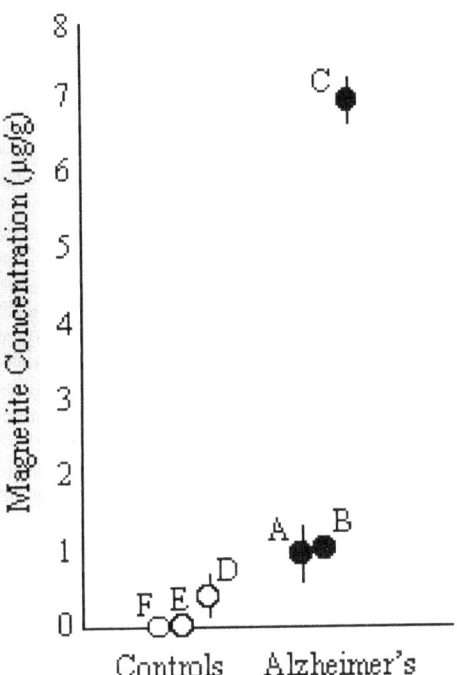

FIGURE 2. Tissue concentration of magnetite as measured using hysteresis methods for control subjects (*open circles*: subjects D–F) and AD subjects (*solid circles*: subjects A–C).

While this technique may not be quite as sensitive as measuring isothermal remanent magnetization (IRM) in the absence of an applied field (due to the fact that there are other nonremanence-carrying components contributing to the signal during hysteresis measurement), it should be, and appears to be, much better as a relative measure when comparing diseased tissue with controls.

We have recently employed this technique to evaluate the magnetic properties of pathogenic tissue from AD samples and compared this with age- and sex-matched controls.[23] The results of our first pilot study on tissue samples from three female AD subjects and three female controls demonstrated a significant increase in the level of magnetite in the AD subjects compared to controls (FIG. 2). One control had magnetite concentrations that were intermediate between the two groups. This subject had no clinical signs of AD, but pathological changes were revealed on autopsy suggestive of a neurodegenerative process (significant plaques and tangles). In addition, the sample with the highest concentration of magnetite came from the subject with the most aggressive and advanced form of AD at the time of death. At present, we are examining tissue from other neurodegenerative disorders, such as HD, PD, and neuroferritinopathy,[24] using these techniques.

POTENTIAL ROLE AND ORIGIN OF BIOGENIC MAGNETITE IN NEURODEGENERATIVE DISEASES

These results raise questions concerning the origin and physiological function of biogenic magnetite as well as the consequences of its presence. As these particles generate strong, local magnetic fields that may have an influence on biochemical reactions and provide a source of toxic Fe^{2+}, the effects of these particles on processes associated with neurodegenerative diseases should be examined. Theoretical studies have shown that magnetic fields can influence reaction yields and also lead to the formation of free radicals through stabilization of the triplet state (e.g., ref. 25). In addition, experimental evidence indicates that the presence of synthetic magnetite particles has a dramatic effect on the production of free radicals *in vitro*.[26,27]

In order to investigate the possible role of biogenic magnetite in neurodegenerative disease, we have begun investigations of magnetic iron in AD plaque material as well as experiments to examine the influence of the local magnetic fields generated by biogenic magnetite on amyloid-β plaque aggregation. Preliminary results indicate that plaque aggregation may be affected by strong, local magnetic fields (FIG. 3), although these experiments are still in progress.[28]

The origin of biogenic magnetite in the human brain is thus far unknown. It has been proposed that it may originate within the core of ferritin, perhaps through a malfunction in the way the protein handles iron,[29] and there is some evidence for this. If the ferritin core becomes overloaded or there is a breakdown in the protein's

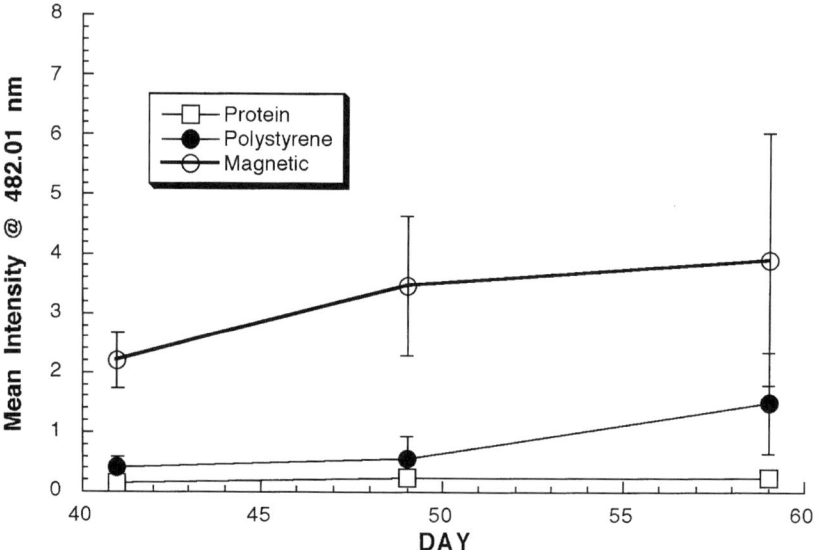

FIGURE 3. Results of thioflavin-T analysis of the effects of encapsulated magnetic particles on β-amyloid peptide aggregation. *Open squares* are solutions of β-amyloid (1–40) only. *Closed circles* represent peptide solution and latex particles. *Open circles* represent solutions of peptide with magnetite particles coated with latex. All concentrations are at 2 μM of β-amyloid in buffer [$n = 9$ per group]. (Figure modified after ref. 28.)

function, a mechanism for Fe^{2+} oxidation is lost. This process could lead to the formation of biogenic magnetite, which is more strongly magnetic than ferrihydrite.

Recent work by Carmen Quintana's group used high-resolution transmission electron microscopy and electron energy loss spectroscopy to examine ferritin in paired helical filaments from AD tissue and ferritin bound to aberrant tau filaments in neurodegenerative progressive supranuclear palsy. The results give a preliminary indication of the presence of a cubic iron oxide within the ferritin protein cage with spectra similar to synthetic magnetite standards.[30] It is possible that ferritin may act as a precursor for the formation of biogenic magnetite in humans, perhaps through excess loading of iron in the core and the breakdown of normal protein function. This core environment provides, under certain circumstances, both a source of ferrous and ferric iron as well as a potential nucleation site for magnetite growth.[31]

It is also possible that magnetite may play a secondary role in neurodegenerative diseases, particularly in relation to amyloid-β, the proteinaceous component of AD plaques. The role of Aβ in neurodegeneration is hotly debated (e.g., ref. 32). Recent evidence suggests that Aβ is toxic primarily through its complex interactions with redox-active iron, which can promote aggregation of the peptide.[33,34] However, it is also possible that, in some instances, Aβ plays a protective role by binding iron and reducing the production of iron-generated free radicals.[35] As we have observed magnetite in increased concentrations in AD tissue, it is possible that it is also forming as a by-product of the interaction of other forms of redox-active iron with Aβ due to the high iron concentrations within the plaque. In this case, as the protein aggregates, the mechanism for iron binding may be gradually lost, resulting in the production of magnetite. This would create a feedback mechanism by which toxic Fe^{2+} would be made available and local magnetic fields generated by the particles could influence free radical production as described previously. We are currently investigating these possibilities.

DISCUSSION

Anomalous iron concentrations are virtually ubiquitous in neurodegenerative diseases; however, its role is almost certainly very complicated and the compounds of iron present are not well known. Our most recent results indicate that biogenic magnetite accounts for at least some of the excess iron present in this diseased tissue. As discussed in this paper, magnetite has the potential to play an active role in disease progression and, in fact, in disease initiation. At this point, however, it is not clear whether it is a causative factor in these diseases or simply a by-product of disrupted iron homeostasis.

Regardless of its role, our recent data indicate that magnetite appears to be forming at the very earliest stages of the disease, at least in the case of AD. This result points to a possible beneficial aspect of magnetite biomineralization in neurodegenerative diseases. Early detection is one of the primary goals of neurodegenerative research efforts and MRI is currently used as a tool for this, but generally only for patients at a more advanced stage of the disease, where atrophy can be imaged as regions of *hyperintensity* (e.g., ref. 36). Although MRI analysis can be successful in identifying neurodegenerative diseases, the disease has normally progressed significantly by the time it is observed.

The presence of anomalous concentrations of biogenic magnetite may make it possible to detect the early stages of neurodegenerative diseases. As biogenic magnetite produces strong local magnetic fields, these fields will have an effect on proton relaxation rates in the surrounding tissue and should lead to *hypointensity* artifacts in MRI scans of patients suspected of AD. Alteration of MRI pulse sequences could be used to detect these accumulations of biogenic magnetite. In fact, these artifacts are well understood and form the basis for the use of biocompatible synthetic magnetite nanoparticles as contrast agents in MRI investigations (e.g., ref. 37). We are currently investigating this tool in gel phantoms and hope to progress to human studies in the near future.[38]

Investigations of biogenic magnetite and other iron compounds associated with neurodegenerative tissue should lead to a better understanding of their roles in these diseases. Such studies could give rise to new treatments as well as new diagnostic techniques as examined in this paper. It should be stressed, however, that our understanding of the role of iron in neurodegenerative diseases is still at an early stage. There is much more work to be done.

ACKNOWLEDGMENTS

The author is supported by a Royal Society–Wolfson Foundation Research Merit Award and would like to acknowledge the support of NIH Grant No. R01 AG02030-01 A1 and The Swiss National Science Funds. He also thanks Q. Pankhurst, D. Hautot, J. Collingwood, C. Batich, M. Davidson, H. G. Wieser, M. Fuller, C. Exley, P. Schultheiss-Grassi, and A. Mikhailov for their contributions to this research.

REFERENCES

1. HARRISON, P.M. & P. AROSIO. 1996. The ferritins: molecular properties, iron storage function, and cellular regulation. Biochim. Biophys. Acta **1275:** 161–203.
2. SMITH, M.A., P.L.R. HARRIS, L.M. SAYRE & G. PERRY. 1997. Iron accumulation in Alzheimer disease is a source of redox-generated free radicals. Proc. Natl. Acad. Sci. USA **94:** 9866–9868.
3. LOVELL, M.A., J.D. ROBERTSON, W.J. TEESDALE *et al.* 1998. Copper, iron, and zinc in Alzheimer's disease senile plaques. J. Neurol. Sci. **158:** 470–502.
4. GOODMAN, L. 1953. Alzheimer's disease—a clinicopathologic analysis of 23 cases with a theory on pathogenesis. J. Nerv. Ment. Dis. **118:** 97–130.
5. FISHER, P., M.E. GOTZ, W. DANIELCZYK *et al.* 1997. Blood transferrin and ferritin in Alzheimer's disease. Life Sci. **60:** 2273–2278.
6. LOEFFLER, D.A., J.R. CONNOR, P.L. JUNEAU *et al.* 1995. Transferrin and iron in normal, Alzheimer's disease, and Parkinson's disease brain regions. J. Neurochem. **65:** 710–716.
7. SAYRE, L.M., G. PERRY, P.L.R. HARRIS *et al.* 2000. *In situ* oxidative catalysis by neurofibrillary tangles and senile plaques in Alzheimer's disease: a central role for bound transition metals. J. Neurochem. **74:** 270–279.
8. KIRSCHVINK, J.L., A. KOBAYASHI-KIRSCHVINK & B.J. WOODFORD. 1992. Magnetite biomineralization in the human brain. Proc. Natl. Acad. Sci. USA **89:** 7683–7687.
9. DUNN, J.R., M. FULLER, J. ZOEGER *et al.* 1995. Magnetic material in the human hippocampus. Brain Res. Bull. **36:** 149–153.
10. DOBSON, J. & P.P. GRASSI. 1996. Magnetic properties of human hippocampal tissue: evaluation of artefact and contamination sources. Brain Res. Bull. **39:** 255–259.

11. WILLMORE, L.J., G.W. SYPERT & J.B. MUNSON. 1978. Recurrent seizures induced by cortical iron injection—model of post-traumatic epilepsy. Ann. Neurol. **4:** 329–336.
12. ENGSTROM, E.R., L. HILLERED, R. FLINK *et al.* 2001. Extracellular amino acid levels measured with intracerebral microdialysis in the model of posttraumatic epilepsy induced by intracortical iron injection. Epilepsy Res. **43:** 135–144.
13. UEDA, Y. & L.J. WILLMORE. 2000. Sequential changes in glutamate transporter protein levels during Fe3+-induced epileptogenesis. Epilepsy Res. **39:** 201–209.
14. IKEDA, M. 2001. Iron overload without the C282Y mutation in patients with epilepsy. J. Neurol. Neurosurg. Psychiatry **70:** 551–553.
15. UEDA, Y., L.J. WILLMORE & W.J. TRIGGS. 1998. Amygdalar injection of $FeCl_3$ causes spontaneous recurrent seizures. Exp. Neurol. **153:** 123–137.
16. SCHULTHEISS-GRASSI, P.P. & J. DOBSON. 1999. Magnetic analysis of human brain tissue. BioMetals **12:** 67–72.
17. FULLER, M.D., J. DOBSON, H.G. WIESER & S. MOSER. 1995. On the sensitivity of the human brain to magnetic fields: evocation of epileptiform activity. Brain Res. Bull. **36:** 155–159.
18. DOBSON, J., T.G. ST. PIERRE, P.P. SCHULTHEISS-GRASSI *et al.* 2000. Magnetic stimulation of epileptiform activity in mesial temporal lobe epilepsy patients. Brain Res. **868:** 386–391.
19. DOBSON, J., T.G. ST. PIERRE, H.G. WIESER & M. FULLER. 2000. Changes in paroxysmal brainwave patterns of epileptics by application of DC magnetic fields. Bioelectromagnetics **21:** 94–99.
20. FULLER, M.D. *et al.* 2003. In preparation.
21. DOBSON, J. 2002. Investigation of age-related variations in biogenic magnetite levels in the human hippocampus. Exp. Brain Res. **144:** 122–126.
22. PARDOE, H. & J. DOBSON. 1999. Magnetic iron biomineralization in rat brains: effects of iron loading. BioMetals **12:** 77–82.
23. HAUTOT, D., Q.A. PANKHURST, N. KAHN & J. DOBSON. 2003. Preliminary evaluation of nanoscale biogenic magnetite and Alzheimer's disease. Proc. R. Soc. B Biol. Lett. **270:** S62–S64.
24. CURTIS, A.R.J. *et al.* 2001. Mutation in the gene encoding ferritin light polypeptide causes dominant adult-onset basal ganglia disease. Nat. Genet. **28:** 350–354.
25. TIMMEL, C.R., U. TILL, B. BROCKLEHURST *et al.* 1998. Effects of weak magnetic fields on free radical recombination reactions. Mol. Phys. **95:** 71–89.
26. SCAIANO, J.C., S. MONAHAN & J. RENAUD. 1997. Dramatic effect of magnetite particles on the dynamics of photogenerated free radicals. Photochem. Photobiol. **65:** 759–762.
27. CHIGNELL, C.F. & R.H. SIK. 1998. Effect of magnetite particles on photoinduced and non-photoinduced free radical processes in human erythrocytes. Photochem. Photobiol. **68:** 598–601.
28. DOBSON, J., B. GROSS, C. EXLEY *et al.* 2001. Biomedical applications of biogenic and biocompatible magnetic nanoparticles. *In* Modern Problems of Cellular and Molecular Biophysics, pp. 121–130. Noyan Tapan/UNESCO. Yerevan.
29. DOBSON, J. 2001. Nanoscale biogenic iron oxides and neurodegenerative disease. FEBS Lett. **496:** 1–5.
30. QUINTANA, C., M. LANCIN, C. MARHIC *et al.* 2000. Preliminary high resolution TEM and electron energy loss spectroscopy studies of ferritin cores extracted from brain in patients with neurodegenerative PSP and Alzheimer diseases. Cell. Mol. Biol. **46:** 807–820.
31. ZHAO, G., F. BOU-ABDALLAH, P. AROSIO *et al.* 2003. Multiple pathways for mineral core formation in mammalian apoferritin: the role of hydrogen peroxide. Biochemistry **42:** 3142–3150.
32. SELKOE, D.J. 2001. Alzheimer's disease: genes, proteins, and therapy. Physiol. Rev. **81:** 741–766.
33. SCHUBERT, D. & M. CHEVION. 1995. The role of iron in beta amyloid toxicity. Biochem. Biophys. Res. Commun. **216:** 702–707.
34. ROTTKAMP, C.A., A.K. RAINA, X. ZHU *et al.* 2001. Redox-active iron mediates amyloid-β toxicity. Free Radical Biol. Med. **30:** 447–450.
35. BISHOP, G.M., S.R. ROBINSON, Q. LIU *et al.* 2002. Iron: a pathological mediator of Alzheimer's disease. Dev. Neurosci. **24:** 184–187.

36. CHETELAT, G. & J.C. BARON. 2003. Early diagnosis of Alzheimer's disease: contribution of structural neuroimaging. Neuroimage **18:** 525–541.
37. SJØGREN, C.E., K. BRILEY-SAEBØ, M. HANSON & C. JOHANSSON. 1994. Magnetic characterization of iron-oxides for magnetic resonance imaging. Magn. Reson. Med. **31:** 268–272.
38. PARDOE, H., W. CHUA-ANUSORN, T.G. ST. PIERRE & J. DOBSON. 2003. Detection limits for ferrimagnetic particle concentrations using magnetic resonance imaging–based proton transverse relaxation rate measurements. Phys. Med. Biol. **48:** 89–95.

The Relevance of Iron in the Pathogenesis of Parkinson's Disease

MARIO E. GÖTZ,[a] KAY DOUBLE,[b] MANFRED GERLACH,[c] MOUSSA B. H. YOUDIM,[d] AND PETER RIEDERER[e]

[a]*Department of Pharmacology and Toxicology, University of Würzburg, Würzburg, Germany*

[b]*Prince of Wales Medical Research Institute, Sydney, Australia*

[c]*Department of Child and Youth Psychiatry and Psychotherapy, University of Würzburg, Würzburg, Germany*

[d]*Eve Topf and National Parkinson Foundation Centers of Excellence for Neurodegenerative Diseases Research, and Department of Pharmacology, Technion-Faculty of Medicine, Haifa, Israel*

[e]*Department of Psychiatry, Clinical Neurochemistry, University of Würzburg, Würzburg, Germany*

ABSTRACT: Investigations that revealed increased levels of iron in postmortem brains from patients with Parkinson's disease (PD) as compared to those from individuals not suffering from neurological disorders are reported. The chemical natures in which iron predominates in the brain and the relevance of neuromelanin for neuronal iron binding are discussed. Major findings have been that iron levels increase with the severity of neuropathological changes in PD, presumably due to increased transport through the blood-brain barrier in late stages of parkinsonism. Glial iron is mainly stored as ferric iron in ferritin, while neuronal iron is predominantly bound to neuromelanin. Iron overload may induce progressive degeneration of nigrostriatal neurons by facilitating the formation of reactive biological intermediates, including reactive oxygen species, and the formation of cytotoxic protein aggregates. There are indications that iron-mediated neuronal death in PD proceeds retrogradely. These results are also discussed with respect to their relevance for disease progression in relation to cytotoxic α-synuclein protofibril formation.

KEYWORDS: Parkinson's disease; substantia nigra; dopamine; iron; transition metal; neuromelanin; α-synuclein; Lewy body; oxidative stress; radicals

TRANSITION METAL EPIDEMIOLOGY IN PARKINSON'S DISEASE

Several epidemiological studies suggest that long-term occupational and dietary metal exposure is associated with the occurrence of Parkinson's disease (PD).[1,2]

Address for correspondence: M. E. Götz, Dept. of Pharmacology, University of Kiel, Hospitalstr. 4, D-24105 Kiel, Germany. Voice: +49-431-597-3521; fax: +49-431-597-3522; *or* P. Riederer, Füchsleinstrasse 15, D-97080 Würzburg, Germany. Voice: +49-931-201-77200; fax: +49-931-201-77220. mgoetz@pharmakologie.uni-kiel.de *or* peter.riederer@mail.uni-wuerzburg.de

Gorell et al.[3,4] reported significant associations of PD with manganese and copper, as well as combinations of lead with copper and iron, and iron with copper, for workers with more than 20 years of occupational contact, and that an increased risk for patients with family history may exist.[5] A combination of manganese, iron, and aluminum might favor the development of PD after 30 years of exposure.[6] The only positive dose-response relationship was found between mercury exposure and PD, but not for other metals.[7,8] In contrast to Logroscino et al.[9] in a recent study, Powers et al.[2] reported a moderate association between iron intake from foods and PD and an apparent joint effect of iron and manganese. These epidemiological findings might point to a role of transition metals in PD, but cannot explain most cases of idiopathic parkinsonism. Thus, transition metal alterations described in the brain of parkinsonian patients and of other neurodegenerative diseases might, in addition to occupational exposure, be caused to a large degree by endogenous dysregulation of iron uptake, transport, distribution, and storage (TABLE 1). This chapter is mainly an overview of how iron is stored in the parkinsonian brain. In addition, we discuss the putative cellular consequences, a topic that has been dealt with in a series of reviews.[10–26]

IRON DEPOSITS IN THE PARKINSONIAN BRAIN

Iron is the most abundant transition metal in the body and has a unique distribution in the brain. A quantitative regional distribution of nonhemin iron in the normal brain was presented in 1958 by Hallgren and Sourander,[27] and confirmed in 1989 by Riederer et al.[28] Accordingly, iron levels range from 1.4 mg/100 g fresh weight in the medulla oblongata up to 21.3 mg/100 g fresh weight in the globus pallidus in individuals older than 29 years. The substantia nigra (SN), one of the most vulnerable brain regions in PD, was reported to contain 18.5 mg/100 g fresh weight. These values were confirmed to be in the same range when analyzed with extended X-ray fine-structure absorption.[29] Iron concentrations in the SN and the globus pallidus are higher than those of the liver, where iron is stored and disposed of to other tissues. Other brain regions where iron is found in high concentrations are the dentate gyrus, interpeduncular nucleus, thalamus, ventral pallidus, nucleus basalis, and red nucleus (FIG. 1).

Increased levels of iron in the brain of parkinsonian patients was demonstrated by Lhermitte et al.,[30] Earle,[31] Riederer et al.,[32] and Dexter et al.[33] The increase in non-hemin chelatable iron III[28,34] was not seen in patients with mild neuropathological changes in the SN. However, total iron content, determined following reduction of iron III to iron II by using the iron II chelator ferrozine and granulated ascorbic acid as a reductant for iron III, increased only in the SN in more severe cases of PD. This was not seen in incidental Lewy body disease, a neuropathologically classified disorder that is assumed to reflect presymptomatic cases of PD since they are devoid of clinical signs of parkinsonism, but exhibit cell loss in the SN and diffuse Lewy bodies postmortem.[35,36] A further study using iron chelators and spectrophotometry reported increased total iron and iron III content in the SN pars compacta from PD subjects, but not from Alzheimer's subjects.[37] Histochemical iron staining of paraffin sections using Perls' stain of iron III after pretreatment with ferrocyanide[38] revealed iron III in astrocytes, microglia, the walls of arterioles, and veins in putamen

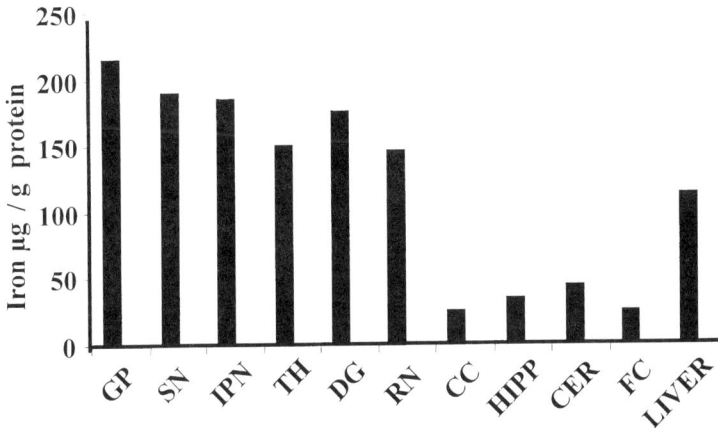

FIGURE 1. Distribution of iron in human brains. Terms: GP, globus pallidus; SN, substantia nigra; IPN, interpeduncular nucleus; TH, thalamus; DG, dentate gyrus; CC, cerebral cortex; HIPP, hippocampus; CER, cerebellum; FC, frontal cortex. Note that a similar type of distribution is also observed for the monkey, rat, and dog brains. (Figure adapted from ref. 120, with kind permission from Elsevier.)

and pallidum, but only rarely in neurons. In the SN pars compacta, iron III was localized in microglia, in astrocytes often localized next to neurons, and in single nonpigmented neurons. The selective increase in total iron in the parkinsonian SN pars compacta was detected as well, with inductively coupled plasma spectroscopy (ICP) and energy dispersive X-ray microanalysis (EDX) (TABLE 1).[33,39–44] Several other research groups, however, could not find significant changes in total iron content in the SN of parkinsonian patients using ICP, atomic absorption, emission (AAS), or Mössbauer spectroscopy (MS) (TABLE 1).[45–48] Explanations for the divergent findings of different groups can be based on the usage of brain tissue of different stages of neurodegeneration at the time of death, as well as different dissection and tissue-handling protocols. Moreover, sensitivities of methods vary considerably. For example, MS only measures ^{57}Fe, which is a low abundance isotope in brain tissue.[49,50] Other methods do not give total levels of iron, but detailed microstructural information such as EDX and laser microprobe mass analysis (LAMMA). Using EDX on an electron microscope working in the scanning transmission mode, neuromelanin-bound iron could be detected only in pigmented neurons of the SN pars compacta of parkinsonian patients.[42,43] Similarly, using the sensitive LAMMA technique, prominent iron and aluminum levels were seen to be associated with neuromelanin granules.[51,52] Probe sites directed to nonmelanized portions of cytoplasm of these cells or to the adjacent neuropil revealed lower concentrations of both metals with EDX and LAMMA. Further, the use of X-ray absorption fine-structure (EXAFS) and cryoelectron transmission microscopy confirmed the increased iron in parkinsonian nigra and the lateral globus pallidus, and revealed that ferritin was more heavily loaded with iron in PD when compared with age-matched controls.[29] When iron levels are correlated with dopamine concentrations, the most significant negative association was found between dopamine and iron III in the putamen and not in the

TABLE 1. Increased localized brain ferritin and iron deposits in neurodegenerative diseases

Aceruloplasminemia
Alzheimer's disease
Amyotrophic lateral sclerosis
Acquired immune deficiency syndrome
Friedreich's ataxia
Hallervorden syndrome
Huntington's chorea
Juvenile Parkinson's disease
Multiple sclerosis
Parkinson's disease
Prion diseases

SOURCE: Adapted from ref. 120, with kind permission from Elsevier.

SN in severe PD,[21] although in putamen total iron levels are not significantly different from controls (TABLE 1). In the globus pallidus, there is a subregional alteration of iron levels in the lateral versus the medial part (TABLE 2).[29,39,40] These findings may be indicative of retrograde degeneration of catecholaminergic neurons in PD. However, unlike the presence of oxidative stress and neurochemical changes reported for SN pars compacta, where iron is increased, the striatum is relatively unaltered biochemically.

The nigral increase in iron levels identified biochemically in the postmortem brain from parkinsonian patients appears to be confirmed and is related to the severity of the disease in the living patient as assessed by magnetic resonance imaging (MRI).[53–56] Inhomogeneities in the magnetic field induced by tissue attributes have an important effect on image contrast of water proton resonance in MRI. The changes in prominent contrast seen on T_2-weighted images correlate with sites of ferric iron that dephase the excited proton spins and thus lead to decreased T_2 relaxation times in MRI experiments. T_2-weighted images demonstrate prominent low-signal areas in the red nucleus and the SN pars reticulata, which may be regarded as indications for high ferric iron content since iron concentrations in the adult are much higher than those of paramagnetic manganese or copper.[57]

Further, several studies have suggested that transcranial ultrasound (TCS) may reflect alterations in metal constituents of the brain since PD patients exhibit a substantially increased echogenicity in the SN.[58] In more than 90% of PD patients, the SN is superimposed by extended white signals, reflecting increased echogenicity, which is mainly contralateral to the clinically more affected body side.[17,18,59–61] A recent postmortem study using brains from normal subjects at different ages suggests a relationship of SN echogenicity with higher levels of iron, L-ferritin, and H-ferritin, and a reduced neuromelanin concentration.[62] This molecular constellation of different iron species may describe a noxious cellular milieu promoting the generation of oxyradicals and cell damage. It may explain the increased susceptibility of subjects with SN hyperechogenicity for nigral injury, as demonstrated by positron emission tomography (PET) studies.

In summary, the reported data on iron deposits suggest that brain iron in PD is related to neurite degeneration proceeding in a retrograde manner to neuronal cell death as the disease progresses. A causative role of iron in neurodegeneration, however, cannot be implicated because results of the analytical approaches described are not sufficient for testing causal relationships.

TABLE 2. Alterations in levels of iron and of parameters proposed to reflect iron content (e.g., T_2 relaxation time in MRI or echogenicity in TCS) determined by different histochemical, biochemical, and physicochemical methods in PD

	SN	Putamen	Globus pallidus
Semiquantitative estimations			
Lhermitte (H)[30]	Normal	—	Increase
Earle (XF)[31]	General, but nonspecific increase in PD		
Jellinger (H)[41]	Increase (S)	(NS)	(NS)
Jellinger (EDX)[42]	Increase (S)	—	—
Gorell (MRI)[53]	Increase (S)	—	—
Ryvlin (MRI)[54]	Increase (S)	Decrease (S)	Decrease (S)
Becker (TCS)[58]		Basal ganglia increase (S)	
Ye (MRI)[55]		Basal ganglia increase (S)	
Quantitative determinations			
Riederer (SP)[32]	1.22 (a)	1.17 (a)	0.87 (a)
Dexter (ICP)[33]	1.35 (S)	—	—
Sofic (SP)[34]	1.89 (S)	0.81 (NS)	1.20 (NS)
Riederer (SP)[28]	1.77 (S)	0.81 (NS)	1.20 (NS)
Uitti (AAS)[45]	1.07 (NS)	—	—
Dexter[39]	1.34 (S)(t)	1.04 (NS)(l)	0.71 (S)(l)
Dexter (ICP)[40]	1.30 (S)(pc)	0.83 (NS)(m)	0.71 (S)(m)
Sofic (SP)[37]	1.50 (S)	—	—
Hirsch (XMA)[44]	3.40 (S)(b)	—	—
Good (LAMMA)[51]	1.45 (S)(c)	—	—
Mann (ICP)[46]	1.56 (S)	—	—
Galazka-Friedman (MS)[48]	0.98 (NS)	—	—
Loeffler (Col)[47]	0.82 (NS)	0.64 (NS)	1.80 (NS)
Griffiths (EXAFS)[29]	2.01 (S)	1.23 (NS)	1.43 (S)(l)
Griffiths & Crossman[129]			0.69 (S)(m)

NOTE: Values express fold-change of arithmetic mean values compared to control means. Terms: (S) statistically significant; (NS) statistically not significant; (a) no statistics calculated; (b) on zones lacking neuromelanin; (c) on neuromelanin granule; (—) not determined; (pc) substantia nigra pars compacta; (l) lateral; (m) medial; (t) total; (AAS) atomic absorption spectroscopy; (Col) colorimetry; (EDX) energy dispersive X-ray analysis; (EXAFS) extended X-ray absorption fine structure; (H) histological staining; (ICP) inductively coupled plasma spectroscopy; (LAMMA) laser microprobe mass analysis; (MRI) magnetic resonance imaging; (MS) Mössbauer spectroscopy; (SP) spectrophotometry; (TCS) transcranial ultrasound; (XF) X-ray fluorescence spectroscopy; (XMA) X-ray microprobe analysis.

SOURCES OF IRON AND ITS INTERACTION WITH NEUROMELANIN

Following transferrin-mediated transport, the highest levels of iron in the brain are bound to ferritin and neuromelanin. Ferritin is a 450-kDa protein with 24 subunits forming a cavity that can store up to 4500 atoms of ferric iron. It is expressed in oligodendrocytes, astrocytes, and microglia, but hardly in neurons.[63,64] In PD, the number of ferritin-immunoreactive microglial cells in the SN increases, with many reactive microglial cells located in close proximity to melanin-containing or degenerating neurons.[41,42,65] Ferritin cores in the SN of PD patients are reported to be denser and contain more iron than those in the SN of healthy subjects.[29] As long as iron is bound to ferritin, cytotoxic reactions are not expected. However, iron can be released from ferritin by various exogenous[66–69] and endogenous substances via reductive mechanisms.[70,71] If released from ferritin, low molecular iron II complexes may undergo redox reactions resulting in cytotoxic damage of proteins, DNA, or lipids.[72,73] *In vitro* experiments revealed that ortho-dihydroxyphenyl compounds such as the endogenous neurotransmitter dopamine potently release iron II from ferritin.[69,74] Since ferritin is located predominantly in glial cells,[63] oxidative stress from glial cells might be a secondary event following high dopamine turnover, neuronal cell dysfunction, and death.

An interesting intraneuronal iron source or iron sink is neuromelanin. Ben-Shachar *et al.*[75] demonstrated for the first time that synthetic dopamine-derived melanin has a capacity to bind a highly significant amount of iron at two sites, and recently this has been confirmed by the studies of Double *et al.*[76] Neuromelanin contains large amounts of iron in the SN of patients with PD,[42,43] a dark pigment produced in catecholaminergic neurons of the human SN and locus coeruleus, and is generally regarded as the result of the reaction of oxidized catechols with a variety of nucleophiles, including thiols from glutathione and proteins. The involvement of enzymes in neuromelanin synthesis is still under debate.[77] This pigment may be different in some respects as compared to epidermal melanocytes and synthetic dopamine melanins.[78] It appears that iron is bound to catecholic groups. EPR studies showed that, in the SN, the ferric iron is bound to neuromelanin as a high-spin complex with an octahedral configuration.[79–81] MS demonstrates that ferric iron is bound to ferritin-like oxyhydroxide clusters.[26,48,50] Neuromelanin is only about 50% saturated with iron III in the SN. The remaining binding capacity can be used for further metal binding or binding of other cationic compounds, such as paraquat and methyl-phenylpyridinium (MPP$^+$), as well as of dopamine oxidation products. Thus, neuromelanin may have a cytoprotective role as long as transition metals are bound to it, or it may trigger cytotoxicity if iron is released within neurons, depending on the reducing cellular environment and probably other yet unknown factors.

The number of pigmented and nonpigmented neurons is constant in normal aging, while in PD there is a preferential loss of pigmented neurons.[82] Precise analysis of the dopaminergic neurons in the midbrain demonstrates that the loss of dopaminergic neurons in the mesencephalon is not uniform. Some dopaminergic cell groups are more vulnerable than others. The degree of cell loss is severe in the SN pars compacta, intermediate in the ventral tegmental area and cell group A8, but nonexistent in the central gray substance. This heterogeneity provides a good paradigm for analyzing the factors implicated in the differential vulnerability of dopaminergic neurons. So far, neurons that degenerate have been shown to contain neuromelanin, high amounts of iron, and no calbindin 28K, and are poorly protected against oxidative stress. By contrast, neurons that survive in PD are free of neuromelanin, are calbindin D28–positive, contain low amounts of iron, and are better protected against oxidative stress.[14]

The absolute concentration of nigral neuromelanin is less than 50% in PD with respect to age-matched controls.[83] This may lead to the conclusion that neuromelanin might be somehow involved in neurotoxicity in PD. On the other hand, it is discussed that, in PD, either not enough neuromelanin is produced or the structure of neuromelanin is changed. Hence, the ability to bind cytotoxic components is reduced and thus neuromelanin would be rather a neuroprotectant than a neurotoxicant in PD brains. This view is supported by neuropathological investigations that show that pigmented cells in PD contain less neuromelanin compared with those in control brains.[84] Furthermore, it was reported[85] that the more vulnerable nigral ventral tier cells contain less neuromelanin than the more heavily pigmented cells in the dorsal tier, suggesting that neuromelanin may confer an advantage upon the cells in which it is found. Further support for the hypothesis that neuromelanin can act as a protective substance comes from *in vitro* studies. Human neuromelanin can significantly reduce both basal and iron-stimulated lipid peroxidation in rat cortical homogenates.[75,86] Thus, neuromelanin may attenuate oxidative damage by directly interacting with, and inactivating, free radical species. The ability of neuromelanin to bind potentially damaging species such as transition metals may represent another mechanism by which it is neuroprotective. However, if high-affinity binding capacity is overcome, low-affinity binding may result in iron overload prone to leak back iron into the cytosol.[75,76] Alternatively, it is possible that iron bound to neuromelanin may remain redox-active. Although ferrous iron bound to synthetic melanin is reported to be not very efficient for decomposing hydrogen peroxide *in vitro*,[87] redox-active iron has recently been reported on neuromelanin granules in both the normal and the PD brain.[88] It is suggested from structural analysis using nuclear magnetic resonance spectroscopy and electron paramagnetic resonance spectroscopy that neuromelanin isolated from parkinsonian brain may have decreased ability to bind iron.[89,90] Consequently, a reduction in iron binding ability of neuromelanin would render catecholaminergic neurons more vulnerable due to increased redox-capable iron levels.[20] In fact, Faucheux *et al.*[91] investigated the redox activity of neuromelanin aggregates in a group of parkinsonian patients, who presented a statistically significant reduction (−70%) in the number of melanized neurons and an increased nonheme iron III content as compared with a group of matched control subjects. The level of redox activity detected in neuromelanin aggregates was significantly increased (+69%) in parkinsonian patients and was highest in patients with the most severe neuronal loss. This change was not observed in tissue in the immediate vicinity of melanized neurons. These results strongly support the hypothesis that neuromelanin-iron interactions are involved in neurodegeneration in PD. However, whether neuromelanin acts as a toxifying or detoxifying component in PD depends on the human SN intracellular environment and, most probably, the stage of the disease.

Additional important clues to the understanding of elevated cerebral iron are provided by the findings of mutated genes relevant for iron metabolism in other neurodegenerative disorders (TABLE 3), such as HFE, the gene most commonly mutated in patients with hereditary hemochromatosis, and TFR2, a mutated transferrin receptor, which is important in non-HFE-associated iron-overload disorders. How these mutated genes operate and interact to induce abnormal brain iron metabolism, trans-

TABLE 3. Disorders of iron metabolism and their expressed mutated genes

19-133	Defect in ferritin accumulation (relevance in PD)
DCYTB	Mutated iron transporter
Ferroportin 1	Iron exporter across membrane
FTL	Ferritin light polypeptide (abnormal ferritin accumulation)
HFE	Hemochromatosis protein (iron metabolism, relevance in PD)
Ireg1	Iron regulated transporter gene
IRP2	Iron regulated protein 2 (degeneration, ataxia, bradykinesia)
MTP1	Mouse transition protein 1
NRAMP1	Iron transporter (relevance in multiple sclerosis)
NRAMP2	Iron importer
P53	Mutations in Wilson's disease, PD, and hemochromatosis
PANK2	Pantothenate kinase defect (PD and Hallervorden syndrome)
Sfxn1	Mutated transmembrane protein (important for iron utilization)
SLC11A3	Ferroprotein gene
Tfc2	Mutated transferrin receptor (relevance in Alzheimer's disease)
Tfr2	Mutated transferrin receptor (linked to neurodegeneration)
YFH1	Friedreich's ataxia (mitochondrial iron accumulation, neurodegeneration)

SOURCE: Adapted from ref. 119, with kind permission from Springer-Verlag, and from ref. 120, with kind permission from Elsevier.

port, and accumulation in the CNS, and how iron in turn interacts with proteins (e.g., α-synuclein), causing them to aggregate into toxic inclusions, are points under current investigation, as emphasized below.

CONSEQUENCES OF IRON OVERLOAD IN THE PARKINSONIAN SN

There are suggestive findings that iron is triggering protein aggregation in Lewy bodies, the major neuropathological hallmark of PD. Lewy bodies contain many lipids and proteins, including ubiquitin, tyrosine hydroxylase, nitrosylated iron regulatory protein 2 (IRP2), and α-synuclein. The relation of mutations in α-synuclein and familial PD[92–94] and the accumulation of α-synuclein in Lewy bodies[95–97] suggest that α-synuclein may be part of the pathogenetic mechanism at least in familial PD. α-Synuclein associated with the presynaptic vesicles tends to self-aggregation; however, as long as it is associated with the presynaptic membrane, it is not toxic. Recent studies showed that it forms stable aggregates in the presence of iron or calcium.[98–100] Iron-catalyzed oxidative reactions (FIG. 2) convert α-helical α-synuclein conformation into β-pleated sheet conformation, which is the form of synuclein found in Lewy bodies and glial cytoplasmic inclusions.[101–104] Further, it has been shown *in vitro* that α-synuclein promotes the production of hydroxyl radicals, as detected with spin trapping agents by electron spin resonance spectroscopy in the presence of iron II.[99,105] Cross-linking of α-synuclein may in addition be promoted by advanced

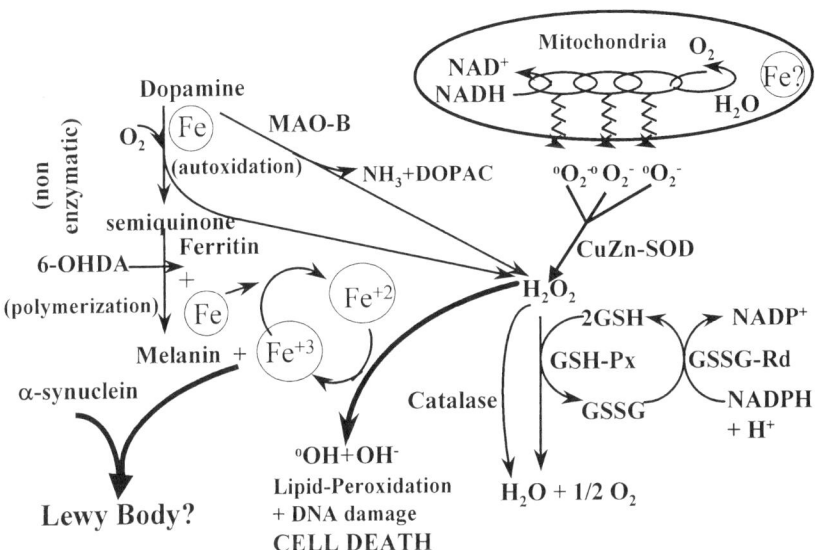

FIGURE 2. The oxidative stress hypothesis of PD. Similar mechanisms involving iron-induced and hydrogen peroxide–induced oxidative stress have been put forward for other neurodegenerative diseases. Terms: MAO-B, monoamine oxidase B; CuZn-SOD, copper and zinc–containing superoxide dismutase; GSH, glutathione; GSH-Px, GSH-peroxidase; GSSG, glutathione disulfide; GSSG-Rd, GSSG-reductase; DOPAC, 3,4-dihydroxyphenyl-acetate; 6-OHDA, 6-hydroxydopamine; NAD, nicotine amide adenine dinucleotide; H_2O_2, hydrogen peroxide; ˙OH, hydroxyl radical; ˙O_2^-, superoxide radical anion; OH^-, hydroxide anion. (Figure adapted from ref. 120, with kind permission from Elsevier.)

glycation end products.[106] Interestingly, formation of advanced glycation end products is accelerated by iron since the reaction products formed (Amadori products) from reducing sugars with primary or secondary amino groups are further oxidized. These advanced glycation end products may be involved in protein cross-linking in various neurodegenerative diseases.[107]

Very interestingly, catecholamines stabilize oligomers by covalent ligation.[108] Thus, iron-catalyzed oxidation of catechols may lead to α-synuclein protofibrils. Protofibrils of α-synuclein interact with membranes of vesicles and mitochondria, leading to transient permeabilization and leakage of transmitters or mitochondrial factors, triggering cell death.[109–111] In addition, α-synuclein may bind iron, leading to aggregation of α-synuclein.[25,112,113] It is mainly in this process that the oligomers, not the mature aggregates of α-synuclein, appear to be selectively toxic to dopaminergic neurons.[114–118]

One of the most significant features of 6-hydroxydopamine and MPTP-induced dopaminergic neurotoxicity is that both these neurotoxins induce a selective increase of iron in SN pars compacta of rats, mice, and monkeys[119–125] (see Youdim et al., this volume). Pretreatment with iron chelators, desferal and VK-28, is neuroprotective against these neurotoxins[123–125] and prevents the appearance of iron, IRP2 (iron responsive protein 2), and α-synuclein in SN pars compacta[119,124] (FIG. 3). These

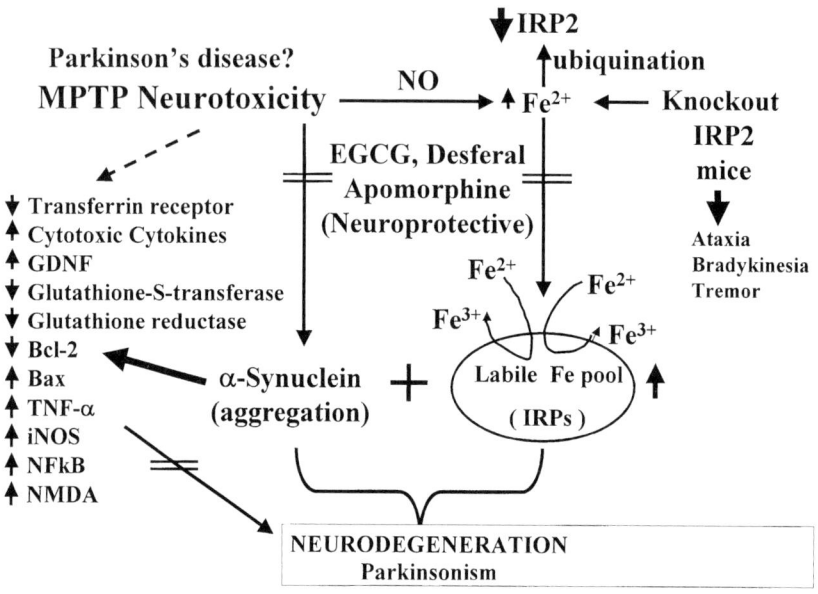

FIGURE 3. Proposed mechanism of nigrostriatal dopamine neuron degeneration following chronic MPTP exposure to mice involving iron metabolism. Induction of iNOS may lead to NO-mediated inactivation of IRP2 by ubiquitination, and consequently to increased cellular iron in a reactive form. This labile iron pool alone or in combination with α-synuclein aggregates may lead to cytotoxicity via oxidative stress and inflammatory response. Terms: IRP, iron regulatory protein; NO, nitric oxide, EGCG, epigallocatechin-3-gallate; GDNF, glia-derived neurotrophic factor; iNOS, inducible nitric oxide synthase; TNF, tumor necrosis factor; NFkB, nuclear factor κB; NMDA, N-methyl-D-aspartate. (Figure adapted from ref. 119, with kind permission from Springer-Verlag, and from ref. 120, with kind permission from Elsevier.)

results might encourage clinical studies in the future, with the aim of slowing down disease progression in PD and possibly other neurodegenerative diseases as well.

In conclusion, there might be interactions of several pathogenetic pathways that are active on the background of normal human brain aging and aggravated by individual susceptibility that lead to the histopathological changes in PD because, so far, complete pathology cannot be mimicked in mice that solely overexpress α-synuclein[126,127] or following intranigral iron injections to rats.[121–125] Future research using viral vectors to increase cerebral overexpression of putatively pathogenetic proteins in rodents may elucidate the complex interactions between iron, oxidative stress, and protein aggregation in neurodegeneration[128] and hopefully pave the way for the development of novel therapeutic strategies to combat neurodegeneration.

SUMMARY

Histopathological, biochemical, and *in vivo* brain imaging techniques, such as MRI and transcranial color-coded real-time sonography, revealed a consistent increase of brain iron in PD. Increased iron deposits in the SN in PD may have several causes that may reside in disturbances of iron uptake, storage, and transport as neurodegeneration progresses. Major iron stores are ferritin and hemosiderin in glial cells as well as neuromelanin in neurons. Age- and disease-dependent overload of iron storage proteins may result in iron release upon reduction. Consequently, the low-molecular-weight chelatable iron complexes may trigger redox reactions, leading to damage of biomolecules. These processes may also promote disease progression by enabling aggregation of susceptible proteins such as α-synuclein. Additionally, upon neurodegeneration, there is strong microglial activation, which can be another source of high iron concentrations in the brain. Although the current evidence suggests that increased brain iron may be a secondary result of neuronal degeneration and reactive gliosis in several neurological disorders affecting the basal ganglia, including progressive supranuclear palsy and multisystem atrophy, the question whether iron is a major progressor of neuronal death can still not be excluded.

REFERENCES

1. LAI, B.C.L., S.A. MARION, K. TESCHKE et al. 2002. Occupational and environmental risk factors for Parkinson's disease. Parkinson's Rel. Dis. **8:** 297–309.
2. POWERS, K.M., T. SMITH-WELLER, G.M. FRANKLIN et al. 2003. Parkinson's disease risks associated with dietary iron, manganese, and other nutrient intakes. Neurology **60:** 1761–1766.
3. GORELL, J.M., C.C. JOHNSON, B.A. RYBICKI et al. 1997. Occupational exposures to metals as risk factors for Parkinson's disease. Neurology **48:** 650–658.
4. GORELL, J.M., C.C. JOHNSON, B.A. RYBICKI et al. 1999. Occupational exposure to manganese, copper, lead, iron, mercury, and zinc and the risk of Parkinson's disease. Neurotoxicology **20:** 239–247.
5. RYBICKI, B.A., C.C. JOHNSON, E.L. PETERSON et al. 1999. A family history of Parkinson's disease and its effect on other PD risk factors. Neuroepidemiology **18:** 270–278.
6. ZAYED, J., S. DUCIC, G. CAMPANELLA et al. 1990. Facteurs environnementaux dans l'etiologie de la maladie de Parkinson. Can. J. Neurol. Sci. **17:** 286–291.
7. NGIM, C.H. & G. DEVATHASAN. 1989. Epidemiologic study on the association between body burden mercury level and idiopathic Parkinson's disease. Neuroepidemiology **8:** 128–141.
8. SEIDLER, A., W. HELLENBRAND, B.P. ROBRA et al. 1996. Possible environmental, occupational, and other etiologic factors for Parkinson's disease: a case control study in Germany. Neurology **46:** 1275–1284.
9. LOGROSCINO, G., K. MARDER, J. GRAZIANO et al. 1998. Dietary iron, animal fats, and risk of Parkinson's disease. Mov. Disord. **13**(suppl. 1): 13–16.
10. YOUDIM, M.B. & A.R. GREEN. 1976. Biogenic monoamine metabolism and functional activity in iron-deficient rats: behavioural correlates. Ciba Found. Symp. **7–9:** 201–225.
11. YOUDIM, M.B.H. 1985. Brain iron metabolism: biochemical and behavioural aspects of iron in relation to dopaminergic neurotransmission. *In* Handbook of Neurochemistry. Volume 10, pp. 731–756. Plenum. New York.
12. YOUDIM, M.B., D. BEN-SHACHAR & P. RIEDERER. 1989. Is Parkinson's disease a progressive siderosis of substantia nigra resulting in iron and melanin induced neurodegeneration? Acta Neurol. Scand. Suppl. **126:** 47–54.
13. YOUDIM, M.B., D. BEN-SHACHAR & P. RIEDERER. 1993. The possible role of iron in the etiopathology of Parkinson's disease. Mov. Disord. **8:** 1–12.
14. HIRSCH, E.C. 1994. Biochemistry of Parkinson's disease with special reference to the dopaminergic systems. Mol. Neurobiol. **9:** 135–142.
15. JELLINGER, K.A. 1999. The role of iron in neurodegeneration: prospects for pharmacotherapy of Parkinson's disease. Drugs Aging **14:** 115–140.

16. JELLINGER, K.A. 2003. General aspects of neurodegeneration. J. Neural Transm. Suppl. **65:** 101–144.
17. BERG, D., M. GERLACH, M.B.H. YOUDIM *et al.* 2001. Brain iron pathways and their relevance to Parkinson's disease. J. Neurochem. **79:** 225–236.
18. BERG, D., W. ROGGENDORF, U. SCHRÖDER *et al.* 2002. Echogenicity of the substantia nigra: association with increased iron content and marker for susceptibility to nigrostriatal injury. Arch. Neurol. **59:** 999–1005.
19. DOUBLE, K.L., M. GERLACH, M.B.H. YOUDIM & P. RIEDERER. 2000. Impaired iron homeostasis in Parkinson's disease. J. Neural Transm. Suppl. **60:** 37–58.
20. DOUBLE, K.L., D. BEN-SHACHAR, M.B.H. YOUDIM *et al.* 2002. Influence of neuromelanin on oxidative pathways within the human substantia nigra. Neurotoxicol. Teratol. **24:** 621–628.
21. GERLACH, M., D. BEN-SHACHAR, P. RIEDERER *et al.* 1994. Altered brain metabolism of iron as a cause of neurodegenerative diseases. J. Neurochem. **63:** 793–807.
22. GERLACH, M., K.L. DOUBLE, D. BEN-SHACHAR *et al.* 2003. Neuromelanin and its interaction with iron as a potential risk factor for dopaminergic neurodegeneration underlying Parkinson's disease. Neurotoxicol. Res. **5:** 35–44.
23. PERRY, G., L.M. SAYRE *et al.* 2002. The role of iron and copper in the aetiology of neurodegenerative disorders: therapeutic implications. CNS Drugs **16:** 339–352.
24. SIPE, J.C., P. LEE & E. BEUTLER. 2002. Brain iron metabolism and neurodegenerative disorders. Dev. Neurosci. **24:** 188–196.
25. WOLOZIN, B. & N. GOLTS. 2002. Iron and Parkinson's disease. Neuroscientist **8:** 22–32.
26. FRIEDMAN, A. & J. GALAZKA-FRIEDMAN. 2001. The current state of free radicals in Parkinson's disease: nigral iron as a trigger of oxidative stress. *In* Parkinson's Disease. Volume 86, pp. 137–142. Lippincott/Williams & Wilkins. Philadelphia.
27. HALLGREN, B. & P. SOURANDER. 1958. The effect of age on the non-haemin iron in the human brain. J. Neurochem. **3:** 41–51.
28. RIEDERER, P., E. SOFIC, W-D. RAUSCH *et al.* 1989. Transition metals, ferritin, glutathione, and ascorbic acid in parkinsonian brains. J. Neurochem. **52:** 515–520.
29. GRIFFITHS, P.D., B.R. DOBSON, G.R. JONES *et al.* 1999. Iron in the basal ganglia in Parkinson's disease: an *in vitro* study using extended X-ray absorption fine structure and cryo-electron microscopy. Brain **122:** 667–673.
30. LHERMITTE, J., W.M. KRAUS & D. MCALPINE. 1924. On the occurrence of abnormal deposits of iron in the brain in parkinsonism with special reference to its localisation. J. Neurol. Psychopathol. **5:** 195–208.
31. EARLE, K.M. 1968. Studies in Parkinson's disease including X-ray fluorescent spectroscopy of formalin-fixed tissues. J. Neuropathol. Exp. Neurol. **27:** 1–14.
32. RIEDERER, P., E. SOFIC, W.D. RAUSCH *et al.* 1985. Dopaminforschung heute und morgen—L-Dopa in der Zunkuft. *In* L-Dopa Substitution der Parkinson-Krankheit, Geschichte-Gegenwart-Zukunft, pp. 127–144. Springer-Verlag. Wien.
33. DEXTER, D.T., F.R. WELLS *et al.* 1987. Increased nigral iron content in post-mortem parkinsonian brain. Lancet **341:** 1219–1220.
34. SOFIC, E., P. RIEDERER, H. HEINSEN *et al.* 1988. Increased iron(III) and total iron content in post-mortem substantia nigra of parkinsonian brain. J. Neural Transm. **74:** 199–205.
35. DEXTER, D.T., J. SIAN, S. ROSE *et al.* 1994. Indices of oxidative stress and mitochondrial function in individuals with incidental Lewy body disease. Ann. Neurol. **35:** 38–44.
36. OWEN, A.D., A.H.V. SCHAPIRA, P. JENNER *et al.* 1997. Indices of oxidative stress in Parkinson's disease, Alzheimer's disease, and dementia with Lewy bodies. J. Neural Transm. Suppl. **51:** 167–173.
37. SOFIC, E., W. PAULUS, K. JELLINGER *et al.* 1991. Selective increase of iron in substantia nigra zona compacta of parkinsonian brains. J. Neurochem. **56:** 978–982.
38. GOMORI, G. 1936. Microtechnical demonstration of iron. Am. J. Pathol. **12:** 655–663.
39. DEXTER, D.T., F.R. WELLS, A.J. LEES *et al.* 1989. Increased nigral iron content and alterations in other metal ions occurring in brain in Parkinson's disease. J. Neurochem. **52:** 1830–136.
40. DEXTER, D.T., A. CARAYON, F. JAVOY-AGID *et al.* 1991. Alterations in the levels of iron, ferritin, and other trace metals in Parkinson's disease and other neurodegenerative diseases affecting the basal ganglia. Brain **114:** 1953–1975.

41. JELLINGER, K., W. PAULUS, I. GRUNDKE-IQBAL et al. 1990. Brain iron and ferritin in Parkinson's and Alzheimer's disease. J. Neural Transm. Parkinson's Dis. Dementia Sect. **2:** 327–340.
42. JELLINGER, K., E. KIENZL, G. RUMPELMAIR et al. 1992. Iron-melanin complex in substantia nigra of parkinsonian brains: an X-ray microanalysis. J. Neurochem. **59:** 1168–1171.
43. JELLINGER, K., E. KIENZL, G. RUMPELMAIR et al. 1993. Iron and ferritin in substantia nigra in Parkinson's disease. In Parkinson's Disease. Volume 60, pp. 267–272. Raven Press. New York.
44. HIRSCH, E.C., J-P. BRANDEL, P. GALLE et al. 1991. Iron and aluminum increase in the substantia nigra of patients with Parkinson's disease: an X-ray microanalysis. J. Neurochem. **56:** 446–451.
45. UITTI, R.J., A.H. RAJPUT, B. ROZDILSKY et al. 1989. Regional metal concentrations in Parkinson's disease, other chronic neurological diseases, and control brains. Can. J. Neurol. Sci. **16:** 310–314.
46. MANN, V.M., J.M. COOPER, S.E. DANIEL et al. 1994. Complex I, iron, and ferritin in Parkinson's disease substantia nigra. Ann. Neurol. **36:** 876–881.
47. LOEFFLER, D.A., J.R. CONNOR et al. 1995. Transferrin and iron in normal, Alzheimer's disease, and Parkinson's disease brain regions. J. Neurochem. **65:** 710–716.
48. GALAZKA-FRIEDMAN, J., E.R. BAUMINGER, A. FRIEDMAN et al. 1996. Iron in parkinsonian and control substantia nigra: a Mössbauer spectroscopy study. Mov. Disord. **11:** 8–16.
49. GERLACH, M., K. DOUBLE, P. RIEDERER et al. 1997. Iron in the parkinsonian substantia nigra. Mov. Disord. **12:** 258–260.
50. GERLACH, M., A.X. TRAUTWEIN, L. ZECCA et al. 1995. Mössbauer spectroscopic studies of human neuromelanin isolated from the substantia nigra. J. Neurochem. **65:** 923–926.
51. GOOD, P., C.W. OLANOW & D.P. PERL. 1992. Neuromelanin-containing neurons of the substantia nigra accumulate iron and aluminum in Parkinson's disease: a LAMMA study. Brain Res. **593:** 343–346.
52. IANCU, T.C., D.P. PERL, I. STERNLIEB et al. 1996. The application of laser microprobe mass analysis to the study of biological material. Biometals **9:** 57–65.
53. GORELL, J.M., R.J. ORDIDGE, G.G. BROWN et al. 1995. Increased iron-related MRI contrast in the substantia nigra in Parkinson's disease. Neurology **45:** 1138–1143.
54. RYVLIN, P., E. BROUSSOLLE, H. PIOLLET et al. 1995. Magnetic resonance imaging evidence of decreased putamenal iron content in idiopathic Parkinson's disease. Arch. Neurol. **52:** 583–588.
55. YE, F.Q., P.S. ALLEN & W.R.W. MARTIN. 1996. Basal ganglia iron content in Parkinson's disease measured with magnetic resonance. Mov. Disord. **11:** 243–249.
56. BARTZOKIS, G., J.L. CUMMINGS, C.H. MARKHAM et al. 1999. MRI evaluation of brain iron in earlier- and later-onset Parkinson's disease and normal subjects. Magn. Reson. Imaging **17:** 213–222.
57. MARTIN, W.R.W. 2001. Magnetic resonance imaging and spectroscopy in Parkinson's disease. In Parkinson's Disease. Volume 86, pp. 197–203. Lippincott/Williams & Wilkins. Philadelphia.
58. BECKER, G., J. SEUFERT, U. BOGDAHN et al. 1995. Degeneration of substantia nigra in chronic Parkinson's disease visualized by transcranial color-coded real-time sonography. Neurology **45:** 182–184.
59. BERG, D., G. BECKER, B. ZEILER et al. 1999. Vulnerability of the nigrostriatal system as detected by transcranial ultrasound. Neurology **53:** 1026–1031.
60. BERG, D., C. SIEFKER, P. RUPRECHT-DÖRFLER et al. 2001. Echo pattern of substantia nigra and its relevance for motor function and motility in elderly subjects. Neurology **56:** 13–17.
61. WALTER, U., M. WITTSTOCK, R. BENECKE et al. 2002. Substantia nigra echogenicity is normal in non-extrapyramidal cerebral disorders, but increased in Parkinson's disease. J. Neural Transm. **109:** 191–196.
62. ZECCA, L., D. BERG, T. ARZBERGER et al. 2004. The in vivo detection of iron and neuromelanin by transcranial sonography: a new approach for early detection of substantia nigra damage. Neurology. Submitted.
63. CONNOR J., S. MENZIES, S. ST. MARTIN et al. 1990. The cellular distribution of transferrin, ferritin, and iron in the human brain. J. Neurosci. Res. **27:** 595–611.

64. ZECCA, L., M. GALLORINI, V. SCHÜNEMANN et al. 2001. Iron, neuromelanin, and ferritin content in the substantia nigra of normal subjects at different ages: consequences for iron storage and neurodegenerative processes. J. Neurochem. **76:** 1766–1773.
65. RIEDERER, P., W.D. RAUSCH, B. SCHMIDT et al. 1988. Biochemical fundamentals in Parkinson's disease. Mt. Sinai J. Med. **55:** 21–28.
66. MONTEIRO, H., G. VILLE & C. WINTERBOURN. 1989. Release of iron from ferritin by semiquinone, anthracycline, bipyridyl, and nitroaromatic radicals. Free Radical Biol. Med. **6:** 587–591.
67. LAPENNA, D., S. DEGIOIA, G. CIOFANI et al. 1995. Captopril induces iron release from ferritin and oxidative stress. J. Pharm. Pharmacol. **47:** 1–6.
68. LINERT, W., E. HERLINGER, R.F. JAMESON et al. 1996. Dopamine, 6-hydroxydopamine, iron, and dioxygen: their mutual interactions and possible implication in the development of Parkinson's disease. Biochim. Biophys. Acta **1316:** 160–168.
69. DOUBLE, K.L., M. MAYWALD, M. SCHMITTEL et al. 1997. In vitro studies of ferritin iron release and neurotoxicity. J. Neurochem. **70:** 2492–2499.
70. MONTEIRO, H. & C. WINTERBOURN. 1988. The superoxide-dependent transfer of iron from ferritin to transferrin and lactoferrin. Biochem. J. **256:** 923–928.
71. BOYER, R., T. GRABILL & R. PETROVICH. 1988. Reactive release of ferritin iron: a kinetic assay. Anal. Biochem. **174:** 17–22.
72. HALLIWELL, B. & J. GUTTERIDGE. 1986. Iron and free radical reactions: two aspects of antioxidant protection. Trends Biol. Sci. **11:** 1372–1375.
73. GÖTZ, M.E., G. KÜNIG, P. RIEDERER et al. 1994. Oxidative stress: free radical production in neural degeneration. Pharmacol. Ther. **63:** 37–122.
74. DOUBLE, K.L., P. RIEDERER & M. GERLACH. 1998. The role of iron in 6-hydroxydopamine neurotoxicity. Adv. Neurol. **80:** 287–296.
75. BEN-SHACHAR, D., P. RIEDERER & M.B.H. YOUDIM. 1991. Iron-melanin interaction and lipid peroxidation: implications for Parkinson's disease. J. Neurochem. **57:** 1609–1614.
76. DOUBLE, K.L., M. GERLACH, V. SCHÜNEMANN et al. 2003. Iron-binding characteristics of neuromelanin of the human substantia nigra. Biochem. Pharmacol. **66:** 489–494.
77. ZECCA, L., D. TAMPELLINI, M. GERLACH et al. 2001. Substantia nigra neuromelanin: structure, synthesis, and molecular behaviour. Mol. Pathol. **54:** 414–418.
78. DOUBLE, K.L., L. ZECCA et al. 2000. Structural characteristics of human substantia nigra neuromelanin and synthetic dopamine melanins. J. Neurochem. **75:** 2583–2589.
79. ZECCA, L. & H.M. SWARTZ. 1993. Total and paramagnetic metals in human substantia nigra and its neuromelanin. J. Neural Transm. Parkinson's Dis. Dementia Sect. **5:** 203–213.
80. ZECCA, L., T. SHIMA, A. STROPPOLO et al. 1996. Interaction of neuromelanin and iron in substantia nigra and other areas of human brain. Neuroscience **73:** 407–415.
81. SHIMA, T., T. SARNA, A. STROPPOLO et al. 1997. Binding of iron to neuromelanin of human substantia nigra and synthetic melanin: an electron paramagnetic resonance spectroscopy study. Free Radical Biol. Med. **23:** 110–119.
82. HIRSCH, E.C., A.M. GRAYBIEL & Y.A. AGID. 1988. Melanized dopaminergic neurons are differentially susceptible to degeneration in Parkinson's disease. Nature **334:** 345–348.
83. ZECCA, L., R. FARIELLO, P. RIEDERER et al. 2002. The absolute concentration of nigral neuromelanin, assayed by a new sensitive method, increases throughout the life and is dramatically decreased in Parkinson's disease. FEBS Lett. **510:** 216–220.
84. KASTNER, A., E. HIRSCH, O. LEJEUNE et al. 1992. Is the vulnerability of neurons in the substantia nigra of patients with Parkinson's disease related to their neuromelanin content? J. Neurochem. **59:** 1080–1089.
85. GIBB, W. 1992. Melanin, tyrosine hydroxylase, calbindin, and substance P in the human midbrain and substantia nigra in relation to nigrostriatal projections and differential neuron susceptibility in Parkinson's disease. Brain Res. **581:** 283–291.
86. DOUBLE, K.L., P. RIEDERER & M. GERLACH. 1999. The significance of neuromelanin in Parkinson's disease. Drug News Dev. **12:** 333–340.
87. PILAS, B., T. SARNA, B. KALYANARAMAN et al. 1988. The effect of melanin on iron associated decomposition of hydrogen peroxide. Free Radical Biol. Med. **4:** 285–293.
88. CASTELLANI, R., S. SIEDLAK, G. PERRY et al. 2000. Sequestration of iron by Lewy bodies in Parkinson's disease. Acta Neuropathol. **100:** 111–114.

89. AIME, S., B. BERGAMASCO, M. CASU et al. 2000. Isolation and ^{13}C-NMR characterization of an insoluble proteinaceous fraction from substantia nigra of patients with Parkinson's disease. Mov. Disord. **15:** 977–981.
90. LOPIANO, L., M. CHIESA, D. DIGILIO et al. 2000. Q-band EPR investigations of neuromelanin in control and Parkinson's disease patients. Biochim. Biophys. Acta **1500:** 306–312.
91. FAUCHEUX, B.A., M.E. MARTIN, C. BEAUMONT et al. 2003. Neuromelanin associated redox-active iron is increased in the substantia nigra of patients with Parkinson's disease. J. Neurochem. **86:** 1142–1148.
92. POLYMEROPOULOS, M., C. LAVEDAN, E. LEROY et al. 1997. Mutation in the α-synuclein gene identified in families with Parkinson's disease. Science **276:** 2045–2047.
93. KRÜGER, R., W. KUHN, T. MÜLLER et al. 1998. Ala30Pro mutation in the gene encoding α-synuclein in Parkinson's disease. Nat. Genet. **18:** 106–108.
94. MIZUNO, Y., N. HATTORI, T. KITADA et al. 2001. Familial Parkinson's disease, α-synuclein, and parkin. *In* Parkinson's Disease. Volume 86, pp. 13–21. Lippincott/Williams & Wilkins. Philadelphia.
95. SPILLANTINI, M.G., M.L. SCHMIDT, V.M. LEE et al. 1997. α-Synuclein in Lewy bodies. Nature **388:** 839–840.
96. SPILLANTINI, M.G., R.A. CROWTHER, R. JAKES et al. 1998. α-Synuclein in filamentous inclusions of Lewy bodies from Parkinson's disease and dementia with Lewy bodies. Proc. Natl. Acad. Sci. USA **95:** 6469–6473.
97. MARKOPOULOU, K., Z. WSZOLEK, R. PFEIFFER et al. 1999. Reduced expression of the G209A α-synuclein allele in familial parkinsonism. Ann. Neurol. **46:** 374–381.
98. OSTREROVA-GOLTS, N., L. PETRUCELLI et al. 2000. The A53T α-synuclein mutation increases iron-dependent aggregation and toxicity. J. Neurosci. **20:** 6048–6054.
99. TURNBULL, S., B.J. TABNER, O.M.A. EL-AGNAF et al. 2001. α-Synuclein implicated in Parkinson's disease catalyses the formation of hydrogen peroxide *in vitro*. Free Radical Biol. Med. **30:** 1163–1170.
100. NIELSEN, M.S., H. VORUM, E. LINDERSSON et al. 2001. Ca^{2+} binding to α-synuclein regulates ligand binding and oligomerization. J. Biol. Chem. **276:** 22680–22684.
101. HASHIMOTO, M., L.J. HSU, Y. XIA et al. 1999. Oxidative stress induces amyloid-like aggregate formation of NACP/α-synuclein *in vitro*. Neuroreport **10:** 717–721.
102. HASHIMOTO, M., E. ROCKENSTEIN & E. MASLIAH. 2003. Transgenic models of α-synuclein pathology. Ann. N.Y. Acad. Sci. **991:** 171–188.
103. PAIK, S.R., H. SHIN, J. LEE et al. 1999. Copper(II)-induced self oligomerization of α-synuclein. Biochem. J. **340:** 821–828.
104. PAIK, S.R., D.Y. LEE, H.J. CHO et al. 2003. Oxidized glutathione stimulated the amyloid formation of α-synuclein. FEBS Lett. **537:** 63–67.
105. TABNER, B.J., S. TURNBULL, O.M.A. EL-AGNAF et al. 2002. Formation of hydrogen peroxide and hydroxyl radicals from Aβ and α-synuclein as a possible mechanism of cell death in Alzheimer's disease and Parkinson's disease. Free Radical Biol. Med. **32:** 1076–1083.
106. MÜNCH, G., H.J. LÜTH, A. WONG et al. 2000. Crosslinking of α-synuclein by advanced glycation endproducts—an early pathophysiological step in Lewy body formation? J. Chem. Neuroanat. **20:** 253–257.
107. MÜNCH, G., J, GASIC-MILENKOVIC & T. ARENDT. 2003. Effect of advanced glycation endproducts on cell cycle and their relevance for Alzheimer's disease. J. Neural Transm. Suppl. **65:** 63-71.
108. CONWAY, K.A., J.C. ROCHET, R.M. BIEGANSKI et al. 2001. Kinetic stabilization of the alpha-synuclein protofibril by a dopamine alpha-synuclein adduct. Science **294:** 1346–1349.
109. TABRIZI, S.J., M. ORTH, J.M. WILKINSON et al. 2001. Expression of mutant α-synuclein causes increased susceptibility to dopamine toxicity. Hum. Mol. Genet. **9:** 2683–2689.
110. HSU, L.J., Y. SAGARRA, A. ARROYO et al. 2001. α-Synuclein promotes mitochondrial deficit and oxidative stress. Am. J. Pathol. **157:** 401–410.
111. SAHA, A.R., N.N. NINKINA, D.P. HANGER et al. 2001. Induction of neuronal death by α-synuclein. Eur. J. Neurosci. **12:** 3073–3077.

112. KIM, K.S., S.Y. CHOI, H.Y. KWON *et al.* 2002. Aggregation of α-synuclein induced by the Cu,Zn–superoxide dismutase and hydrogen peroxide system. Free Radical Biol. Med. **32:** 544–550.
113. YOUDIM, M.B.H., G. STEPHENSEN & D. BEN SHACHAR. 2004. Ironing iron out in Parkinson's disease and other neurodegenerative diseases with iron chelators: a lesson from 6-hydroxydopamine and iron chelators, desferal and VK-28. This volume.
114. FORLONI, G., I. BERTANI, A.M. CALELLA *et al.* 2001. α-Synuclein and Parkinson's disease: selective neurodegenerative effect of α-synuclein fragment on dopaminergic neurons *in vitro* and *in vivo*. Ann. Neurol. **47:** 632–640.
115. LO BIANCO, C., J.L. RIDET, B.L. SCHNEIDER *et al.* 2002. α-Synucleinopathy and selective dopaminergic loss in a rat lentiviral-based model of Parkinson's disease. Proc. Natl. Acad. Sci. USA **99:** 10813–10818.
116. VOLLES, M.J., S.J. LEE, J-C. ROCHET *et al.* 2001. Vesicle permeabilization by protofibrillar α-synuclein: implications for the pathogenesis and treatment of Parkinson's disease. Biochemistry **40:** 7812–7819.
117. VOLLES, M.J. & P.T. LANSBURY, JR. 2002. Vesicle permeabilization by protofibrillar α-synuclein is sensitive to Parkinson's disease–linked mutations and occurs by a pore-like mechanism. Biochemistry **41:** 4595–4602.
118. VOLLES, M.J. & P.T. LANSBURY, JR. 2003. Zeroing in on the pathogenic form of alpha-synuclein and its mechanism of neurotoxicity in Parkinson's disease. Biochemistry **42:** 7871–7878.
119. YOUDIM, M.B.H. 2003. What have we learnt from cDNA microarray gene expression studies about the role of iron in MPTP-induced neurodegeneration and Parkinson's disease. J. Neural Transm. Suppl. **65:** 73–88.
120. YOUDIM, M.B.H. & P. RIEDERER. 1999. Iron in the brain, normal and pathological. *In* Elsevier's Encyclopedia of Neuroscience. Second edition, pp. 984–987. Elsevier. Amsterdam.
121. OESTREICHER, E., G.J. SENGSTOCK, P. RIEDERER *et al.* 1994. Degeneration of nigrostriatal dopaminergic neurons increases iron within the substantia nigra: a histochemical and neurochemical study. Brain Res. **660:** 8–18.
122. SENGSTOCK, G.J., N.H. ZAWIA, C.W. OLANOW *et al.* 1997. Intranigral iron infusion in the rat: acute elevations in nigral lipid peroxidation and striatal dopaminergic markers with ensuing nigral degeneration. Biol. Trace Elem. Res. **58:** 177–195.
123. BEN-SHACHAR, D., G. ESHEL, J.P. FINBERG *et al.* 1991. The iron chelator desferrioxamine (Desferal) retards 6-hydroxydopamine-induced degeneration of nigrostriatal dopamine neurons. J. Neurochem. **56:** 1441–1444.
124. BEN-SHACHAR, D., N. KAHANA, L. KAMPEL *et al.* 2004. Neuroprotection by a novel brain permeable iron chelator, VK-28, against 6-hydroxydopamine lesion in rats. Neuropharmacology **46:** 254–263.
125. BEN-SHACHAR, D. & M.B. YOUDIM. 1991. Intranigral iron injection induces behavioural and biochemical "parkinsonism" in rats. J. Neurochem. **57:** 2133–2135.
126. MATSUOKA, Y., M. VILA, S. LINCOLN *et al.* 2001. Lack of nigral pathology in transgenic mice expressing human α-synuclein driven by the tyrosine hydroxylase promoter. Neurobiol. Dis. **8:** 535–539.
127. DONG, Z., B. FERGER, J. FELDON *et al.* 2002. Overexpression of Parkinson's disease associated alpha-synuclein A53T by recombinant adeno-associated virus in mice does not increase the vulnerability of dopaminergic neurons to MPTP. J. Neurobiol. **53:** 1–10.
128. BJÖRKLUND, A. & D. KIRIK. 2003. Modeling CNS neurodegeneration by overexpression of disease-causing proteins using viral vectors. TINS **26:** 386–392.
129. GRIFFITHS, P.D. & A.R. CROSSMAN. 1993. Distribution of iron in the basal ganglia and neocortex in postmortem tissue in Parkinson's disease and Alzheimer's disease. Dementia **4:** 61–65.

Manganese-Induced Parkinsonism and Parkinson's Disease

C. W. OLANOW

Department of Neurology, Mount Sinai School of Medicine, New York, New York, USA

ABSTRACT: It has long been appreciated that manganese exposure can cause neurotoxicity and a neurologic syndrome that resembles Parkinson's disease (PD). Current evidence indicates that manganese-induced parkinsonism can be differentiated from PD because of its predilection to accumulate in and damage the pallidum and striatum rather than the SNc. The clinical syndrome, response to levodopa, imaging studies with MRI and PET, and pathologic features all help to distinguish these two conditions and permit the correct diagnosis to be established. This is of particular relevance in differentiating patients with parkinsonism due to manganese intoxication from patients with idiopathic PD who have incidental manganese exposure.

KEYWORDS: Parkinson's disease (PD); parkinsonism; manganese; features; clinical; imaging; pathologic; levodopa

INTRODUCTION

Manganese is a paramagnetic heavy metal that is widely distributed in the environment, being present in air, water, and food. It is the 12th most common element in the Earth's crust and the 4th most widely used metal in the world. More that 8 million tons of manganese metal are extracted annually and more that 90% is employed in the manufacture of steel, where it imparts hardness to the metal. Manganese is also employed in the manufacture of batteries, in water purification, in bactericidal and fungicide agents, and as an antiknock agent in gasoline. It has long been appreciated that manganese exposure can cause neurotoxicity and a neurologic syndrome that resembles Parkinson's disease (PD). There has, however, been confusion in trying to sort out patients with parkinsonism induced by manganese from patients with idiopathic PD and incidental exposure to manganese. There is now, though, considerable evidence indicating that manganese preferentially damages different areas of the brain from those that are affected in PD and that these entities can be readily differentiated based on the consequent clinical, pharmacologic, imaging, and pathologic features.[1]

Address for correspondence: C. W. Olanow, M.D., FRCPC, Professor and Chair, Department of Neurology, Mount Sinai School of Medicine, 1 Gustave Levy Place, Annenberg 14-94, New York, NY 10029. Voice: 212-241-8435; fax: 212-987-7635.
warren.olanow@mssm.edu

MANGANESE NEUROTOXICITY: HISTORY

Manganese neurotoxicity was first described in 1837 in 5 patients who worked in a manganese ore crushing plant in France.[2] These patients were reported to have muscle weakness, bent posture, whispering speech, limb tremor, and salivation. Several cases of manganese neurointoxication were subsequently reported over the next 150 years, particularly in miners, smelters, and workers involved in the manufacture of dry batteries.[3–21] Manganese toxicity has also been described in patients receiving long-term parental nutrition[22] and following potassium permanganate ingestion.[23]

The clinical features of manganese neurotoxicity include psychiatric features, parkinsonism, and dystonia. Patients with extreme exposure are reported to have suffered acute behavioral disturbances, hallucinations, and psychoses, referred to as "manganese madness" or "locura manganica".[10,13] Extrapyramidal features are the more common manifestation and, in well-defined cases, include gait dysfunction with a propensity to fall backward, postural instability, bradykinesia, rigidity, micrographia, masked facies, and speech disturbances.[20,21,24] Tremor is less common and, when present, tends to be postural or kinetic rather than resting as seen in PD.[5,6,13,20,21,24] Patients with manganese-induced parkinsonism also frequently experience characteristic forms of dystonia consisting of facial grimacing and/or plantar flexion of the foot, which interferes with gait and is known as "coq au pied" or "cock-walk". In general, there is little or no response to levodopa.[20,21,24] While this is the typical clinical picture, the problem of diagnosis is confused by reports of manganese-exposed individuals who have a clinical picture that more closely resembles PD and includes resting tremor and a good response to levodopa.[25–27] There was thus confusion in trying to ascertain whether these patients did in fact have parkinsonism due to manganese intoxication or PD with incidental exposure to manganese. Differentiating these conditions is an important practical problem as there are many individuals with exposure to manganese in the workplace, and PD is a common disorder that affects 1–2% of adults independent of whether or not they have been exposed to manganese.

The ability to distinguish manganese-induced parkinsonism from PD has recently been clarified because of (a) delineation of the clinical syndromes associated with damage to the substantia nigra pars compacta (SNc) and damage to the striatum and globus pallidus (GP); (b) clinical, pharmacologic, and brain imaging studies in a cohort of Taiwanese patients with well-defined manganese-induced parkinsonism; (c) behavioral, pharmacological, imaging, and pathological studies of manganese intoxication in nonhuman primates; and (d) studies of manganese-induced parkinsonism in patients with liver failure. This paper will review these various features and propose criteria for distinguishing manganese-induced parkinsonism from PD.

PD AND MANGANESE-INDUCED PARKINSONISM

Clinicopathologic Correlations (Nigral vs. Pallidal Parkinsonism)

A pathologic review of brains from 100 patients diagnosed during life as having idiopathic PD detected a misdiagnosis rate of approximately 25%.[28] In this study, classical PD pathology with degeneration of dopamine neurons in the SNc and intracellular inclusions or Lewy bodies was most accurately predicted during life by a clinical picture characterized by resting tremor, asymmetry, and a good response to levodopa.[29] In contrast, patients with atypical parkinsonism with damage primarily localized to the pallidum and striatum have a clinical picture characterized by early involvement of speech, gait, and balance; a relative absence of resting tremor; lack of asymmetry; and a poor response to levodopa.[30] MRI studies similarly suggest that patients with PD, who have primary involvement of the SNc, could be differentiated from patients with atypical parkinsonism based on the presence of resting tremor, asymmetry, and a good response to levodopa.[31] In these studies, parkinsonian patients with MRI studies indicative of striatal/pallidal degeneration had a syndrome marked by the early appearance of gait and balance disturbance, speech impairment, absence of resting tremor, and little or no response to levodopa. Indeed, prospective studies demonstrate that MRI abnormalities indicative of degeneration of the striatum predict a poor response to levodopa and the evolution of atypical parkinsonism rather than PD.[32] These observations indicate that the clinical picture can differentiate patients with PD where there is degeneration of SNc dopaminergic neurons from those with parkinsonism related to damage primarily affecting the striatum and/or pallidum. Differences in the response to levodopa in these two groups likely relate to the fact that PD patients have degeneration of dopaminergic neurons in the nigra, but preservation of dopamine receptors on striatal neurons, and are thus capable of responding to levodopa. In contrast, patients with parkinsonism due to degeneration of the striatum or pallidum have damage to downstream neurons and receptors that preclude a satisfactory response to levodopa.

Degeneration of dopamine neurons in the SNc coupled with intracytoplasmic Lewy bodies and a loss of striatal dopamine are the pathologic hallmarks of PD.[33] PD is also associated with degeneration and Lewy bodies in other regions including the locus coeruleus (LC), the nucleus basalis of Meynert (NBM), the dorsal motor nucleus of the vagus (DMNV), as well as selected neurons of the cerebral cortex, spinal cord, and peripheral nervous system.[34] In contrast, there is substantial evidence that the primary sites of damage following manganese neurotoxicity are the GP and the substantia nigra pars reticularis (SNr), while the nigrostriatal system and other regions affected in PD are relatively spared. Manganese intoxication in rodents, primates, and humans is associated with neuronal loss and gliosis, which are most prominent in the GP (especially the medial segment), the SNr, and to a lesser extent the striatum, with sparing of SNc neurons and striatal dopamine.[35–41] Indeed, pallidal neurons have been shown to be selectively vulnerable to manganese intoxication,[42] while manganese does not increase nigrostriatal damage or dopamine depletion in MPTP-lesioned mice or 6-OHDA-lesioned rats.[42,43] Additionally, damage caused by manganese is not associated with Lewy body formation, the characteristic histologic feature of PD, nor does manganese cause damage to the LC, NBM, DMNV, or other sites that are specifically affected in PD. Manganese levels are highest in the GP in normal individuals[44,45] and manganese preferentially accumulates in the pallidum following systemic administration to monkeys.[46,47] MRI studies in manganese-intoxicated patients demonstrate a characteristic signal abnormality on T1-weighted studies that is not seen in normal individuals or in patients with PD or other forms of parkinsonism (FIG. 1). MRI studies following manganese administration in monkeys similarly note signal changes that are primarily located in the GP

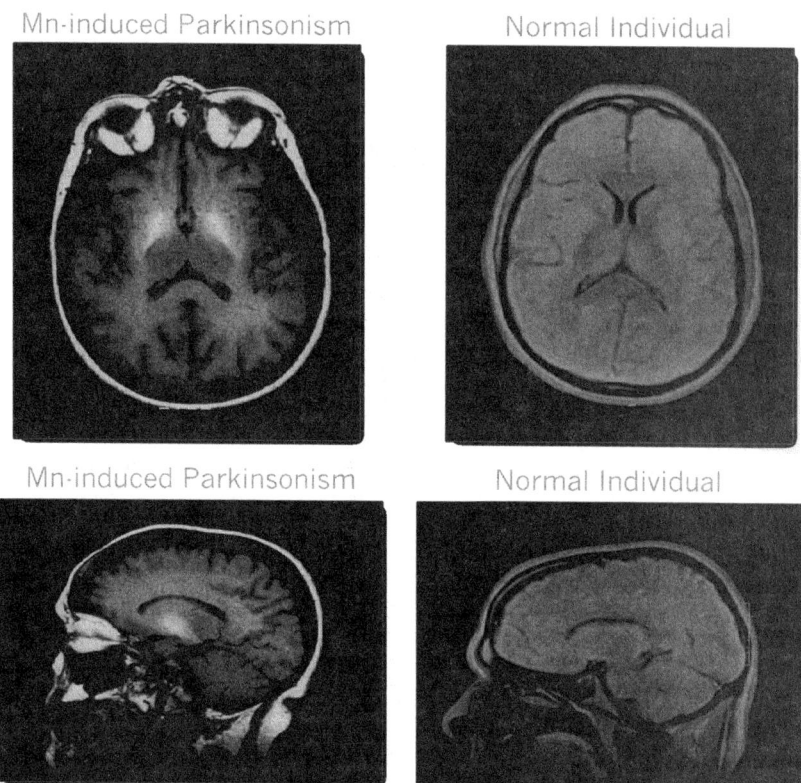

FIGURE 1. T1-weighted MRI studies in a normal patient and a patient with manganese-induced parkinsonism: **(top)** transverse sections and **(bottom)** sagittal sections. Note the bilateral, symmetric, high signal abnormalities in the globus pallidus in the patient with manganese-induced parkinsonism. Similar changes are also noted in the substantia nigra pars reticularis. This is a characteristic signature of manganese accumulation. Patients with PD and other forms of parkinsonism have normal T1-weighted MRI studies and do not show the signal abnormalities seen with manganese accumulation.

and striatum.[48,49] In these studies, no pathologic changes were noted in the SNc and striatal dopamine levels were preserved.[40] Finally, striatal fluorodopa (FD) uptake on positron emission tomography (PET), a measure of the integrity of the nigrostriatal system, is consistently reduced in PD, especially in the posterior putamen,[50,51] but is normal in instances when parkinsonism is induced by manganese toxicity (FIG. 2).[52–55]

These findings demonstrate that in contrast to PD, which preferentially damages dopamine neurons in the SNc, manganese preferentially accumulates within and damages the pallidum and striatum, while sparing the nigrostriatal system. This difference in the pattern of neuronal degeneration in patients with PD and those with basal ganglia damage due to manganese neurotoxicity suggests that these conditions can be differentiated based on their clinical picture, response to levodopa, imaging

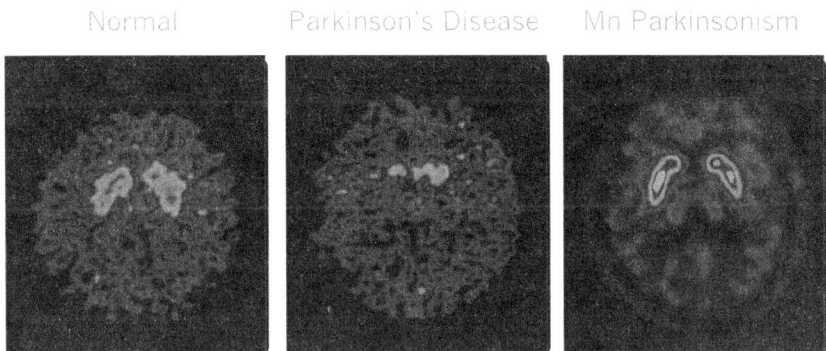

FIGURE 2. PET measure of striatal fluorodopa (FD) uptake in a normal individual, a patient with PD, and a patient with parkinsonism due to manganese intoxication. Note that PD is associated with a marked reduction in striatal FD uptake, which primarily affects the posterior putamen. This reflects the loss of dopamine neurons in the SNc and their striatal terminals where dopamine is stored. In contrast, striatal FD uptake is normal in patients with manganese-induced parkinsonism, reflecting the preservation of the nigrostriatal system. [Figure courtesy of Donald Calne.]

studies, and findings at pathology. Patients with PD who have primary degeneration in SNc neurons would be expected to have a parkinsonian picture characterized by resting tremor and a good response to levodopa. In contrast, patients with manganese-induced parkinsonism where damage is primarily localized to the pallidum and striatum would be expected to have a clinical picture characterized by early gait and balance impairment, speech dysfunction, absence of resting tremor, and a poor response to levodopa.

The Taiwanese Cohort

To test this hypothesis, we had the opportunity to examine a cohort of patients who worked in a manganese smelting plant in Taiwan and who developed parkinsonism that was almost certainly due to manganese intoxication.[19,24,54] Six of 13 individuals chronically exposed to a very high ambient concentration of manganese (in excess of 27 mg/m^3) developed a basal ganglia syndrome characterized by gait dysfunction with particular difficulty walking backwards, bradykinesia, micrographia, and hypophonia. Five had dystonia manifest as facial grimace and/or coq au pied. Postural tremor was intermittently observed in 3, but none had resting tremor. Subjective improvement with levodopa was initially recorded in some patients,[19] but benefits did not persist and the drug was eventually discontinued in all patients for lack of efficacy.[24] Further, no levodopa benefit was detected in a double-blind placebo-controlled study.[56] In addition, no patient developed dyskinesia or motor fluctuations while on levodopa, a side effect that is common in PD.[57] MRI studies demonstrated bilateral signal hyperintensities in the pallidum and striatum on T1-weighted images, indicative of manganese accumulation. FD-PET studies were also performed to assess the integrity of the nigrostriatal system. As indicated above, PD is characterized by a reduction in striatal FD uptake on PET, particularly in the posterior putamen,[50,51] consistent with the 40–80% reduction in SNc dopaminergic neurons and brain dopamine that is thought to already be present by the time of the emergence of the first features of clinical dysfunction.[25,58,59] Indeed, striatal FD uptake is reduced in asymptomatic individuals who are known to have an increased risk of developing PD and might thus have a preclinical form of the disorder.[60,61] In contrast to these findings in PD, each patient studied in the Taiwanese cohort who had parkinsonism related to manganese neurotoxicity had normal striatal FD uptake on PET.[52] Indeed, FD-PET studies performed many years later in patients from this same cohort remained normal,[53] despite the fact that the parkinsonism had persisted and even progressed.[54] Kim *et al.* similarly described a normal FD-PET scan in a 51-year-old patient with parkinsonism secondary to manganese exposure who had a propensity to fall backwards, no resting tremor, cock-walk, and a poor response to levodopa.[55] These findings indicate that the nigrostriatal system is relatively spared in patients with manganese-induced parkinsonism. In contrast, in patients with manganese-induced parkinsonism, PET studies demonstrated reduced F-18 deoxyglucose utilization in the pallidum and decreased raclopride binding in the striatum, indices of dysfunction in these brain regions.[53]

These cases support the notion that the clinical picture, response to levodopa, and imaging studies can differentiate PD and manganese-induced parkinsonism based on their propensity to preferentially damage the SNc and the pallidum, respectively.

Manganese-Induced Parkinsonism in Rhesus Monkeys

Animal models of manganese toxicity have further helped to clarify the features of manganese-induced parkinsonism. Mella described a parkinsonian syndrome coupled with dystonia in monkeys chronically treated with manganese chloride for 18 months.[38] At postmortem, degeneration was primarily localized to the pallidum and striatum. We also intoxicated adult rhesus monkeys with weekly intravenous injections of manganese chloride and described a parkinsonian syndrome characterized by gait dysfunction, bradykinesia, rigidity, and facial grimacing, but not tremor.[40] None of these animals responded to levodopa and none developed dyskinesia, which are routinely observed in MPTP-lesioned monkeys where the nigrostriatal system is damaged as it is in PD.[62] MRI studies demonstrated high signal abnormalities bilaterally in the striatum and pallidum on T1-weighted scans in manganese-lesioned animals. In contrast, striatal FD uptake on PET and striatal dopamine levels were normal. PET studies further showed that manganese intoxication was associated with decreased striatal raclopride binding, suggestive of local damage, in contrast to nigrostriatal lesions that are associated with denervation receptor supersensitivity. At postmortem, neuronal loss and gliosis most prominently affected the medial segment of the pallidum, there was no evidence of damage to SNc neurons, and dopamine levels were normal. Further, there were no Lewy bodies as is typically found in PD. Interestingly, numerous Alzheimer's type II astrocytes were detected throughout the GP. These histopathologic features have not been described in PD and are discussed in more detail below. These findings illustrate that manganese intoxication is associated with relative sparing of the nigrostriatal system, despite the fact that it induces a parkinsonian syndrome.

These studies provide further information on the clinical, pharmacologic, imaging, and pathologic features that characterize manganese-induced parkinsonism and illustrate the differences between this condition and PD based on their predilection to cause damage to neurons in the pallidum and nigrostriatal system, respectively.

Manganese-Induced Parkinsonism in Patients with Liver Failure

Additional clarification of the parkinsonian syndrome associated with manganese intoxication comes from studies of patients with chronic liver failure. It has long been appreciated that patients with chronic liver disease can develop a form of parkinsonism sometimes referred to as non-Wilsonian hepatolenticular degeneration.[63,64] There are also several reports noting that liver failure can be associated with bilateral symmetric high signal changes in the GP on T1-weighted MRI,[65–67] a pattern strikingly similar to that seen with manganese accumulation. Manganese is present in high concentration in many foods and liquids,[68] and approximately 98% of the dietary manganese load is cleared by the liver.[69–71] Based on these observations, we have postulated that the MRI changes in patients with liver failure are due to the accumulation of manganese as a result of the failure of the liver to adequately clear dietary manganese, and further that what has been referred to as non-Wilsonian hepatolenticular degeneration in the older literature is in fact a form of manganese-induced parkinsonism.[72] Indeed, we have demonstrated that plasma manganese levels are increased in patients with liver failure in comparison to controls and that there is a correlation between plasma manganese levels and the severity of signal change on T1-weighted MRI.[73] We further demonstrated that Alzheimer type II astrocytes, a characteristic finding in the pallidum of patients with non-Wilsonian hepatolenticular degeneration,[74] are also found in the brains of monkeys that have been intoxicated with manganese.[40]

Collectively, these observations indicate that the parkinsonism associated with liver failure is due to manganese intoxication and thus represents a good model for studying manganese-induced parkinsonism. Surveys of patients with chronic liver failure suggest that as many as 20–50% of these patients have moderate to severe parkinsonism.[75,76] Parkinsonism in these patients is characterized by early development of gait and balance dysfunction, speech impairment, bradykinesia, rigidity, relative absence of resting tremor, occasional dystonia, and generally a poor response to levodopa. The syndrome thus mirrors the parkinsonism described secondary to manganese intoxication and differs from PD. Interestingly, in our study, all patients with liver failure who had parkinsonism had the typical MRI changes, but not all patients with MRI changes had parkinsonian features. Kim *et al.* similarly noted MRI abnormalities in asymptomatic manganese-exposed workers.[77] These findings suggest that manganese-related alterations on MRI represent a threshold that must be exceeded before the emergence of motor dysfunction due to manganese intoxication. PET imaging of FD uptake and SPECT assessments of dopamine transporters in the liver failure cohort are ongoing.

Distinguishing PD from Manganese-Induced Parkinsonism

It is becoming increasingly clear that PD and manganese-induced parkinsonism are distinct entities that can be differentiated based on their clinical features, response to levodopa, imaging studies, and brain pathology[1] (see TABLE 1). PD primarily affects the nigrostriatal system and is characterized clinically by resting

TABLE 1. Features of PD vs. manganese-induced parkinsonism

Feature	PD	Manganese-induced parkinsonism
Clinical	Resting tremor; asymmetry	Early speech and balance dysfunction; symmetric impairment; relative absence of tremor; specific dystonia (grimace, coq au pied)
Response to levodopa	Good response	Poor response
Levodopa-induced motor complications	Very common, especially in young-onset cases	Never observed
MRI (T1-weighted images)	Normal	High signal change in GP, striatum, and SNr bilaterally
FD-PET	Decreased striatal uptake, especially in posterior putamen	Normal
Pathology	Degeneration of neurons in SNc, LC, NBM, DMNV, cortex, spinal cord, peripheral nervous system; Lewy bodies	Degeneration of neurons in GP; no Lewy bodies

tremor, asymmetry, and a good response to levodopa. MRI studies are normal and FD-PET demonstrates decreased striatal uptake, with the most severe changes being in the posterior putamen, even in patients in the earliest stages of the disease. Pathologically, there is degeneration of SNc dopamine neurons, Lewy bodies, and a reduction in striatal dopamine. Degeneration and Lewy body formation also occur in other specific brain regions, including the LC, NBM, and DMNV. In contrast, manganese-induced parkinsonism reflects pallidal degeneration and is characterized by gait and balance dysfunction, speech impairment, no asymmetry, absence of resting tremor, and little (if any) response to levodopa. MRI studies show a characteristic bilateral high signal abnormality in the pallidum on T1-weighted scans, and FD-PET studies are normal. At pathology, damage primarily affects the pallidum, and the nigrostriatal system is spared. On the other hand, pathology is not observed in the SNc, LC, NBM, or DMNV and there are no Lewy bodies. Based on these criteria, it is now possible to differentiate patients with manganese-induced parkinsonism from patients with PD who have incidental exposure to manganese.

Is Manganese a Risk Factor for PD?

An additional question relates to the possibility that exposure to manganese could serve as a risk factor for the development of PD, independent of its capacity to cause parkinsonism through direct damage to the pallidum. Gorell *et al.* performed a case-control study that suggested an increased risk of PD in patients with occupational exposure to manganese for greater than 20 years.[78] However, there was no correlation with the degree of exposure and this observation was not confirmed in a multiple variate logistic regression analysis of this same cohort.[79] Further, other epidemiologic studies have failed to identify a correlation between PD and manganese.[80–82] Hundreds of different toxins, exposures, diets, occupations, and lifestyles have been identified in epidemiologic studies as having a possible relationship to PD,[83–93] but it has proven difficult to determine if these are false associations or etiologic clues.

There have also been a few patients reported with clinical and imaging features consistent with the diagnosis of PD, but who also had a history of exposure to manganese in the workplace. Racette *et al*. described patients derived from their clinical practice who worked as welders and had clinical features of PD with good responses to levodopa and abnormal FD-PET studies.[94] They speculated that manganese exposure from the welding fume may have been a risk factor for the development of PD, but provided no specific evidence to support this hypothesis. Kim *et al*. similarly described two cases of workers exposed to manganese who developed a PD syndrome with typical imaging changes.[95,96] Initially, they argued that these patients had PD with incidental exposure to manganese;[95] however, they subsequently raised the possibility that manganese might have been a risk factor for the development of PD,[96] although they could not exclude the possibility that these cases simply had idiopathic PD with incidental exposure to manganese. PD is a common disorder affecting approximately 1 million persons in the United States and is associated with a 1–2% lifetime risk that an individual will develop this condition.[97,98] It is thus reasonable to consider that a large number of manganese-exposed individuals would be expected to develop PD by chance alone. To assess the frequency of PD in welders who may have been exposed to manganese in welding fumes, we performed a retrospective chart review of approximately 1500 PD patients and gathered information on their profession at the time of diagnosis and on their major lifetime profession. This study was performed at 3 major PD centers in different geographic regions of the United States. In these studies, only 1 patient indicated that they had been working as a welder at the time of diagnosis and no patient reported welding as their major lifetime profession (Olanow, unpublished data). Tanner *et al*. similarly reviewed the occupations of PD patients and found no evidence that having worked as a welder increased the likelihood of developing PD.[82] These studies have the advantage that they are not biased, do not seek out any one profession, and reflect information put into charts by independent observers unrelated to the study. They suffer from being retrospective, lacking a control group, and not providing complete information on all professions and tasks performed by an individual during their lifetime and the level of exposure. To better assess the role of manganese exposure in the development of PD, we have instituted a multicenter case-control study in which complete occupational histories with detailed accounts of tasks and exposures will be obtained from a large sample of PD patients and appropriate controls.

Finally, a number of studies have reported "preclinical" motor abnormalities in workers who have been exposed to manganese in the workplace.[99–108] However, there studies suffer from numerous methodologic limitations. Most were not prospective or blinded, end points were not predetermined, relatively small sample sizes were employed, motor assessments used were not validated, and almost none of the studies controlled for alcohol consumption, liver damage, drug use, or cognitive function. Further, the magnitude of change between groups in the various tests was small, corrections were not made for multiple comparisons, and abnormalities did not correlate with degree of manganese exposure. More importantly, none of these patients had clinical abnormalities indicative of PD and none developed PD during follow-up.

Laboratory studies have not been particularly helpful in assessing the potential of manganese to cause PD and, for that matter, it is not completely clear how manganese causes toxicity. Manganese is a transition metal that was originally thought to cause damage by inducing oxidative stress.[109] However, manganese does not participate in the Fenton reaction and some studies suggest that it acts by interfering with energy production and promoting excitotoxicity.[110] It has also been shown that manganese may be neuroprotective under some circumstances.[111,112] These issues may be moot as the question of whether or not manganese induces toxicity may be a function of its distribution. Further, while manganese may be toxic to dopamine neurons if injected directly into the nigra,[113] all studies performed to date in humans and primates indicate that the metal preferentially accumulates in and causes damage to the pallidum, with sparing of the nigrostriatal system.[35–40,44–49,53,55,66,72,73]

SUMMARY

In summary, current evidence indicates that manganese-induced parkinsonism can be differentiated from PD because of its predilection to accumulate in and damage the pallidum and striatum rather than the SNc. The clinical syndrome, response to levodopa, imaging studies with MRI and PET, and pathologic features all help to distinguish these two conditions and permit the correct diagnosis to be established. This is of particular relevance in differentiating patients with parkinsonism due to manganese intoxication from patients with idiopathic PD who have incidental manganese exposure.

REFERENCES

1. CALNE, D.B., N.S. CHU, C.C. HUNG et al. 1999. Manganism and idiopathic parkinsonism: similarities and differences. Neurology **44:** 1583–1586.
2. COUPER, J. 1837. On the effects of black oxide of manganese which inhaled into the lungs. Br. Ann. Med. Pharm. **1:** 41–42.
3. EMBDEN, H. 1901. Zur kentniss dre metallischen nervengifte. Deutsch. Med. Wochenschr. **27:** 795–796.
4. VON JAKSCH, R. 1907. Uber mangantoxikosen und maganophobie. Munchen Med. Wochenschr. **54:** 969–972.
5. CASAMAJOR, L. 1913. An unusual form of mineral poisoning affecting the nervous system: manganese? JAMA **60:** 646–649.
6. EDSALL, D.L., F.P. WILBUR & C.K. DRINKER. 1919. The occurence, course, and prevention of chronic manganese poisoning. J. Ind. Hyg **1:** 183–193.
7. ASHIZAWA, R. 1927. Uber einen Sektionsfall von chronischer manganvergiftung. Jpn. J. Med. Sci. Trans. Intern. Med. Pediatr. Psychiatr. **1:** 173–191.
8. CANAVAN, M.M., S. COBB & C.K. DRINKER. 1934. Chronic manganese poisoning. Psychiatry **32:** 501–512.
9. FLINN, R.H., P.A. NEAL & W.B. FULTON. 1941. Industrial manganese poisoning. J. Ind. Hyg. Toxicol. **23:** 374–387.
10. RODIER, J. 1955. Manganese poisoning in Moroccan miners. Vrit. J. Ind. Med. **12:** 21–35.
11. SCHULER, P., H. OYANGUREN, V. MATURANA et al. 1957. Manganese poisoning: environmental and medical study at Chilean mine. Int. Med. Surg. **26:** 167–173.
12. ABD, E.L., S. NABY & M. HASSANEIN. 1965. Neuropsychiatric manifestation of chronic manganese poisoning. J. Neurol. Neurosurg. Psychiatry **28:** 282–288.

13. MENA, I., O. MARIN, S. FUENZALIDA & G.C. COTIZAS. 1967. Chronic manganese poisoning: clinical picture and manganese turnover. Neurology **17:** 128–136.
14. TANAKA, S. & J. LIEBEN. 1969. Manganese poisoning and exposure in Pennsylvania. Arch. Environ. Health **19:** 674–684.
15. EMARA, A.M., S.H. AL-SHAWABI, O.I. MADKOUR & G.H. EL-SAMRA. 1971. Chronic manganese poisoning in the dry battery industry. Br. J. Ind. Med. **28:** 78–82.
16. SARIC, M., A. MARKICERVIC & O. HRUSTIC. 1977. Occupational exposure to manganese Br. J. Ind. Med. **34:** 114–118.
17. FERRAZ, H.B., P.H.F. BERTOLUCCI, J.S. PEREIRA et al. 1988. Chronic exposure to the fungicide Maneb may produce symptoms and signs of CNS manganese intoxication. Neurology **38:** 550–553.
18. SMYTH, L.T., R.C. RUHF, N.E. WHITMAN et al. 1973. Clinical manganism and exposure to manganese in the production and processing of ferromanganese alloy. J. Occup. Med. **15:** 101–109.
19. HUANG, C.C., N.S. CHU, C.S. LU et al. 1989. Chronic manganese intoxication. Arch. Neurol. **46:** 1104–1106.
20. COOK, D.G., S. FAHN & K.A. BRAIT. 1974. Chronic manganese intoxication. Arch. Neurol. **30:** 59–64.
21. GREENHOUSE, A.H. 1971. Manganese intoxication in the United States. Trans. Am. Neurol. Assoc. **96:** 248–249.
22. EJIMA, A., T. IMAMURA, S. NAKAMURA et al. 1992. Manganese intoxication during total parenteral nutrition. Lancet **ii:** 426.
23. HOLZGRAEFE, R., W. POSER, H. KIJEWSKI & W. BEEUCHE. 1986. Chronic poisoning caused by potassium permanganate: a case report. Clin. Toxicol. **24:** 235–244.
24. HUANG, C.C., C.S. LU, N.S. CHU et al. 1993. Progression after chronic manganese exposure. Neurology **43:** 1479–1483.
25. BERNHEIMER, H., W. BIRKMAYER, O. HORNYKIEWICZ et al. 1973. Brain dopamine and the syndromes of Parkinson and Huntington. J. Neurol. Sci. **20:** 415–455
26. ROSENSTOCK, H.A., D.G. SIMONS & J.S. MEYER. 1971. Chronic manganism: neurologic and laboratory studies during treatment with levodopa. JAMA **217:** 1354–1358.
27. MENA, I. et al. 1970. Modification of chronic manganese poisoning. N. Engl. J. Med. **282:** 5–10.
28. HUGHES, A.J., S.E. DANIEL, L. KILFORD & A.J. LEES. 1992. Accuracy of clinical diagnosis of idiopathic Parkinson's disease: a clinico-pathologic study of 100 cases. J. Neurol. Neurosurg. Psychiatry **55:** 181–184.
29. HUGHES, A.J., Y. BEN-SHLOMO, S.E. DANIEL & A.J. LEES. 1992. What features improve the accuracy of clinical diagnosis in Parkinson's disease: a clinicopathologic study. Neurology **42:** 1142–1146.
30. WENNING, G.K., Y. BEN-SHLOMO, A. HUGHES et al. 2000. What clinical features are most useful to distinguish definite multiple system atrophy from Parkinson's disease? J. Neurol. Neurosurg. Psychiatry **68:** 434–440.
31. OLANOW, C.W. 1992. Magnetic resonance imaging in parkinsonism. Neurol. Clin. North Am. **10:** 405–420.
32. OLANOW, C.W., M. ALBERTS, W. DJANG & J. STAJICH. 1990. MR imaging of putamenal iron predicts response to dopaminergic therapy in parkinsonian patients. In Early Markers in Parkinson's and Alzheimer's Diseases, pp. 99–109. Springer-Verlag. Berlin/New York.
33. FORNO, L.S. 1996. Neuropathology of Parkinson's disease. J. Neuropathol. Exp. Neurol. **55:** 259–272.
34. BRAAK, H. et al. 2003. Staging of brain pathology related to sporadic Parkinson's disease. Neurobiol. Aging **24:** 197–211.
35. YAMADA, M., S. OHNO, I. OKAYASU et al. 1986. Chronic manganese poisoning: a neuropathological study with determination of manganese distribution in the brain. Acta Neuropathol. (Berlin) **70:** 273–278.
36. ASHIZAWA, R. 1927. Uber einen Sektionsfall von chronischer manganvergiftung. Jpn. J. Med. Sci. Trans. Intern. Med. Pediatr. Psychiatr. **1:** 173–191.
37. CANAVAN, M., S. COBB & C.K. DRINKER. 1934. Chronic manganese poisoning: report of a case with autopsy. Arch. Neurol. Psychiatry **32:** 501–512.

38. MELLA, H. 1924. The experimental production of basal ganglion symptomatology in Macacus Rhesus. Arch. Neurol. Psychiatry **11:** 405–417.
39. PENTSCHEW, A., F. EBNER & R. KOVATCH. 1963. Experimental manganese encephalopathy in monkeys: a preliminary report. J. Neuropathol. Exp. Neurol. **22:** 488–499.
40. OLANOW, C.W., P.F. GOOD, H. SHINOTOH et al. 1996. Manganese intoxication in the rhesus monkey: a clinical, pathologic, and biochemical study. Neurology **46:** 492–498.
41. SPADONI, F., A. STEFANI, M. MORELLO et al. 2000. Selective vulnerability of pallidal neurons in the early phases of manganese intoxication. Exp. Brain Res. **135:** 544–551.
42. GWIAZDA, R.H., D. LEE, J. SHERIDAN & D.R. SMITH. 2002. Low cumulative manganese exposure affects striatal GABA, but not dopamine. Neurotoxicology **23:** 69–76.
43. BAEK, S.Y., M.J. LEE, H.S. JUNG et al. 2003. Effect of manganese exposure on MPTP neurotoxicities. Neurotoxicology **24:** 657–665.
44. LARSEN, N.A., H. PAKKENBERG, E. DAMSGAARD & K. HEYDORN. 1979. Topographical distribution of arsenic, manganese, and selenium in the normal human brain. J. Neurol. Sci. **42:** 407–416.
45. BONILLA, E., E. SALAZAR, J. JOAQUIN et al. 1982. The regional distribution of manganese in the normal human brain. Neurochem. Res. **7:** 221–227.
46. SUZUKI, Y., Y. MOURI, K. SUZUKI et al. 1975. Study of subacute toxicity of manganese dioxide in monkeys. Tokoshima J. Exp. **22:** 5–10.
47. DASTUR, D.K., D.K. MANGAHANIA & K.V. RAGHAVENDRAN. 1968. Distribution and fate of ^{54}Mn in the monkey: studies of different parts of the central nervous system and other organs. J. Clin. Invest. **50:** 9–20.
48. NEWLAND, M.C., T.L. CECKLER, J.H. KORDOWER et al. 1989. Visualizing manganese in the primate basal ganglia with magnetic resonance imaging. Exp. Neurol. **106:** 251–258.
49. SHINOTOH, H., B.J. SNOW, K.A. HEWITT et al. 1995. MRI and PET studies of manganese-intoxicated monkeys. Neurology **45:** 1199–1204.
50. BROOKS, D.J. 1998. The early diagnosis of Parkinson's disease. Ann. Neurol. **44**(suppl. 1): 10–18.
51. LEENDERS, K.L., E.P. SALMON, P. TYRRELL et al. 1990. The nigrostriatal dopaminergic system assessed *in vivo* by positron emission tomography in healthy volunteer subjects and patients with Parkinson's disease. Arch. Neurol. **47:** 1290–1297.
52. WOLTERS, E.C.H., C.C. HUANG, C. CLARK et al. 1989. Positron emission tomography in manganese intoxication. Ann. Neurol. **26:** 647–651.
53. SHINOTOH, H., B.J. SNOW, N.S. CHU et al. 1997. Presynaptic and postsynaptic striatal dopaminergic function in patients with manganese intoxication: a positron emission tomography study. Neurology **48:** 1053–1056.
54. HUANG, C.C., N.S. CHU, C.S. LU et al. 1998. Long-term progression in chronic manganism: ten years of follow-up. Neurology **50:** 698–700.
55. KIM, J.W., Y. KIM, H.K. CHEONG & K. ITO. 1998. Manganese induced parkinsonism: a case report. J. Korean Med Sci. **13:** 437–439.
56. LU, C.S., C.C. HUANG, N.S. CHU & D.B. CALNE. 1994. Levodopa failure in chronic manganism. Neurology **44:** 1600–1602.
57. LANG, A.P. & A.E. LOZANO. 1998. Parkinson's disease. N. Engl. J. Med. **339:** 1044–1053.
58. FEARNLEY, J.M. & A.J. LEES. 1991. Ageing and Parkinson's disease: substantia nigra regional selectivity. Brain **114:** 2283–2301.
59. MORRISH, P.K., G.V. SAWLE & D.J. BROOKS. 1995. Clinical and [^{18}F]dopa PET findings in early Parkinson's disease. J. Neurol. Neurosurg. Psychiatry **59:** 597–600.
60. PICCINI, P., P.K. MORRISH, N. TURJANSKI et al. 1997. Dopaminergic function in familial Parkinson's disease: a clinical and ^{18}F-dopa PET study. Ann. Neurol. **41:** 222–229.
61. BURN, D.J., M.H. MARK, E.D. PLAYFORD et al. 1992. Parkinson's disease in twins studied with ^{18}F-dopa and positron emission tomography. Neurology **42:** 1894–1900.
62. PEARCE, R.K.B., M. JACKSON, L. SMITH et al. 1995. Chronic L-dopa administration induces dyskinesia in the 1-methyl-4-phenyl-1,2,3,6-tetrahydropyridine-treated common marmoset (*Callithrix jacchus*). Mov. Disord. **10:** 731–740.
63. VICTOR, M., R.D. ADAMS & M. COLE. 1965. The acquired (non-Wilsonian) type of chronic hepatocerebral degeneration. Medicine **44**(5): 345–396.
64. SHERLOCK, S., W.H.J. SUMMERSKILL, L.P. WHITE & E.A. PHEAR. 1954. Portal-systemic encephalopathy: neurological complications of liver disease. Lancet **267:** 453–457.

65. INOUE, E., H. SHINICHI, Y. NARUMI et al. 1991. Portal-systemic encephalopathy: presence of basal ganglia lesions with high signal intensity on MR images. Radiology **179:** 551–555.
66. BRUNBERG, J.A., E. KANAL, W. HIRSCH & D.H. VAN THIEL. 1991. Chronic acquired hepatic failure: MR imaging of the brain at 1.5 T. AJNR **12:** 909–914.
67. KULISEVSKY, J., J. PUJOL, C. JUNQUE et al. 1993. MRI pallidal hyperintensity and brain atrophy in cirrhotic patients: two different MRI patterns of clinical deterioration. Neurology **43:** 2570–2573.
68. GREGER, J.L., C.D. DAVIS, J.W. SUTTIE & B.J. LYLE. 1990. Intake, serum concentrations, and urinary excretion of manganese by adult males. Am. J. Clin. Nutr. **51:** 457–461.
69. GREENBERG, D.M., D.H. COPP & E.M. CUTHBERSTON. 1943. Studies in mineral metabolism with aid of artificial radioactive isotopes: distribution and excretion, particularly by way of bile, of iron, cobalt, and manganese. J. Biol. Chem. **147:** 749–756.
70. POLLACK, S., J.N. GEORGE, R.C. REBA et al. 1965. The absorption of nonferrous metals in iron deficiency. J. Clin. Invest. **44:** 1470–1473.
71. KLAASSEN, C.D. 1974. Biliary excretion of manganese in rats, rabbits, and dogs. Toxicol. Appl. Pharmacol. **29:** 458–468.
72. HAUSER, R.A., T.A. ZESIEWICZ, A.S. ROSEMURGY et al. 1994. Manganese intoxication and chronic liver failure. Ann. Neurol. **36:** 871–875.
73. HAUSER, R.A., T.A. ZESIEWICZ, C. MARTINEZ et al. 1996. Blood manganese concentrations are increased and correlate with T-1 weighted signal hyperintensity on magnetic resonance imaging of the brain in patients with hepatic cirrhosis. Can. J. Neurol. Sci. **23:** 1–4.
74. NORENBERG, M.D. 1981. The astrocyte in liver disease. Adv. Cell. Neurobiol. **2:** 303.
75. BRODSKY, M.A., M.L. SCHILSKY, D.L. BRONSTER et al. 2002. Parkinsonism, manganese, and brain imaging in liver failure. Mov. Disord. **17**(suppl. 5)**:** 258.
76. BURKHARD, P.R., J. DELAVELLE, R. DU PASQUIER & L. SPAHR. 2003. Chronic parkinsonism associated with cirrhosis: a distinct subset of acquired hepatocerebral degeneration. Arch. Neurol. **60:** 521–528.
77. KIM, Y., K.S. KIM, J.S. YANG et al. 1999. Increase in signal intensities on T1-weighted magnetic resonance images in asymptomatic manganese-exposed workers. Neurotoxicology **20:** 901–907.
78. GORELL, J.M., C.C. JOHNSON, B.A. RYBICKI et al. 1999. Occupational exposure to manganese, copper, lead, iron, mercury, and zinc and the risk of Parkinson's disease. Neurotoxicology **20:** 239–247.
79. GORELL, J.M., B.A. RYBICKI, C.C. JOHNSON & E.L. PETERSON. 1999. Assessment of the multifactorial risk of Parkinson's disease. Neurology **52:** A429.
80. SEMCHUK, K.M., E.J. LOVE & R.G. LEE. 1993. Parkinson's disease: a test of the multifactorial etiologic hypothesis. Neurology **43:** 1173–1180.
81. HERTZMAN, C., M. WIENS, D. BOWERING et al. 1990. Parkinson's disease: a case-control study of occupational and environmental risk factors. Am. J. Ind. Med. **17:** 349–355.
82. TANNER, C.M., S.M. GOLDMAN, P. QUINLAN et al. 2003. Occupation and risk of Parkinson's disease (PD): a preliminary investigation of standard occupational codes (SOC) in twins discordant for disease. Neurology **60:** A415.
83. RAJPUT, A.H., A. RAJPUT & M. RAJPUT. 2003. Epidemiology of parkinsonism. *In* Handbook of Parkinson's Disease, pp. 17–42. Dekker. New York.
84. MENEGON, A., P.G. BOARD, A.C. BLACKBURN et al. 1998. Parkinson's disease, pesticides, and glutathione transferase polymorphisms. Lancet **352:** 1344–1346.
85. GORELL, J.M., C.C. JOHNSON, B.A. RYBICKI et al. 1998. The risk of Parkinson's disease with exposure to pesticides, farming, well water, and rural living. Neurology **50:** 1346–1350.
86. SEIDLER, A., W. HELLENBRAND, B.P. ROBRA et al. 1996. Possible environmental, occupational, and other etiologic factors for Parkinson's disease: a case-control study in Germany. Neurology **46:** 1275–1284.
87. BUTTERFIELD, P.G., B.G. VALANIS, P.S. SPENCER et al. 1993. Environmental antecedents of young-onset Parkinson's disease. Neurology **43:** 1150–1158.
88. MARDER, K., G. LOGROSCINO, B. ALFARO et al. 1998. Environmental risk factors for Parkinson's disease in an urban multiethnic community. Neurology **50:** 279–281.

89. PRIYADARSHI, A., S.A. KHUDER, E.A. SCHAUB & S.S. PRIYADARSHI. 2001. Environmental risk factors and Parkinson's disease: a metaanalysis. Environ. Res. **86:** 122–127.
90. ZORZON, M., L. CAPUS, A. PELLEGRINO *et al.* 2002. Familial and environmental risk factors in Parkinson's disease: a case-control study in north-east Italy. Acta Neurol. Scand. **105:** 77–82.
91. ROSS, G.W., R.D. ABBOTT, H. PETROVITCH *et al.* 2000. Association of coffee and caffeine intake with the risk of Parkinson disease. JAMA **283:** 2674–2679.
92. MORENS, D.M., A. GRANDINETTI, D. REED *et al.* 1995. Cigarette smoking and protection from Parkinson's disease: false association or etiologic clue? Neurology **45:** 1041–1051.
93. TANNER, C.M. & D.A. ASTON. 2000. Epidemiology of Parkinson's disease and akinetic syndromes. Curr. Opin. Neurol. **13:** 427–430.
94. RACETTE, B.A., L. MCGEE-MINNICH, S.M. MOERLEIN *et al.* 2001. Welding-related parkinsonism: clinical features, treatment, and pathophysiology. Neurology **56:** 8–13.
95. KIM, Y., J.W. KIM & K. ITO. 1999. Idiopathic parkinsonism with superimposed manganese exposure: utility of positron emission tomography. Neurotoxicology **20:** 249–252.
96. KIM, Y., J.M. KIM & J.W. KIM. 2002. Dopamine transporter density is decreased in parkinsonian patients with a history of manganese exposure: what does it mean? Mov. Disord. **17:** 568–575.
97. BOWER, J.H., J.M. MARAGANORE, S.K. MCDONNELL & W.A. ROCCA. 1999. Incidence and prevalence of parkinsonism in Olmstead County, Minnesota, 1979–1990. Neurology **52:** 1214–1220.
98. TWELVES, D., K.S. PERKINS & C. COUNSELL. 2003. Systematic review of incidence studies of Parkinson's disease. Mov. Disord. **18:** 19–31.
99. ROELS, H.A., P. GHYSELEN, J.P. BUCHET *et al.* 1992. Assessment of the permissible exposure level to manganese in workers exposed to manganese dioxide dust. Br. J. Ind. Med. **49:** 25–34.
100. ROELS, H., R. LAUWERYS, J.P. BUCHET *et al.* 1987. Epidemiological survey among workers exposed to manganese: effects on lung, central nervous system, and some biological indices. Am. J. Ind. Med. **11:** 307–327.
101. WENNBERG, A., A. IREGREN, G. STRUWE *et al.* 1991. Manganese exposure in steel smelters: a health hazard to the nervous system. Scand. J. Work Environ. Health **17:** 255–262.
102. WENNBERG, A., M. HAGMAN & L. JOHANSSON. 1992. Preclinical neurophysiological signs of parkinsonism in occupational manganese exposure. Neurotoxicology **13:** 271–274.
103. CHIA, S.E., J. GOH, G. LEE *et al.* 1993. Use of a computerized postural sway measurement system for assessing workers exposed to manganese. Clin. Exp. Pharmacol. Physiol. **20:** 549–553.
104. IREGREN, A. 1999. Manganese neurotoxicity in industrial exposures: proof of effects, critical exposure level, and sensitive tests. Neurotoxicology **20:** 315–323.
105. LUCCHINI, R., P. APOSTOLI, C. PERRONE *et al.* 1999. Long-term exposure to "low levels" of manganese oxides and neurofunctional changes in ferroalloy workers. Neurotoxicology **20:** 287–297.
106. MERGLER, D., G. HUEL, R. BOWLER *et al.* 1994. Nervous system dysfunction among workers with long-term exposure to manganese. Environ. Res. **64:** 151–180.
107. SJOGREN, B., A. IREGREN, W. FRECH *et al.* 1996. Effects on the nervous system among welders exposed to aluminum and manganese. Occup. Environ. Med. **53:** 32–40.
108. HOCHBERG, F., G. MILLER, R. VALENZUELA *et al.* 1996. Late motor deficits of Chilean manganese miners: a blinded control study. Neurology **47:** 788–795.
109. DONALDSON, J., F.S. LABELLA & D. GESSER. 1981. Enhanced autooxidation of dopamine as a possible basis of manganese neurotoxicity. Neurotoxicology **2:** 53–64.
110. BROUILLET, B.P., L. SHINOBU, U. MCGARVEY *et al.* 1993. Manganese injection into the striatum produces excitotoxic lesions by impairing energy metabolism. Exp. Neurol. **120:** 89–94.

111. SZIRAKI, I., K.P. MOHANAKUMAR, P. RAUHALA et al. 1998. Manganese: a transition metal protects nigrostriatal neurons from oxidative stress in the iron-induced animal model of parkinsonism. Neuroscience **85:** 1101–1111.
112. TSAI, S.S., A.Y. SUN, H.D. KIM & G.Y. SUN. 1993. Manganese exposure to PC-12 cells alters triacylglycerol metabolism and promotes neurite outgrowth. Life Sci. **52:** 1567–1575.
113. DANIELS, A.J. & J. ABARCA. 1991. Effect of intranigral Mn^{2+} on striatal and nigral synthesis and levels of dopamine and cofactor. Neurotoxicol. Teratol. **13:** 483–487.

Brain Ferritin Iron as a Risk Factor for Age at Onset in Neurodegenerative Diseases

GEORGE BARTZOKIS,[a,b,c,d] TODD A. TISHLER,[c,e] IL-SEON SHIN,[a,f] PO H. LU,[a,g] AND JEFFREY L. CUMMINGS[a,g]

[a]*Department of Neurology, UCLA, Los Angeles, California, USA*

[b]*Laboratory of Neuroimaging, Department of Neurology, Division of Brain Mapping, UCLA, Los Angeles, California, USA*

[c]*Greater Los Angeles VA Healthcare System, Department of Psychiatry, West Los Angeles, California, USA*

[d]*Department of Psychiatry, Charles R. Drew University of Medicine and Science, Los Angeles, California, USA*

[e]*Neuroscience Interdepartmental Graduate Program, UCLA, Los Angeles, California, USA*

[f]*Department of Psychiatry, Chonnam National University Medical School, Kwangju, Korea*

[g]*Department of Psychiatry and Biobehavioral Sciences, UCLA, Los Angeles, California, USA*

ABSTRACT: Tissue iron can promote oxidative damage. Brain iron increases with age and is abnormally elevated early in the disease process in several neurodegenerative disorders, including Alzheimer's disease (AD) and Parkinson's disease (PD). Higher iron levels in males may contribute to higher risk for younger-onset PD and recent studies have linked the presence of the hemochromatosis gene with a younger age at onset of AD. We examined whether age at onset of PD and AD was associated with increased brain ferritin iron. Ferritin iron can be measured with specificity *in vivo* with MRI utilizing the field-dependent relaxation rate increase (FDRI) method. FDRI was assessed in three basal ganglia regions (caudate, putamen, and globus pallidus) and frontal lobe white matter for younger- and older-onset male PD and AD patients and healthy controls. Significant increases in basal ganglia FDRI levels were observed in the younger-onset groups of both diseases compared to their respective control groups, but were absent in the older-onset patients. The results support the suggestion that elevated ferritin iron and its associated toxicity is a risk factor for age at onset of neurodegenerative diseases such as AD and PD. Clinical phenomena such as gender-associated risk of developing neurodegenerative diseases and the age at onset of such diseases may be associated with brain iron levels. *In vivo* MRI can measure and track brain ferritin iron levels and provides an opportunity to design therapeutic interventions that target high-risk populations early in the course of illness, possibly even before symptoms appear.

Address for correspondence: George Bartzokis, M.D., UCLA Alzheimer's Disease Center, 710 Westwood Plaza, Room 2-238, Los Angeles, CA 90095-1769. Fax: 310-268-3266. gbar@ucla.edu

KEYWORDS: iron; ferritin; risk factor; age at onset; aging; Alzheimer's disease; Parkinson's disease; dementia; sex; gender; brain development; myelin; oligodendrocytes; degeneration; amyloid; tau; α-synuclein; free radicals; Lewy body; neuritic plaques; synucleinopathy; proteinopathy; tauopathy; chelation; treatment; hemochromatosis

INTRODUCTION

The human brain is unique in its protracted process of development, which continues until late in the fifth decade.[1] This protracted developmental process is driven in large part by the complex process of myelination (for review, see refs. 2 and 3). Adequate iron levels are essential for normal myelination. Oligodendrocytes (the cells that produce myelin) contain the highest levels of iron of any brain cell,[4,5] their differentiation into myelin-producing cells may be dependent on iron availability,[6–8] and inadequate iron levels result in poor myelination and mental deficits.[4,9] Increasing levels of brain tissue iron are an integral part of brain development, with iron accumulation in normal individuals[10–14] continuing into the fifth decade and possibly beyond in some brain regions.[13–15]

Both iron deposition and myelination are highly heterochronic, with different regions developing at different rates (for review, see refs. 2, 3, 15, and 16). Oligodendrocytes are unique in their capacity to obtain their iron directly through binding ferritin[17] and may be directly involved in brain iron regulation.[18] During development, brain ferritin binding coincides with the onset and progression of myelination.[17] Oligodendrocytes and especially their precursors are also some of the most vulnerable cells in the brain during development, as well as during aging and degeneration; their dysfunction and/or loss[19,20] may have multiple devastating consequences, including the release of their iron stores (for review, see refs. 2 and 3).

Although essential for cell function, tissue iron can promote oxidative damage by catalyzing free radical reactions, resulting in the formation of hydroxyl radicals (the most reactive free radical species), which denature protein and DNA and initiate lipid peroxidation.[21] Our group documented increasing brain iron levels during normal aging.[11,15] We also demonstrated in several neurodegenerative diseases that iron levels are abnormally elevated early in the disease process and suggested that high iron levels may be contributing to the age risk factor of neurodegenerative diseases.[12,22–24] Tissue iron (as well as other metals such as zinc and copper) contributes to the development of proteinopathies (abnormal deposits of proteins) associated with several neurodegenerative diseases such as Alzheimer's disease (AD), Parkinson's disease (PD), and Dementia with Lewy Bodies (DLB) (for review, see refs. 2 and 3). Through its extensive process of myelination and the associated brain changes it produces, including increases in iron levels, the normal development of the human brain may establish the conditions for the most important risk factor for many of the prevalent neurodegenerative diseases such as AD, PD, and DLB: age (for review, see refs. 2 and 3).

The contribution of brain iron deposition to the age risk factor for AD has recently been assessed in studies of the hemochromatosis gene[25–27] that increases body iron load. The presence of this gene was found to be associated with the risk of developing AD in individuals with younger onset (before age 70),[26] especially in males.[25] This iron-associated risk is amenable to indirect *in vivo* measurement using magnetic resonance imaging (MRI) through the effect of iron on transverse relaxation rates (R_2).[22,28]

The bulk of brain iron is stored in ferritin molecules.[29,30] An *in vivo* MRI method called field-dependent R_2 increase (FDRI) can obtain specific measures of the iron content inside ferritin molecules.[10,15,28] The method takes advantage of the fact that ferritin increases R_2 linearly with the field strength of the MRI instrument to produce highly specific and reproducible measures of tissue iron deposits.[10,15,28] Briefly, FDRI is the difference in measures of brain R_2 obtained with two different field-strength MRI instruments. In the presence of ferritin, R_2 increases with increasing magnetic field-strength.[10,11,15,28,31–34] This field-dependent R_2 increase is specifically associated with the total iron contained in ferritin molecules[28,32] and has been shown to be independent of the amount of iron loading (number of iron atoms per ferritin molecule) of ferritin[33] and to increase linearly with field-strength.[32–34] Thus, FDRI is a specific measure of the total iron contained in ferric oxyhydroxide particles that form the mineral core of ferritin molecules. In human tissue, ferritin and its breakdown product (hemosiderin) are the only known physiologic sources of such particles.[28,31,32,35] The FDRI measure of the iron content of these particles will therefore be referred to as ferritin iron.[12,22–24]

In a previous study, we assessed the impact of iron levels on the age at onset of males with PD.[22] The focus on males was driven in part by evidence that studies examining younger PD subjects reported gender differences,[36–38] with prevalence of PD being higher in younger males than females, while those excluding younger subjects do not observe this difference (for review, see ref. 39). We therefore collected a bimodal sample composed of younger- and older-onset male PD subjects with a 20-year difference in mean age at onset and respective groups of age-matched normal controls. We observed that male subjects with PD with a younger onset of symptoms had increased ferritin iron levels in several basal ganglia regions, suggesting that ferritin iron may be a risk factor for earlier-onset PD (TABLE 1A). We also observed that older-onset patients had low levels of ferritin iron (TABLE 1B), but had evidence that free iron content (iron not in ferritin) was elevated compared to age-matched normal controls.[22] The MRI results on our older-onset sample were consistent with the interpretation made by Dexter *et al.*[40] and Connor *et al.*,[16] who suggested that, in their older age-range postmortem samples, decreased basal ganglia ferritin levels and increased free iron levels are present in PD subjects, concluding that elevated basal ganglia iron may be a contributing risk factor for PD.

In addition to the increased likelihood that younger-onset patients with PD are male,[36–38,41] younger-onset PD patients also experience more extrapyramidal motor side effects when treated,[36,41–45] suggesting an elevated susceptibility of the basal ganglia to treatment-associated toxicity. Like PD, DLB also has a male predominance in younger age cohorts (under 70)[46] and the presence of extrapyramidal symptoms is one of the diagnostic criteria used in diagnosing DLB. The pathology findings in DLB (the presence of Lewy bodies at brain autopsy) are very similar to those in PD, although the regional distribution of these lesions differs.[47] Often, DLB pathology occurs with AD pathology (neuritic plaques and neurofibrillary tangles) and the clinical manifestations of the two diseases can often not be clearly distinguished from each other.[46,47]

Postmortem studies have shown reduced ferritin levels with increased iron levels in older samples of subjects with AD as well as PD.[16] Our group reported increased

TABLE 1A. Data for younger-onset age (<60): PD

Region	PD FDRI (n = 6)		N FDRI (n = 6)		PD vs. N	
	Mean	SD	Mean	SD	t	p
C	2.27	0.42	2.11	0.51	−0.15	.57
G	5.73	0.54	4.88	0.68	−2.38	.039
P	2.91	0.55	2.26	0.42	−2.29	.045
W	1.32	0.46	1.15	0.32	−0.72	.49

NOTE: Data are presented as the unadjusted mean (standard deviation). Field-dependent relaxation rate (R_2) increase (FDRI) is an estimate of ferritin iron and is defined as the difference in the R_2 values of each brain region obtained using high (1.5 T) and low (0.5 T) field-strength MRI instruments. All statistical tests were two-tailed and the a priori alpha level was set to .05. A t test was used to determine whether PD subjects had increased iron in the basal ganglia. Separate t tests were computed in each region. Age was not included as a covariate in PD analyses because each pair of groups (PD and normals [N]) were matched with respect to age.

TABLE 1B. Data for older-onset age (>60): PD

Region	PD FDRI (n = 6)		N FDRI (n = 8)		PD vs. N	
	Mean	SD	Mean	SD	t	p
C	2.62	0.70	2.60	0.17	−0.06	.96
G	5.24	0.84	5.70	1.16	0.81	.43
P	2.63	0.68	3.03	0.55	1.23	.24
W	1.57	0.56	1.51	0.42	−0.20	.84

NOTE: Data are presented as the unadjusted mean (standard deviation). Field-dependent relaxation rate (R_2) increase (FDRI) is an estimate of ferritin iron and is defined as the difference in the R_2 values of each brain region obtained using high (1.5 T) and low (0.5 T) field-strength MRI instruments. All statistical tests were two-tailed and the a priori alpha level was set to .05. A t test was used to determine whether PD subjects had increased iron in the basal ganglia. Separate t tests were computed in each region. Age was not included as a covariate in PD analyses because each pair of groups (PD and normals [N]) were matched with respect to age.

basal ganglia ferritin iron levels in AD.[24] We therefore reexamined our AD data to assess whether, as was the case in PD, the iron risk factor is also associated with the age at onset in male subjects with AD.

METHODS

PD Study Samples

To study the effect of age at onset, we recruited a bimodal sample of 6 younger (under 65 years of age) and 6 older (over 65) male PD patients (mean age 56.0 versus 71.3, $t = -5.65$, $p = .0002$). The 2 groups had markedly different age at illness onset (mean 47.5 for the younger group versus 67.3 for the older group, $t = -5.61$, $p = .0002$), which did not overlap between groups (younger group all had illness onset before 60 years old; older group all had illness onset after 60 years old). None of these subjects were demented and all had Mini-Mental State Examination[48] scores in the normal range (27 or above).[22]

The healthy control group consisted of 14 male subjects, matched to the 2 PD groups in age. The 2 PD samples (younger- and older-onset) did not differ in age from their respective subgroups of 6 normal control subjects under 65 and 8 normal control subjects over 65.[22]

AD Study Samples

The male AD population ($n = 14$) ranged in age from 68 to 85 (mean 79.0, SD 5.5) and 12 met NINCDS-ADRDA[49] criteria for Probable AD and 2 for Possible AD. All had mild to severe AD with Mini-Mental State Examination[48] scores ranging between 8 and 26 (mean 19.4, SD 5.4).[24] For the purposes of the current analysis, this sample was segregated into subjects with onset before and after age 70. The 4 younger-onset subjects had a mean age of 72.6 with SD of 4.3, while the 10 older subjects had a mean age of 81.6 with SD of 3.4 ($t = -4.16$, df = 12, $p = .0013$). The 2 groups had markedly different age at illness onset (mean 67.5 with SD 1.9 for the younger group versus 75.7 with SD 2.7 for the older group) ($t = -5.43$, df = 12, $p = .0002$).

The healthy comparison group for the AD study was drawn from a large pool of subjects participating in a study on normal aging. Because brain iron increases with age in a nonlinear fashion (linear increase in adults and plateauing in the older age groups[13]), the normal comparison group was first chosen to have the same age range as the patient groups. Thus, unlike the PD study described above[22] where bimodal samples of patients and controls were recruited, the control sample for the AD group was recruited based on the age range of the AD group.[24] The mean ages did not differ between the younger-onset AD group and the healthy control group ($t = 1.56$, df = 38, $p = .13$). However, the older-onset AD group was older than their respective control group (AD: mean age 81.6, SD 3.4; controls: mean age 68.7, SD 4.9; $t = 7.86$, df = 44, $p < .0001$). In this older age range, correlations between age and FDRI did not reach statistical significance for either the control or AD groups in any of the four structures ($p > .07$). However, in order to eliminate any effect of age on iron levels in comparisons between groups, age was included as a covariate in statistical analyses of AD samples (TABLES 2A and 2B).

MRI Protocol

The methods have been described in detail elsewhere[15] and will only be summarized here. The subjects from each study were all scanned using two MR instruments (1.5 T and 0.5 T Picker instruments, Cleveland, OH), and the two scans were done within an hour of each other using the same imaging protocol. Coronal and sagittal pilot scans were first obtained to specify the location and spatial orientation of the head and the position of the axial image acquisition grid. The axial image acquisition sequence acquired interleaved contiguous slices using a Carr Purcell Meiboom Gill dual spin-echo sequence (2500/20,90/2, 3-mm slice thickness, 192 gradient steps,

TABLE 2A. Data for younger-onset age (<70): AD

Region	AD FDRI ($n = 4$)		N FDRI ($n = 36$)		AD vs. N	
	Mean	SD	Mean	SD	F	p
C	3.16	0.34	2.55	0.44	6.34	.016
G	5.85	0.98	4.67	0.67	8.84	.0052
P	4.30	0.60	3.05	0.57	13.79	.0007
W	1.39	0.33	1.64	0.28	2.24	.14

NOTE: Data are presented as the unadjusted mean (standard deviation). Field-dependent relaxation rate (R_2) increase (FDRI) is an estimate of ferritin iron and is defined as the difference in the R_2 values of each brain region obtained using high (1.5 T) and low (0.5 T) field-strength MRI instruments. All statistical tests were two-tailed and the a priori alpha level was set to .05. An analysis of covariance (ANCOVA) design, with diagnosis (AD versus normal controls [N]) as the independent variable and age as a covariate, was used to determine whether AD subjects had increased iron in the basal ganglia. Separate ANCOVAs were computed in each region. Age was included as a covariate in all analyses because of the known effects of age on brain ferritin iron levels.[13] Simple t tests (uncorrected for age) result in the same significant findings as the ANCOVA results reported. For comparison between tables, the F value is equal to the square of the t value reported in the PD tables above.

TABLE 2B. Data for older-onset age (>70): AD

Region	AD FDRI ($n = 10$)		N FDRI ($n = 36$)		AD vs. N	
	Mean	SD	Mean	SD	F	p
C	2.81	0.55	2.55	0.44	0.11	.74
G	4.43	1.17	4.67	0.67	0.55	.46
P	3.43	1.12	3.05	0.57	0.00	.97
W	1.47	0.30	1.64	0.28	0.30	.58

NOTE: Data are presented as the unadjusted mean (standard deviation). Field-dependent relaxation rate (R_2) increase (FDRI) is an estimate of ferritin iron and is defined as the difference in the R_2 values of each brain region obtained using high (1.5 T) and low (0.5 T) field-strength MRI instruments. All statistical tests were two-tailed and the a priori alpha level was set to .05. An analysis of covariance (ANCOVA) design, with diagnosis (AD versus normal controls [N]) as the independent variable and age as a covariate, was used to determine whether AD subjects had increased iron in the basal ganglia. Separate ANCOVAs were computed in each region. Age was included as a covariate in all analyses because of the known effects of age on brain ferritin iron levels.[13] Simple t tests (uncorrected for age) result in the same significant findings as the ANCOVA results reported. For comparison between tables, the F value is equal to the square of the t value reported in the PD tables above.

and 25-cm field of view). The coronal and sagittal pilot scans were used to determine the alignment and accuracy of head repositioning in the second MRI instrument. To consistently position the actual image slices identically within the brain and thus sample the same volume of tissue, the axial slice-select grid was adjusted so that the anterior commissure was contained within the same slice in both high and low field-strength instruments. For increased consistency, all subsequent measures were referenced to this slice.[15,28]

Image Analysis

Transverse relaxation times (T_2) were calculated for each voxel by an automated algorithm from the two signal intensities (TE = 20 and 90) of the dual spin-echo sequence to produce gray-scale-encoded T_2 maps of the brain.[15] The T_2 measures were extracted using an Apple Macintosh–configured image analysis workstation. T_2 data for each of the ROIs were obtained from contiguous pairs of slices. The relaxation rate (R_2) was calculated as the reciprocal of $T_2 \times 1000$ ms/s. The average R_2 values of the two slices from both hemispheres were the final measures used in the subsequent analyses. The FDRI measure was calculated as the difference in R_2 (high field R_2 – low field R_2). Test/retest reliability for FDRI measures was very high, with intraclass correlation coefficients ranging from 0.88 to 0.99 ($p < .0023$).[15,28]

RESULTS

Ferritin iron (FDRI) data are presented for three basal ganglia regions (caudate, putamen, and globus pallidus) and frontal lobe white matter for all four subgroups (TABLES 1A, 1B, 2A, 2B). Striking increases in basal ganglia FDRI levels were seen in both younger-onset PD and AD groups compared to their respective control groups (TABLES 1A, 2A). These differences were attenuated in all regions in the older-onset PD and AD groups and were reversed (smaller than the normal group) in globus pallidus and putamen regions of the PD group, and globus pallidus of the AD group (TABLES 1B, 2B). The results were not meaningfully different (significance of findings or direction of differences) if the younger-onset AD sample was increased to an n of 6 by including 2 additional subjects (data not shown). We present the sample split at age 70 to make it more comparable to other published reports on age at onset of AD.[26,46]

DISCUSSION

The FDRI measure of ferritin iron suggests that younger-onset male patients with AD and PD have increased ferritin iron levels in basal ganglia compared to healthy controls. This finding is consistent with the suggestion that increased ferritin iron is a risk factor for these diseases[12,22–24] and may be impacting their clinical manifestations, including their age at onset.[22,23] This suggestion is consistent with recent evidence that the highly prevalent hemochromatosis gene is associated with a younger age at onset of AD,[25–27] that this gene is associated with increased oxidative stress and stage of the disease,[50] and that this association may be especially marked in males.[25] Increased risk for developing motor disorders such as PD seems to be similarly associated with the presence of the hemochromatosis gene.[51,52]

Tissue metals including iron are essential in developing proteinopathies that result in the pathognomonic lesions of many neurodegenerative diseases (for review, see refs. 2 and 3). Normal males have higher brain ferritin iron levels than females across the life span (Bartzokis, unpublished data). Increased ferritin iron could represent one of the general risk factors for neurodegenerative diseases and may be contributing to clinical phenomena such as the male predominance observed in DLB and younger-onset PD, as well as influencing the age at onset of neurodegenerative diseases (for review, see refs. 2 and 3).

The data from the older-onset patients show nonsignificant, but substantial reductions in ferritin iron in the globus pallidus (TABLES 1B and 2B). The globus pallidus region also had the greatest reductions in ferritin immunoreactivity[16] in postmortem brain samples of PD and AD whose mean ages (74 and 80, respectively) were similar to the ages of our *in vivo* older-onset samples (71 and 82, respectively). These same investigators[16] also observed significant, albeit smaller reductions of ferritin immunoreactivity in the other two basal ganglia regions where ferritin iron was reduced in our samples only in the putamen of the PD group (TABLES 1B and 2B). Thus, our ferritin iron measures in the older-onset samples of AD and PD are consistent in direction, but not in magnitude and pervasiveness with postmortem data of ferritin levels.[16] One may speculate that these differences could be due to longer lengths of illness in the postmortem samples[16] compared to our *in vivo* patient samples.

Together, the *in vivo* and postmortem data could be interpreted in a single developmental model that includes illness length. The *in vivo* data suggest that elevated ferritin iron and its associated toxicity (for review, see refs. 2 and 3) are associated with a younger onset of disease. Subsequently, the disease process itself could result in the destruction of ferritin molecules (or destruction of high-ferritin-containing oligodendrocytes) and cause the release of toxic "free" iron that culminates in the postmortem findings of low ferritin levels, but continued abnormally high iron levels.[16] Such a developmental (time-dependent) model will be difficult to unambiguously demonstrate in cross-sectional studies.[53] In our samples, for example, length of illness was highly confounded with age and an adequate assessment of the impact that length of illness had on ferritin iron was not possible. This developmental interpretation of the FDRI and postmortem results remains speculative until prospective studies are performed. Nonetheless, the pattern of *in vivo* and postmortem findings supports the supposition that complex changes in iron metabolism may be involved in both diseases (for review, see refs. 2 and 3).

The *in vivo* measures of basal ganglia ferritin iron must be considered in the context of the interaction between normal brain development and the known pathology of each disease state. The basal ganglia myelinate early in development,[54] reach the highest iron levels of all brain regions,[13–15] and may therefore be intrinsically more resistant to iron-associated toxicity than other brain regions such as association cortices (for review, see refs. 2 and 3). Increased iron levels in basal ganglia regions are present early in the disease process in multiple neurodegenerative disorders.[12,22,23] The abnormal basal ganglia ferritin iron increases may thus be a good indicator of iron elevations or iron metabolism abnormalities and toxicity in other more vulnerable brain regions that are more directly impacted by disease (for review, see refs. 2, 3, and 22).

Normal ferritin, a spherical protein in which upwards of 90% of tissue nonheme iron is stored,[29,30] can sequester iron and other transition metals and may function as a general metal detoxicant.[55] Oligodendrocytes have the highest iron content of all brain cell types[4,5] and as much as 70% of brain iron is associated with myelin.[56] Many normal as well as pathological processes (anoxia, oxidative stress, etc.) that have been shown to damage oligodendrocytes and especially their precursors (for review, see refs. 2 and 3) can also release iron from ferritin.[21,57–60] Oligodendrocytes may be more vulnerable than other cells to such iron releases since, in addition to containing the highest iron stores, their particular ferritin subunit composition makes iron available with greater ease than in other cells.[4,61]

The importance of metal metabolism and iron in particular has been highlighted by recent studies suggesting central roles for such metals in the pathophysiology of amyloid β (Aβ)– and α-synuclein (αSyn)–associated lesions such as neuritic plaques in AD and Lewy bodies in PD and DLB (for review, see refs. 2 and 3). For example, elevated iron levels increase the production of amyloid precursor protein (APP) through an untranslated region that contains an iron-responsive element,[62] and soluble Aβ (the initial Aβ form produced from APP cleavage) can act as an iron chelator and oppose the age-associated elevation of brain iron levels.[63,64] Thus, the normal function of APP and Aβ increases may be part of the mechanisms to protect against iron-mediated toxicity. However, iron and other transition metals such as copper and zinc can also promote Aβ oligomerization.[63,65–68] Oligomerized Aβ is the toxic entity associated with free radical damage and AD.[66,69,70] In addition, the abnormal aggregation of αSyn (a cytoskeletal protein whose precipitation results in Lewy bodies) can also be promoted by iron.[71] Furthermore, intracellular tau and αSyn can promote each other's fibrillization[72] and are often observed to coexist in many neurodegenerative disorders.[46,47,73,74] In the context of developmental increases in brain iron levels, the homeostatic mechanism protecting cells from iron's deleterious effects may represent a crucial element of the pathophysiology of neurodegenerative disorders (for review, see refs. 2 and 3).

The data on the impact of ferritin iron on the age at onset of degenerative diseases are compatible with the hypothesis that elevated iron levels are one of the risk factors for neurodegenerative disorders (for review, see refs. 2, 3, 12, and 22–24). Important clinical phenomena such as the gender-associated risk of developing neurodegenerative diseases and the age at onset of such diseases may be associated with brain iron levels. The complex findings and the pathophysiologic implications of increased iron levels support the notion that tissue iron measures should be considered as an integral part of research efforts for brain diseases that demonstrate developmental (age-related) patterns of clinical manifestations or free radical involvement. *In vivo* assessments of ferritin iron represent a unique opportunity to more clearly define the development of such disorders through prospective studies that are difficult to perform with postmortem studies. *In vivo* studies can also provide an opportunity to design therapeutic interventions, including chelation[75,76] targeting susceptible populations with elevated ferritin iron, early in the course of illness or possibly even before symptoms appear (for review, see refs. 2 and 3).

ACKNOWLEDGMENTS

This work was supported in part by an NIA Alzheimer's Disease Center Grant (AG 16570), an Alzheimer's Disease Research Center of California Grant, the Sidell-Kagam Foundation (MH635701A1), and a Merit Review Grant from the Department of Veterans Affairs.

REFERENCES

1. BARTZOKIS, G., M. BECKSON, P.H. LU *et al.* 2001. Age-related changes in frontal and temporal lobe volumes in men: a magnetic resonance imaging study. Arch. Gen. Psychiatry **58:** 461–465.

2. BARTZOKIS, G. 2004. Age-related myelin breakdown: a developmental model of cognitive decline and Alzheimer's disease. Neurobiol. Aging **25:** 5–18.
3. BARTZOKIS, G. 2004. Quadratic trajectories of brain myelin content: unifying construct for neuropsychiatric disorders. Neurobiol. Aging **25:** 49–62.
4. CONNOR, J.R. & S.L. MENZIES. 1996. Relationship of iron to oligodendrocytes and myelination. Glia **17:** 83–93.
5. ERB, G.L., D.L. OSTERBUR & S.M. LEVINE. 1996. The distribution of iron in the brain: a phylogenetic analysis using iron histochemistry. Brain Res. Dev. Brain Res. **93:** 120–128.
6. MORATH, D.J. & M. MAYER-PROSCHEL. 2001. Iron modulates the differentiation of a distinct population of glial precursor cells into oligodendrocytes. Dev. Biol. **237:** 232–243.
7. POWER, J., M. MAYER-PROSCHEL, J. SMITH et al. 2002. Oligodendrocyte precursor cells from different brain regions express divergent properties consistent with the differing time courses of myelination in these regions. Dev. Biol. **245:** 362–375.
8. SMITH, M.A., C.A. ROTTKAMP, A. NUNOMURA et al. 2000. Oxidative stress in Alzheimer's disease. Biochim. Biophys. Acta **1502:** 139–144.
9. RONCAGLIOLO, M., M. GARRIDO, T. WALTER et al. 1998. Evidence of altered central nervous system development in infants with iron deficiency anemia at 6 mo: delayed maturation of auditory brainstem responses. Am. J. Clin. Nutr. **68:** 683–690.
10. BARTZOKIS, G., D. SULTZER, J. MINTZ et al. 1994. In vivo evaluation of brain iron in Alzheimer's disease and normal subjects using MRI. Biol. Psychiatry **35:** 480–487.
11. BARTZOKIS, G., M. BECKSON, D.B. HANCE et al. 1997. MR evaluation of age-related increase of brain iron in young adult and older normal males. Magn. Reson. Imaging **15:** 29–35.
12. BARTZOKIS, G., D. SULTZER, B.J. CUMMINGS et al. 2000. In vivo evaluation of brain iron in Alzheimer's disease and normal controls using magnetic resonance imaging. Arch. Gen. Psychiatry **57:** 47–53.
13. HALLGREN, B. & P. SOURANDER. 1958. The effect of age on the non-haemin iron in the human brain. J. Neurochem. **3:** 41–51.
14. KLINTWORTH, G.K. 1973. Huntington's chorea—morphologic contributions of a century. In Advances in Neurology. Vol. 1: Huntington's Chorea, 1872–1972, pp. 353–368. Raven Press. New York.
15. BARTZOKIS, G., J. MINTZ, D. SULTZER et al. 1994. In vivo MR evaluation of age-related increases in brain iron. Am. J. Neuroradiol. **15:** 1129–1138.
16. CONNOR, J.R., B.S. SNYDER, P. AROSIO et al. 1995. A quantitative analysis of iso-ferritins in select regions of aged, parkinsonian, and Alzheimer's diseased brains. J. Neurochem. **65:** 717–724.
17. HULET, S.W., S. MENZIES & J.R. CONNOR. 2002. Ferritin binding in the developing mouse brain follows a pattern similar to myelination and is unaffected by the jimpy mutation. Dev. Neurosci. **24:** 208–213.
18. GERBER, M.R. & J.R. CONNOR. 1989. Do oligodendrocytes mediate iron regulation in the human brain? Ann. Neurol. **26:** 95–98.
19. TANG, Y., J.R. NYENGAARD, B. PAKKENBERG et al. 1997. Age-induced white matter changes in the human brain: a stereological investigation. Neurobiol. Aging **18:** 609–615.
20. MARNER, L., J.R. NYENGAARD, Y. TANG et al. 2003. Marked loss of myelinated nerve fibers in the human brain with age. J. Comp. Neurol. **462:** 144–152.
21. HALLIWELL, B. & J.M.C. GUTTERIDGE. 1988. Iron as biological pro-oxidant. ISI Atlas Sci. Biochem. **1:** 48–52.
22. BARTZOKIS, G., J.L. CUMMINGS, C.H. MARKHAM et al. 1999. MRI evaluation of brain iron in earlier- and later-onset Parkinson's disease and normal subjects. Magn. Reson. Imaging **17:** 213–222.
23. BARTZOKIS, G., J. CUMMINGS, S. PERLMAN et al. 1999. Increased basal ganglia iron levels in Huntington disease. Arch. Neurol. **56:** 569–574.
24. BARTZOKIS, G. & T.A. TISHLER. 2000. MRI evaluation of basal ganglia ferritin iron and neurotoxicity in Alzheimer's and Huntington's disease. Cell. Mol. Biol. **46:** 821–833.
25. MOALEM, S., M.E. PERCY, D.F. ANDREWS et al. 2000. Are hereditary hemochromatosis mutations involved in Alzheimer disease? Am. J. Med. Genet. **93:** 58–66.

26. SAMPIETRO, M., L. CAPUTO, A. CASATTA et al. 2001. The hemochromatosis gene affects the age of onset of sporadic Alzheimer's disease. Neurobiol. Aging. **22:** 563–568.
27. COMBARROS, O., M. GARCIA-ROMAN, A. FONTALBA et al. 2003. Interaction of the h63d mutation in the hemochromatosis gene with the apolipoprotein E epsilon 4 allele modulates age at onset of Alzheimer's disease. Dement. Geriatr. Cogn. Disord. **15:** 151–154.
28. BARTZOKIS, G., M. ARAVAGIRI, W.H. OLDENDORF et al. 1993. Field dependent transverse relaxation rate increase may be a specific measure of tissue iron stores. Magn. Reson. Med. **29:** 459–464.
29. FLOYD, R.A. & J.M. CARNEY. 1993. The role of metal ions in oxidative processes and aging. Toxicol. Ind. Health **9:** 197–214.
30. MORRIS, C.M., J.M. CANDY, A.E. OAKLEY et al. 1992. Histochemical distribution of non-haem iron in the human brain. Acta Anat. (Basel) **144:** 235–257.
31. VYMAZAL, J., M. HAJEK, N. PATRONAS et al. 1995. The quantitative relation between T1-weighted and T2-weighted MRI of normal gray matter and iron concentration. J. Magn. Reson. Imaging **5:** 554–560.
32. VYMAZAL, J., R.A. BROOKS, C. BAUMGARNER et al. 1996. The relation between brain iron and NMR relaxation times: an *in vitro* study. Magn. Reson. Med. **35:** 56–61.
33. VYMAZAL, J., O. ZAK, J.W. BULTE et al. 1996. T1 and T2 of ferritin solutions: effect of loading factor. Magn. Reson. Med. **36:** 61–65.
34. VYMAZAL, J., R.A. BROOKS, N. PATRONAS et al. 1995. Magnetic resonance imaging of brain iron in health and disease. J. Neurol. Sci. **134**(suppl.)**:** 19–26.
35. BULTE, J.W., G.F. MILLER, J. VYMAZAL et al. 1997. Hepatic hemosiderosis in non-human primates: quantification of liver iron using different field strengths. Magn. Reson. Med. **37:** 530–536.
36. PANTELATOS, A. & F. FORNADI. 1993. Clinical features and medical treatment of Parkinson's disease in patient groups selected in accordance with age at onset. Adv. Neurol. **60:** 690–697.
37. KOSTIC, V.S., S.R. FILIPOVIC, D. LECIC et al. 1994. Effect of age at onset on frequency of depression in Parkinson's disease. J. Neurol. Neurosurg. Psychiatry **57:** 1265–1267.
38. MAYEUX, R., K. MARDER, L.J. COTE et al. 1995. The frequency of idiopathic Parkinson's disease by age, ethnic group, and sex in northern Manhattan, 1988–1993 [see comments]. Am. J. Epidemiol. **142:** 820–827.
39. DE RIJK, M.C., C. TZOURIO, M.M. BRETELER et al. 1997. Prevalence of parkinsonism and Parkinson's disease in Europe—The EuroParkinson Collaborative Study: European community concerted action on the epidemiology of Parkinson's disease. J. Neurol. Neurosurg. Psychiatry **62:** 10–15.
40. DEXTER, D.T., J. SIAN, P. JENNER et al. 1993. Implications of alterations in trace element levels in brain in Parkinson's disease and other neurological disorders affecting the basal ganglia. *In* Parkinson's Disease: From Basic Research to Treatment, pp. 273–281. Raven Press. New York.
41. FRIEDMAN, A. 1994. Old-onset Parkinson's disease compared with young-onset disease: clinical differences and similarities. Acta Neurol. Scand. **89:** 258–261.
42. WU, R.M., H.C. CHIU, M. WANG et al. 1993. Risk factors on the occurrence of response fluctuations and dyskinesias in Parkinson's disease. J. Neural Transm. Parkinson's Dis. Dement. Sect. **5:** 127–133.
43. KOSTIC, V.S., S. PRZEDBORSKI, E. FLASTER et al. 1991. Early development of levodopa-induced dyskinesias and response fluctuations in young-onset Parkinson's disease. Neurology **41:** 202–205.
44. GOLBE, L.I. 1991. Young-onset Parkinson's disease: a clinical review. Neurology **41:** 168–173.
45. WAGNER, M.L., M.N. FEDAK, J.I. SAGE et al. 1996. Complications of disease and therapy: a comparison of younger and older patients with Parkinson's disease. Ann. Clin. Lab. Sci. **26:** 389–395.
46. BARKER, W.W., C.A. LUIS, A. KASHUBA et al. 2002. Relative frequencies of Alzheimer disease, Lewy body, vascular and frontotemporal dementia, and hippocampal sclerosis in the State of Florida Brain Bank. Alzheimer's Dis. Assoc. Disord. **16:** 203–212.

47. COLOSIMO, C., A.J. HUGHES, L. KILFORD et al. 2003. Lewy body cortical involvement may not always predict dementia in Parkinson's disease. J. Neurol. Neurosurg. Psychiatry **74:** 852–856.
48. FOLSTEIN, M.F., S.E. FOLSTEIN & P.R. MCHUGH. 1975. "Mini-mental state": a practical method for grading the cognitive state of patients for the clinician. J. Psychiatr. Res. **12:** 189–198.
49. MCKHANN, G., D. DRACHMAN, M. FOLSTEIN et al. 1984. Clinical diagnosis of Alzheimer's disease: report of the NINCDS-ADRDA work group under the auspices of Department of Health and Human Services Task Force on Alzheimer's Disease. Neurology **34:** 939–944.
50. PULLIAM, J.F., C.D. JENNINGS, R.J. KRYSCIO et al. 2003. Association of hfe mutations with neurodegeneration and oxidative stress in Alzheimer's disease and correlation with apoE. Am. J. Med. Genet. **119B:** 48–53.
51. DEMARQUAY, G., A. SETIEY, Y. MOREL et al. 2000. Clinical report of three patients with hereditary hemochromatosis and movement disorders. Mov. Disord. **15:** 1204–1209.
52. DEKKER, M.C., P.C. GIESBERGEN, O.T. NJAJOU et al. 2003. Mutations in the hemochromatosis gene (hfe), Parkinson's disease, and parkinsonism. Neurosci. Lett. **348:** 117–119.
53. KRAEMER, H.C., J.A. YESAVAGE, J.L. TAYLOR et al. 2000. How can we learn about developmental processes from cross-sectional studies, or can we? Am. J. Psychiatry **157:** 163–171.
54. YAKOVLEV, P.I. & A.R. LECOURS. 1967. Regional development of the brain in early life. Blackwell Sci. Pub. Boston.
55. JOSHI, J.G., S.R. SCZEKAN & J.T. FLEMING. 1989. Ferritin—a general metal detoxicant. Biol. Trace Elem. Res. **21:** 105–110.
56. DE LOS MONTEROS, A.E., R.A. KORSAK, T. TRAN et al. 2000. Dietary iron and the integrity of the developing rat brain: a study with the artificially-reared rat pup. Cell. Mol. Biol. (Noisy-Le-Grand) **46:** 501–515.
57. ANGGARD, E. 1994. Nitric oxide: mediator, murderer, and medicine [see comments]. Lancet **343:** 1199–1206.
58. DAI, L., P.G. WINYARD, Z. ZHANG et al. 1996. Ascorbate promotes low density lipoprotein oxidation in the presence of ferritin. Biochim. Biophys. Acta **1304:** 223–228.
59. DOUBLE, K.L., M. MAYWALD, M. SCHMITTEL et al. 1998. In vitro studies of ferritin iron release and neurotoxicity. J. Neurochem. **70:** 2492–2499.
60. HALLIWELL, B. & J.M. GUTTERIDGE. 1986. Oxygen free radicals and iron in relation to biology and medicine: some problems and concepts. Arch. Biochem. Biophys. **246:** 501–514.
61. BLISSMAN, G., S. MENZIES, J. BEARD et al. 1996. The expression of ferritin subunits and iron in oligodendrocytes in neonatal porcine brains. Dev. Neurosci. **18:** 274–281.
62. ROGERS, J.T., J.D. RANDALL, P.S. EDER et al. 2002. Alzheimer's disease drug discovery targeted to the APP mRNA 5′ untranslated region. J. Mol. Neurosci. **19:** 77–82.
63. MAYNARD, C.J., R. CAPPAI, I. VOLITAKIS et al. 2002. Overexpression of Alzheimer's disease amyloid-beta opposes the age-dependent elevations of brain copper and iron. J. Biol. Chem. **277:** 44670–44676.
64. ZOU, K., J.S. GONG, K. YANAGISAWA et al. 2002. A novel function of monomeric amyloid beta-protein serving as an antioxidant molecule against metal-induced oxidative damage. J. Neurosci. **22:** 4833–4841.
65. ATWOOD, C.S., R.D. MOIR, X. HUANG et al. 1998. Dramatic aggregation of Alzheimer Aβ by Cu(II) is induced by conditions representing physiological acidosis. J. Biol. Chem. **273:** 12817–12826.
66. CURTAIN, C.C., F. ALI, I. VOLITAKIS et al. 2001. Alzheimer's disease amyloid-beta binds copper and zinc to generate an allosterically ordered membrane-penetrating structure containing superoxide dismutase–like subunits. J. Biol. Chem. **276:** 20466–20473.
67. GARZON-RODRIGUEZ, W., A.K. YATSIMIRSKY & C.G. GLABE. 1999. Binding of Zn(II), Cu(II), and Fe(II) ions to Alzheimer's Aβ peptide studied by fluorescence. Bioorg. Med. Chem. Lett. **9:** 2243–2248.
68. LOSKE, C., A. GERDEMANN, W. SCHEPL et al. 2000. Transition metal-mediated glycoxidation accelerates cross-linking of beta-amyloid peptide. Eur. J. Biochem. **267:** 4171–4178.

69. KIURU, S., O. SALONEN & M. HALTIA. 1999. Gelsolin-related spinal and cerebral amyloid angiopathy. Ann. Neurol. **45:** 305–311.
70. TORP, R., E. HEAD & C.W. COTMAN. 2000. Ultrastructural analyses of beta-amyloid in the aged dog brain: neuronal beta-amyloid is localized to the plasma membrane. Prog. Neuropsychopharmacol. Biol. Psychiatry **24:** 801–810.
71. GOLTS, N., H. SNYDER, M. FRASIER *et al.* 2002. Magnesium inhibits spontaneous and iron-induced aggregation of alpha-synuclein. J. Biol. Chem. **277:** 16116–16123.
72. GIASSON, B.I., M.S. FORMAN, M. HIGUCHI *et al.* 2003. Initiation and synergistic fibrillization of tau and alpha-synuclein. Science **300:** 636–640.
73. ARIMA, K., S. HIRAI, N. SUNOHARA *et al.* 1999. Cellular co-localization of phosphorylated tau- and NACP/alpha-synuclein-epitopes in Lewy bodies in sporadic Parkinson's disease and in dementia with Lewy bodies. Brain Res. **843:** 53–61.
74. PIAO, Y.S., S. HAYASHI, M. HASEGAWA *et al.* 2001. Co-localization of alpha-synuclein and phosphorylated tau in neuronal and glial cytoplasmic inclusions in a patient with multiple system atrophy of long duration. Acta Neuropathol. (Berlin) **101:** 285–293.
75. CHERNY, R.A., C.S. ATWOOD, M.E. XILINAS *et al.* 2001. Treatment with a copper-zinc chelator markedly and rapidly inhibits beta-amyloid accumulation in Alzheimer's disease transgenic mice. Neuron **30:** 665–676.
76. MCLACHLAN, D.R.C., A.J. DALTON, T.P. KRUCK *et al.* 1991. Intramuscular desferrioxamine in patients with Alzheimer's disease. Lancet **337:** 1304–1308.

Hematoma Removal, Heme, and Heme Oxygenase Following Hemorrhagic Stroke

KENNETH R. WAGNER[a] AND BARNEY E. DWYER[b]

[a]*Medical Research Service, Department of Veterans Affairs Medical Center, and Department of Neurology, University of Cincinnati College of Medicine, Cincinnati, Ohio, USA*

[b]*Medical Research Service, Department of Veterans Affairs Medical Center, White River Junction, Vermont, and Department of Medicine (Neurology), Dartmouth Medical School, Hanover, New Hampshire, USA*

ABSTRACT: The hemorrhagic strokes, intracerebral (ICH) and subarachnoid hemorrhage (SAH), often have poor outcomes. Indeed, the most common hemorrhagic stroke, ICH, has the highest mortality and morbidity rates of any stroke subtype. In this report, we discuss the evidence for the staging of red blood cell removal after ICH and the significance of control of this process. The protective effects of clinically relevant metalloporphyrin heme oxygenase inhibitors in experimental models of ICH and in superficial siderosis are also discussed. We also examine literature paradoxes related to both heme and heme oxygenase in various disorders of the central nervous system. Last, new data are presented that support the concept that heme, although primarily a pro-oxidant, can also have antioxidant properties.

KEYWORDS: brain hemorrhage; oxidative stress; heme; heme oxygenase; stroke

INTRODUCTION

Hemorrhagic stroke is a broad classification of several stroke subtypes including spontaneous intracerebral hemorrhage (ICH), subarachnoid hemorrhage (SAH), and hemorrhagic infarct conversion after ischemic stroke. For all of these hemorrhagic stroke subtypes, mortality and morbidity are significant. Indeed, the incidences of death and neurological deficits are greater following intracerebral bleeds as compared to ischemic stroke.[1,2] The current use of tissue plasminogen activator to treat ischemic stroke and myocardial infarction can itself induce ICH and increase bleeding into ischemic infarcts.[3] Morbidity and mortality from ICH are the result of three factors: (1) mechanical damage to the brain parenchyma, (2) mass effect from the hematoma and/or edema development, leading to increased intracranial pressure, and (3) the toxicity of the blood components on adjacent brain tissue.[1,4–11] Mechanical

Address for correspondence: Kenneth R. Wagner, Ph.D., Research Service (151), Department of Veterans Affairs Medical Center, 3200 Vine Street, Cincinnati, OH 45220. Voice: 513-861-3100, ext. 4339; fax: 513-475-6415.
wagnerkr@email.uc.edu

Ann. N.Y. Acad. Sci. 1012: 237–251 (2004). © 2004 New York Academy of Sciences.
doi: 10.1196/annals.1306.020

damage to brain tissue at the site of the bleed is unavoidable; it presents a difficult therapeutic target unless it can be anticipated (e.g., during surgery). Surgical removal of the hematoma shows promise for reducing the potentially lethal mass effect of expanding edema.[12,13] Therapeutic strategies to reduce the toxicity of blood products in brain parenchyma are potentially important, but at present there are no accepted pharmacologic treatments for hemorrhagic stroke. This report focuses on the natural course of hematoma removal, the roles of heme and heme oxygenase in ICH, and novel therapeutic approaches suggested by studies of experimental ICH. Due to the large volume of original literature that exists in this general area, we have often cited review papers to reduce the number of references.

INTRACEREBRAL HEMATOMA AND THE FATE OF HEMOGLOBIN AND HEME

Oxyhemoglobin Conversion and Red Blood Cell Lysis

Within hours following an intracerebral bleed, the red cells in the hematoma center become deoxygenated, resulting in the conversion of oxy- to deoxyhemoglobin.[14,15] Deoxyhemoglobin spontaneously and nonenzymatically oxidizes to methemoglobin as its iron is converted from the ferrous to the ferric state.[16] Bradley[14] defined five stages of magnetic resonance imaging changes as the hematoma ages in which hemoglobin passes through several forms (oxyhemoglobin, deoxyhemoglobin, and methemoglobin) prior to red cell lysis.

Rapid hemolysis does not occur following an intracerebral bleed. Rather, red blood cells and their contents are removed slowly over several days to weeks. This process appears to be carefully controlled. Red cells are known to maintain their biconcave shape for 2–3 days in rats, 4–8 days in dogs, and 5–10 days in humans.[17,18] Red cell lysis *in vitro* also occurs slowly, with hemoglobin release starting after 2 days.[19] In dogs, hemoglobin concentrations peak in cerebrospinal fluid (CSF) by day 2 following intracisternal blood injections.[20] The time course of hemolysis following SAH in humans is similar.[19] In our porcine ICH model,[21,22] we have observed that hematoma volumes remain unchanged for the first 7 days. The majority of the hematoma is then removed by 14 days.[8] Recently, it was reported that erythrocyte lysis and hemoglobin release following ICH may be mediated by activation of complement and the formation of a membrane attack complex.[23]

How red cells are lysed may be less important than timing. This is because the timing of red cell lysis after ICH may determine where heme degradation occurs. Macrophages and reactive microglia become abundant in days after ICH. Both cell types express heme oxygenase and are well suited for heme disposal. Lysis of red cells within the hematoma before sufficient phagocytic cells are present could result in significant uptake of hemoglobin (and heme) by neurons and astrocytes.[24] The toxicity of hemoglobin in brain tissue has been well described,[25] and its toxicity in neurons in culture has been abundantly established.[26,27] In this regard, an important determinant of outcome after ICH may be the amount of blood that lyses during the initial bleed.

The Heme Oxygenase–Biliverdin Reductase System

The conversion of deoxyhemoglobin to methemoglobin appears to facilitate the release of heme from hemoglobin. Heme could be released into the extracellular space after ICH and into the CSF after SAH. Extracellular heme is lipophilic and could enter cells after direct insertion into cell membranes. Heme binding to serum proteins might facilitate cellular uptake. Hemopexin, for example, is a serum protein that binds free heme with high affinity and transports it to liver for degradation.[28] In addition, heme may be released directly into neurons, astrocytes, and microglia if these cells internalize hemoglobin from lysed red cells. The details of these processes in the central nervous system (CNS) are largely unknown. Regardless of the cell type, heme is degraded by the heme oxygenase–biliverdin reductase (HO-BVR) system. HO (EC 1.14.99.3) catalyzes oxidative cleavage of the porphyrin ring of heme, the rate-limiting step in heme catabolism; carbon monoxide and iron are released and the bile pigment biliverdin is formed.[29] Two functional HO isoenzymes, designated HO-1 and HO-2, are known.[30] HO-1 is largely absent from normal brain, whereas HO-2 is abundantly expressed in neurons.[31–38] HO-1 (HSP-32) is a heat shock or stress protein that is inducible by a variety of noxious stimuli and associated with acquired resistance to cellular injury. Interestingly, HO-1 has the largest number of diverse inducers of any known enzyme.[30] HO-2 is generally considered refractory to induction, except by glucocorticoids.[39–41] However, HO-2 induction has also been reported after chronic morphine exposure,[42] spinal nerve lesioning,[43] and spinal cord injury.[44] A third isoenzyme, designated HO-3, has been found in the CNS,[45,46] but its significance is unknown. Biliverdin, which is the product of oxidative cleavage of heme by HO, is subsequently reduced to bilirubin by biliverdin IXα reductase (BVR, EC 1.3.1.24).[47] BVR is present in brain tissue, where it appears to be coexpressed with HO isoforms,[48] and is inducible by cerebral ischemia.[49]

HO, METALLOPORPHYRINS, AND BRAIN PROTECTION AFTER HEMORRHAGE

The role of the HO-BVR system in brain cell injury is controversial. In part, this is because heme and its degradation products (iron, carbon monoxide, and biliverdin/bilirubin) are all biologically active molecules, yet the functional consequences of their elaboration often seem contradictory (see ref. 35 and discussion below). Beyond regulating heme degradation, the HO-BVR system may have as yet unrecognized cytoprotective properties. For example, inducible HO-1 and constitutive HO-2 may subserve different protective functions beyond their catalytic function. Thus, a human monoblastic lymphoma cell line, which was transfected with a mutant human HO-1 devoid of catalytic activity, was protected from oxidative stress.[50] HO-2 (but not HO-1) contains a "heme regulatory motif" and may help to detoxify nitric oxide.[51] Moreover, BVR was localized to the nucleus of rat kidney cells after stress.[52] Subsequent studies in COS cells characterized BVR as a leucine zipper-like DNA binding protein that regulated HO-1 induction.[53] Metalloporphyrin HO inhibitors may also have unexpected effects. A potentially important if overlooked property of some porphyrin compounds, zinc-protoporphyrin IX (ZnPP), for example, is their anti-inflammatory effect.[54,55] It is noteworthy that Kadoya and coworkers,[56] who reported that ZnPP treatment reduced infarct size and edema formation in a rat model of transient cerebral ischemia, used ZnPP because it was an interleukin-1 antagonist and not because it was an HO inhibitor.

The abundance of heme after cerebral bleeding (see ref. 16) suggested to us that the HO-BVR system would be an important therapeutic target in ICH. The clinically relevant question was whether HO should be inhibited or upregulated (HO-1). Our initial studies focused on HO inhibitors because metalloporphyrins, which are structural analogues of heme and potent HO inhibitors, had demonstrated neuroprotective properties, that is, treatment with metalloporphyrins reduced infarct size and edema in experimental models of transient forebrain ischemia.[56,57] *In vitro*, metalloporphyrins reduced injury in rat hippocampal slices caused by hydrogen peroxide, hypoxia, and trauma.[58] Metalloporphyrins reduced myoglobin toxicity in cultured human proximal tubular cells (HK-2),[59] and they reduced hydrogen peroxide toxicity in monkey renal epithelial cells (BSC-1)[60] and in a rat astroglial cell line.[35] The studies briefly reviewed below present a consistent picture, namely, that inhibition of the HO-BVR system can be beneficial in acute ICH.

Metalloporphyrins Are Neuroprotective after Experimental ICH

In our initial studies, we reported that tin-mesoporphyrin (SnMP), a potent HO inhibitor, when mixed with blood immediately before intracerebral injection, markedly reduced clot size and edema formation in a porcine ICH model.[61] Oxidative stress appears to be important in the pathophysiology of edema formation after ICH.[8,11,62–65] However, no evidence has been obtained to suggest SnMP reduced oxidative stress. Thus, SnMP did not inhibit reductive ferritin iron release or prevent iron-stimulated lipid peroxidation (TBARS) *in vitro*. Previously, SnMP reduced hydrogen peroxide toxicity in a rat astroglial cell line.[35] However, we found no evidence that SnMP was a superoxide dismutase (SOD) mimetic or a catalase mimetic (e.g., as are some manganese-centered metalloporphyrins). Moreover, no effect of SnMP on blood clotting *in vitro* was found.[61] Delayed clotting might lessen injury by allowing backflow of blood from the injection site. At present, the presumed protective mechanism of SnMP is HO inhibition.

Huang *et al.*[66] reported that intracerebral injection of hemoglobin, hemin, bilirubin, and iron chloride produced significant edema in the ipsilateral cortex and basal ganglia of rats at 24 h. The HO inhibitor, tin-protoporphyrin IX (SnPP), when coinjected with hemoglobin significantly reduced cortical edema. The protective mechanism of SnPP is uncertain. It is noteworthy that two products of heme degradation by HO-BVR (iron and bilirubin) also produced edema and that the iron chelator, deferoxamine, albeit only at high doses, reduced edema. HO inhibition (reducing iron and bilirubin) as the protective mechanism of SnPP is consistent with these data. However, HO-1 induction was reported in adjacent tissue. How widely distributed SnPP (or other metalloporphyrins) becomes after coinjection with blood is unknown. Most metalloporphyrins, including SnPP, are poorly soluble in neutral aqueous solution and bind to serum proteins. We speculate that they may not penetrate very far into normal tissue and have limited uptake into cells. Thus, it may prove to be true that optimal antiedema effect of metalloporphyrins requires local HO inhibition (close to the site of hemorrhage) and upregulated HO activity in more distant brain tissue.

Superficial Siderosis

In human superficial siderosis, small continuous or recurrent SAHs lead to the deposition of iron (derived from heme) as hemosiderin at the surface of the brain and to neurological dysfunction in advanced stages of the disease.[67] Surgical intervention to repair the source of bleeding is potentially useful, but often the source of bleeding is unknown. Koeppen and coworkers[67] tested the hypothesis that inhibitors of HO would reduce iron deposition in an experimental model of superficial siderosis. Washed autologous red blood cells (with or without added SnPP) were injected into the cisterna magna of rabbits weekly for 16 weeks. Iron deposition in cerebellum (a major site of superficial siderosis in humans) was shown dramatically by histochemistry (enhanced Perls' stain), with the crests of cerebellar folia being prominently involved. Addition of SnPP to the red cell injectate dramatically reduced iron deposition in cerebellum. SnPP was reported to accumulate in brain tissue after successive injections, suggesting HO inhibition was in fact the protective mechanism of SnPP.

SAH and Cerebral Vasospasm

Cerebral vasospasm caused by SAH is a major medical complication of ruptured intracranial aneurysms and is associated with high levels of morbidity and mortality.[19,68,69] No consistently effective therapy exists at present. Oxyhemoglobin has been implicated as a causative factor, possibly by generating free radicals and lipid peroxides, but the mechanism(s) is uncertain.[19] Two products of heme degradation by HO-BVR (iron and bilirubin) might aggravate cerebral vasospasm. Thus, iron chelators have had some success in attenuating experimental systems.[70–72] Moreover, bilirubin oxidation products (BOXes) that are generated in an oxidative stress environment are potent vasospastic agents *in vitro* and *in vivo*.[73,74] Interestingly, expression of HO-1 has been reported to be beneficial for preventing vasospasm in experimental systems.[75–77] Possibly, the HO-BVR system both promotes and prevents vasospasm. As suggested earlier, iron and bilirubin may be vasospastic agents, while carbon monoxide, which is released in equimolar amounts with bilirubin and iron during heme degradation, may antagonize vasoconstriction (see refs. 78 and 79).

HEME AND NEUROPROTECTION

Heme is generally considered a pro-oxidant molecule, and potentially harmful effects of excess heme (which might be mimicked after ICH) have been documented (see refs. 26, 27, and 80–83). Nevertheless, beneficial effects of heme infusion in nonneural injury have been reported (see ref. 84). Moreover, heme and heme-arginate are used to treat acute intermittent porphyria, a metabolic disease caused by a deficiency in the heme biosynthetic pathway.[85] Heme infusion was also reported to be protective in brain injury models. Takizawa *et al.*[86] reported that infusion of heme, but not SnMP, reduced neuronal injury after transient forebrain ischemia in striatum and parietal cortex (but not in hippocampal CA1 or CA3 regions). Each compound (30 μmol/kg) was administered intraperitoneally at the time of reperfusion and then daily for 2 additional days. Protection was attributed to HO-1 induction. However, HO-1 induction was predominantly glial, and the mechanistic relationship between glial HO-1 induction and neuronal sparing is unclear. Although hyperbilirubinemia was protective in focal ischemia,[87] whether bilirubin was elevated in heme-treated rats was not determined.

In another neurological disease model, experimental allergic encephalomyelitis (EAE), the clinical severity was reduced by heme treatment.[88] Lewis rats were injected with heme (40 µmol/kg) once a day, on days 7 to 17, after they were immunized with guinea pig myelin basic protein. In contrast, the HO inhibitor, SnMP (40 µmol/kg), exacerbated EAE. These results suggest a protective role for HO, possibly because bilirubin production is increased.[89] In contrast, the clinical severity of EAE in SJL mice (immunized with proteolipid protein peptide) was reduced by injections of SnMP (200 µmol/kg injected 8–19 days after immunization).[90] These results suggest that HO activity amplified the disease process. Why these studies came to different conclusions is unclear, but may reflect differences in the animal model used and the amount of SnMP injected. Also, in the SJL mouse model of EAE, extravasated blood is present in the CNS, potentially leading to increased iron levels following HO-1 degradation of heme.[90]

ANTIOXIDANT PROPERTIES OF HEME AND SOME METALLOPORPHYRINS

The beneficial effects of HO are generally attributed to the antioxidant effects of increased synthesis of bilirubin and ferritin (secondary to iron release). Carbon monoxide may act as a vasodilator and also may have a direct cytoprotective effect in some tissues (see ref. 91), but its role in brain injury is unclear. Moreover, there is ample evidence that heme is a pro-oxidant molecule,[92] and degradation of free heme is generally considered beneficial. However, preliminary data obtained during preparation of reference 61 showed that heme could reduce lipid peroxidation in rat brain homogenates. Those studies have been expanded and data are shown below.

Lipid peroxidation in rat brain homogenates (with and without iron supplementation) was estimated by the formation of TBARS (thiobarbituric acid reacting substances), which was measured by fluorescence spectrophotometry, as previously described.[61] One-hour incubation at 37°C increased TBARS about 10-fold in non-iron-supplemented rat brain homogenates (TABLE 1, expt. 1). The addition of 50 µM heme to homogenates significantly reduced TBARS formation as early as 10 min after the start of the incubation and by over 75% after 1 h (FIG. 1 and TABLE 1, expt. 1). Addition of 200 µM $FeSO_4$ to homogenates stimulated TBARS formation as much as 8-fold (FIG. 2 and TABLE 1, expt. 2). In this case, addition of 50 µM heme reduced TBARS formation, albeit the results were more variable: by about 65% after a 30-min incubation (FIG. 2) and by about 35% in homogenates incubated 1 h (TABLE 1, expt. 2). This variability was likely due to technical factors, but together the results suggested diminution of heme's antioxidant effect in iron-supplemented homogenates. This was confirmed in a second experiment (FIG. 3). It is noteworthy that the concentrations of heme used here might be expected after the injection of neuroprotective doses of heme.[57,86] A single intraperitoneal injection of heme (30 µmol/kg) given to rats transiently increased serum heme, measured by the pyridine hemochromogen assay: control, 15.2 ± 7.0 µM ($n = 6$); heme + 1 h, 60.9 ± 7.2 µM ($n = 5$); and heme + 24 h, 16.1 ± 4.6 µM ($n = 6$). SnPP did not reduce TBARS formation in rat brain homoge-

TABLE 1. Effect of heme and porphyrins on lipid peroxidation in rat brain homogenates

Control	TBARS, µmol/L (n)	%
Expt. 1: Complete ($t = 0$)	1.51 ± 0.17 (8)	—
Complete ($t = 1$ h)	10.05 ± 1.35 (8)	100
+ DMSO (2.2 µL/tube)	8.47 ± 0.50 (8)	84*
+ 50 µM heme	2.45 ± 0.39 (8)	24*
+ 50 µM SnPP	12.20 ± 1.20 (8)	120*
+ 50 µM heme + 50 µM SnPP	3.04 ± 0.22 (8)	30*
Expt. 2: Complete + 200 µM FeSO$_4$ ($t = 1$ h)	80.6 ± 3.20 (8)	100
+ DMSO (2.2 µL/tube)	75.7 ± 3.70 (8)	94
+ 50 µM heme	52.0 ± 10.4 (8)	65*
+ 50 µM SnPP	77.6 ± 3.60 (8)	96
+ 50 µM heme + 50 µM SnPP	71.0 ± 6.70 (8)	88*†
Expt. 3: Complete ($t = 1$ h)		
+ DMSO (2.2 µL/tube)	11.9 ± 2.3 (8)	100
+ 50 µM ZnPP	3.2 ± 1.0 (8)	27*
+ 50 µM PPIX	10.4 ± 2.3 (8)	88
+ 100 µM ZnCl$_2$	14.7 ± 1.4 (8)	124*
Expt. 4: Complete + 200 µM FeSO$_4$ ($t = 1$ h)		
+ DMSO (2.2 µL/tube)	67.5 ± 3.6 (8)	100
+ 50 µM ZnPP	29.4 ± 9.9 (8)	43*
+ 50 µM PPIX	68.0 ± 2.2 (8)	101
+ 100 µM ZnCl$_2$	69.8 ± 3.3 (8)	103

NOTE: The effect of heme, SnPP, ZnPP, protoporphyrin IX (PPIX), and ZnCl$_2$ on TBARS formation in rat brain homogenates, without or with iron supplementation, was measured as described in the legends of FIGURES 1 and 2, respectively. Incubations were at 37°C for 1 h, except "complete" ($t = 0$), which was the zero time control. Data expressed as the mean ± SD were analyzed by one-way ANOVA followed by Tukey's test for multiple comparisons: *$p < .05$ vs. complete and vehicle (DMSO) control; †$p < .05$, SnPP + heme vs. heme alone.

nates (TABLE 1, expts. 1 and 2). Previously, we showed SnMP did not reduce TBARS formation in similar homogenates.[61] In contrast, ZnPP did reduce TBARS accumulation in rat brain homogenates (TABLE 1, expts. 3 and 4). Heme is generally considered a pro-oxidant molecule, so we considered the possibility that heme had antioxidant properties at low concentration, but became a pro-oxidant molecule at high concentration. We found no evidence for this. TBARS accumulation in non-iron-supplemented rat brain homogenates was reduced by 64% when 50 µM heme was present and by 77% when 500 µM heme was present (FIG. 4). In iron-supplemented homogenates, heme was less effective. Nevertheless, 500 µM heme reduced TBARS formation in this case by 64% (FIG. 5).

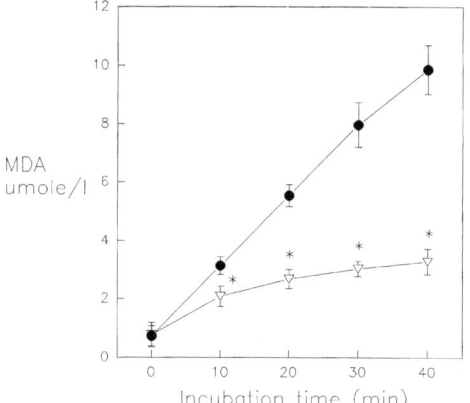

FIGURE 1. TBARS formation in rat brain homogenates in the presence of 50 μM heme (▽) or vehicle (●) was measured by fluorescence spectrophotometry.[61] Briefly, homogenates (100 μL aliquots) were kept on ice, and heme (hemin chloride, no. 3741, Calbiochem) or vehicle (DMSO) was added. Reaction tubes were incubated at 37°C and the reaction was terminated with 11 volumes of 10% TCA at the time indicated. TBARS in the acid-soluble extract and in an identically treated malondialdehyde standard curve was measured as described.[61] Data are expressed as malondialdehyde equivalents (MDA, μmol/L). Each data point is the mean ± SD of 5 independent samples. Data were analyzed by one-way ANOVA followed by Tukey's test for multiple comparisons using SigmaStat software: *$p < .05$ for heme-treated vs. vehicle-treated homogenates.

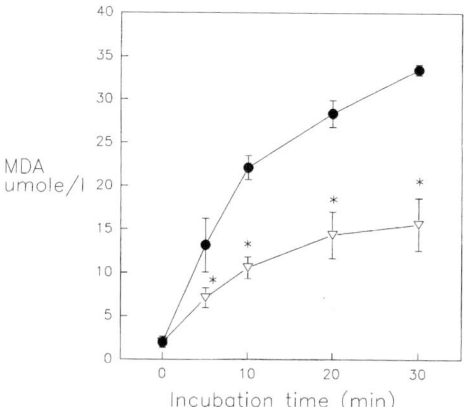

FIGURE 2. TBARS formation in iron-supplemented rat brain homogenates in the presence of 100 μM heme (▽) or vehicle (●) was measured as described in FIGURE 1. Heme and vehicle were added to homogenates on ice, and the reaction was initiated by the addition of 200 μM FeSO$_4$. Reaction tubes were incubated at 37°C for the time indicated. Based on preliminary studies, higher heme concentration and shorter incubation times were used to better depict the effect of heme on iron-stimulated TBARS formation. Each data point, expressed as MDA, μmol/L, is the mean ± SD of 4 independent samples. Data were analyzed by one-way ANOVA as described in FIGURE 1: *$p < .05$ for heme-treated vs. vehicle-treated homogenates.

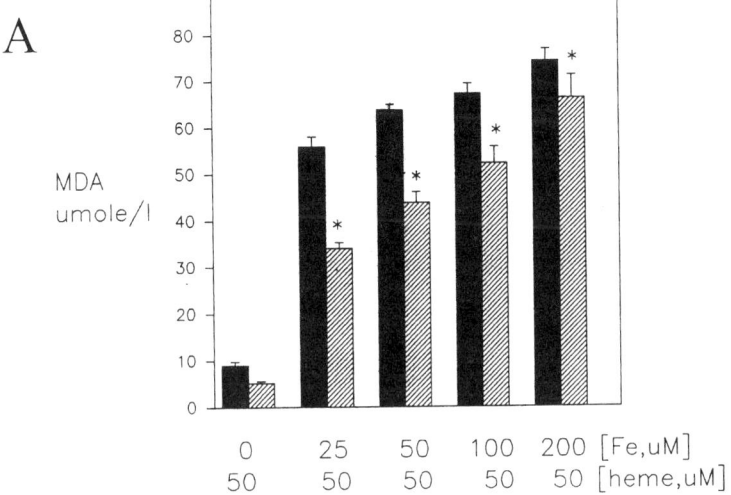

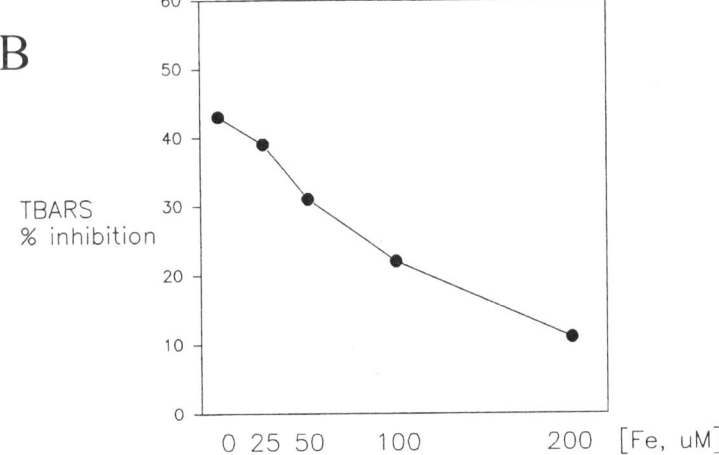

FIGURE 3. Iron dependence of heme's antioxidant effect. TBARS formation in iron-supplemented rat brain homogenates in the presence of 50 µM heme (*striped bars*) or vehicle (*solid bars*) was measured as described in FIGURE 1. Heme was added to homogenates and the reaction was initiated by the addition of $FeSO_4$ (at the concentrations indicated). The reaction tubes were incubated for 1 h at 37°C. **(A)** Each data point, expressed as MDA, µmol/L, is the mean ± SD of 4 independent samples: *$p < .05$ for heme-treated vs. vehicle-treated homogenates. **(B)** The % inhibition of TBARS formation by 50 µM heme plotted vs. the iron concentration.

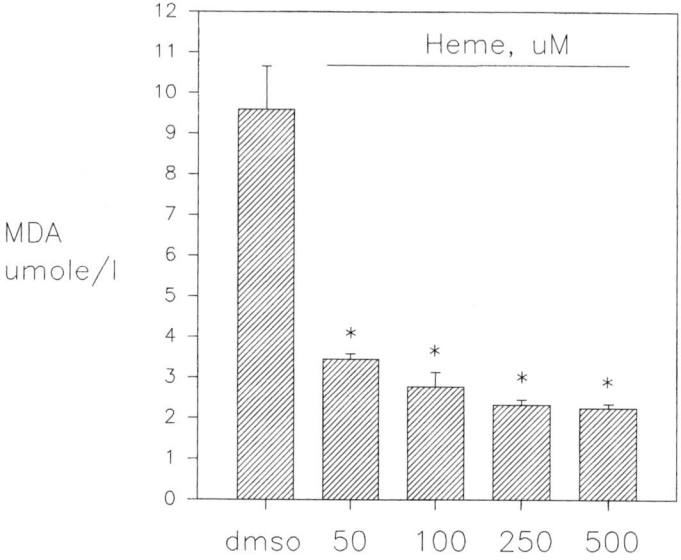

FIGURE 4. Effect of heme concentration on TBARS formation in rat brain homogenates measured as described in FIGURE 1. Heme was added to homogenates at the concentrations indicated and the reaction mixture was incubated for 1 h at 37°C. Each data point, expressed as MDA, μmol/L, is the mean ± SD of 4 independent samples: $*p < .05$ for heme-treated groups vs. the vehicle-treated control.

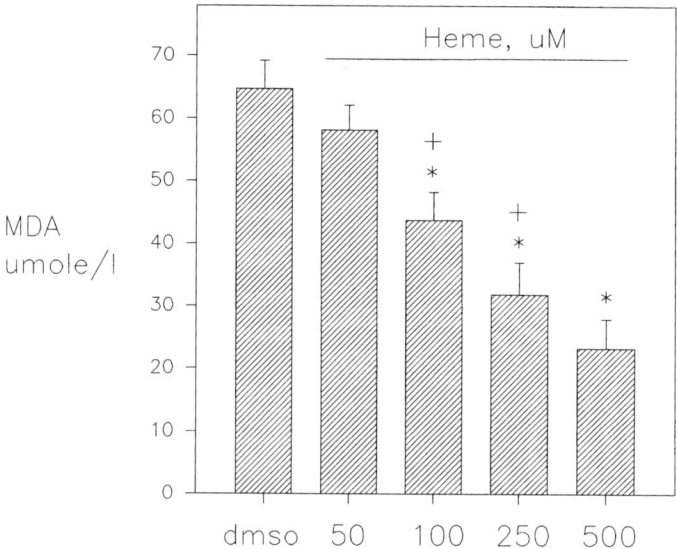

FIGURE 5. *See following page for legend.*

The results obtained were not surprising since several porphyrin compounds, including heme, inhibited iron-stimulated lipid peroxidation in liver homogenates.[93] Moreover, several synthetic porphyrin compounds developed as stable catalytic antioxidants inhibit lipid peroxidation.[94] Lipid hydroperoxides, which form as a result of free radical processes, disrupt cell membrane function. Heme is a lipophilic molecule, and its presence within membranes is the ideal location to influence lipid peroxidation reactions. Hemoglobin undergoes one-electron oxidation and reduction and can act as a source or sink of free radicals.[95] Heme in membranes may act in a similar way. The significance of heme as an antioxidant remains to be clarified, but should be considered when excessive amounts of heme are present.

CONCLUSIONS

As detailed in this chapter, various findings regarding blood removal after ICH suggest that it occurs through a controlled process. However, since the degree of morbidity in ICH survivors is considerably greater than after ischemic stroke, it is possible that this removal process may be inefficient, resulting in the exposure of brain to elevated hemoglobin/heme in an untimely manner. Furthermore, mechanical lysis of blood could occur with traumatic brain injury and contusion or during ICH treatment (aspiration of clots by stereotactic approaches through small cannulae), potentially resulting in elevated hemoglobin/heme concentrations in the tissue prior to "priming" of the brain for the removal process. In the treatment setting, stabilizing red blood cells within hematoma might be valuable strategy. Some metalloporphyrins, including ZnPP and synthetic porphyrins, which have been developed as catalytic antioxidants, may be protective because they have antioxidant effects. Some like SnMP and SnPP may be protective because they inhibit heme oxygenase (this needs to be rigorously tested). The role of heme in the pathophysiology of brain injury in ICH may need to be reevaluated. We suggest that metalloporphyrins might be a useful adjunct treatment for ICH by reducing direct tissue injury caused by oxidative stress and heme degradation products, and possibly by reducing conversion of heme to potentially injurious bilirubin oxidation products, as observed in SAH.

ACKNOWLEDGMENTS

Our studies referred to in this manuscript were supported by NIH Grant No. R01-NS30652 and funds from the Office of Research and Development, Medical Research Service, Department of Veterans Affairs.

FIGURE 5. Effect of heme concentration on TBARS formation in iron-supplemented rat brain homogenates measured as described in FIGURE 1. Heme was added to homogenates and the reaction was initiated by addition of 200 μM FeSO$_4$. The reaction mixture was incubated for 1 h at 37°C. Each data point, expressed as MDA, μmol//L, is the mean ± SD of 4 independent samples: *$p < .05$ for heme-treated groups vs. the vehicle-treated control; $^+p < .05$ for heme-treated group vs. next highest heme dose. Note: Failure to detect significant TBARS reduction in homogenates treated with 50 μM heme was unique to this experiment.

REFERENCES

1. QURESHI, A.I., S. TUHRIM, J.P. BRODERICK *et al.* 2001. Spontaneous intracerebral hemorrhage. N. Engl. J. Med. **344:** 1450–1460.
2. WOO, D. & J.P. BRODERICK. 2002. Spontaneous intracerebral hemorrhage: epidemiology and clinical presentation. Neurosurg. Clin. N. Am. **13:** 265–279.
3. THE NATIONAL INSTITUTE OF NEUROLOGICAL DISORDERS AND STROKE RT-PA STROKE STUDY GROUP. 1995. Tissue plasminogen activator for acute ischemic stroke. N. Engl. J. Med. **333:** 1581–1587.
4. SUZUKI, J. & T. EBINA. 1980. Sequential changes in tissue surrounding ICH. *In* Spontaneous Intracerebral Hematomas, pp. 121–128. Springer-Verlag. Berlin.
5. MENDELOW, A.D. 1993. Mechanisms of ischemic brain damage with intracerebral hemorrhage. Stroke **24:** I115–I117.
6. KOEPPEN, A.H., A.C. DICKSON & J.A. MCEVOY. 1995. The cellular reactions to experimental intracerebral hemorrhage. J. Neurol Sci. **134**(suppl.)**:** 102–112.
7. XI, G., K.R. WAGNER, R.F. KEEP *et al.* 1998. Role of blood clot formation on early edema development after experimental intracerebral hemorrhage. Stroke **29:** 2580–2586.
8. WAGNER, K.R. & J.P. BRODERICK. 2001. Hemorrhagic stroke: pathophysiological mechanisms and neuroprotective treatments. *In* Neuroprotection, pp. 471–508. Prominent Press. Scottsdale, AZ.
9. XI, G., R.F. KEEP & J.T. HOFF. 2002. Pathophysiology of brain edema formation. Neurosurg. Clin. N. Am. **13:** 371–383.
10. XI, G., G. REISER & R.F. KEEP. 2003. The role of thrombin and thrombin receptors in ischemic, hemorrhagic, and traumatic brain injury: deleterious or protective? J. Neurochem. **84:** 3–9.
11. WAGNER, K.R., F.R. SHARP, T.D. ARDIZZONE *et al.* 2003. Heme and iron metabolism: role in cerebral hemorrhage. J. Cereb. Blood Flow Metab. **23:** 629–652.
12. FAYAD, P.B. & I.A. AWAD. 1998. Surgery for intracerebral hemorrhage. Neurology **51:** 69–73.
13. ZUCCARELLO, M., N. ANDALUZ & K.R. WAGNER. 2002. Minimally invasive therapy for intracerebral hematomas. Neurosurg. Clin. N. Am. **13:** 349–354.
14. BRADLEY, W.G., JR. 1993. MR appearance of hemorrhage in the brain. Radiology **189:** 15–26.
15. ATLAS, S.W. & K.R. THULBORN. 1998. MR detection of hyperacute parenchymal hemorrhage of the brain. Am. J. Neuroradiol. **19:** 1471–1477.
16. LETARTE, P.B., K. LIEBERMAN, K. NAGATANI *et al.* 1993. Hemin: levels in experimental subarachnoid hematoma and effects on dissociated vascular smooth-muscle cells. J. Neurosurg. **79:** 252–255.
17. DARROW, V.C., E.C. ALVORD, JR., L.A. MACK *et al.* 1988. Histologic evolution of the reactions to hemorrhage in the premature human infant's brain: a combined ultrasound and autopsy study and a comparison with the reaction in adults. Am. J. Pathol. **130:** 44–58.
18. XI, G., R.F. KEEP & J.T. HOFF. 1998. Erythrocytes and delayed brain edema formation following intracerebral hemorrhage in rats. J. Neurosurg. **89:** 991–996.
19. MACDONALD, R.L. & B.K. WEIR. 1991. A review of hemoglobin and the pathogenesis of cerebral vasospasm. Stroke **22:** 971–982.
20. MARLET, J.M. & J.P. BARRETO FONSECA. 1982. Experimental determination of time of intracranial hemorrhage by spectrophotometric analysis of cerebrospinal fluid. J. Forensic Sci. **27:** 880–888.
21. WAGNER, K.R., G. XI, Y. HUA *et al.* 1996. Lobar intracerebral hemorrhage model in pigs: rapid edema development in perihematomal white matter. Stroke **27:** 490–497.
22. ANDALUZ, N., M. ZUCCARELLO & K.R. WAGNER. 2002. Experimental animal models of intracerebral hemorrhage. Neurosurg. Clin. N. Am. **13:** 385–393.
23. HUA, Y., G. XI, R.F. KEEP *et al.* 2000. Complement activation in the brain after experimental intracerebral hemorrhage. J. Neurosurg. **92:** 1016–1022.
24. MATZ, P.G., P.R. WEINSTEIN & F.R. SHARP. 1997. Heme oxygenase-1 and heat shock protein 70 induction in glia and neurons throughout rat brain after experimental intracerebral hemorrhage. Neurosurgery **40:** 152–160.
25. SADRZADEH, S.M., D.K. ANDERSON, S.S. PANTER *et al.* 1987. Hemoglobin potentiates central nervous system damage. J. Clin. Invest. **79:** 662–664.

26. REGAN, R.F. & S.S. PANTER. 1993. Neurotoxicity of hemoglobin in cortical cell culture. Neurosci. Lett. **153:** 219–222.
27. REGAN, R.F. & S.S. PANTER. 1996. Hemoglobin potentiates excitotoxic injury in cortical cell culture. J. Neurotrauma **13:** 223–231.
28. TOLOSANO, E. & F. ALTRUDA. 2002. Hemopexin: structure, function, and regulation. DNA Cell Biol. **21:** 297–306.
29. ORTIZ DE MONTELLANO, P.R. & A. WILKS. 2001. Heme oxygenase structure and mechanism. Adv. Inorg. Chem. **51:** 359–405.
30. MAINES, M.D. 1997. The heme oxygenase system: a regulator of second messenger gases. Annu. Rev. Pharmacol. Toxicol. **37:** 517–554.
31. TRAKSHEL, G.M., R.K. KUTTY & M.D. MAINES. 1988. Resolution of the rat brain heme oxygenase activity: absence of a detectable amount of the inducible form (HO-1). Arch. Biochem. Biophys. **260:** 732–739.
32. EWING, J.F., S.N. HABER & M.D. MAINES. 1992. Normal and heat-induced patterns of expression of heme oxygenase-1 (HSP32) in rat brain: hyperthermia causes rapid induction of mRNA and protein. J. Neurochem. **58:** 1140–1149.
33. DWYER, B.E., R.N. NISHIMURA, J. DE VELLIS et al. 1992. Heme oxygenase is a heat shock protein and PEST protein in rat astroglial cells. Glia **5:** 300–305.
34. DWYER, B.E., R.N. NISHIMURA, S.Y. LU et al. 1996. Transient induction of heme oxygenase after cortical stab wound injury. Brain Res. Mol. Brain Res. **38:** 251–259.
35. DWYER, B.E., S.Y. LU, J.T. LAITINEN et al. 1998. Protective properties of tin- and manganese-centered porphyrins against hydrogen peroxide–mediated injury in rat astroglial cells. J. Neurochem. **71:** 2497–2504.
36. VERMA, A., D.J. HIRSCH, C.E. GLATT et al. 1993. Carbon monoxide: a putative neural messenger. Science **259:** 381–384.
37. DWYER, B.E., R.N. NISHIMURA & S.Y. LU. 1995. Differential expression of heme oxygenase-1 in cultured cortical neurons and astrocytes determined by the aid of a new heme oxygenase antibody: response to oxidative stress. Brain Res. Mol. Brain Res. **30:** 37–47.
38. DWYER, B.E., R.N. NISHIMURA & S.Y. LU. 1995. Differential localization of heme oxygenase and NADPH-diaphorase in spinal cord neurons. Neuroreport **6:** 973–976.
39. WEBER, C.M., B.C. EKE & M.D. MAINES. 1994. Corticosterone regulates heme oxygenase-2 and NO synthase transcription and protein expression in rat brain. J. Neurochem. **63:** 953–962.
40. RAJU, V.S., W.K. MCCOUBREY, JR. & M.D. MAINES. 1997. Regulation of heme oxygenase-2 by glucocorticoids in neonatal rat brain: characterization of a functional glucocorticoid response element. Biochim. Biophys. Acta **1351:** 89–104.
41. DENNERY, P.A. 2000. Regulation and role of heme oxygenase in oxidative injury. Curr. Top. Cell. Regul. **36:** 181–199.
42. LI, X. & J.D. CLARK. 2000. Chronic morphine exposure and the expression of heme oxygenase type 2. Mol. Brain Res. **75:** 179–184.
43. GORDH, T., H.S. SHARMA, M. AZIZI et al. 2000. Spinal nerve lesion induces upregulation of constitutive isoform of heme oxygenase in the spinal cord: an immunohistochemical investigation in the rat. Amino Acids **19:** 373–381.
44. PANAHIAN, N. & M.D. MAINES. 2001. Site of injury-directed induction of heme oxygenase-1 and -2 in experimental spinal cord injury: differential functions in neuronal defense mechanisms? J. Neurochem. **76:** 539–554.
45. MCCOUBREY, W.K., JR., T.J. HUANG & M.D. MAINES. 1997. Isolation and characterization of a cDNA from the rat brain that encodes hemoprotein heme oxygenase-3. Eur. J. Biochem. **247:** 725–732.
46. SCAPAGNINI, G., R. FORESTI, V. CALABRESE et al. 2002. Caffeic acid phenethyl ester and curcumin: a novel class of heme oxygenase-1 inducers. Mol. Pharmacol. **61:** 554–561.
47. MCDONAGH, A.F. 2001. Turning green to gold. Nat. Struct. Biol. **8:** 198–200.
48. EWING, J.F., C.M. WEBER & M.D. MAINES. 1993. Biliverdin reductase is heat resistant and coexpressed with constitutive and heat shock forms of heme oxygenase in brain. J. Neurochem. **61:** 1015–1023.
49. PANAHIAN, N., T. HUANG & M.D. MAINES. 1999. Enhanced neuronal expression of the oxidoreductase—biliverdin reductase—after permanent focal cerebral ischemia. Brain Res. **850:** 1–13.

50. HORI, R., M. KASHIBA, T. TOMA et al. 2002. Gene transfection of H25A mutant heme oxygenase-1 protects cells against hydroperoxide-induced cytotoxicity. J. Biol. Chem. **277:** 10712–10718.
51. MAINES, M.D. & N. PANAHIAN. 2001. The heme oxygenase system and cellular defense mechanisms. *In* From Genes to Bedside, pp. 249–272. Kluwer/Plenum. New York.
52. MAINES, M.D., J.F. EWING, T.J. HUANG et al. 2001. Nuclear localization of biliverdin reductase in the rat kidney: response to nephrotoxins that induce heme oxygenase-1. J. Pharmacol. Exp. Ther. **296:** 1091–1097.
53. AHMAD, Z., M. SALIM & M.D. MAINES. 2002. Human biliverdin reductase is a leucine zipper-like DNA-binding protein and functions in transcriptional activation of heme oxygenase-1 by oxidative stress. J. Biol. Chem. **277:** 9226–9232.
54. NAGAI, H., K. KITAGAKI, K. KUWABARA et al. 1992. Anti-inflammatory properties of zinc protoporphyrin disodium (Zn-PP-2Na). Agents Actions **37:** 273–283.
55. NAGAI, H., Y. TAKAOKA, H. MORI et al. 1996. The effects of mesoporphyrin on experimental arthritis in mice. Inflammation Res. **45:** 293–298.
56. KADOYA, C., E.F. DOMINO, G.Y. YANG et al. 1995. Preischemic, but not postischemic zinc protoporphyrin treatment reduces infarct size and edema accumulation after temporary focal cerebral ischemia in rats. Stroke **26:** 1035–1038.
57. LIN, T.N., S.F. CHEN, H. LIN et al. 1999. Role of heme oxygenase-1 (HO-1) in cerebral ischemia-reperfusion. Soc. Neurosci. Abstr. **25**(part 2)**:** 1587.
58. PANIZZON, K.L., B.E. DWYER, R.N. NISHIMURA et al. 1996. Neuroprotection against CA1 injury with metalloporphyrins. Neuroreport **7:** 662–666.
59. ZAGER, R.A. & K. BURKHART. 1997. Myoglobin toxicity in proximal human kidney cells: roles of Fe, Ca^{2+}, H_2O_2, and terminal mitochondrial electron transport. Kidney Int. **51:** 728–738.
60. DA SILVA, J.L., T. MORISHITA, B. ESCALANTE et al. 1996. Dual role of heme oxygenase in epithelial cell injury: contrasting effects of short-term and long-term exposure to oxidant stress. J. Lab. Clin. Med. **128:** 290–296.
61. WAGNER, K.R., Y. HUA et al. 2000. Tin-mesoporphyrin, a potent heme oxygenase inhibitor, for treatment of intracerebral hemorrhage: *in vivo* and *in vitro* studies. Cell. Mol. Biol. (Noisy-Le-Grand) **46:** 597–608.
62. PEELING, J., H.J. YAN, S.G. CHEN et al. 1998. Protective effects of free radical inhibitors in intracerebral hemorrhage in rat. Brain Res. **795:** 63–70.
63. PEELING, J., M.R. DEL BIGIO, D. CORBETT et al. 2001. Efficacy of disodium 4-[(*tert*-butylimino)methyl]benzene-1,3-disulfonate *N*-oxide (NXY-059), a free radical trapping agent, in a rat model of hemorrhagic stroke. Neuropharmacology **40:** 433–439.
64. WAGNER, K.R., B.A. PACKARD, C.L. HALL et al. 2002. Protein oxidation and heme oxygenase-1 induction in porcine white matter following intracerebral infusions of whole blood or plasma. Dev. Neurosci. **24:** 154–160.
65. WU, J., Y. HUA, R.F. KEEP et al. 2002. Oxidative brain injury from extravasated erythrocytes after intracerebral hemorrhage. Brain Res. **953:** 45–52.
66. HUANG, F.P., G. XI, R.F. KEEP et al. 2002. Brain edema after experimental intracerebral hemorrhage: role of hemoglobin degradation products. J. Neurosurg. **96:** 287–293.
67. KOEPPEN, A.H. & A.C. DICKSON. 2002. Tin-protoporphyrin prevents experimental superficial siderosis in rabbits. J. Neuropathol. Exp. Neurol. **61:** 689–701.
68. TREGGIARI-VENZI, M.M., P.M. SUTER & J.A. ROMAND. 2001. Review of medical prevention of vasospasm after aneurysmal subarachnoid hemorrhage: a problem of neurointensive care. Neurosurgery **48:** 249–261.
69. DUMONT, A.S., R.J. DUMONT, M.M. CHOW et al. 2003. Cerebral vasospasm after subarachnoid hemorrhage: putative role of inflammation. Neurosurgery **53:** 123–133.
70. VOLLMER, D.G., K. HONGO, H. OGAWA et al. 1991. A study of the effectiveness of the iron-chelating agent deferoxamine as vasospasm prophylaxis in a rabbit model of subarachnoid hemorrhage. Neurosurgery **28:** 27–32.
71. ARTHUR, A.S., A.H. FERGUS, G. LANZINO et al. 1997. Systemic administration of the iron chelator deferiprone attenuates subarachnoid hemorrhage–induced cerebral vasospasm in the rabbit. Neurosurgery **41:** 1385–1391.
72. HORKY, L.L., R.M. PLUTA, R.J. BOOCK et al. 1998. Role of ferrous iron chelator 2,2′-dipyridyl in preventing delayed vasospasm in a primate model of subarachnoid hemorrhage. J. Neurosurg. **88:** 298–303.

73. KRANC, K.R., G.J. PYNE, L. TAO et al. 2000. Oxidative degradation of bilirubin produces vasoactive compounds. Eur. J. Biochem. **267:** 7094–7101.
74. CLARK, J.F., M. REILLY & F.R. SHARP. 2002. Oxidation of bilirubin produces compounds that cause prolonged vasospasm of rat cerebral vessels: a contributor to subarachnoid hemorrhage–induced vasospasm. J. Cereb. Blood Flow Metab. **22:** 472–478.
75. SUZUKI, H., K. KANAMARU, H. TSUNODA et al. 1999. Heme oxygenase-1 gene induction as an intrinsic regulation against delayed cerebral vasospasm in rats. J. Clin. Invest. **104:** 59–66.
76. SUZUKI, H., K. KANAMARU, H. TSUNODA et al. 2001. The functional significance of heme oxygenase-1 gene induction in a rat vasospasm model. Acta Neurochir. Suppl. **77:** 89–91.
77. ONO, S., T. KOMURO & R.L. MACDONALD. 2002. Heme oxygenase-1 gene therapy for prevention of vasospasm in rats. J. Neurosurg. **96:** 1094–1102.
78. SCHWARTZ, S.M. 2001. A protective player in the vascular response to injury. Nat. Med. **7:** 656–657.
79. WINESTONE, J.S., C. BONNER & C.W. LEFFLER. 2003. Carbon monoxide as an attenuator of vasoconstriction in piglet cerebral arterioles. Exp. Biol. Med. (Maywood) **228:** 46–50.
80. WAGENER, F.A., A. EGGERT, O.C. BOERMAN et al. 2001. Heme is a potent inducer of inflammation in mice and is counteracted by heme oxygenase. Blood **98:** 1802–1811.
81. WAGENER, F.A., H.D. VOLK, D. WILLIS et al. 2003. Different faces of the heme–heme oxygenase system in inflammation. Pharmacol. Rev. **55:** 551–571.
82. GRACA-SOUZA, A.V., M.A. ARRUDA, M.S. DE FREITAS et al. 2002. Neutrophil activation by heme: implications for inflammatory processes. Blood **99:** 4160–4165.
83. JENEY, V., J. BALLA, A. YACHIE et al. 2002. Pro-oxidant and cytotoxic effects of circulating heme. Blood **100:** 879–887.
84. MOSLEY, K., D.E. WEMBRIDGE, V. CATTELL et al. 1998. Heme oxygenase is induced in nephrotoxic nephritis and hemin, a stimulator of heme oxygenase synthesis, ameliorates disease. Kidney Int. **53:** 672–678.
85. ANDERSON, K.E. 2003. Approaches to treatment and prevention of human porphyrias. *In* The Porphyrin Handbook: Medical Aspects of Porphyrias, pp. 247–284. Elsevier. Amsterdam/New York.
86. TAKIZAWA, S., H. HIRABAYASHI, K. MATSUSHIMA et al. 1998. Induction of heme oxygenase protein protects neurons in cortex and striatum, but not in hippocampus, against transient forebrain ischemia. J. Cereb. Blood Flow Metab. **18:** 559–569.
87. KITAMURA, Y., Y. ISHIDA, K. TAKATA et al. 2003. Hyperbilirubinemia protects against focal ischemia in rats. J. Neurosci. Res. **71:** 544–550.
88. LIU, Y., B. ZHU, L. LUO et al. 2001. Heme oxygenase-1 plays an important protective role in experimental autoimmune encephalomyelitis. Neuroreport **12:** 1841–1845.
89. LIU, Y., B. ZHU, X. WANG et al. 2003. Bilirubin as a potent antioxidant suppresses experimental autoimmune encephalomyelitis: implications for the role of oxidative stress in the development of multiple sclerosis. J. Neuroimmunol. **139:** 27–35.
90. CHAKRABARTY, A., M.R. EMERSON & S.M. LEVINE. 2003. Heme oxygenase-1 in SJL mice with experimental allergic encephalomyelitis. Mult. Scler. **9:** 372–381.
91. ZHANG, X., P. SHAN, L.E. OTTERBEIN et al. 2003. Carbon monoxide inhibition of apoptosis during ischemia-reperfusion lung injury is dependent on the p38 mitogen-activated protein kinase pathway and involves caspase 3. J. Biol. Chem. **278:** 1248–1258.
92. RYTER, S.W. & R.M. TYRRELL. 2000. The heme synthesis and degradation pathways: role in oxidant sensitivity—heme oxygenase has both pro- and antioxidant properties. Free Radical Biol. Med. **28:** 289–309.
93. IMAI, K., T. AIMOTO, M. SATO et al. 1990. Antioxidative effect of several porphyrins on lipid peroxidation in rat liver homogenates. Chem. Pharm. Bull. (Tokyo) **38:** 258–260.
94. PATEL, M. & B.J. DAY. 1999. Metalloporphyrin class of therapeutic catalytic antioxidants. Trends Pharmacol. Sci. **20:** 359–364.
95. WINTERBOURN, C.C. 1990. Oxidative reactions of hemoglobin. Methods Enzymol. **186:** 265–272.

The Role of Iron in the Pathogenesis of Experimental Allergic Encephalomyelitis and Multiple Sclerosis

STEVEN M. LeVINE AND ANURADHA CHAKRABARTY

Department of Molecular and Integrative Physiology, Mental Retardation and Human Development Center, University of Kansas Medical Center, Kansas City, Kansas, USA

ABSTRACT: Multiple sclerosis (MS) and its animal model, experimental allergic encephalomyelitis (EAE), are autoimmune disorders resulting in demyelination in the central nervous system (CNS). Pathologically, the blood-brain barrier becomes damaged, macrophages and T cells enter into the CNS, oligodendrocytes and myelin are destroyed, astrocytes and microglia undergo gliosis, and axons become transected. Data from several biochemical and pharmacological studies indicate that free radicals participate in the pathogenesis of EAE, and iron has been implicated as the catalyst leading to their formation. The primary focus of this article is the examination of the role of iron in the pathogenesis of MS and EAE. Particular attention will be paid to the role and distribution of iron and proteins involved with iron metabolism (e.g., transferrin, ferritin, heme oxygenase-1, etc.) in normal and disease states of myelin. Furthermore, therapeutic interventions aimed at iron, iron-binding proteins, and substrates or products of iron-catalyzed reactions leading to free radical production will be discussed.

KEYWORDS: iron; myelin; oligodendrocyte; multiple sclerosis; experimental allergic encephalomyelitis; oxidative damage; heme oxygenase; ferritin; transferrin

INTRODUCTION

In humans, lipids represent ~33% of the dry weight of gray matter, ~55% of the dry weight of white matter, and ~70% of the dry weight of myelin.[1] The oligodendrocyte is responsible for producing massive quantities of lipids that become incorporated into the multilamellar structure of myelin, and each oligodendrocyte can produce up to 50 or more myelin segments. This "lipid factory" requires sufficient enzymatic machinery for the biosynthetic steps required for this high level of lipid production, and many of these enzymes utilize iron as part of their catalytic center. It has been suggested that the high concentration of iron observed within the oligo-

---

Address for correspondence: Steven M. LeVine, Ph.D., Department of Molecular and Integrative Physiology, Mental Retardation and Human Development Center, University of Kansas Medical Center, 3901 Rainbow Boulevard, Kansas City, KS 66160. Voice: 913-588-7420; fax: 913-588-5677.

slevine@kumc.edu

dendrocyte is due to an elevated expression of enzymes involved with myelin production, although other explanations have been put forth, such as the oligodendrocyte being a center for iron distribution to the rest of the central nervous system (CNS). In addition to partaking in normal physiological processes within oligodendrocytes, the high levels of iron have been suggested to promote pathogenesis during disease states such as multiple sclerosis (MS) and its animal model, experimental allergic encephalomyelitis (EAE), due to the ability of iron to catalyze reactions that lead to oxidative tissue damage. The function of high levels of iron within oligodendrocytes in healthy states and the role of iron in demyelinating diseases of the CNS are the focus of this paper.

IRON, TRANSFERRIN, AND FERRITIN IN OLIGODENDROCYTES AND MYELIN

A large number of studies have examined the distribution of iron in the CNS by histochemical staining procedures. The Perls' histochemical stain has been used in many investigations, especially in combination with 3,3'-diaminobenzidine enhancement of the ferric ferrocyanide reaction product.[2] Other modifications include the utilization of permeabilization steps to increase the penetration of histochemical reagents into densely myelinated areas[3,4] and changes in fixatives and/or incubation times.[5] A consensus among many studies is that iron is enriched within oligodendrocytes and myelin.[3,4,6-12] Electron microscopic studies revealed iron deposits in the cytoplasm of oligodendrocytes[11] and within the inner and outer loops of myelin,[6] and it is possible that compact myelin also contains appreciable amounts of iron.[4] In addition to iron histochemical staining, substantial concentrations of iron also have been detected by atomic absorption in myelin fractions of brain homogenates.[13]

The role that high levels of iron perform within oligodendrocytes is not fully established. The high concentration of iron in oligodendrocytes has been suggested to be associated with biosynthetic enzymes that are involved with the high metabolic demands of myelinogenesis.[3,14] However, phylogenetic studies revealed that oligodendrocytes in the fish and frog did not have high iron levels as detected by histochemistry,[15] suggesting that high iron levels are not essential for the formation and/or maintenance of myelin. On the other hand, iron deficiency during early postnatal life causes a reduction in myelination,[16,17] indicating that the oligodendrocyte is sensitive to low iron levels. An alternate suggestion was that, in species with iron-enriched oligodendrocytes, the iron serves as a storage depot to be tapped for delivery of iron to other cells in the CNS. In support of this idea is the observation that oligodendrocytes synthesize transferrin,[12,18-20] and this transferrin may deliver iron to other cells in the CNS.[21] In one study, cultured oligodendrocytes were shown to synthesize and secrete transferrin,[22] but this was not confirmed in a follow-up study.[23] Thus, it is uncertain whether iron delivery by oligodendrocyte transferrin contributes significantly to the transport of iron in the CNS. Oligodendrocyte transferrin also has been suggested to serve in an autocrine capacity to help oligodendrocytes accumulate iron.[24] However, transferrin receptors are absent or present in low abundance in white matter,[25,26] and mechanisms to sequester iron other than involving transferrin can be used by glial cells.[27] Transferrin has been shown to serve as a growth factor[28] and it was found to be important for the maturation and function of oligodendrocytes,[29–31] suggesting that the receptor may be present in oligodendrocyte progenitors. However, the role of transferrin as a growth factor would not account for the large accumulation of iron that is observed within oligodendrocytes.

Unlike transferrin receptors, ferritin receptors are present in high concentrations in white matter,[32,33] and cultured oligodendrocytes bind and internalize ferritin.[34] Furthermore, due to the large binding capacity of ferritin for iron, there is the potential for a greater delivery of iron by ferritin than transferrin.[32] In addition to binding ferritin, oligodendrocytes express ferritin.[12,35–37] There are two subunits of ferritin, heavy (H) and light (L), and both subunits are expressed by oligodendrocytes.[38–41] Neurons, microglia, and astrocytes also express ferritin similar to oligodendrocytes,[40,42] yet these cells are not routinely stained by iron histochemistry, suggesting that the large accumulation of iron observed in oligodendrocytes is not necessarily due solely to the presence of ferritin. Moreover, immunohistochemical staining of ferritin or transferrin fails to reveal staining of myelin. This is in contrast to findings with iron histochemistry where myelin staining has been clearly documented.[3,4,6–12] Thus, it is likely that there is no one protein that accounts for the large majority of iron binding in oligodendrocytes. Transferrin, ferritin, and iron-containing enzymes involved with myelinogenesis probably all contribute to sites of iron localization within oligodendrocytes. The proteins that bind iron within myelin are less clear. However, iron enrichment within both oligodendrocytes and myelin raises the possibility that an imbalance in the management of iron during disease could lead to the production of iron-catalyzed free radicals that result in oxidative damage.

IRON, FERRITIN, AND TRANSFERRIN IN EAE AND MS

Histochemical staining of iron in CNS tissue from SJL mice with EAE revealed iron deposits that were not present in the CNS of normal animals. For example, during clinically active disease, there was histochemical staining of iron within macrophages and extravasated RBCs, and granular staining was present in extracellular sites and possibly within some astrocytes.[43] During the recovery phase of disease, staining persisted in macrophages and granular deposits.[43] In tissue from MS patients, an initial report by Craelius et al.[44] revealed abnormal iron deposits in 5 out of 5 MS patients, but these findings were not fully confirmed in two subsequent studies.[45,46] These three studies on MS tissue did not include any steps to enhance the permeability of the tissue to the histochemical reagents, and they were carried out on paraffin sections where the processing steps could facilitate the leaching of iron from the tissue. When Vibratome sections were utilized together with permeabilization steps, iron deposits were observed in macrophages in tissue from 5 out of 5 MS patients, and labeled reactive microglia and ameboid macrophages were observed in 3 out of 5 tissues.[47] In tissue from 1 MS patient, labeling of axons was present.[47] Craelius et al. also noted axonal staining in their report.[44] It is possible that the axonal staining revealed axons that were recently transected since axonal transection is a predominant pathological feature of MS.[48–51] In addition to labeled axons, punctate iron deposits were observed within some neurons of patients with MS similar to that observed for neurons in CNS tissue from patients affected with Alzheimer's disease.[47] These deposits within neurons likely represent cells undergoing degeneration since neuronal loss is also a pathological feature of MS.[52] In addition to neurons, punctate deposits were observed within some oligodendrocytes in MS tissue.[47] Mitochondria are possible sites of these punctate deposits within oligodendrocytes and neurons since two proinflammatory cytokines found in MS, TNF-α and IL-1β, have been shown to lead to the accumulation of iron within mitochondria in astrocyte cultures.[53]

During stress, such as hypoxia, oligodendrocytes increase their synthesis of ferritin,[54,55] and ferritin levels are increased in the CNS of EAE animals compared to control animals.[56] Ferritin levels, but not transferrin or iron levels,[57,58] were found to be significantly elevated within the CSF of MS patients with chronic progressive active disease, but not relapsing remitting disease, compared to levels in the CSF of control patients.[58] In other conditions, the upregulation of ferritin expression is thought to be associated with the protection of cells against oxidative damage[59–62] and/or the inhibition of cell-mediated immunity.[63–67] Cell-mediated immunity, that is, T cells and macrophages, is the major contributor to pathology in EAE and MS. Thus, the elevated levels of ferritin in EAE and MS may be a protective mechanism to limit the toxic effects of iron during ongoing pathogenesis.

In MS tissue, there is an absence of ferritin binding sites in and adjacent to lesion sites, which is likely due to the loss of oligodendrocytes in this disease, while in the normal brain there is a relatively high concentration of ferritin receptors in white matter compared to gray matter.[33] Unlike ferritin, transferrin can bind to periplaque regions and to occasional plaques in MS tissue,[33] indicating that the receptors accounting for transferrin binding were present in cells other than oligodendrocytes.

Natural resistance–associated macrophage protein-1 (Nramp1) modulates iron metabolism in macrophages and is thought to play an important role in macrophage activation.[68] Since the macrophage is critical for the pathogenic development of EAE and MS, Nramp1 has been suggested to be involved with CNS demyelinating diseases. Although far from proven, an allele of this gene has been suggested to be associated with MS susceptibility in South African Caucasians[69] and alleles in this gene might influence the susceptibility and/or severity of other autoimmune diseases such as rheumatoid arthritis.[70,71] Thus, the management of iron in the CNS may be a precipitating factor for the onset and/or progression of MS.

FREE RADICAL DAMAGE TO OLIGODENDROCYTES/MYELIN

The abnormal iron deposits observed in EAE and MS tissues indicate that the normal homeostasis of iron is disrupted, and iron is likely released from the proteins that it normally binds. Released iron will quickly bind to neighboring molecules, and iron that is loosely bound, or in a free state, can catalyze reactions that lead to the production of reactive oxygen intermediates (ROI). ROI can promote cellular damage at many levels, for example, proteins, DNA, lipids, mitochondrial function, etc. Data from a variety of studies indicate that oxidative tissue damage occurs in EAE and MS. For example, lipid peroxidation products were observed in EAE and MS tissue,[72,73] the production rates of ROI from inflammatory cells were increased from EAE mice compared to control mice,[74,75] and pharmacological interventions aimed at disrupting oxidative damage have therapeutic value in EAE and possibly MS (discussed below).

During EAE or MS, cells in the CNS respond to inflammation by inducing the expression of stress response proteins.[76,77] One stress response protein related to iron metabolism is heme oxygenase-1 (HO-1). HO-1 expression can be induced by many factors including heme, metals, glutathione depletion, nitric oxide, cytokines, etc., and several of these stressors are present in EAE and MS. For example, abnormal iron deposits are present in EAE and MS tissues,[43,47] glutathione depletion occurs in EAE and MS,[56,78,79] nitric oxide and its products are increased in EAE and MS,[80,81] and proinflammatory cytokines are enhanced in MS.[82,83]

HO-1 acts in association with NADPH cytochrome P450 reductase, which provides reducing equivalents, to convert heme into biliverdin, carbon monoxide, and iron. The various products of HO-1 activity have pro- or antioxidative properties.[84] For example, biliverdin and bilirubin (which is rapidly generated from biliverdin by biliverdin reductase) are both antioxidants. The released iron is a pro-oxidant if it is not properly sequestered. Cells protect themselves from the toxic effects of iron by responding to an increased iron load by downregulating the transferrin receptor[85] and by upregulating ferritin expression.[86,87] Lower levels of transferrin receptor will restrict the entry of additional iron into the cell, while the increased expression of ferritin will bind and store the released iron. Iron-responsive proteins (IRPs) bind iron-responsive elements on ferritin mRNA, preventing its translation.[86,87] When IRPs sense an increase in cytosolic iron levels, they allow ferritin mRNA to undergo translation, which results in the sequestering of iron by ferritin. Disruption of this regulation can lead to an enhanced accumulation of iron and ubiquitin-containing inclusions within oligodendrocytes.[88] Furthermore, mice deficient in H-ferritin have increased evidence of oxidative stress in their CNS.[89] Thus, a coordinated response is required to limit the potential toxic effects of iron that is liberated by HO-1.

HO-1 expression has been observed to be increased in EAE[90,91] and MS[53] tissues, and immunohistochemical studies revealed that HO-1 was expressed predominantly by macrophages and some astrocytes in EAE tissue[90,91] and in astrocytes in MS tissue.[53] The administration of the HO-1 inhibitor, tin-protoporphyrin IX, to EAE mice resulted in the enhanced induction of HO-1 levels above the levels already increased by the EAE disease.[56] HO-1 induction was not observed in control animals given tin-protoporphyrin IX, suggesting that the breakdown of the blood-brain barrier that occurs in EAE[92,93] accounts for the greater access of drugs to the CNS in EAE animals compared to control animals,[94] and this would allow tin-protoporphyrin IX to induce the expression of HO-1 in CNS cells, similar to what has been observed in the liver.[95] Tin-protoporphyrin IX appeared to enhance the expression of HO-1 in astrocytes and microglia since these cells were only occasionally observed in EAE animals given vehicle, but they were observed more frequently in EAE animals given tin-protoporphyrin IX (FIGS. 1A–C). In addition to greater expression in astrocytes and microglia, induction was also enhanced in radial glia (FIG. 1D) and possible other cells as well. Besides inducing HO-1 expression, tin-protoporphyrin IX is known to inhibit HO-1 activity, especially at higher doses (discussed below).

During EAE, the oligodendrocyte is exposed to stress due to the inflammatory response directed at its antigens. However, HO-1 staining was not evident in this cell even after the administration of tin-protoporphyrin IX to EAE animals. Even though HO-1 staining was not clearly observed within oligodendrocytes, all cells need to metabolize heme. Thus, oligodendrocytes should have the capacity to express one or more types of HO, which could facilitate the deposition of iron in ferritin, whose expression is elevated in this cell type[38–41] and whose levels increase in the CNS during EAE.[56] The inability to detect HO-1 staining in oligodendrocytes during

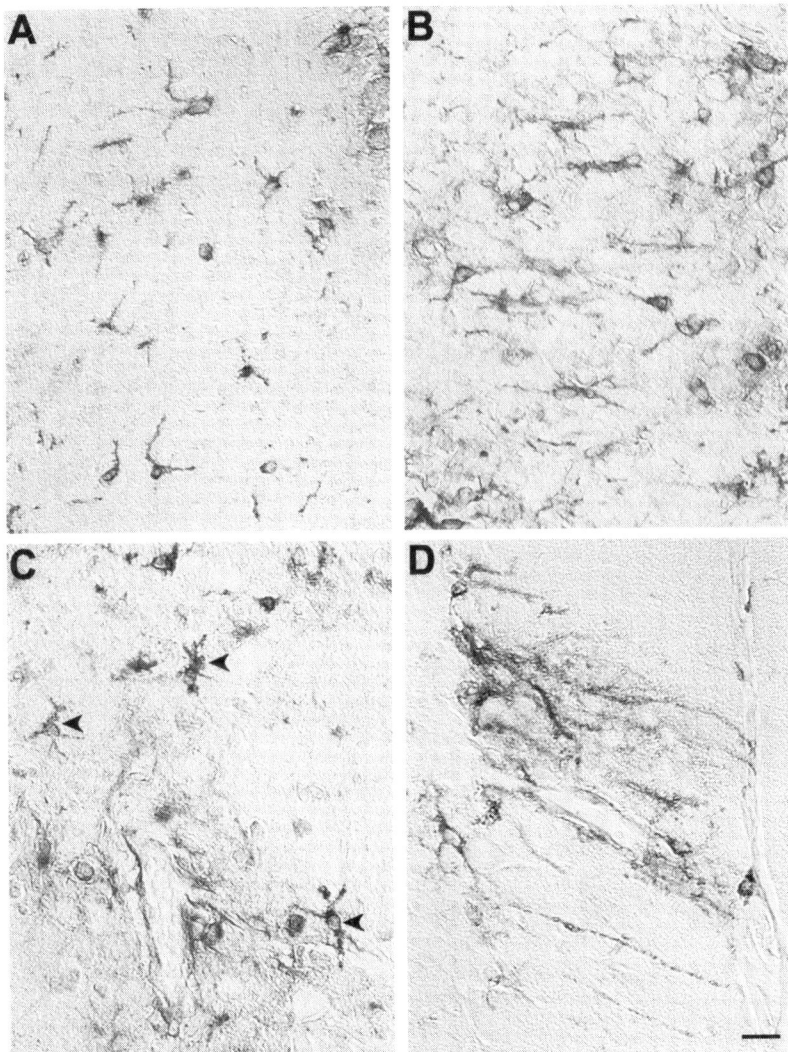

FIGURE 1. HO-1 immunohistochemical staining on formalin-fixed, paraffin sections utilizing 1:6000 rabbit anti-mouse HO-1 (StressGen Biotechnologies, Victoria, British Columbia, Canada), horseradish peroxidase–labeled goat anti-rabbit IgG, and 3,3'-diaminobenzidine. The frequency of labeling in **(A)** reactive microglia, **(B)** amoeboid microglia, **(C)** astrocytes (*arrowheads*), and **(D)** radial glia appeared greater in EAE SJL mice given 50 µmol/kg (C, D) or 200 µmol/kg (A, B) tin-protoporphyrin IX than EAE mice given vehicle, suggesting that tin-protoporphyrin IX could induce HO-1 expression in the CNS of EAE animals. Labeled infiltrating, round macrophages were abundant in EAE animals given vehicle or tin-protoporphyrin IX (not shown). Bar: 20 µm.

disease could be due to several factors. For example, the levels of expression could be well below that for other cells, such as macrophages, which would make it difficult to optimize staining conditions that clearly reveal staining in both cell types. Alternatively, a different form of HO than HO-1, for example, HO-2, could be expressed by oligodendrocytes or the expression of HO-1 could last only minutes or hours and thus missed in a disease lasting several days for EAE and many years for MS patients.

As HO-1 expression increases during disease, the overall expression of NADPH cytochrome P450 reductase expression decreases in EAE tissue.[90] The reduction of NADPH cytochrome P450 reductase would appear to be inconsistent with the increase of HO-1 in the CNS of EAE animals; however, this enzyme is also used by other enzymes such as the cytochrome P450s. The activity of a cytochrome P450 was found to be reduced in the CNS of EAE animals,[96] which would parallel the reduction of NADPH cytochrome P450 reductase,[90] and other cytochrome P450s also may be reduced. Furthermore, the reduction in NADPH cytochrome P450 reductase levels is consistent with its pattern of expression in other models of stress.[97,98] Thus, while NADPH cytochrome P450 reductase is likely associating with HO-1 during EAE, on balance its levels throughout the brain are reduced during disease.

THERAPEUTIC APPROACHES TARGETING PATHOGENIC MECHANISM INVOLVING IRON

Various therapeutic interventions targeting iron, iron management, or iron-catalyzed free radicals have been explored for the treatment of EAE and MS. The most direct approach has been the utilization of iron chelation therapy. In 1984, Bowern et al.[99] administered the iron chelator, Desferal (also known as desferrioxamine and deferoxamine), to Lewis rats given guinea pig spinal cord homogenates as the encephalitogen. Both the duration and severity of disease were reduced in the treated groups. In a subsequent study, Desferal failed to reduce disease severity in Lewis rats given myelin basic protein (MBP) as the encephalitogen, but the drug was administered only from days 1–7 postencephalitogen injection, while disease onset was day 11.[100] In an effort to clarify the discrepancy between these two studies, a third study was performed on SJL mice given MBP as the encephalitogen.[94] Desferal was given during the clinical stage of disease rather than the preclinical period as was the case in the second negative study.[100] Treatment with Desferal resulted in disease suppression in this third study, and immunohistochemical staining of Desferal revealed its presence in the CNS of EAE animals.[94] Administration of Desferal during the active stages of experimental uveitis was also found to suppress lipid peroxidation in the retina.[101] Thus, taken together, these data support the notion that Desferal acts to suppress the active stage of disease by limiting iron-catalyzed free radical tissue damage.

Due to the ability to inhibit free radical tissue injury, Desferal was tested for therapeutic value in three studies on MS patients. In the first study, 12 MS patients were given 2 g/day for 5 days/week for 3 months.[102] At the end of the study, 7 of the 12 patients showed improvement, 4 patients were unchanged, and 1 was worse. The second study gave Desferal at 2 g/day for 7 days followed by 1 g/day for an additional 7 days. At 3 months following treatment, 9/18 patients showed improvement, 7/18 were unchanged, and 2/18 showed worsening; however, as time progressed, the patients displayed a trend to have a worsening of disease.[103] In the third study, Desferal was given at 2 g/day for 7 days followed by 1 g/day for an additional 7 days and this was repeated every 3 months for 2 years. Out of 9 patients, 1 showed improvement, 3 were unchanged, and 5 worsened by 0.5 points on the Kurtzke expanded disability status scale.[104] Taken together, the results are inconclusive about whether Desferal has therapeutic value for the treatment of MS, and a larger, double-blind trial needs to be performed to resolve this question. Although the patients appeared to tolerate Desferal reasonably well, a serious drawback is that this drug is usually administered by a subcutaneous pump over several hours, which is a difficult and cumbersome method for drug administration. A more promising approach may be the administration of an iron chelator that can be given by an oral route when a suitable one becomes available.

Although there is debate as to whether HO-1 serves a protective or pathogenic role during disease,[84,105] interventions aimed at HO-1 have been pursued for the treatment of EAE. Hemin (40 µmol/kg), an inducer of HO-1 expression, was found to ameliorate EAE in Lewis rats, while tin-mesoporphyrin (40 µmol/kg), an inhibitor of HO-1, was found to worsen disease.[106] The authors suggest that HO-1 suppresses disease by the production of biliverdin, which is converted to bilirubin, and/or carbon monoxide.[106] Bilirubin serves as an antioxidant,[85,107,108] while carbon monoxide is thought to be an anti-inflammatory agent.[109] In a follow-up study, bilirubin was administered (50, 100, 200 mg/kg) to Lewis rats during the active stage of EAE and it was found to suppress disease in a dose-dependent manner.[110] However, caution should be exercised with respect to advancing this form of therapy for a chronic condition like MS since high levels of bilirubin can adversely affect the nervous system.

A second study examined the role of HO-1 in EAE.[56] In this study, the SJL mouse model was used together with the HO-1 inhibitor, tin-protoporphyrin IX (50 and 200 µmol/kg). The high dose of inhibitor was found to suppress clinical and pathological evidence of disease, and oxidative stress was reduced.[56] In the SJL model, there is extravasation of RBCs into the CNS during disease,[43,111] and the release of iron from heme by HO-1 has been suggested to account for the pathogenic effects of HO-1 in this model, which may be similar to pathogenic mechanisms of HO-1 in stroke,[112] traumatic brain injury,[113] and cerebral ischemia.[114] Thus, the different results between studies on Lewis rats and SJL mice may be due to variations in pathological features between these models or the different doses of inhibitors. Since pathological studies indicate variations of pathological mechanisms in MS,[115] and extravasation of RBCs has been suggested in some MS patients,[46,116] it is unclear whether HO-1 serves to advance or attenuate pathology during the course of this disease.

The propathogenic mechanism of HO-1 in SJL mice with EAE was suggested to be related to a large release of iron from heme and a failure of ferritin to adequately sequester the iron in a timely and/or complete manner.[56] Thus, to test this possibility, apoferritin was administered to SJL mice with EAE. Apoferritin was found to suppress disease activity in EAE mice, while injections of iron, which increased serum ferritin levels, failed to ameliorate the disease course.[117] It was suggested that the ferritin synthesized in response to iron injections quickly acquired the injected iron and lost some or most of its therapeutic potential since ferritin loaded with iron can release iron, especially when exposed to superoxide anion radical or nitric oxide.[118,119] The therapeutic action of apoferritin was suggested to be due to the sequestering of the exogenous iron that occurs in this disease,[43] and this mechanism would be similar to that suggested for Desferal described above.

Other therapies aimed at reducing the substrates or products of iron-catalyzed reactions leading to ROI have been tested in EAE. The administration of catalase, but not superoxide dismutase, to Lewis rats with EAE resulted in the suppression of disease activity.[75] High doses of uric acid, a scavenger of peroxynitrite, which is produced from superoxide anion radical and nitric oxide, resulted in suppression of disease in PLSJL mice.[120] Scavengers of oxygen radicals such as α-lipoic acid,[121] butylated hydroxyanisole,[122] EUK-8,[123] melatonin,[124] N-acetyl-L-cysteine,[125] thymoquinone,[126] etc., resulted in disease suppression. Thus, there is a growing body of evidence indicating that interventions aimed at iron or at the substrates or products of iron-catalyzed reactions that produce ROI ameliorate EAE disease. These studies suggest that this pathogenic mechanism may hold potential as a target for therapeutic intervention for MS.

SUMMARY

The massive quantities of lipids produced by oligodendrocytes may be responsible, in part, for the high accumulation of iron in this cell type. Some proteins that are involved with iron management, for example, ferritin and transferrin, are also expressed in abundance within this cell. Disruption of iron metabolism within oligodendrocytes, or in other cells within the CNS, could help to precipitate or advance MS, and interventions targeting iron-catalyzed reactions warrant further exploration for the treatment of MS.

ACKNOWLEDGMENTS

Studies by S. M. LeVine were supported by research grants from the National Multiple Sclerosis Society (RG 3188A3), NINDS (NS 33596), and NICHD (Mental Retardation Research Center Grant HD 02528).

REFERENCES

1. MORELL, P., R.H. QUARLES & W.T. NORTON. 1993. Myelin formation, structure, and biochemistry. *In* Basic Neurochemistry, pp. 117–143. Raven Press. New York.
2. NGUYEN-LEGROS, J., J. BIZOT, M. BOLESSE & J.P. PULICANI. 1980. "Diaminobenzidine black" as a new histochemical demonstration of exogenous iron. Histochemistry **66:** 239–244.
3. LEVINE, S.M. & W.B. MACKLIN. 1990. Iron-enriched oligodendrocytes: a reexamination of their spatial distribution. J. Neurosci. Res. **26:** 508–512.
4. LEVINE, S.M. 1991. Oligodendrocytes and myelin sheaths in normal, quaking, and shiverer brains are enriched in iron. J. Neurosci. Res. **29:** 413–419.

5. SMITH, M.A., P.L. HARRIS, L.M. SAYRE & G. PERRY. 1997. Iron accumulation in Alzheimer disease is a source of redox-generated free radicals. Proc. Natl. Acad. Sci. USA **94:** 9866–9868.
6. FRANCOIS, C., J. NGUYEN-LEGROS & G. PERCHERON. 1981. Topographical and cytological localization of iron in rat and monkey brains. Brain Res. **215:** 317–322.
7. HILL, J.M. & R.C. SWITZER III. 1984. The regional distribution and cellular localization of iron in the rat brain. Neuroscience **11:** 595–603.
8. HILL, J.M., M.R. RUFF, R.J. WEBER & C.B. PERT. 1985. Transferrin receptors in rat brain: neuropeptide-like pattern and relationship to iron distribution. Proc. Natl. Acad. Sci. USA **82:** 4553–4557.
9. DWORK, A.J., E.A. SCHON & J. HERBERT. 1988. Nonidentical distribution of transferrin and ferric iron in human brain. Neuroscience **27:** 333–345.
10. GERBER, M.R. & J.R. CONNOR. 1989. Do oligodendrocytes mediate iron regulation in the human brain? Ann. Neurol. **26:** 95–98.
11. CONNOR, J.R. & S.L. MENZIES. 1990. Altered cellular distribution of iron in the central nervous system of myelin deficient rats. Neuroscience **34:** 265–271.
12. CONNOR, J.R., S.L. MENZIES, S.M. ST. MARTIN & E.J. MUFSON. 1990. Cellular distribution of transferrin, ferritin, and iron in normal and aged human brains. J. Neurosci. Res. **27:** 595–611.
13. RAJAN, K.S., R.W. COLBURN & J.M. DAVIS. 1976. Distribution of metal ions in the subcellular fractions of several rat brain areas. Life Sci. **18:** 423–431.
14. CONNOR, J.R., G. PAVLICK, D. KARLI et al. 1995. A histochemical study of iron-positive cells in the developing rat brain. J. Comp. Neurol. **355:** 111–123.
15. ERB, G.L., D.L. OSTERBUR & S.M. LEVINE. 1996. The distribution of iron in the brain: a phylogenetic analysis using iron histochemistry. Brain Res. Dev. Brain Res. **93:** 120–128.
16. YU, G.S., T.M. STEINKIRCHNER, G.A. RAO & E.C. LARKIN. 1986. Effect of prenatal iron deficiency on myelination in rat pups. Am. J. Pathol. **125:** 620–624.
17. BEARD, J.L., J.A. WIESINGER & J.R. CONNOR. 2003. Pre- and postweaning iron deficiency alters myelination in Sprague-Dawley rats. Dev. Neurosci. **25:** 308–315.
18. BLOCH, B., T. POPOVICI, M.J. LEVIN et al. 1985. Transferrin gene expression visualized in oligodendrocytes of the rat brain by using in situ hybridization and immunohistochemistry. Proc. Natl. Acad. Sci. USA **82:** 6706–6710.
19. CONNOR, J.R. & R.E. FINE. 1987. Development of transferrin-positive oligodendrocytes in the rat central nervous system. J. Neurosci. Res. **17:** 51–59.
20. URBAN, K., M. HEWICKER-TRAUTWEIN & G. TRAUTWEIN. 1998. Immunohistochemical localization of transferrin in the pre- and postnatal bovine brain. Anat. Histol. Embryol. **27:** 45–49.
21. ESPINOSA DE LOS MONTEROS, A., L. PENA & J. DE VELLIS. 1989. Does transferrin have a special role in the nervous system? J. Neurosci. Res. **24:** 125–136.
22. ESPINOSA DE LOS MONTEROS, A., S. KUMAR, S. SCULLY et al. 1990. Transferrin gene expression and secretion by rat brain cells in vitro. J. Neurosci. Res. **25:** 576–580.
23. DE ARRIBA ZERPA, G.A., M.C. SALEH, P.M. FERNANDEZ et al. 2000. Alternative splicing prevents transferrin secretion during differentiation of a human oligodendrocyte cell line. J. Neurosci. Res. **61:** 388–395.
24. CONNOR, J.R. & S.L. MENZIES. 1996. Relationship of iron to oligodendrocytes and myelination. Glia **17:** 83–93.
25. MOOS, T. 1996. Immunohistochemical localization of intraneuronal transferrin receptor immunoreactivity in the adult mouse central nervous system. J. Comp. Neurol. **375:** 675–692.
26. DICKINSON, T.K. & J.R. CONNOR. 1998. Immunohistochemical analysis of transferrin receptor: regional and cellular distribution in the hypotransferrinemic (hpx) mouse brain. Brain Res. **801:** 171–181.
27. TAKEDA, A., A. DEVENYI & J.R. CONNOR. 1998. Evidence for non-transferrin-mediated uptake and release of iron and manganese in glial cell cultures from hypotransferrinemic mice. J. Neurosci. Res. **51:** 454–462.
28. MESCHER, A.L. & S.I. MUNAIM. 1988. Transferrin and the growth-promoting effect of nerves. Int. Rev. Cytol. **110:** 1–26.

29. ESCOBAR CABRERA, O.E., E.R. BONGARZONE, E.F. SOTO & J.M. PASQUINI. 1994. Single intracerebral injection of apotransferrin in young rats induces increased myelination. Dev. Neurosci. **16:** 248–254.
30. ESPINOSA DE LOS MONTEROS, A., S. KUMAR, P. ZHAO et al. 1999. Transferrin is an essential factor for myelination. Neurochem. Res. **24:** 235–248.
31. MARTA, C.B., O.E. ESCOBAR CABRERA, C.I. GARCIA et al. 2000. Oligodendroglial cell differentiation in rat brain is accelerated by the intracranial injection of apotransferrin. Cell. Mol. Biol. (Noisy-Le-Grand) **46:** 529–539.
32. HULET, S.W., E.J. HESS, W. DEBINSKI et al. 1999. Characterization and distribution of ferritin binding sites in the adult mouse brain. J. Neurochem. **72:** 868–874.
33. HULET, S.W., S. POWERS & J.R. CONNOR. 1999. Distribution of transferrin and ferritin binding in normal and multiple sclerotic human brains. J. Neurol. Sci. **165:** 48–55.
34. HULET, S.W., S.O. HEYLIGER, S. POWERS & J.R. CONNOR. 2000. Oligodendrocyte progenitor cells internalize ferritin via clathrin-dependent receptor mediated endocytosis. J. Neurosci. Res. **61:** 52–60.
35. OZAWA, H., A. NISHIDA, T. MITO & S. TAKASHIMA. 1994. Development of ferritin-positive cells in cerebrum of human brain. Pediatr. Neurol. **10:** 44–48.
36. IIDA, K., S. TAKASHIMA & K. UEDA. 1995. Immunohistochemical study of myelination and oligodendrocyte in infants with periventricular leukomalacia. Pediatr. Neurol. **13:** 296–304.
37. CHEEPSUNTHORN, P., C. PALMER & J.R. CONNOR. 1998. Cellular distribution of ferritin subunits in postnatal rat brain. J. Comp. Neurol. **400:** 73–86.
38. SANYAL, B., P.E. POLAK & S. SZUCHET. 1996. Differential expression of the heavy-chain ferritin gene in non-adhered and adhered oligodendrocytes. J. Neurosci. Res. **46:** 187–197.
39. BLISSMAN, G., S. MENZIES, J. BEARD et al. 1996. The expression of ferritin subunits and iron in oligodendrocytes in neonatal porcine brains. Dev. Neurosci. **18:** 274–281.
40. HAN, J., J.R. DAY, J.R. CONNOR & J.L. BEARD. 2002. H and L ferritin subunit mRNA expression differs in brains of control and iron-deficient rats. J. Nutr. **132:** 2769–2774.
41. THOMPSON, K., S. MENZIES, M. MUCKENTHALER et al. 2003. Mouse brains deficient in H-ferritin have normal iron concentration, but a protein profile of iron deficiency and increased evidence of oxidative stress. J. Neurosci. Res. **71:** 46–63.
42. DICKINSON, T.K. & J.R. CONNOR. 1995. Cellular distribution of iron, transferrin, and ferritin in the hypotransferrinemic (Hp) mouse brain. J. Comp. Neurol. **355:** 67–80.
43. FORGE, J.K., T.V. PEDCHENKO & S.M. LEVINE. 1998. Iron deposits in the central nervous system of SJL mice with experimental allergic encephalomyelitis. Life Sci. **63:** 2271–2284.
44. CRAELIUS, W., M.W. MIGDAL, C.P. LUESSENHOP et al. 1982. Iron deposits surrounding multiple sclerosis plaques. Arch. Pathol. Lab. Med. **106:** 397–399.
45. WALTON, J.C. & J.C. KAUFMANN. 1984. Iron deposits and multiple sclerosis. Arch. Pathol. Lab. Med. **108:** 755–756.
46. ADAMS, C.W. 1988. Perivascular iron deposition and other vascular damage in multiple sclerosis. J. Neurol. Neurosurg. Psychiatry **51:** 260–265.
47. LEVINE, S.M. 1997. Iron deposits in multiple sclerosis and Alzheimer's disease brains. Brain Res. **760:** 298–303.
48. ARNOLD, D.L., G.T. RIESS, P.M. MATTHEWS et al. 1994. Use of proton magnetic resonance spectroscopy for monitoring disease progression in multiple sclerosis. Ann. Neurol. **36:** 76–82.
49. TRAPP, B.D., J. PETERSON, R.M. RANSOHOFF et al. 1998. Axonal transection in the lesions of multiple sclerosis. N. Engl. J. Med. **338:** 278–285.
50. TRAPP, B.D., L. BO, S. MORK & A. CHANG. 1999. Pathogenesis of tissue injury in MS lesions. J. Neuroimmunol. **98:** 49–56.
51. BJARTMAR, C., G. KIDD, S. MORK et al. 2000. Neurological disability correlates with spinal cord axonal loss and reduced N-acetyl aspartate in chronic multiple sclerosis patients. Ann. Neurol. **48:** 893–901.
52. PETERSON, J.W., L. BO, S. MORK et al. 2001. Transected neurites, apoptotic neurons, and reduced inflammation in cortical multiple sclerosis lesions. Ann. Neurol. **50:** 389–400.

53. MEHINDATE, K., D.J. SAHLAS, D. FRANKEL et al. 2001. Proinflammatory cytokines promote glial heme oxygenase-1 expression and mitochondrial iron deposition: implications for multiple sclerosis. J. Neurochem. **77:** 1386–1395.
54. QI, Y. & G. DAWSON. 1994. Hypoxia specifically and reversibly induces the synthesis of ferritin in oligodendrocytes and human oligodendrogliomas. J. Neurochem. **63:** 1485–1490.
55. QI, Y., T.M. JAMINDAR & G. DAWSON. 1995. Hypoxia alters iron homeostasis and induces ferritin synthesis in oligodendrocytes. J. Neurochem. **64:** 2458–2464.
56. CHAKRABARTY, A., M.R. EMERSON & S.M. LEVINE. 2003. Heme oxygenase-1 in SJL mice with experimental allergic encephalomyelitis. Mult. Scler. **9:** 372–381.
57. BAARK, J.P., O. BERG & L. HEMMINGSEN. 1983. Immunonephelometric determination of proteins in cerebrospinal fluid in various neurological disorders. Clin. Chim. Acta **127:** 271–277.
58. LEVINE, S.M., S.G. LYNCH, C.N. OU et al. 1999. Ferritin, transferrin, and iron concentrations in the cerebrospinal fluid of multiple sclerosis patients. Brain Res. **821:** 511–515.
59. BALLA, G., H.S. JACOB, J. BALLA et al. 1992. Ferritin: a cytoprotective antioxidant strategem of endothelium. J. Biol. Chem. **267:** 18148–18153.
60. HERBERT, V., S. SHAW, E. JAYATILLEKE & T. STOPLER-KASDAN. 1994. Most free-radical injury is iron-related: it is promoted by iron, hemin, holoferritin, and vitamin C, and inhibited by deferoxamine and apoferritin. Stem Cells **12:** 289–303.
61. JUCKETT, M.B., J. BALLA, G. BALLA et al. 1995. Ferritin protects endothelial cells from oxidized low density lipoprotein in vitro. Am. J. Pathol. **147:** 782–789.
62. VOGT, B.A., J. ALAM, A.J. CROATT et al. 1995. Acquired resistance to acute oxidative stress: possible role of heme oxygenase and ferritin. Lab. Invest. **72:** 474–483.
63. MATZNER, Y., C. HERSHKO, A. POLLIACK et al. 1979. Suppressive effect of ferritin on in vitro lymphocyte function. Br. J. Haematol. **42:** 345–353.
64. KEOWN, P. & B. DESCAMPS-LATSCHA. 1983. In vitro suppression of cell-mediated immunity by ferroproteins and ferric salts. Cell. Immunol. **80:** 257–266.
65. MATZNER, Y., A.M. KONIJN, Z. SHLOMAI & B.H. BEN. 1985. Differential effect of isolated placental isoferritins on in vitro T-lymphocyte function. Br. J. Haematol. **59:** 443–448.
66. HARADA, T., M. BABA, I. TORII & S. MORIKAWA. 1987. Ferritin selectively suppresses delayed-type hypersensitivity responses at induction or effector phase. Cell. Immunol. **109:** 75–88.
67. HANN, H.W., M.W. STAHLHUT, S. LEE et al. 1989. Effects of isoferritins on human granulocytes. Cancer **12:** 2492–2496.
68. WYLLIE, S., P. SEU & J.A. GOSS. 2002. The natural resistance–associated macrophage protein 1 Slc11a1 (formerly Nramp1) and iron metabolism in macrophages. Microb. Infect. **4:** 351–359.
69. KOTZE, M.J., J.N. DE VILLIERS, R.N. ROONEY et al. 2001. Analysis of the NRAMP1 gene implicated in iron transport: association with multiple sclerosis and age effects. Blood Cells Mol. Dis. **27:** 44–53.
70. YANG, Y.S., S.J. KIM, J.W. KIM & E.M. KOH. 2000. NRAMP1 gene polymorphisms in patients with rheumatoid arthritis in Koreans. J. Korean Med. Sci. **15:** 83–87.
71. RODRIGUEZ, M.R., M.F. GONZALEZ-ESCRIBANO, F. AGUILAR et al. 2002. Association of NRAMP1 promoter gene polymorphism with the susceptibility and radiological severity of rheumatoid arthritis. Tissue Antigens **59:** 311–315.
72. BRETT, R. & M.G. RUMSBY. 1993. Evidence of free radical damage in the central nervous system of guinea-pigs at the prolonged acute and early relapse stages of chronic relapsing experimental allergic encephalomyelitis. Neurochem. Int. **23:** 35–44.
73. LEVINE, S.M. & D.L. WETZEL. 1998. Chemical analysis of multiple sclerosis lesions by FT-IR microspectroscopy. Free Radical Biol. Med. **25:** 33–41.
74. MACMICKING, J.D., D.O. WILLENBORG, M.J. WEIDEMANN et al. 1992. Elevated secretion of reactive nitrogen and oxygen intermediates by inflammatory leukocytes in hyper-acute experimental autoimmune encephalomyelitis: enhancement by the soluble products of encephalitogenic T cells. J. Exp. Med. **176:** 303–307.
75. RUULS, S.R., J. BAUER, K. SONTROP et al. 1995. Reactive oxygen species are involved in the pathogenesis of experimental allergic encephalomyelitis in Lewis rats. J. Neuroimmunol. **56:** 207–217.

76. AQUINO, D.A., A.A. KLIPFEL, C.F. BROSNAN & W.T. NORTON. 1993. The 70-kDa heat shock cognate protein (HSC70) is a major constituent of the central nervous system and is up-regulated only at the mRNA level in acute experimental autoimmune encephalomyelitis. J. Neurochem. **61:** 1340–1348.
77. DUZHAK, T., M.R. EMERSON, A. CHAKRABARTY et al. 2003. Analysis of protein induction in the CNS of SJL mice with experimental allergic encephalomyelitis by proteomic screening and immunohistochemistry. Cell. Mol. Biol. (Noisy-Le-Grand) **49:** 723–732.
78. HONEGGER, C.G., W. KRENGER & H. LANGEMANN. 1989. Measurement of free radical scavengers in the spinal cord of rats with experimental autoimmune encephalomyelitis. Neurosci. Lett. **98:** 327–332.
79. CALABRESE, V., G. SCAPAGNINI, A. RAVAGNA et al. 2003. Disruption of thiol homeostasis and nitrosative stress in the cerebrospinal fluid of patients with active multiple sclerosis: evidence for a protective role of acetylcarnitine. Neurochem. Res. **28:** 1321–1328.
80. CROSS, A.H., P.T. MANNING, M.K. STERN & T.P. MISKO. 1997. Evidence for the production of peroxynitrite in inflammatory CNS demyelination. J. Neuroimmunol. **80:** 121–130.
81. VAN DER VEEN, R.C., D.R. HINTON, F. INCARDONNA & F.M. HOFMAN. 1997. Extensive peroxynitrite activity during progressive stages of central nervous system inflammation. J. Neuroimmunol. **77:** 1–7.
82. LINK, H. 1998. The cytokine storm in multiple sclerosis. Mult. Scler. **4:** 12–15.
83. NAVIKAS, V. & H. LINK. 1996. Review: cytokines and the pathogenesis of multiple sclerosis. J. Neurosci. Res. **45:** 322–333.
84. RYTER, S.W. & R.M. TYRRELL. 2000. The heme synthesis and degradation pathways: role in oxidant sensitivity—heme oxygenase has both pro- and antioxidant properties. Free Radical Biol. Med. **28:** 289–309.
85. ELBIRT, K.K. & H.L. BONKOVSKY. 1999. Heme oxygenase: recent advances in understanding its regulation and role. Proc. Assoc. Am. Physicians **111:** 438–447.
86. ROUAULT, T. & R. KLAUSNER. 1997. Regulation of iron metabolism in eukaryotes. Curr. Top. Cell. Regul. **35:** 1–19.
87. SCHNEIDER, B.D. & E.A. LEIBOLD. 2000. Regulation of mammalian iron homeostasis. Curr. Opin. Clin. Nutr. Metab. Care **3:** 267–273.
88. LAVAUTE, T., S. SMITH, S. COOPERMAN et al. 2001. Targeted deletion of the gene encoding iron regulatory protein-2 causes misregulation of iron metabolism and neurodegenerative disease in mice. Nat. Genet. **27:** 209–214.
89. THOMPSON, K., S. MENZIES, M. MUCKENTHALER et al. 2003. Mouse brains deficient in H-ferritin have normal iron concentration, but a protein profile of iron deficiency and increased evidence of oxidative stress. J. Neurosci. Res. **71:** 46–63.
90. EMERSON, M.R. & S.M. LEVINE. 2000. Heme oxygenase-1 and NADPH cytochrome P450 reductase expression in experimental allergic encephalomyelitis: an expanded view of the stress response. J. Neurochem. **75:** 2555–2562.
91. SCHLUESENER, H.J. & K. SEID. 2000. Heme oxygenase-1 in lesions of rat experimental autoimmune encephalomyelitis and neuritis. J. Neuroimmunol. **110:** 114–120.
92. GOLDMUNTZ, E.A., C.F. BROSNAN & W.T. NORTON. 1986. Prazosin treatment suppresses increased vascular permeability in both acute and passively transferred experimental autoimmune encephalomyelitis in the Lewis rat. J. Immunol. **137:** 3444–3450.
93. CLAUDIO, L., Y. KRESS, J. FACTOR & C.F. BROSNAN. 1990. Mechanisms of edema formation in experimental autoimmune encephalomyelitis: the contribution of inflammatory cells. Am. J. Pathol. **137:** 1033–1045.
94. PEDCHENKO, T.V. & S.M. LEVINE. 1998. Desferrioxamine suppresses experimental allergic encephalomyelitis induced by MBP in SJL mice. J. Neuroimmunol. **84:** 188–197.
95. SARDANA, M.K. & A. KAPPAS. 1987. Dual control mechanism for heme oxygenase: tin(IV)-protoporphyrin potently inhibits enzyme activity while markedly increasing content of enzyme protein in liver. Proc. Natl. Acad. Sci. USA **84:** 2464–2468.
96. MONSHOUWER, M., D. AGNELLO, P. GHEZZI & P. VILLA. 2000. Decrease in brain cytochrome P450 enzyme activities during infection and inflammation of the central nervous system. Neuroimmunomodulation **8:** 142–147.

97. SEWERYNEK, E., M. ABE, R.J. REITER et al. 1995. Melatonin administration prevents lipopolysaccharide-induced oxidative damage in phenobarbital-treated animals. J. Cell. Biochem. **58:** 436–444.
98. SCHOLZ, R.W., P.V. REDDY, A.D. LIKEN & C.C. REDDY. 1996. Inhibition of rat liver microsomal NADPH cytochrome P450 reductase by glutathione and glutathione disulfide. Biochem. Biophys. Res. Commun. **226:** 475–480.
99. BOWERN, N., I.A. RAMSHAW, I.A. CLARK & P.C. DOHERTY. 1984. Inhibition of autoimmune neuropathological process by treatment with an iron-chelating agent. J. Exp. Med. **160:** 1532–1543.
100. WILLENBORG, D.O., N.A. BOWERN, G. DANTA & P.C. DOHERTY. 1988. Inhibition of allergic encephalomyelitis by the iron chelating agent desferrioxamine: differential effect depending on type of sensitizing encephalitogen. J. Neuroimmunol. **17:** 127–135.
101. WU, G.S., J. WALKER & N.A. RAO. 1993. Effect of deferoxamine on retinal lipid peroxidation in experimental uveitis. Invest. Ophthalmol. Visual Sci. **34:** 3084–3089.
102. NORSTRAND, I.F. & W. CRAELIUS. 1989. A trial of deferoxamine (Desferal) in the treatment of multiple sclerosis: a pilot study. Clin. Trials J. **26:** 365–369.
103. LYNCH, S.G., K. PETERS & S.M. LEVINE. 1996. Desferrioxamine in chronic progressive multiple sclerosis: a pilot study. Mult. Scler. **2:** 157–160.
104. LYNCH, S.G., T. FONSECA & S.M. LEVINE. 2000. A multiple course trial of desferrioxamine in chronic progressive multiple sclerosis. Cell. Mol. Biol. (Noisy-Le-Grand) **46:** 865–869.
105. MATSUOKA, Y., M. OKAZAKI & Y. KITAMURA. 1999. Induction of inducible heme oxygenase (HO-1) in the central nervous system: is HO-1 helpful or harmful? Neurotox. Res. **1:** 113–117.
106. LIU, Y., B. ZHU, L. LUO et al. 2001. Heme oxygenase-1 plays an important protective role in experimental autoimmune encephalomyelitis. Neuroreport **12:** 1841–1845.
107. STOCKER, R., Y. YAMAMOTO, A.F. MCDONAGH et al. 1987. Bilirubin is an antioxidant of possible physiological importance. Science **235:** 1043–1046.
108. BARANANO, D.E., M. RAO, C.D. FERRIS & S.H. SNYDER. 2002. Biliverdin reductase: a major physiologic cytoprotectant. Proc. Natl. Acad. Sci. USA **99:** 10693–10698.
109. OTTERBEIN, L.E., F.H. BACH, J. ALAM et al. 2000. Carbon monoxide has anti-inflammatory effects involving the mitogen-activated protein kinase pathway. Nat. Med. **6:** 422–428.
110. LIU, Y., B. ZHU, X. WANG et al. 2003. Bilirubin as a potent antioxidant suppresses experimental autoimmune encephalomyelitis: implications for the role of oxidative stress in the development of multiple sclerosis. J. Neuroimmunol. **139:** 27–35.
111. RAINE, C.S., L.B. BARNETT, A. BROWN et al. 1980. Neuropathology of experimental allergic encephalomyelitis in inbred strains of mice. Lab. Invest. **43:** 150–157.
112. WAGNER, K.R., Y. HUA, G.M. DE COURTEN-MYERS et al. 2000. Tin-mesoporphyrin, a potent heme oxygenase inhibitor, for treatment of intracerebral hemorrhage: in vivo and in vitro studies. Cell. Mol. Biol. (Noisy-Le-Grand) **46:** 597–608.
113. PANIZZON, K.L., B.E. DWYER, R.N. NISHIMURA & R.A. WALLIS. 1996. Neuroprotection against CA1 injury with metalloporphyrins. Neuroreport **7:** 662–666.
114. KADOYA, C., E.F. DOMINO, G.Y. YANG et al. 1995. Preischemic but not postischemic zinc protoporphyrin treatment reduces infarct size and edema accumulation after temporary focal cerebral ischemia in rats. Stroke **26:** 1035–1038.
115. LUCCHINETTI, C., W. BRUCK, J. PARISI et al. 2000. Heterogeneity of multiple sclerosis lesions: implications for the pathogenesis of demyelination. Ann. Neurol. **47:** 707–717.
116. ADAMS, C.W., R.N. POSTON, S.J. BUK et al. 1985. Inflammatory vasculitis in multiple sclerosis. J. Neurol. Sci. **69:** 269–283.
117. LEVINE, S.M., S. MAITI, M.R. EMERSON & T.V. PEDCHENKO. 2002. Apoferritin attenuates experimental allergic encephalomyelitis in SJL mice. Dev. Neurosci. **24:** 177–183.
118. REIF, D.W. & R.D. SIMMONS. 1990. Nitric oxide mediates iron release from ferritin. Arch. Biochem. Biophys. **283:** 537–541.
119. REIF, D.W. 1992. Ferritin as a source of iron for oxidative damage. Free Radical Biol. Med. **12:** 417–427.

120. HOOPER, D.C., S. SPITSIN, R.B. KEAN et al. 1998. Uric acid, a natural scavenger of peroxynitrite, in experimental allergic encephalomyelitis and multiple sclerosis. Proc. Natl. Acad. Sci. USA **95:** 675–680.
121. MARRACCI, G.H., R.E. JONES, G.P. MCKEON & D.N. BOURDETTE. 2002. Alpha lipoic acid inhibits T cell migration into the spinal cord and suppresses and treats experimental autoimmune encephalomyelitis. J. Neuroimmunol. **131:** 104–114.
122. HANSEN, L.A., D.O. WILLENBORG & W.B. COWDEN. 1995. Suppression of hyperacute and passively transferred experimental autoimmune encephalomyelitis by the antioxidant, butylated hydroxyanisole. J. Neuroimmunol. **62:** 69–77.
123. MALFROY, B., S.R. DOCTROW, P.L. ORR et al. 1997. Prevention and suppression of autoimmune encephalomyelitis by EUK-8, a synthetic catalytic scavenger of oxygen-reactive metabolites. Cell. Immunol. **177:** 62–68.
124. KANG, J.C., M. AHN, Y.S. KIM et al. 2001. Melatonin ameliorates autoimmune encephalomyelitis through suppression of intercellular adhesion molecule-1. J. Vet. Sci. **2:** 85–89.
125. LEHMANN, D., D. KARUSSIS, R. MISRACHI-KOLL et al. 1994. Oral administration of the oxidant-scavenger N-acetyl-L-cysteine inhibits acute experimental autoimmune encephalomyelitis. J. Neuroimmunol. **50:** 35–42.
126. MOHAMED, A., A. SHOKER, F. BENDJELLOUL et al. 2003. Improvement of experimental allergic encephalomyelitis (EAE) by thymoquinone, an oxidative stress inhibitor. Biomed. Sci. Instrum. **39:** 440–445.

Hereditary Causes of Disturbed Iron Homeostasis in the Central Nervous System

PREM PONKA

Lady Davis Institute for Medical Research, Sir Mortimer B. Davis Jewish General Hospital, Department of Physiology and Medicine, McGill University, Montreal, Quebec, Canada

ABSTRACT: Iron is essential for oxidation-reduction catalysis and bioenergetics; however, unless appropriately shielded, this metal plays a crucial role in the formation of toxic oxygen radicals that can attack all biological molecules. Organisms are equipped with specific proteins designed for iron acquisition, export and transport, and storage, as well as with sophisticated mechanisms that maintain the intracellular labile iron pool at an appropriate level. Despite these homeostatic mechanisms, organisms often face the threat of either iron deficiency or iron overload. This review describes several hereditary iron-overloading conditions that are confined to the brain. Recently, a mutation in the L-subunit of ferritin has been described that causes the formation of aberrant L-ferritin with an altered C-terminus. Individuals with this mutation in one allele of L-ferritin have abnormal aggregates of ferritin and iron in the brain, primarily in the globus pallidus. Patients with this dominantly inherited late-onset disease present with symptoms of extrapyramidal dysfunction. Mice with a targeted disruption of a gene for iron regulatory protein 2 (IRP2), a translational repressor of ferritin, misregulate iron metabolism in the intestinal mucosa and the central nervous system. Significant amounts of ferritin and iron accumulate in white matter tracts and nuclei, and adult IRP2-deficient mice develop a movement disorder consisting of ataxia, bradykinesia, and tremor. Mutations in the frataxin gene are responsible for Friedreich's ataxia, the most common of the inherited ataxias. Frataxin appears to regulate mitochondrial iron-sulfur cluster formation, and the neurologic and cardiac manifestations of Friedreich's ataxia are due to iron-mediated mitochondrial toxicity. Patients with Hallervorden-Spatz syndrome, an autosomal recessive, progressive neurodegenerative disorder, have mutations in a novel pantothenate kinase gene (PANK2). The cardinal feature of this extrapyramidal disease is pathologic iron accumulation in the globus pallidus. The defect in PANK2 is predicted to cause the accumulation of cysteine, which binds iron and causes oxidative stress in the iron-rich globus pallidus. Finally, aceruloplasminemia is an autosomal recessive disorder of iron metabolism caused by loss-of-function mutations in ceruloplasmin gene that leads to misregulation of both systemic and central nervous system iron trafficking. Affected individuals suffer from extrapyramidal signs, cerebellar ataxia, progressive neurodegeneration of retina, and diabetes mellitus. Excessive iron depositions are found in the brain, liver, pancreas, and other parenchymal cells, but plasma iron concentrations

Address for correspondence: Dr. Prem Ponka, Lady Davis Institute for Medical Research, LDI, Room 202, 3755 Cote St. Catherine Road, Montreal, Quebec H3T 1E2, Canada. Voice: 514-340-8260; fax: 514-340-7502.

prem.ponka@mcgill.ca

are decreased. These conditions are not common, but awareness about them is important for differential diagnosis of various neurodegenerative disorders.

KEYWORDS: iron; ferritin; neuroferritinopathy; iron regulatory protein 2 (IRP2); IRP2 deficiency; Friedreich's ataxia; frataxin; pantothenate kinase 2 deficiency; ceruloplasmin; aceruloplasminemia

INTRODUCTION

Iron is indispensable for life, serving as metal cofactor for many proteins and enzymes. It is involved in a broad spectrum of essential biological functions such as oxygen transport (hemoglobin), electron transfer (mitochondrial heme and nonheme iron proteins essential for energy production), and DNA synthesis (ribonucleotide reductase), to name just a few. However, the chemical properties of iron that allow this versatility also lead to the paradoxical situation that acquisition by the organism of the fourth most abundant element in the Earth's crust is exceedingly difficult. At pH ~7.4 and physiological oxygen tension, the relatively soluble ferrous iron is readily oxidized to ferric ion, which is susceptible to hydrolysis, forming virtually insoluble ferric hydroxides. Moreover, unless bound to specific ligands, iron plays a key role in the formation of harmful oxygen radicals, which ultimately cause peroxidative damage to vital cell structures.[1,2] Because of this virtual insolubility and potential toxicity, specialized mechanisms and molecules for the acquisition, transport, and storage of iron in a soluble nontoxic form have evolved to meet cellular and organismal iron requirements. In addition, organisms are equipped with sophisticated mechanisms that prevent the expansion of catalytically active intracellular iron pool, while maintaining sufficient concentrations of the metal for metabolic use.[3–6]

Iron, in both its heme and nonheme forms, is present and exerts its function in every cell in the body, and most cells, except for highly differentiated cells, have the capacity to acquire Fe. However, cellular iron acquisition and its proper intracellular targeting into functional iron proteins depend on an array of other proteins. "Traditional" proteins involved in iron metabolism include transferrin, transferrin receptor, and ferritin, but a remarkable flurry of activity in the past several years has identified a large number of novel genes whose products emerge as important players in iron metabolism (TABLE 1). Importantly, many proteins encoded by these genes are expressed in the brain,[45–48] where they undoubtedly play an essential role in maintaining iron homeostasis.

With some notable exceptions (e.g., enterocytes), physiologically, virtually all the cells in the organism take up iron from transferrin. Delivery of iron to cells occurs following the binding of transferrin to transferrin receptors on the cell membrane. The transferrin receptor complexes are then internalized by endocytosis and iron is released from transferrin by a process involving endosomal acidification.[3–6] Iron is then transported through the endosomal membrane by the Fe^{2+} transporter Nramp2/DMT1. Since iron bound to transferrin is in its oxidized (ferric, Fe^{3+}) form and since the substrate for Nramp2/DMT1 is ferrous ion (Fe^{2+}), reduction of Fe^{3+} must occur in endosomes, but nothing is known about the mechanism of this process. McKie et al.[21] recently isolated a cDNA that encodes a plasma membrane diheme protein present in mouse duodenal cells that exhibits ferric reductase activity. This protein (Dcytb) belongs to the cytochrome b 561 family of plasma membrane

TABLE 1. Some proteins involved in iron metabolism

Protein	Function	Result of deficiency	Ref.
Transferrin (Tf)	Fe(III) carrier in plasma	Severe Fe-deficient anemia; generalized Fe overload including the brain	3–6
TfR	Membrane receptor for Fe_2-Tf	Embryonic lethality	7, 8
TfR2	As "classical" TfR (?); different tissue distribution	Hemochromatosis (H) type 3	9–11
HFE	Unknown; binds TfR; restricted expression (enterocytes, macrophages)	H type 1	12, 13
Ferritin (H and L)	Cellular Fe storage	H: embryonic lethality	4, 14, 15
IRP (1 and 2)	Fe "sensors"; binding to IREs	IRP2: brain Fe overload	3–6, 16–18
DMT1/DCT1/Nramp2	Membrane transporter for Fe^{2+}	Hypochromic microcytic anemia	19, 20
Duodenal cytochrome b (Dcytb)	Ferric reductase (provides Fe^{2+} for DMT1 in duodenum)	Unknown	21
Ferroportin 1/Ireg1/MTP1	Fe export from cells	H type 4	22–25
Ceruloplasmin (Cp)	Regulation of Fe export from cells	Hypochromic microcytic anemia	26, 27
Hephaestin	Regulation of Fe export from enterocytes (membrane-bound Cu homologue)	Hypochromic microcytic anemia	28, 29
Frataxin	Control of [Fe-S] cluster synthesis	Friedreich's ataxia	30, 31
Ferrochelatase	The last enzyme of heme synthesis; Fe^{2+} insertion into protoporphyrin IX	Erythropoietic protoporphyria	32
Mitochondrial ferritin	Mitochondrial Fe storage (?); high expression in testes	Unknown; high expression in "ring" sideroblasts	33, 34
Heme oxygenase 1 (HO1)	Recycling of hemoglobin Fe	Severe anemia and inflammation	35–38
HO2	Neuroprotection (?)	Ejaculatory abnormalities	39–40
Hepcidin	Plasma peptide that appears to inhibit Fe absorption	Fe overload	41–44

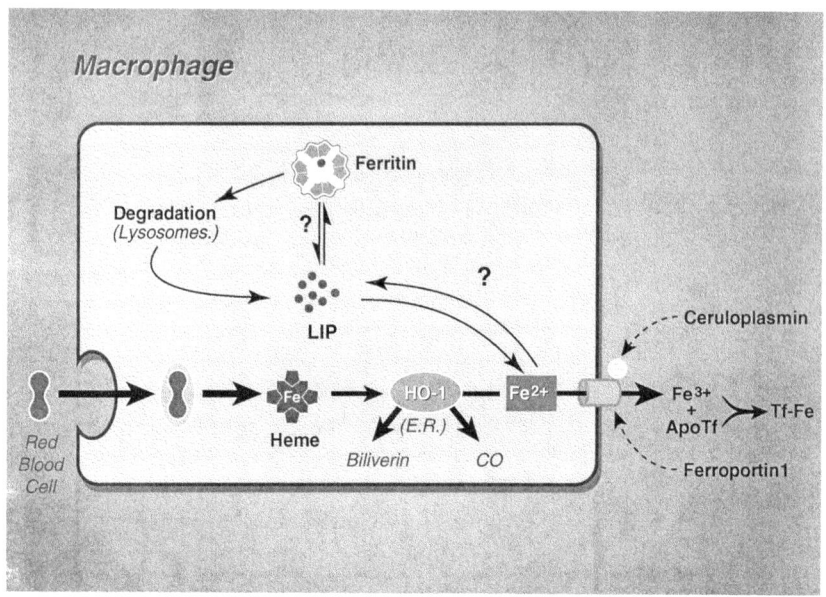

FIGURE 1. Scheme of possible iron pathways in reticuloendothelial macrophages involved in the recycling of hemoglobin iron (Fe). After phagocytosis of senescent red blood cells, the erythrocyte membrane is lysed and heme is transported to the endoplasmic reticulum (E.R.) to be degraded by heme oxygenase 1 (HO-1). Most of the Fe derived from hemoglobin catabolism is promptly returned to the circulation, likely being transported across the plasma membrane by ferroportin 1. Ferroxidase activity of ceruloplasmin oxidizes iron at the external side of the cell membrane, which "drives" the transport of Fe^{2+} and makes it available for binding to plasma transferrin. (Reprinted from ref. 96, with permission.)

reductases and it would seem important to examine whether this or similar b-type cytochrome is involved in Fe^{3+} reduction within endosomes. Following its escape from endosomes, iron is transported to intracellular sites of use and/or storage in ferritin, but this aspect of Fe metabolism, including the nature of the elusive intermediary pool of Fe and its cellular trafficking, remains enigmatic.

Iron release from "donor cells" (primarily enterocytes, macrophages, and likely astrocytes) to plasma transferrin is poorly understood, but a number of recent studies have provided new clues in this important area of iron metabolism. A likely candidate for iron export from the cells is ferroportin 1, also known as Ireg1 or MTP1,[22–24] and the ferroxidase activity of hephaestin[28] and ceruloplasmin[26,27] facilitates the movement of iron across the membranes of enterocytes and macrophages, respectively. FIGURE 1 schematically shows the involvement of both ferroportin 1 and ceruloplasmin in iron release from macrophages; these proteins have similar functions in duodenal enterocytes, astrocytes, and probably other cells, but obviously the substrate for ferroportin 1 in these cells is not iron derived from hemoglobin.

Normally, the body iron content in humans is maintained within narrow limits by the regulation of intestinal iron absorption.[49] Both heme and inorganic iron are absorbed through the brush border of the upper small intestine. Heme is more readily available for absorption, but usually constitutes only a small fraction of dietary iron.

Heme is taken up intact, probably via specific high-affinity heme-binding sites in the mucosal brush border.[50,51] After entering the intestinal epithelial cells, Fe is enzymatically released from heme by heme oxygenase. Nramp2/DMT1, which is involved in Fe transport across the endosomal membrane (see earlier), is also the principal transporter of inorganic iron in the intestine. Nramp2/DMT1 transports only Fe^{2+} and this explains why reducing agents enhance iron absorption. Moreover, the duodenal brush border contains a ferric reductase (Dcytb),[21] which generates Fe^{2+} prior to its transport into the enterocyte. Ferroportin 1,[22–24] with assistance provided by hephaestin,[28,29] is probably involved in the exit of iron from enterocytes into the circulation.

Cells are also equipped with a remarkable regulatory system that tightly controls iron levels in the "labile iron pool" (LIP), that is, iron-in-transit among various intracellular compartments. Sensitive control mechanisms exist that monitor iron levels in the LIP and prevent its expansion, while still making the metal available for iron-dependent proteins and enzymes. In general, enlargement of the LIP leads to a stimulation of ferritin synthesis and to a decrease in the expression of transferrin receptors, and the opposite scenario develops when this pool is depleted of iron. Pivotal players in this regulation are iron regulatory proteins (IRP1 and IRP2), which "sense" iron levels in the LIP. IRPs control the synthesis of transferrin receptors and ferritin by binding to iron-responsive elements (IREs) that are located in the 3'-untranslated region (UTR) and the 5'-UTR of their respective mRNAs. When cellular iron is scarce, IRPs bind to IREs and this association blocks the translation of ferritin mRNA and stabilizes transferrin receptor mRNA. The expansion of the LIP inactivates IRP1 binding to the IREs and leads to a degradation of IRP2, resulting in an efficient translation of ferritin and rapid degradation of transferrin receptor mRNA.[3–6,16,17] Although the IRE/IRP system has probably evolved to sense the iron-in-transit and to maintain it at appropriate levels, iron is not the only factor that modulates IRPs. Recently, nitric oxide and oxidative stress were shown to dramatically affect the binding of IRPs to IREs and, consequently, cellular iron metabolism.[52–56] Moreover, the mRNAs for ferritin and transferrin receptors are not the only transcripts containing IREs. mRNAs for both Nramp2/DMT1 and ferroportin contain IREs in their untranslated regions and it seems likely that the expression of the above proteins is controlled, at least in part, by the IRE/IRP system.

As already alluded to, iron is transported within the body between sites of absorption, storage, and utilization by the plasma glycoprotein transferrin, which binds ferric iron very tightly, but reversibly. The turnover of transferrin iron is roughly 30 mg/24 h and, normally, about 80% of this iron is transported to the bone marrow for hemoglobin synthesis in developing erythroid cells. Senescent erythrocytes are phagocytosed by macrophages of the reticuloendothelial system, where the heme moiety is split from hemoglobin and catabolized enzymatically via heme oxygenase 1.[35] Iron, which is liberated from its confinement within the protoporphyrin ring inside macrophages, is returned almost quantitatively to the circulation (see FIG. 1). The remaining 5 mg of the daily plasma iron turnover is exchanged with nonerythroid tissues, namely, the liver. About 1 mg of dietary iron is absorbed per 24 h and the total organismal iron balance is maintained by a daily loss of 1 mg via nonspecific mechanisms (mostly cell desquamation).[3–6]

Some tissues such as the brain require strict constancy of their environment and do not allow free passage of large molecules, including transferrin and ceruloplasmin (see below). The endothelial cells of the brain comprise the blood-brain barrier and are known to express transferrin receptors (reviewed in ref. 3). Differic transferrin binds to the receptors and is taken up at the luminal membrane of the blood capillaries. The iron dissociates from transferrin in the endosomal compartment (see earlier) and iron-free apotransferrin is recycled to the blood. The liberated iron is then released into the intestinal space, and ferroportin is a likely candidate for iron transport across the abluminal membrane. Further details on iron transport to and within the central nervous system (CNS) can be found in the article by Moos and Morgan in this issue.

Iron uptake into the brain is maximal during rapid brain growth, but brain iron uptake continues throughout life.[57] However, the actual daily rate of iron exchange between transferrin in plasma and the CNS is unknown. This value is probably significantly less than 3 mg Fe/24 h (representing ~10% of daily plasma iron turnover) since close to 90% of transferrin iron turnover is with erythroid cells/macrophage system and the liver (see earlier). This does not necessarily mean that iron turnover between the cerebrospinal fluid and neurons and other cells in the CNS is also low. Unfortunately, very little is known about rates of iron trafficking inside the brain and nothing is known about mechanisms and controls of iron release from the CNS to the systemic circulation. Nevertheless, it seems reasonable to propose that the exchange of iron between the systemic circulation and plasma is highly regulated since neither patients with hereditary hemochromatosis nor those with secondary iron overload develop iron overload in the brain. As will be discussed later, one of the conditions that leads to a significant iron accumulation in the CNS is aceruloplasminemia. Brain iron accumulation in patients with aceruloplasminemia can be explained by the fact that ceruloplasmin is involved in iron release from astrocytes and likely some other brain cells. This phenotype provides clear evidence that iron exchange between plasma and the brain occurs in adults and also that such an exchange is rather slow since clinical symptoms in patients with aceruloplasminemia develop relatively late in life. In summary, there seems to be a cycle of iron in the brain that is separate from that present in the circulation, but that exhibits some similarities with it.

GENETICALLY BASED MISREGULATION OF IRON METABOLISM IN THE BRAIN

Neuroferritinopathy

Ferritin[4,14] is a ubiquitous protein whose only clearly defined function is the sequestration and storage of iron. It has been detected in almost all animal and plant cells as well as fungi and bacteria. Mammalian ferritin consists of a protein shell that can accommodate up to 4500 atoms of iron in its internal cavity. The molecular structure of ferritin has been well characterized. The protein shell by itself has a molecular mass between 430 and 460 kDa, is approximately 25 Å thick, and is made up of 24 symmetrically related subunits of two types, a light subunit (L-subunit) of about 19 kDa and a heavy subunit (H-subunit) of about 21 kDa. The amino acid sequences of the H- and L-subunits differ by about 50%. Relatively soluble ferrous iron, which is incorporated into the shell much more efficiently than ferric iron, is oxidized and deposited after its association with the inner surface of the subunits. The H-subunit has a ferroxidase activity, whereas the L-subunit of ferritin has a higher capacity to induce iron-core nucleation. Although the different rations of H- to L-chains in various tissues are determined mainly at a transcriptional level, the iron-dependent regulation of H- and L-ferritin expression occurs mainly at the translational level via interactions of IREs with IRPs whose binding affinities are modulated by iron levels in the LIP (see earlier).

Recently, Curtis *et al.*[58] have described a previously unknown, dominantly inherited basal ganglia disease that is associated with iron accumulation in various parts of the brain. The affected patients present at 40–55 years of age with involuntary movements. Symptoms of extrapyramidal dysfunction include choreoathetosis, dystonia, spasticity, and rigidity, but there is no cognitive decline or cerebellar involvement. All affected individuals (5 patients in total) were from the Cumbrian region of Northern England. By linkage analysis, the disorder was mapped to 19q13.3, which contains the gene for the L-chain of ferritin. All the affected individuals have a heterozygous adenine insertion at position 460–461 in the L-ferritin gene. This insertion is predicted to alter 22 C-terminal residues of L-ferritin and extend this chain by 4 additional amino acids.[58]

Brain histochemistry disclosed abundant accumulation of iron and ferritin, primarily in the neurons of the globus pallidus.[58] Iron- and ferritin-positive inclusions were also present in the forebrain and cerebellum, mainly extracellularly, but also in microglia and oligodendrocytes. The patients did not have symptoms of iron overload and diabetes, and surprisingly their serum ferritin levels were abnormally low. Although liver function tests were normal,[58] biopsy revealed unique inclusions in the patients' hepatocytes (John Burn, personal communication).

Curtis *et al.*[58] have suggested that the mutation found in patients with neuroferritinopathy disrupts the C-terminus of the L-chain of ferritin, affecting its stability and function. It has also been proposed[59] that the C-terminus of the mutated L-chain might interfere with the formation of the holoferritin, causing inappropriate release of iron from iron-laden ferritin. It is also possible that mutated holoferritin has an impaired capacity to take up iron, leaving it unshielded in the cytosol. In any case, it seems likely that "free" iron, with its capacity to catalyze the formation of devastatingly toxic free radicals, is responsible for the pathology seen in patients with neuroferritinopathy. However, it is not totally excluded that the mutated L-ferritin develops a new function.[58] This interesting case would seem to suggest that L-ferritin mutations should be sought in patients with extrapyramidal dysfunction unrelated to known neurological disorders.

IRP2 Deficiency Causes Misregulation of Iron Metabolism in the Brain

Recently, LaVaute *et al.*[18] generated mice with a targeted disruption of the gene encoding IRP2 (*Ireb2*). These mice have a gross misregulation of iron metabolism in the intestinal mucosa and CNS. It seems highly likely that the overexpression of both ferritin subunits, due to the lack of the translational repressor IRP2, plays a major role in the progression of this mouse disease. At the age of 6 months, $Ireb2^{-/-}$ mice developed a progressive neurodegenerative disease characterized initially by an unsteady, wide-based gait, followed by the gradual development of ataxia, vestibular dysfunction, tremor, bradykinesia, and postural abnormalities; $Ireb2^{+/-}$ mice showed an intermediate degree of impairment. Serum ferritin levels were significantly elevated after 6 months of age in $Ireb2^{-/-}$ mice, but serum iron and transferrin levels as well as liver function tests were normal. Total amount of iron in the liver and duodenum was increased by 70% and 50%, respectively.[18] Moreover, the absence of IRP2 was associated with abnormally high expression levels of ferritin, together with increased expressions of DMT1, which transports ferrous ions (Fe^{2+}) across the apical membrane, and ferroportin 1, which exports iron from enterocytes to the circulation. It is intriguing that the IRP1 is unable to repress ferritin mRNA translation in the absence of IRP2; this scenario bears similarity to a recent report from this author's laboratory demonstrating that, in a macrophage-like cell line, the degradation of IRP2 leads to a stimulation of ferritin synthesis despite the presence of a highly active IRP1.[60] In the brain of IRP2-deficient mice, iron accumulated prominently in cerebellar white matter, caudate putamen, thalamus, and colliculi.[18] Iron accumulation preceded axonal degeneration, the earliest pathological event identified in neurons that overexpress ferritin. Abnormal accumulation of ferritin colocalized with iron accumulation in populations of neurons that degenerate. However, it may not be iron per se that is directly responsible for neuronal damage in IRP2-deficient mice. As suggested by LaVaute et al.,[18] a potential consequence of ferritin and ferroportin 1 overexpression would be functional iron deficiency, depriving iron-dependent proteins of this valuable element. In spite of uncertainties regarding molecular pathogenesis of this disease, it is important to bear in mind that some human movement disorders with unidentified etiology may be caused by IRP2 deficiency.

Friedreich's Ataxia

Friedreich's ataxia is an autosomal recessive neurodegenerative disease that is the most common of the inherited ataxias. The onset is before the age of 25, and clinical symptoms include progressive limb and gait ataxia, absent tendon reflexes in the legs, dysarthria, areflexia, sensory loss, and cardiomyopathy. The defective gene in Friedreich's ataxia is *FRDA*, a nuclear gene that encodes a mitochondrial protein known as frataxin. About 98% of mutant alleles have an expansion of GAA trinucleotide repeat in intron 1 of the gene and this mutation leads to a severe reduction in frataxin levels.[30] Frataxin is highly expressed in tissues that are rich in mitochondria, such as heart, liver, and skeletal muscle, and cells from Friedreich's ataxia patients show defective mitochondrial respiration.[61] Hence, it appears that mitochondrial dysfunction is responsible for the pathophysiology of Friedreich's ataxia. Affected individuals develop myocardial fibrosis and degeneration of the cardiac muscle associated with iron deposits in myocardial cells.[62,63] Iron probably accumulates in mitochondria in cardiomyocytes (Massimo Pandolfo, personal communication) and the evidence for this localization has recently been provided by the finding of intramitochondrial iron accumulation in frataxin-deficient mice.[31,64] It is pertinent to point out that mitochondria do not accumulate nonheme iron even in severely iron-overloaded individuals. Sideroblastic anemia is the only other condition, apart from Friedreich's ataxia, associated with intramitochondrial iron accumulation.[65] Importantly, Friedreich's ataxia erythroblasts do not contain iron-overloaded mitochondria. In this context, it is of interest to mention that the induction of erythroid differentiation in murine erythroleukemia cells results in a marked decrease in *FRDA* expression and frataxin levels,[66] suggesting that the highly active heme synthesis in mitochondria requires the absence or low levels of frataxin.

Important clues to the role of frataxin in iron metabolism came from studies on the yeast *FRDA* orthologue (*YFH1*), exploiting *Saccharomyces cerevisiae*.[67] Mitochondria in *yfh* mutants accumulate approximately 10-fold more iron than do mitochondria of wild-type cells.[68] This accumulation is reversible since reintroduction of wild-type *YFH1* results in the rapid export of accumulated mitochondrial iron into the cytosol as free, nonheme-bound iron.[69] Frataxin does not have any transmembrane domain, making it unlikely that it is a membrane transporter. It has been proposed that frataxin may function to form or export iron-sulfur clusters from mitochondria or is involved in mitochondrial iron storage (reviewed in ref. 67). Recent reports by Lill and coworkers[70,71] have provided strong evidence that frataxin is an essential component of a complex machinery that is involved in the synthesis of iron-sulfur clusters in mitochondria.

Neurodegeneration Caused by Pantothenate Kinase 2 (PANK2) Defect

PANK2 defect, formerly known as Hallervorden-Spatz syndrome, is a rare autosomal recessive neurodegenerative disorder associated with a dramatic iron accumulation in the brain.[72,73] This disease with onset in the first two decades is characterized by extrapyramidal dysfunction, manifest as rigidity, dystonia, and choreoathetosis. A massive accumulation of iron in the globus pallidus and the pars reticulata of the substantia nigra can be seen in autopsied brains. Zhou *et al.*[74] recently reported that PANK2 deficiency is caused by a defect in a novel gene that encodes pantothenate kinase that appears to be specifically expressed in the brain. This enzyme is essential in coenzyme A biosynthesis, catalyzing the phosphorylation of pantothenate (vitamin B_5). The new designation for PANK2 deficiency is pantothenate kinase–associated neurodegeneration or PKAN.[73]

Phosphopantothenate, the product of pantothenate kinase, normally condenses with cysteine in the next step of coenzyme A synthesis. Hence, PANK2 deficiency should be expected to be associated with cysteine accumulation and, indeed, high cysteine concentration has been found in the globus pallidus of patients with PKAN.[75] Consequently, the accumulation of cysteine, which has iron-chelating properties, might account for the regional iron accumulation seen in patients with PKAN, and cysteine-bound iron may be responsible for oxidative damage in these regions.[73,74] Indeed, cysteine undergoes rapid auto-oxidation in the presence of iron that results in the generation of free radicals.[76] Further research is needed to confirm whether the cysteine-iron partnership or another mechanism is involved in brain damage in patients with PKAN defect.

Aceruloplasminemia

The involvement of copper in iron metabolism has been known for more than 70 years[77] and, since the mid-1960s,[78] it is becoming increasingly evident that a copper-containing protein, ceruloplasmin, plays a crucial role in this function. Ceruloplasmin is an abundant plasma α_2-glycoprotein that has sky-blue color in its purified form and is synthesized primarily, but not exclusively (see below), in the liver. Ceruloplasmin is composed of 1046 amino acids and contains 6 copper atoms incorporated into the protein during its synthesis. Three of these copper atoms form a trinuclear cluster that activates oxygen needed for the catalytic activity of ceruloplasmin. The substrate for ceruloplasmin is ferrous ion (Fe^{2+}), which becomes oxidized with the concomitant reduction of molecular oxygen. Importantly, the oxidation of Fe^{2+} ions leads to the complete reduction of oxygen that is not accompanied by the production of toxic oxygen free radicals. Although ceruloplasmin can oxidize multiple different substrates *in vitro*, physiologically, the only clearly defined function of ceruloplasmin is its ferroxidase activity.[26,27]

Ferroxidase activity of ceruloplasmin was first noted in 1960, and several years later Frieden and colleagues[78] proposed that ferroxidase activity of ceruloplasmin is needed for an efficient rate of iron insertion into transferrin, which is known to contain ferric iron. The idea that ceruloplasmin is somehow involved in iron release from tissues was supported by subsequent studies using perfused livers[79] and, in particular, exploiting ceruloplasmin knockout mice.[80] However, the exact mechanism by which ceruloplasmin promotes iron efflux from macrophages and other cells is unknown. The most common opinion is that ceruloplasmin plays a role in providing the oxidized form of iron (ferric) for transferrin. However, it should be pointed out that the ferroxidase activity of ceruloplasmin does not seem to be necessary for iron binding to transferrin. Transferrin has its own ferroxidase activity and can oxidize Fe^{2+} to Fe^{3+} followed by the binding of the metal to the protein.[4] Alternatively, it is possible that the ferroxidase activity of ceruloplasmin may be needed for oxidation of ferrous ions following their transport to the cell surface by ferroportin; such an oxidation may be essential for removing iron from the cell membrane and for "driving" the transport of ferrous ions from intracellular milieu (see FIG. 1). In this context, it is pertinent to mention that studies on *Saccharomyces cerevisiae* identified a yeast multicopper oxidase, Fet3, that is homologous to ceruloplasmin.[81] However, in contrast to ceruloplasmin, which is involved in iron release from vertebrate cells, Fet3 is required for the oxidation of ferrous iron present in the extracellular environment prior to the uptake of this metal by yeast membrane permease Ftr1.[82]

As already alluded to, the major site of ceruloplasmin synthesis is the liver. Nevertheless, ceruloplasmin is also expressed in the CNS, where high expression of this protein was noted in astrocytes, particularly in perivascular astrocytes and also those surrounding dopaminergic melanized neurons in the substantia migra. The epithelial cells of the pia, the ependymal cells lining the ventricles, and the choroid plexus also express ceruloplasmin.[83,84] Importantly, David and coworkers recently identified the membrane-bound form of ceruloplasmin and demonstrated that this form of ceruloplasmin is directly anchored to the surface of astrocytes by a glycosylphosphatidylinositol (GPI) anchor.[85] Moreover, this group identified a novel alternatively spliced transcript that codes for the GPI-anchored form of ceruloplasmin and demonstrated that this is the major form of ceruloplasmin in the brain.[86] Furthermore, recently Jeong and David[87] provided some preliminary evidence that astrocytes isolated from ceruloplasmin-deficient mice have decreased capacity to purge themselves of iron, suggesting that GPI-anchored ceruloplasmin plays an important role in maintaining iron homeostasis in the CNS.

The most compelling evidence that ceruloplasmin plays an important role in both systemic and CNS iron trafficking comes from studies of mice and patients with aceruloplasminemia. Ceruloplasmin knockout mice develop iron overload in hepatic macrophages that is caused by a failure to efficiently recycle hemoglobin iron.[80] When ceruloplasmin is administered to such knockout mice, iron is mobilized out of the liver with concomitant increase in plasma iron levels.[80] Adult aceruloplasminemia mice have iron deposits in several regions of the CNS, including the cerebellum, spinal cord, and retina. Signs of increased lipid peroxidation can also be seen in some regions of the CNS.[88]

The first case of aceruloplasminemia in humans was reported by Miyajima *et al.*,[89] who described a 52-year-old Japanese woman with diabetes mellitus, retinal degeneration, and basal ganglia symptoms in association with an absence of ceruloplasmin in the circulation. She had elevated plasma ferritin concentrations, low plasma iron levels, a mild anemia, severe hepatic iron overload, and iron accumulation within the basal ganglia revealed by signal attenuation on T2-weighted magnetic resonance imaging. Subsequently, nucleotide sequence analysis demonstrated a loss-of-function mutation in her ceruloplasmin gene.[90] Since then, multiple distinct mutations in the ceruloplasmin gene have been identified,[91-95] and in all of them the amino acid ligands in the C-terminal region essential for the formation of the trinuclear copper cluster are eliminated. All patients thus far identified have symptoms reflecting serious parenchymal iron overload and progressive neurological disease that is likely caused by iron accumulation in neurons and glia throughout the brain.

In summary, aceruloplasminemia is an autosomal recessive disorder of iron homeostasis caused by mutations in ceruloplasmin gene, characterized by progressive neurodegeneration of the retina and basal ganglia. In patients with aceruloplasminemia, the abnormalities in iron metabolism may be recognized before neurologic symptoms become obvious. In such a case, type 1 hereditary hemochromatosis (see TABLE 1) may be diagnosed and the patient subjected to phlebotomies; however, such a treatment would be totally inappropriate for a patient with aceruloplasminemia since it would exacerbate the anemia. These two diseases can be easily distinguished by measurements of plasma transferrin saturation, which is high in hemochromatosis, but low in aceruloplasminemia.[27] Patients with Wilson's disease also have basal ganglia symptoms and often low plasma ceruloplasmin levels. However, in contrast to Wilson's disease, patients with aceruloplasminemia do not have increased hepatic copper levels and their parents or siblings have half-normal plasma ceruloplasmin concentrations (reviewed in ref. 27). The only other disease, in addition to aceruloplasminemia, PKAN, and IRP2 deficiency (thus far only in mice), in which iron accumulation in the brain has been described is neuroferritinopathy (see earlier). This disease, which has also symptoms of extrapyramidal dysfunction, can be distinguished from hereditary aceruloplasminemia by the lack of systemic iron overload.[27,58]

ACKNOWLEDGMENTS

This work was supported in part by the Canadian Institutes of Health Research. The author thanks Miriam Rosenzweig for editorial assistance.

REFERENCES

1. McCord, J.M. 1998. Iron, free radicals, and oxidative injury. Semin. Hematol. **35:** 5–12.
2. Halliwell, B. 2001. Role of free radicals in the neurodegenerative diseases: therapeutic implications for antioxidant treatment. Drugs Aging **18:** 685–716.
3. Richardson, D.R. & P. Ponka. 1997. The molecular mechanisms of the metabolism and transport of iron in normal and neoplastic cells. Biochim. Biophys. Acta **1331:** 1–40.

4. PONKA, P., C. BEAUMONT & D.R. RICHARDSON. 1998. Function and regulation of transferrin and ferritin. Semin. Hematol. **35:** 35–54.
5. KLAUSNER, R.D. & T. ROUAULT. 1998. The molecular basis of iron metabolism. *In* The Harvey Lectures. Series 92, p. 99.
6. AISEN, P., C. ENNS & M. WESSLING-RESNICK. 2001. Chemistry and biology of eukaryotic iron metabolism. Int. J. Biochem. Cell Biol. **33:** 940–959.
7. PONKA, P. & C.N. LOK. 1999. The transferrin receptor: role in health and disease. Int. J. Biochem. Cell Biol. **31:** 1111–1137.
8. LEVY, J.E., O. JIN, Y. FUKIWARA *et al.* 1999. Transferrin receptor is necessary for development of erythrocytes and the nervous system. Nat. Genet. **21:** 396–399.
9. KAWABATA, H., R. YANG, T. HIRAMA *et al.* 1999. Molecular cloning of transferrin receptor 2: a new member of the transferrin receptor–like family. J. Biol. Chem. **274:** 20826–20832.
10. ROETTO, A., F. DARAIO, F. ALBERTI *et al.* 2002. Hemochromatosis due to mutations in transferrin receptor. Blood Cells Mol. Dis. **29:** 465–470.
11. FLEMING, R.E., J.R. AHMANN, M.C. MIGAS *et al.* 2002. Targeted mutagenesis of the murine transferrin receptor-2 gene products hemochromatosis. Proc. Natl. Acad. Sci. USA **99:** 10653–10658.
12. FEDER, J.N., A. GNIRKE, W. THOMAS *et al.* 1996. A novel MHC class I–like gene is mutated in patients with hereditary haemochromatosis. Nat. Genet. **4:** 399–407.
13. AJIOKA, R.S. & J.P. KUSHNER. 2002. Hereditary hemochromatosis. Semin. Hematol. **39:** 235–241.
14. AROSIO, P. & Z. LEVI. 2002. Ferritin, iron homeostasis, and oxidative damage. Free Radical Biol. Med. **33:** 457–463.
15. FERREIRA, C., D. BUCCHINI, M.E. MARTIN *et al.* 2000. Early embryonic lethality of H ferritin gene deletion in mice. J. Biol. Chem. **275:** 3021–3024.
16. CAIRO, G. & A. PIETRANGELO. 2000. Iron regulatory proteins in pathobiology. Biochem. J. **352:** 241–250.
17. MIKULITS, W., M. SCHRANZHOFER, H. BEUG *et al.* 1999. Post-transcriptional control via iron-responsive elements: the impact of aberrations in hereditary disease. Mutat. Res. **437:** 219–230.
18. LAVAUTE, T., S. SMITH, S. COOPERMAN *et al.* 2001. Targeted deletion of the gene encoding iron regulatory protein-2 causes misregulation of iron metabolism and neurodegenerative disease in mice. Nat. Genet. **27:** 209–214.
19. FLEMING, M.D., C.C. TRENOR III, M.A. SU *et al.* 1997. Microcytic anaemia mice have a mutation in Nramp2, a candidate iron transporter gene. Nat. Genet. **16:** 383–386.
20. GUNSHIN, H., B. MACKENZIE, U.V. BERGER *et al.* 1997. Cloning and characterization of a mammalian proton-coupled metal ion transporter. Nature **388:** 482–488.
21. MCKIE, A.T., D. BARROW, G.O. LATUNDE-DADA *et al.* 2001. An iron-regulated ferric reductase associated with the absorption of dietary iron. Science **291:** 1755–1759.
22. DONOVAN, A., A. BROWNLIE, Y. ZHOU *et al.* 2000. Positional cloning of zebrafish ferroportin 1 identifies a conserved vertebrate iron exporter. Nature **403:** 776–781.
23. MCKIE, A.T., P. MARCIANA, A. ROLFS *et al.* 2000. A novel duodenal iron-regulated transporter, IREG1, implicated in the basolateral transfer of iron to the circulation. Mol. Cell **4:** 299–309.
24. ABBOUD, S. & D.J. HAILE. 2000. A novel mammalian iron-regulated protein involved in intracellular iron metabolism. J. Biol. Chem. **275:** 19906–19912.
25. MONTOSI, G., A. DONOVAN, A. TOTARO *et al.* 2001. Autosomal-dominant hemochromatosis is associated with a mutation in the ferroportin (SLC11A3) gene. J. Clin. Invest. **108:** 619–623.
26. HELLMAN, N.E. & J.D. GITLIN. 2002. Ceruloplasmin metabolism and function. Annu. Rev. Nutr. **22:** 439–458.
27. NITTIS, T. & J.D. GITLIN. 2002. The copper-iron connection: hereditary aceruloplasminemia. Semin. Hematol. **39:** 282–289.
28. VULPE, C.D., Y.M. KUO, T.L. MURPHY *et al.* 1999. Hephaestin, a ceruloplasmin homologue implicated in intestinal iron transport, is defective in the Sla mouse. Nat. Genet. **21:** 195–199.

29. ANDERSON, G.J., D.M. FRAZER, A.T. MCKIE et al. 2002. The ceruloplasmin homolog hephaestin and the control of intestinal iron absorption. Blood Cells Mol. Dis. **29:** 367–375.
30. PANDOLFO, M. 1999. Friedreich's ataxia: clinical aspects and pathogenesis. Semin. Neurol. **19:** 311–321.
31. PUCCIO, H. & M. KOENIG. 2002. Friedreich ataxia: a paradigm for mitochondrial diseases. Curr. Opin. Genet. Dev. **12:** 272–277.
32. COX, T.M. 1997. Erythropoietic protoporphyria. J. Inherit. Metab. Dis. **20:** 258–269.
33. DRYSDALE, J., P. AROSIO, R. INVERNIZZI et al. 2002. Mitochondrial ferritin: a new player in iron metabolism. Blood Cells Mol. Dis. **29:** 376–383.
34. CAZZOLA, M., R. INVERNIZZI, G. BERGAMASCHI et al. 2003. Mitochondrial ferritin expression in erythroid cells from patients with sideroblastic anemia. Blood **101:** 1996–2000.
35. MAINES, M.D. 1997. The heme oxygenase system: a regulator of second messenger gases. Annu. Rev. Pharmacol. Toxicol. **37:** 517–554.
36. POSS, K.D. & S. TONEGAWA. 1997. Heme oxygenase 1 is required for mammalian iron reutilization. Proc. Natl. Acad. Sci. USA **94:** 10919–10924.
37. POSS, K.D. & S. TONEGAWA. 1997. Reduced stress defense in heme oxygenase 1–deficient cells. Proc. Natl. Acad. Sci. USA **94:** 10925–10930.
38. YACHIE, A., Y. NIIDA, T. WADA et al. 1999. Oxidative stress causes enhanced endothelial cell injury in human heme oxygenase-1 deficiency. J. Clin. Invest. **103:** 129–135.
39. DORE, S., K. GOTO, S. SAMPEI et al. 2000. Heme oxygenase-2 acts to prevent neuronal death in brain cultures and following transient cerebral ischemia. Neuroscience **99:** 587–592.
40. BURNETT, A.L., D.G. DEVIN, J.L. KRIEGSFELD et al. 1998. Ejaculatory abnormalities in mice with targeted disruption of the gene for heme oxygenase-2. Nat. Med. **4:** 84–87.
41. PARK, C.H., E.V. VALORE, A.J. WARING & T. GANZ. 2001. Hepcidin, a urinary antimicrobial peptide synthesized in the liver. J. Biol. Chem. **276:** 7806–7810.
42. NICOLAS, G., M. BENNOUN, I. DEVAUX et al. 2001. Lack of hepcidin gene expression and severe tissue iron overload in upstream stimulatory factor 2 (USF2) knockout mice. Proc. Natl. Acad. Sci. USA **98:** 8780–8785.
43. NICOLAS, G., M. BENNOUN, A. PORTEAU et al. 2002. Severe iron deficiency anemia in transgenic mice expressing liver hepcidin. Proc. Natl. Acad. Sci. USA **99:** 4596–4601.
44. NICOLAS, G., L. VIATTE, M. BENNOUN et al. 2002. Hepcidin, a new iron regulatory peptide. Blood Cells Mol. Dis. **29:** 327–335.
45. LEIBOLD, E.A., L.C. GAHRING & S.W. ROGERS. 2001. Immunolocalization of iron regulatory protein expression in the murine central nervous system. Histochem. Cell. Biol. **114:** 195–203.
46. BURDO, J.R., S.L. MENZIES, I.A. SIMPSON et al. 2001. Distribution of divalent metal transporter 1 and metal transport protein 1 in the normal and Belgrade rat. J. Neurosci. Res. **66:** 1198–1207.
47. SIDDAPPA, A.J., R.B. RAO, J.D. WOBKEN et al. 2002. Developmental changes in the expression of iron regulatory proteins and iron transport proteins in the perinatal rat brain. J. Neurosci. Res. **68:** 761–765.
48. CONNOR, J.R. 2003. Iron transport proteins in the diseased brain. J. Neurosci. **207:** 112–113.
49. MIRET, S., R.J. SIMPSON & A.T. MCKIE. 2003. Physiology and molecular biology of dietary iron absorption. Annu. Rev. Nutr. **23:** 283–301.
50. GRASBECK, R., R. MAJURI, I. KOUVONEN & R. TENHUNEN. 1982. Spectral and other studies on the intestinal haem receptor of the pig. Biochim. Biophys. Acta **700:** 137–142.
51. WORTHINGTON, M.T., S.M. COHN, S.K. MILLER et al. 2001. Characterization of a human plasma membrane heme transporter in intestinal and hepatocyte cell lines. Am. J. Physiol. Gastrointest. Liver Physiol. **280:** G1172–G1177.
52. HENTZE, M.W. & L.C. KUHN. 1996. Molecular control of vertebrate iron metabolism: mRNA-based regulatory circuits operated by iron, nitric oxide, and oxidative stress. Proc. Natl. Acad. Sci. USA **93:** 8175–8182.
53. PANTOPOULOS, K. & M.W. HENTZE. 1995. Rapid responses to oxidative stress mediated by iron regulatory protein. EMBO J. **14:** 2917–2924.

54. PANTOPOULOS, K., G. WEISS & M.W. HENTZE. 1996. Nitric oxide and oxidative stress (H_2O_2) control mammalian iron metabolism by different pathways. Mol. Cell. Biol. **16:** 3781–3788.
55. KIM, S. & P. PONKA. 1999. Control of transferrin receptor expression via nitric oxide–mediated modulation of iron-regulatory protein 2. J. Biol. Chem. **274:** 33035–33042.
56. KIM, S. & P. PONKA. 2000. Effects of interferon-γ and lipopolysaccharide on macrophage iron metabolism are mediated by nitric oxide–induced degradation of iron regulatory protein 2. J. Biol. Chem. **275:** 6220–6226.
57. MORGAN, E.H. & T. MOOS. 2002. Mechanism and developmental changes in iron transport across the blood-brain barrier. Dev. Neurosci. **24:** 106–113.
58. CURTIS, A.R., C. FEY, C.M. MORRIS et al. 2001. Mutation in the gene encoding ferritin light polypeptide causes dominant adult-onset basal ganglia disease. Nat. Genet. **28:** 350–354.
59. ROUAULT, T.A. 2001. Iron on the brain. Nat. Genet. **28:** 299–300.
60. KIM, S. & P. PONKA. 2002. Nitrogen monoxide–mediated control of ferritin synthesis: implications for macrophage iron homeostasis. Proc. Natl. Acad. Sci. USA **99:** 12214–12219.
61. BECKER, E. & D.R. RICHARDSON. 2001. Frataxin: its role in iron metabolism and the pathogenesis of Friedreich's ataxia. Int. J. Biochem. Cell Biol. **33:** 1–10.
62. LAMARCHE, J.B., M. COTE & B. LEMIEUX. 1980. The cardiomyopathy of Friedreich's ataxia: morphological observations in 3 cases. Can. J. Neurol. Sci. **7:** 389–396.
63. SANCHEZ-CASIS, G., M. COTE & A. BARBEAU. 1976. Pathology of the heart in Friedreich's ataxia: review of the literature and report of one case. Can. J. Neurol. Sci. **3:** 349–354.
64. PUCCIO, H., D. SIMON, M. COSSEE et al. 2001. Mouse models for Friedreich ataxia exhibit cardiomyopathy, sensory nerve defect, and Fe-S enzyme deficiency followed by intramitochondrial iron deposits. Nat. Genet. **27:** 181–186.
65. PONKA, P. 1997. Tissue-specific regulation of iron metabolism and heme synthesis: distinct control mechanisms in erythroid cells. Blood **89:** 1–25.
66. BECKER, E.M., J.M. GREER, P. PONKA & D.R. RICHARDSON. 2002. Erythroid differentiation and protoporphyrin IX down-regulate frataxin expression in Friend cells: characterization of frataxin expression compared to molecules involved in iron metabolism and hemoglobinization. Blood **99:** 3813–3822.
67. AARDON, O., J. KAPLAN & B.D. MARTIN. 2002. Iron uptake in yeast. *In* Molecular and Cellular Iron Transport, pp. 375–393. Dekker. New York.
68. BABCOCK, M., D. DE SILVA, R. OAKS et al. 1997. Regulation of mitochondrial iron accumulation by Yfh1p, a putative homolog of frataxin. Science **276:** 1709–1712.
69. RADISKY, D.C., M.C. BABCOCK & J. KAPLAN. 1999. The yeast frataxin homologue mediates mitochondrial iron efflux: evidence for a mitochondrial iron cycle. J. Biol. Chem. **274:** 4497–4499.
70. MUHLENHOFF, U., J. GERBER, N. RICHHARDT & R. LILL. 2003. Components involved in assembly and dislocation of iron-sulfur clusters on the scaffold protein Isu1p. EMBO J. **22:** 4815–4825.
71. GERBER, J., U. MUHLENHOFF & R. LILL. 2003. An interaction between frataxin and Isu1/Nfs1 that is crucial for Fe/S cluster synthesis on Isu1. EMBO Rep. **4:** 906–911.
72. SWAIMAN, K.F. 2001. Hallervorden-Spatz syndrome. Pediatr. Neurol. **25:** 102–108.
73. HAYFLICK, S.J. 2003. Unraveling the Hallervorden-Spatz syndrome: pantothenate kinase–associated neurodegeneration is the name. Curr. Opin. Pediatrics **15:** 572–577.
74. ZHOU, B., S.K. WESTAWAY, B. LEVINSON et al. 2001. A novel panthothenate kinase gene (PANK2) is defective in Hallervorden-Spatz syndrome. Nat. Genet. **28:** 345–349.
75. PERRY, T.L., M.G. NORMAN, V.W. YONG et al. 1985. Hallervorden-Spatz disease: cysteine accumulation and cysteine dioxygenase deficiency in the globus pallidus. Ann. Neurol. **18:** 482–489.
76. YOON, S.J., Y.Y. KOH, R.A. FLOYD & J.W. PARK. 2000. Copper, zinc superoxide dismutase enhances DNA damage and mutagenicity induced by cysteine/iron. Mutat. Res. **448:** 97–104.
77. FOX, P.L. The copper-iron chronicles: the story of an intimate relationship. Biometals **16:** 9–40.

78. OSAKI, S., D.A. JOHNSON & E. FRIEDEN. 1971. The mobilization of iron from the perfused mammalian liver by a serum copper enzyme, ferroxidase I. J. Biol. Chem. **246:** 3018–3023.
79. OSAKI, S. & D.A. JOHNSON. 1969. Mobilization of liver iron by ferroxidase (ceruloplasmin). J. Biol. Chem. **244:** 5757–5765.
80. HARRIS, Z.L., A.P. DURLEY, T.K. MAN & J.D. GITLIN. 1999. Targeted gene disruption reveals an essential role for ceruloplasmin in cellular iron efflux. Proc. Natl. Acad. Sci. USA **96:** 10812–10817.
81. ASKWITH, C., D. EIDE, A. VANHO et al. 1994. The FET3 gene of S. cerevisiae encodes a multicopper oxidase required for ferrous iron uptake. Cell **76:** 403–410.
82. STEARMAN, R., D.S. YUAN, Y. YAMAGUCHI-IWA et al. 1996. A permease-oxidase complex involved in high-affinity iron uptake in yeast. Science **271:** 1552–1557.
83. KLOMP, L.W.J., Z.S. FARHANGRAZI, L.L. DUGAN et al. 1996. Ceruloplasmin gene expression in the murine central nervous system. J. Clin. Invest. **98:** 207–215.
84. KLOMP, L.W. & J.D. GITLIN. 1996. Expression of the ceruloplasmin gene in the human retina and brain: implications for a pathogenic model in aceruloplasminemia. Hum. Mol. Genet. **5:** 1989–1996.
85. PATEL, B.N. & S. DAVID. 1997. A novel glycosylphosphatidylinositol-anchored form of ceruloplasmin is expressed by mammalian astrocytes. J. Biol. Chem. **272:** 20185–20190.
86. PATEL, B.N., R.J. DUNN & S. DAVID. 2000. Alternative RNA splicing generates a glycosylphosphatidylinositol-anchored form of ceruloplasmin in mammalian brain. J. Biol. Chem. **275:** 4305–4310.
87. JEONG, S.Y. & S. DAVID. 2003. Glycosylphosphatidylinositol-anchored ceruloplasmin is required for iron efflux from cells in the central nervous system. J. Biol. Chem. **278:** 27144–27148.
88. PATEL, B.N., R.J. DUNN, S.Y. JEONG et al. 2002. Ceruloplasmin regulates iron levels in the CNS and prevents free radical injury. J. Neurosci. **22:** 6578–6586.
89. MIYAJIMA, H., Y. NISHIMURA, K. MIZOGUCHI et al. 1987. Familial apoceruloplasmin deficiency associated with blepharospasm and retinal degeneration. Neurology **37:** 761–767.
90. HARRIS, Z.L., Y. TAKAHASHI, H. MIYAJIMA et al. 1995. Aceruloplasminemia: molecular characterization of this disorder of iron metabolism. Proc. Natl. Acad. Sci. USA **92:** 2539–2543.
91. YOSHIDA, K., K. FURIHATA, S. TAKEDA et al. 1995. A mutation in the ceruloplasmin gene is associated with systemic hemosiderosis in humans. Nat. Genet. **9:** 267–272.
92. DAIMON, M., T. KATO, T. KAWANAMI et al. 1995. A nonsense mutation of the ceruloplasmin gene in hereditary ceruloplasmin deficiency with diabetes mellitus. Biochem. Biophys. Res. Commun. **217:** 89–95.
93. HARRIS, Z.L., M.D. MIGAS, A.E. HUGHES et al. 1996. Familial dementia due to a frameshift mutation in the ceruloplasmin gene. Q. J. Med. **89:** 355–359.
94. OKAMOTO, N., S. WADA, T. OGA et al. 1996. Hereditary ceruloplasmin deficiency with hemosiderosis. Hum. Genet. **97:** 755–758.
95. HELLMAN, N.E., S. KONO, H. MIYAJIMA & J.D. GITLIN. 2002. Biochemical analysis of a missense mutation in aceruloplasminemia. J. Biol. Chem. **277:** 1375–1380.
96. PONKA, P. 2003. Recent advances in cellular iron metabolism. J. Trace Elem. Exp. Med. **16:** 201–217.

Mitochondrial Localization of Human PANK2 and Hypotheses of Secondary Iron Accumulation in Pantothenate Kinase–Associated Neurodegeneration

MONIQUE A. JOHNSON,[a] YIEN MING KUO,[b] SHAWN K. WESTAWAY,[a] SUSAN M. PARKER,[a] KATHERINE H. L. CHING,[a] JANE GITSCHIER,[b] AND SUSAN J. HAYFLICK[a,c]

[a]*Department of Molecular and Medical Genetics, School of Medicine, Oregon Health and Science University, Portland, Oregon, USA*

[b]*Howard Hughes Medical Institute and Departments of Medicine and Pediatrics, University of California, San Francisco, California, USA*

[c]*Departments of Pediatrics and Neurology, School of Medicine, Oregon Health and Science University, Portland, Oregon, USA*

ABSTRACT: Mutations in the pantothenate kinase 2 gene (*PANK2*) lead to pantothenate kinase–associated neurodegeneration (PKAN, formerly Hallervorden-Spatz syndrome). This neurodegenerative disorder is characterized by iron accumulation in the basal ganglia. Pantothenate kinase is the first enzyme in the biosynthesis of coenzyme A from pantothenate (vitamin B_5). *PANK2*, one of four human pantothenate kinase genes, is uniquely predicted to be targeted to mitochondria. We demonstrate mitochondrial localization of PANK2 and speculate on mechanisms of secondary iron accumulation in PKAN. Furthermore, *PANK2* uses an unconventional translational start codon, CUG, which is polymorphic in the general population. The variant sequence, CAG (allele frequency: 0.05), leads to skipping of the mitochondrial targeting signal and cytosolic localization of PANK2. This common variant may cause mitochondrial dysfunction and impart susceptibility to late-onset neurodegenerative disorders with brain iron accumulation, including Parkinson's disease.

KEYWORDS: pantothenate kinase; PKAN; Hallervorden-Spatz syndrome; mitochondrial localization; CoA; brain iron

INTRODUCTION

Pantothenate kinase–associated neurodegeneration (PKAN, formerly Hallervorden-Spatz syndrome) is an autosomal recessive neurodegenerative disease with iron deposition in the basal ganglia, pigmentary retinopathy, and movement and

Address for correspondence: Susan J. Hayflick, Department of Molecular and Medical Genetics, School of Medicine, Oregon Health and Science University, Portland, OR 97239. Voice: 503-494-6866; fax: 503-494-4411.
hayflick@ohsu.edu

speech defects.[1] Recently, we demonstrated that this disease results from mutations in the gene encoding pantothenate kinase 2 (*PANK2*) and proposed hypotheses of pathogenesis.[2] This discovery has led to the previously unappreciated involvement of a well-studied metabolic pathway in neurodegeneration.

Pantothenate kinase, an essential regulatory enzyme in coenzyme A (CoA) biosynthesis, catalyzes the phosphorylation of pantothenate (vitamin B_5), *N*-pantothenoyl-cysteine, and pantetheine.[3] Pantothenate kinase activity in rat has been reported to be cytosolic.[4,5] Although CoA is required in mitochondria, no mitochondrial activity of the first three enzymes in the CoA synthetic pathway (pantothenate kinase, *p*-pantothenate-cysteine synthase, and *p*-pantothenoyl-cysteine decarboxylase) has been found.[6] The final two enzymes in this cascade (pantetheine-*p*-adenylyl transferase and dephospho-CoA kinase) have been associated with mitochondria, either within the matrix[7] or on the cytosolic membrane face.[8]

Humans have four genes predicted to encode a pantothenate kinase. Three human paralogs, *PANK1*, *PANK2*, and *PANK3*, have six nearly identical core exons preceded by significantly divergent sequences in exon 1.[2,9] A fourth human paralog, *PANK4*, is more similar in sequence to its counterparts in *Saccharomyces cerevisiae* and *Caenorhabditis elegans*.[2]

Among these sequences, only PANK2 is predicted to localize to mitochondria, with a strong 29-amino-acid mitochondrial targeting sequence (MTS) at the amino terminus (Predotar, http://www.inra.fr/predotar/; TargetP V1.0, http://www.cbs.dtu.dk/services/TargetP/; and MitoProt, http://www.mips.biochem.mpg.de/cgi-bin/proj/medgen/mitofilter).[10,11] Mitochondrial localization of PANK2, along with its relatively abundant expression in brain,[2] may help to explain the pathophysiology of PKAN.

In this study, we use an antibody against human PANK2 and an antibody against the mouse PANK2, as well as a set of *myc*-HIS-tagged human *PANK2* constructs, to investigate subcellular localization in various cell types. We show that PANK2 is localized to mitochondria, with a small cytosolic fraction. Additionally, we demonstrate that localization is dependent on alternative translational start sites. Finally, we show that a common translation initiation variant produces predominantly cytosolic PANK2 and we propose that this may be a susceptibility allele to common neurodegenerative disorders in which brain iron accumulation has been observed.

EXPERIMENTAL PROCEDURES

Assembly of PANK2 cDNA

Human *PANK2* cDNA, from the predicted translational start to stop codons, was assembled via a three-part ligation. Fragment 1 was generated as follows: exon 1C was amplified by polymerase chain reaction (PCR) using genomic DNA and the primers 5'GTCCCCAGCCTCGTCGGATTGGCTT and 5'ACAGTGTCCGCGCG-CGGGAAAAAC (GenBank accession no. AL353194, nucleotides 28180–28204 and 28699–28676, respectively). The resulting amplicon extended 88 nucleotides 5' of the predicted translational start and 104 base pairs 3' of the end of exon 1C. The final 8 nucleotides of exon 1C are a unique *Not*I site. Fragment 2 was generated as follows: human fetal brain and/or testes cDNA was amplified using primers AP1 and gene specific primer 5'TCTCTGCCCATTTGAATAAAAGCAGGCA by rapid amplification of cDNA ends (RACE) and subcloned into pT-Adv for sequencing according to manufacturer's instructions (Clontech). The 5' end of fragment 2 includes overlapping sequence with the 3' end of fragment 1, containing the same *Not*I site, and additionally an *Apa*LI site 175 bp downstream. Fragment 3 was cloned as discussed for fragment 2, but using the following primers: 5'CGCGGATC-CTTTCCATGGTTTGGACTGGAT and 5'CGAAGATCTCGGGATCTTCAACAGC-TCAAG. The 5' end of fragment 3 includes sequence overlapping with the 3' end of fragment 2, containing the same *Apa*LI site, and *PANK2* sequence, in its entirety, 3' of that site. Fragment 1 was digested with *Not*I, fragment 2 with *Not*I and *Apa*LI, and fragment 3 with *Apa*LI. The fragments were ligated in a stepwise fashion; amplified by PCR using primers 1F (5'CGCGGATCCACGCTGGGGGGCTTGCTCGGGCGG) and 1R (5'CCGGAATTCCGGGATCTTCAACAGCTCAAG), resulting in *PANK2* from translational start to the nucleotide immediately 5' of the stop codon (nucleotide no. 17-1396 of GenBank accession no. BK000010); and subcloned into pBSKS+ (Stratagene) at the *Bam*HI and *Eco*RI sites. This construct was named phPANK2.

Building of myc-*HIS-Tagged cDNA Constructs*

A series of *PANK2*-derived carboxy-terminus *myc*-HIS-tagged constructs were made according to standard recombinant DNA methodologies.[12] Using phPANK2 (this study) as a template, and primer pairs as described below, derivatives of hPANK2 were generated by PCR as follows: primers 1F and 1R (described above) resulted in pCTGcodon1; primers 2F (5'CGCGGATCCGCCACCATGGGGGGCT-TGCTCGGGCGGCAG) and 1R resulted in pATGcodon1; primers 3F (5'CGCG-GATCCACGCAGGGGGGCTTGCTCGGGCGG) and 1R resulted in pCAGcodon1; primers 4F (5'CGCGGATCCGCCACCATGGGAGGGGCCGGCTC) and 1R generated pATGcodon14; primers 5F (5'CGCGGATCCGCCACCATGTCCGC-CACCTCCGTCTCG) and 1R resulted in pΔMTS; primers 1F and 2R (5'CCG-GAATTCAGCCCTGCCGTGGCGCTCCAT) resulted in pCTGMTSonly. These amplicons were then subcloned into pBSKS+ (Stratagene) at the *Bam*HI and *Eco*RI sites and sequenced in both directions. Finally, using *Bam*HI and *Eco*RI, each fragment was excised and subcloned into the human expression vector pcDNA3.1/*myc*-HIS (Invitrogen), resulting in carboxy-terminus *myc*-HIS tags. All junctions of the constructs were sequenced. An overview of all constructs can be found in TABLE 1.

Generation of Anti-PANK2 Antibody

A polyclonal chicken anti-human PANK2 antibody was raised by Aves Labs against a 14-mer synthetic peptide (RRQEPLRRRASSAS), corresponding to amino acids 48–61 (GenBank accession no. DAA00004), located downstream of the MTS. The antibody was then affinity-purified with this 14-mer peptide using Reduce-Imm Reducing Kit and SulfoLink Kit according to the manufacturer's instructions (Pierce). After concentrating, using an Amicon Ultra (Millipore), the antibody was used at 1:5000 dilution for Western blots and 1:1000 for immunofluorescence. As a control for specificity, the antibody was adsorbed for 48 h at 4°C at 10^{-5} M to the above 14-mer and, separately, to an 11-mer (RDVHLELKDLT) containing none of the 14-mer sequence. Only binding to the 14-mer was detected. Additionally, the antibody was shown to bind, as detected by Western blot, to *E. coli*–expressed

TABLE 1. Summary of C-terminus *myc*-HIS-tagged human *PANK2* cDNA constructs

Construct	Description	Primers	Primary localization
pCTGcodon1	Full-length native PANK2	1F/1R	Mitochondria
pΔMTS	N-terminal truncated PANK2, starting at codon 29 (substituted by ATG) and excluding all of the MTS	5F/1R	Cytosol
pCTGMTSonly	PANK2 MTS only	1F/2R	Mitochondria
pATGcodon14	N-terminal truncated PANK2, starting at ATG of codon 14 and excluding half of MTS	4F/1R	Cytosol
pATGcodon1	Full-length PANK2 with CTG start replaced by ATG	2F/1R	Mitochondria
pCAGcodon1	Full-length PANK2 with CTG start replaced by CAG	3F/1R	Cytosol

NOTE: Primer sequences are described in EXPERIMENTAL PROCEDURES. For convenience of comparison, constructs are listed in the order presented in the text and in FIGURE 3.

human PANK2 containing the 14-mer epitope, but not to PANK2 deleted of that epitope (Johnson and Westaway, unpublished data).

Expression in HeLa Cells and COS Cells, Immunohistochemistry, and Immunofluorescence Microscopy

HeLa cells were cultured on Dulbecco's Modified Eagle Medium (Gibco-BRL) supplemented with 10% fetal bovine serum (Gibco-BRL) and grown at 37°C under 5% CO_2. One day prior to transfection, the cells were split using 0.05% trypsin (Gibco-BRL), seeded on coverslips in 6-well plates, and cultured as described above. Some cells were then transfected with various *myc*-HIS-tagged cDNA constructs (described in TABLE 1) using Lipofectamine (Invitrogen) with Opti-MEM I-Reduced (Gibco-BRL) according to the manufacturer's recommendations using empirically derived optimized parameters. Forty-eight hours after transfection, the cells were aspirated and treated with MitoTracker Red CMXRos (Molecular Probes) according to the manufacturer's instructions. For the remainder of the protocol, care was taken to shield the cells from as much light as possible. The cells were then washed with phosphate-buffered saline (PBS) and fixed in 5% (w/v) freshly prepared paraformaldehyde (Electron Microscopy Science) for 15 min at room temperature, followed by washing in PBS. The cells were then blocked in 2.5 mg/mL PBS bovine serum albumin (Sigma) while being permeabilized in 0.1% Triton X-100 (Sigma) for 15 min at room temperature. Following washing in PBS, the cells were incubated for 60 min at room temperature with monoclonal anti-*myc* at a dilution of 1:400 (Upstate Biotechnology), monoclonal anti-HIS at a dilution of 1:400 (Invitrogen), or native anti-PANK2 antibodies (1:1000).

The cells were again washed in PBS and then incubated for 60 min at room temperature with anti-mouse Alexa Fluor 488 at a dilution of 1:600 (Molecular Probes) or fluorescein-labeled goat anti-chicken antibodies at 1:500 (Aves Labs).

Following washing in PBS, the cells were mounted in ProLong Antifade (Molecular Probes) on glass slides according to the manufacturer's instructions. The coverslips were then sealed with clear fingernail polish and dried for 20 min. The fluorescent staining pattern was viewed at 100× using a Leika DMRXA microscope. The images were digitally recorded and then colorized using Adobe Photoshop.

COS7 cells were grown in Lab-TekII chamber slides and incubated in 200 nM MitoTracker Red CMXRos (Molecular Probes) for 30 min at 37°C. Cells were then fixed for 20 min in 4% paraformaldehyde. A polyclonal rabbit anti-mouse PANK2 was raised against a 15-mer peptide (CSGEAESVRRERPGS) and was affinity-purified with the peptide. The antibody (1:200 dilution) was visualized by using a goat Alexa Fluor 488 anti-rabbit IgG antibody at 1:150 dilution (Molecular Probes). Images were visualized at 100× magnification using a Nikon E800 microscope. The images were digitally recorded and then colorized using Adobe Photoshop.

SDS-PAGE and Immunoblots

Protein assays were done using a BCA Protein Assay Reagent Kit (Pierce) according to the manufacturer's instructions. Lysates or crude mitochondrial pellets were analyzed by 7.5% or 10% SDS-PAGE, and immunoblotting was performed according to Sambrook[12] and Bio-Rad instructions. Chemiluminescent immunoblot detection was performed according to Western-Star Systems instructions (Applied Biosystems) or ECL Western Blotting Analysis System instructions (Amersham Pharmacia Biotech). Antibodies were used at the following dilutions: native anti-PANK2 1:5000, anti-*myc* 1:500 (Upstate Biotechnology), anti-HIS 1:5000 (Invitrogen), anti-CPTI 1:3000 (gift from Gebretateos Woldegiorgis, OGI-OHSU), anti-SKL motif 1:1000 (gift from Rick Rachubinski, University of Alberta), anti-BiP/GRP78 1:250 (BD Biosciences), EEA1 1:1000 (BD Biosciences), GM130 1:250 (BD Biosciences), Lamp-1 1:250 (BD Biosciences), MAP2B 1:1000 (BD Biosciences), Nucleoporin p62 1:1000 (BD Biosciences), anti-rabbit 1:6500 (Tropix), anti-chicken 1:7500 (Jackson ImmunoResearch Laboratories), and anti-mouse 1:5000 (ECL, Amersham Pharmacia Biotech). For further descriptions of antibodies, see below.

Isolation of Crude Mitochondrial Pellet and Generation of Percoll Gradient–Purified Mitochondrial Fraction

HeLa cells were cultured as described above and grown to confluency on 100 mM tissue culture plates. To collect a crude mitochondrion pellet, cells were harvested with 0.05% trypsin and transferred to a cold Falcon tube, spun at 4°C, 250*g*, for 2 min in a Sorvall tabletop centrifuge. The supernatant was decanted and the cells washed twice in cold PBS. The cells were resuspended in 1 mL cold buffer A (20 mM potassium HEPES, pH 7.5; 250 mM sucrose; 3 mM EDTA plus protease inhibitor cocktail [Complete, Mini, EDTA-free from Roche]) as directed by the manufacturer. The cells were transferred to a cold ground glass tissue grinder (Kontes) and were lysed by 25 strokes. Following transfer to a cold Falcon tube, the total volume was increased to 10 mL with cold buffer A and spun at 4°C at 1000*g* for 10 min in a Sorvall tabletop centrifuge. The supernatant (I) was saved on ice and the pellet was resuspended in 1 mL cold buffer A, transferred to the Kontes ground glass tissue grinder and lysed by 15 strokes, and spun again as described above. The resulting supernatant was combined with supernatant I and spun at 4°C at 12,000g for 10 min in a Beckman centrifuge. The resulting supernatant was concentrated using an Amicon Ultra 10,000 MWCO (Millipore) and saved for comparison on an immunoblot. The resulting crude mitochondrial pellet was resuspended in 100 µL cold buffer A, assayed by BCA Protein Assay Reagent Kit (Pierce), and stored at 4°C.

For purposes of quantifying percent protein (native PANK2 or *myc*-tagged PANK2) present in crude mitochondria pellet versus resulting supernatant, an equivalent percent of total volume of crude mitochondrial pellet fraction and resulting supernatant fraction was loaded on an SDS-PAGE gel and immunoblotted. Quantification of band densities, corrected for background, was done using LabWorks Image Acquisition and Analysis software (UVP BioImaging Systems).

The crude mitochondrial pellet (1450 µg) was loaded on a Percoll (Amersham Pharmacia Biotech) gradient. The gradient was made as follows: 6.5 mL of 40% Percoll was overlayed with 8 mL 20% Percoll in 18-mm centrifuge tubes (Nalge). The sample was spun at 50,000g for 60 min at 4°C. Ten 1-mL fractions, collected from the top of the tube, were diluted in 2× volume buffer A, spun at 14,000 rpm for 10 min in microfuge at 4°C, pellets collected and resuspended in 100 µL 1× SDS-PAGE sample buffer, and boiled. Ten-µL samples of each fraction were loaded on each of nine SDS-PAGE gels as well as a total cell lysate sample and a control protein. Western blots were performed and probed with the following antibodies against native hPANK2, CPTI (mitochondria), GM130 (Golgi), MAP2B (microtubule), Nucleoporin p62 (nucleus), BiP/GRP78 (endoplasmic reticulum), EEA1 (endosome), and Lamp-1 (lysosome).

Building of EGFP Constructs

pEGFP-N1 (Clontech) (also referred to as pEGFP in FIGURE 5 later) expresses enhanced green fluorescent protein under the CMV promoter. pEGFPΔATG was built as a negative control with respect to EGFP expression as follows: EGFP with the ATG start sequence deleted was amplified by PCR from pEGFP-N1 using primer pairs 5′-CTTCGAATTCTGCAGTCGACCAGCAAGGGCGAGGAGCTG-3′ and 5′-ATGATCTAGAGTCGCGGCCGCTTTACTTGTACAGCTCG-3′ and substituted back into pEGFP-N1 at the *Sal*I and *Not*I sites. pCTGonATGon, in which both the CTG and ATG (nucleotides 31–33 and 97–99 of *PANK2*, GenBank accession no. BK000010) are in frame with EGFP, was constructed by PCR amplification of nucleotides 2–227 of *PANK2* from previously amplified exon 1C PCR product and cloning into pEGFPΔATG at *Sac*I and *Sal*I sites upstream of EGFP using the following primers: forward 5′-ACTCAGATCTCGAGCTCCCGCGGAGGAGGCGAGAAGGA-3′ and reverse 5′-CCGACCGCGGGCGTCGACGCGCTGCTCGCCCG-3′. pCTGoffATGoff was constructed as described above for pCTGonATGon, except using reverse primer 5′-CCGACCGCGGGCGTCGACCGCTGCTCGCCCG-3′, which resulted in both CTG and ATG being out of frame with EGFP. pCTGoff-ATGon, in which the CTG is out of frame and the ATG is in frame with EGFP, was constructed by digesting pCTGonATGon with *Ava*I (cuts between CTG and ATG), filling in the 4-bp overhang, and religating. This created a frameshift, moving CTG out of frame with EGFP. pCTGonATGoff, in which the CTG is in frame with EGFP and ATG is out of frame with EGFP, was digested, filled in, and religated as above, but beginning with pCTGoffATGoff resulting in CTG shifted back into frame with EGFP. All primers were from Gibco-BRL. All constructs were sequenced.

The constructs were transfected into HeLa cells using X-tremeGeneQ2 (Roche) according to the manufacturer's instructions and cultured as described above. The cells were viewed after 24 h and images captured at 10× using a Leika DMRXA microscope at excitation wavelength of 470 nm. The EGFP images were colorized using Adobe Photoshop (Adobe Systems).

RESULTS

Predominant Localization of PANK2 to Mitochondria

Two immunofluorescence experiments demonstrated that PANK2 localizes primarily to the mitochondria. First, using an anti-human PANK2 antibody raised against a peptide corresponding to a sequence downstream of the human PANK2 MTS, we find localization of endogenous PANK2 in HeLa cells to be predominantly mitochondrial and, to a much lesser extent, cytosolic (FIGS. 1A–C). Second, with an antibody raised against a murine PANK2-specific peptide corresponding to a similar sequence downstream of the MTS, fluorescence microscopy revealed staining in the

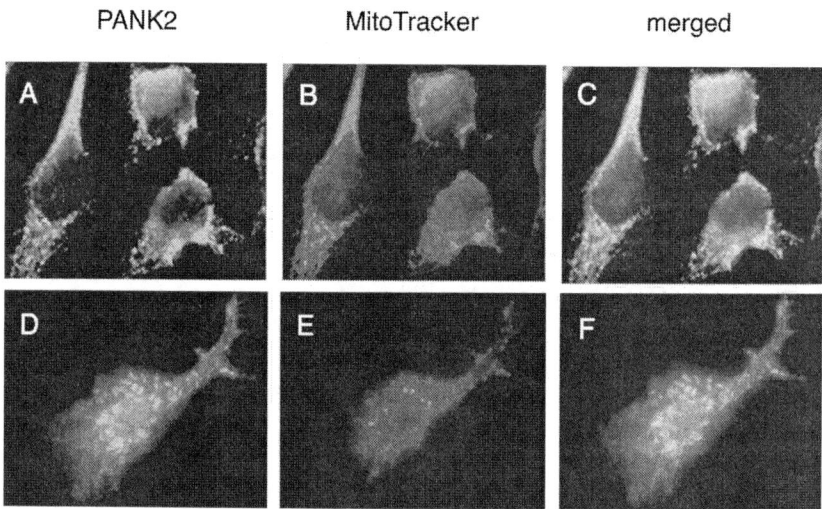

FIGURE 1. Mitochondrial PANK2 distribution detected by indirect immunofluorescence using a PANK2 antibody. HeLa (**A–C**) and COS7 (**D–F**) cells were cultured and prepared for immunofluorescence as described in EXPERIMENTAL PROCEDURES. Cells in panel A were incubated with anti-human PANK2 antibodies and labeled with fluorescein-labeled antichicken antibodies. Cells in panel D were incubated with an antibody against murine PANK2 and binding was visualized with Alexa Fluor 488 Green antibodies. Cells in panels B and E were incubated with MitoTracker Red CMXRos. Panels C and F are merged images. [Images shown in black and white.]

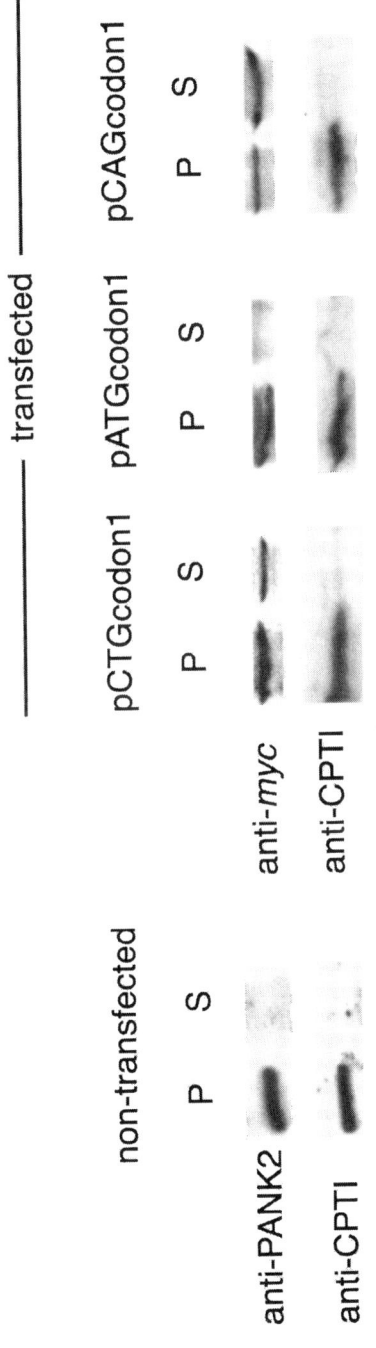

FIGURE 2. Compartmentalization by immunoblot of endogenous and *myc*-tagged PANK2. Equivalent percent volumes of mitochondrial pellet (P) and resulting supernatant (S) were loaded. Anti-PANK2 is antibody against human PANK2. Anti-*myc* is antibody used to decorate transfected *myc*-tagged PANK2 expressed by pCTGcodon1, pATGcodon1, and pCAGcodon1 (see TABLE 1). Anti-CPTI is antibody to mitochondrial marker protein, CPTI. Transfected and nontransfected refer to HeLa cells. Bands were quantitated *in silico* (see EXPERIMENTAL PROCEDURES).

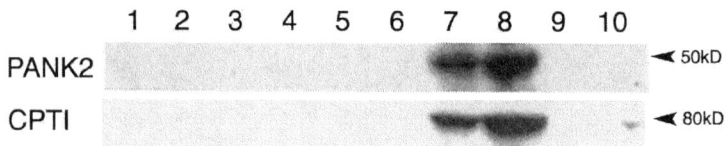

FIGURE 3. HeLa cell mitochondrial localization of PANK2 by immunoblot using Percoll gradient fractions and a native antibody. Equal volumes of Percoll gradient fractions were loaded on SDS-PAGE gels and immunoblotted. The panel labeled PANK2 was incubated with anti-human PANK2 antibody, and the panel labeled CPTI with anti-CPT1 antibody. CPT1 is a mitochondrial marker protein.

mitochondria in COS7 cells, as shown in FIGURES 1D–F. The signal is absent in the control preimmune sera as well as in sera preadsorbed with the peptide (data not shown). As in the transfected HeLa cells, trace cytoplasmic staining was again observed. A similar pattern was seen in HEK and CACO cells with this antibody (data not shown).

We quantitated the relative compartmentalization of endogenous PANK2 by immunoblotting the equivalent percent of total volume of crude mitochondrial pellet fraction and resulting supernatant fraction with anti-human PANK2 antibodies (FIG. 2). In agreement with the immunofluorescence results, the vast majority of native PANK2, 95%, localized to the mitochondrial pellet with 5% in the supernatant. Immunoblotting for CPTI, a mitochondrial marker protein, as a control, showed 100% of CPTI to be in the crude mitochondrial pellet.

Mitochondrial localization was confirmed by studying Percoll gradient–purified mitochondria. Using Percoll gradient fractions, PANK2 was found in, and only in, the same fractions as was CPTI (FIG. 3). To rule out inclusion of organelles other than mitochondria in fractions of interest, immunoblots were done, beginning with the same Percoll gradient fractions used for the anti-PANK2 and anti-CPTI immunoblotting, using antibodies against GM130 (Golgi), MAP2B (microtubule), Nucleoporin p62 (nucleus), BiP/GRP78 (endoplasmic reticulum), EEA1 (endosome), and Lamp-1 (lysosome). With the possible exception of endoplasmic reticulum, no other organelle appears to be comigrating with mitochondria (data not shown).

A series of transfection experiments were undertaken to clarify the role of the predicted leucine codon for translation initiation and resulting localization. We generated *myc*-HIS-tagged constructs derived from human *PANK2* cDNA and performed transfection experiments in HeLa cells (TABLE 1). FIGURES 4A–C demonstrate that a construct of full-length human PANK2 (pCTGcodon1), initiated with the native CUG codon, results in predominantly mitochondrial localization, with a small fraction present in the cytosol. Quantifying this localization was done as previously described for endogenous PANK2, except here pCTGcodon1 transfected HeLa cells were used to generate the crude mitochondrial pellet and resulting supernatant, and immunoblotting was done with anti-*myc* antibodies (FIG. 2). We found 77% of the pCTGcodon1 protein localized to the mitochondrial pellet fraction, with 23% in the resulting supernatant fraction. Control CPTI was found to be entirely in the mitochondrial pellet fraction. Thus, immunofluoresence and quantitation observations on *myc*-tagged, expressed PANK2 and on endogenous PANK2 show localization to be predominantly mitochondrial, with a small cytosolic fraction. Further evidence is presented below supporting leucine as a translational start codon in PANK2.

Evidence for a Mitochondrial Targeting Signal

Deletion of the entire predicted mitochondrial sequence (pΔMTS, lacking the 29 amino-terminal residues L/MGGLLGRQRLLLRMGGGRLGAPMERHGRA) causes PANK2 to be primarily cytosolic (FIGS. 4D–F). In a reciprocal experiment, the predicted MTS from PANK2 (pCTGMTSonly) is sufficient for mitochondrial targeting of the *myc*-HIS tag (FIGS. 4G–I). Use of anti-HIS antibodies for localization confirmed the results and controlled for the possibility that the anti-*myc* antibodies were cross-reacting with the cell's native *myc* (data not shown). From these data, we conclude that PANK2's predicted MTS is authentic.

Evidence for an Alternative Translational Start Site

The presence of a small fraction of PANK2 in the cytosol suggests an alternative translational start site that excludes part or all of the MTS, a possibility for which there is precedence.[13] In PANK2, the first in-frame AUG codon occurs 42 bases downstream of the CUG; translation initiation here would exclude half of the MTS. Indeed, for pATGcodon14, a construct in which the first 13 codons are deleted, the majority of PANK2 is cytosolic (FIGS. 4J–L).

This model would further predict that translation initiation may be less efficient with the CUG codon than with an AUG. To test this possibility, we replaced the initiating unconventional codon in pCTGcodon1 with an ATG (pATGcodon1) and found all of the resulting protein to be mitochondrial (FIGS. 4M–O). By Western blotting, as described above for quantification of pCTGcodon1 protein, but here beginning with pATGcodon1 transfected cells, 100% of resulting protein was found to be in the mitochondrial pellet fraction (FIG. 2). CPTI control results were as described above. These data, taken together, suggest that the cytosolic fraction of PANK2 may be a result of the use of a downstream AUG and that CUG is an authentic translational start site.

An Allelic Variant Associated with Mislocalization of PANK2

During our initial analysis of the *PANK2* gene, a variant in the initiation codon came to light. One of 15 human ESTs (expressed sequence tags) in the public database contained a glutamine codon (CAG) instead of the leucine/methionine codon (CUG) at residue 1. During sequence analysis of our patient population, we also observed a CAG allele frequency of 3.1% (7 of 224 chromosomes). However, additional deleterious mutations could be found in the PKAN patients to account for disease. Subsequent analysis of DNA from CEPH samples confirmed the polymorphism: namely, 5% (9 of 176) of chromosomes from unrelated grandparents in this study sample were found to be heterozygous for the CAG variant. In the 100 CEPH samples tested, no one was found to be homozygous for the CAG allele.

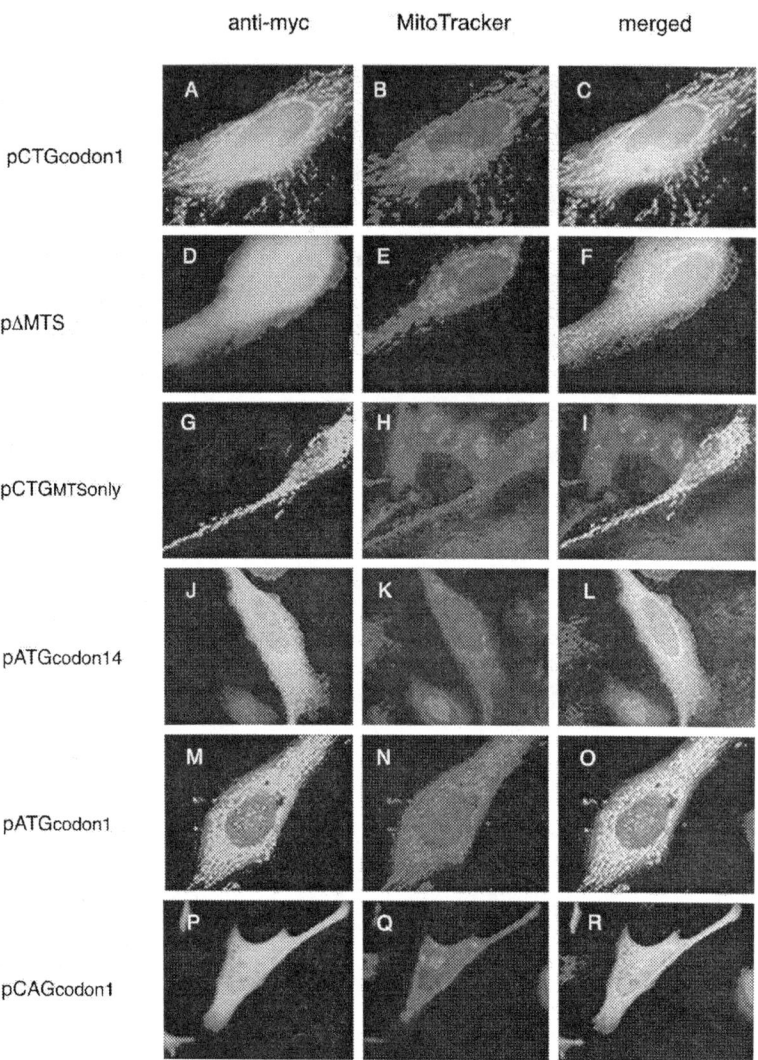

FIGURE 4. Intracellular PANK2 distribution and identification of a mitochondrial targeting sequence as detected by indirect immunofluorescence. HeLa cells were transfected by the plasmids indicated, cultured, and prepared for immunofluorescence as described in EXPERIMENTAL PROCEDURES. Cells in panels A, D, G, J, M, and P were incubated with anti-*myc*. Alexa Fluor 488 Green secondary antibodies were used for fluorescent labeling. Cells in panels B, E, H, K, N, and Q were incubated with MitoTracker Red CMXRos. Panels C, F, I, L, O, and R are merged images. See TABLE 1 for plasmid descriptions. [Images shown in black and white.]

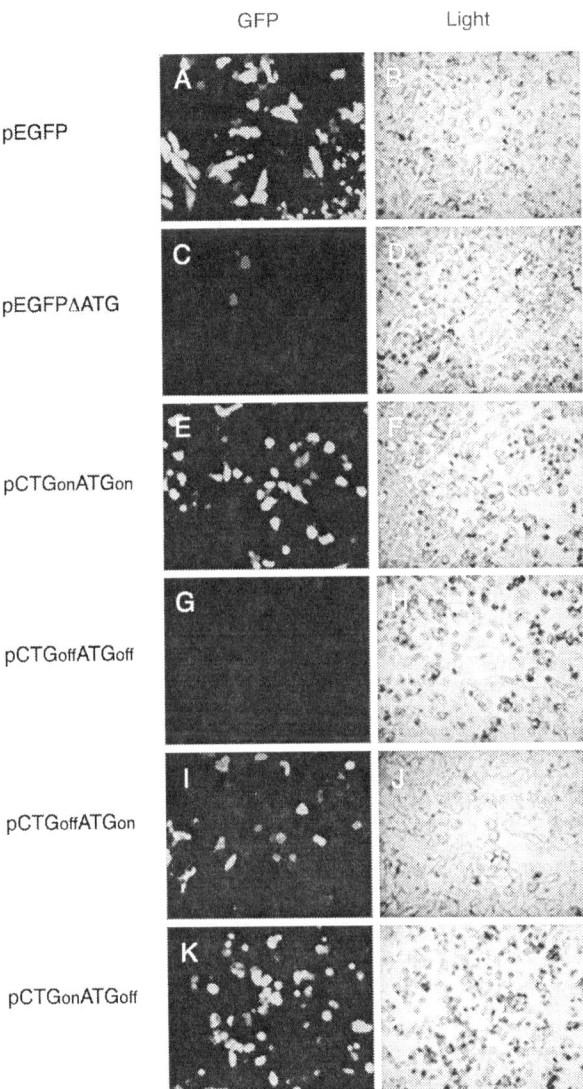

FIGURE 5. Initiation of translation at a CTG as evaluated by an EGFP reporter. Cells in panels A, C, E, G, I, and K were viewed using fluorescent light at excitation wavelength of 470 nm. Cells in panels B, D, F, H, J, and L were viewed using visible light and correspond to panels to their left. pEGFP is the parental vector containing only *EGFP*, driven by the CMV promoter. pEGFPΔATG is the parental vector with the *ATG* of *EGFP* removed. pCTGonATGon is pEGFPΔATG plus the first 212 nucleotides of *PANK2* fused to the amino terminus of *EGFP*, with both the *ATG* and downstream *CTG* of *PANK2* in frame with *EGFP*. pCTGoffATGoff is as in pCTGonATGon, except both the *CTG* and *ATG* are out of frame with *EGFP*. pCTGoffATGon is as in pCTGonATGon, except the *CTG* is out of frame with *EGFP*. pCTGonATGoff is as in pCTGonATGon, except the *ATG* is out of frame with *EGFP*. [Images shown in black and white.]

The CAG codon would not be predicted to serve as a start site for protein translation, and we anticipate that protein encoded by this variant would initiate at methionine codon 14 (see above) and be localized primarily to the cytosol. To test this hypothesis, we transformed HeLa cells with a *myc*-HIS-tagged *PANK2* construct in which the initiating CTG was replaced with a CAG (pCAGcodon1). As shown in FIGURES 4P–R, this variant of PANK2 localizes mainly to the cytosol. The localization was quantified by immunoblot as described above (FIG. 2): 79% of the resulting protein was found to be in the supernatant fraction, yet 100% of the mitochondrial control protein, CPTI, was in the mitochondrial fraction. Based on this mislocalization, we could envision that carriers of the CAG allele might be deficient in mitochondrial PANK2 and might suffer a clinical consequence with time.

Evidence for a CUG Translational Initiation Codon

The localization experiments described above support CUG as a translational initiation codon and, through a set of experiments using a reporter gene initiated by the native PANK2 sequence, we provide further evidence for CUG as a translational initiation codon.

HeLa cells were transiently transfected with a series of *PANK2-EGFP* vectors and tested for EGFP fluorescence (FIG. 5). Cells transfected with the parental vector containing only *EGFP* driven by the CMV promoter (see EXPERIMENTAL PROCEDURES) strongly fluoresce and serve as a positive control. Panel C is as in panel A, except the ATG start codon was removed from *EGFP* and serves as a negative control. In panels E–L, the first 212 nucleotides of *PANK2* were fused to the amino terminus of *EGFP* and, as in the negative control above, the ATG of *EGFP* was removed. Both the native *PANK2 CTG* and downstream *ATG* are in frame with the *EGFP* in panel E, serving as another positive control. Conversely, in panel G, both the native *PANK2 CTG* and *ATG* are out of frame with *EGFP*, demonstrating that there are no alternative initiation codons in the construct. Panel I shows cells in which only the downstream *ATG* is in frame with *EGFP* and we see somewhat reduced fluorescence as compared with the positive controls. Only the *CTG* of *PANK2* is in frame with *EGFP* in panel K and strong fluorescence is observed. Panels B, D, F, H, J, and L show corresponding views of cells as viewed with visible light. These data suggest that CUG in the limited native context of *PANK2* is an effective translation initiation codon.

DISCUSSION

Human PANK2 localizes primarily to mitochondria, with a small fraction found in cytosol. The *PANK2* gene encodes an authentic signal that is both necessary and sufficient for mitochondrial targeting. Two alternate translational start sites in *PANK2* likely account for the dual localization. Our data support a less efficient translation initiation with CUG rather than AUG, with the ribosome at times bypassing this codon and employing a downstream AUG for the translational start that disrupts the MTS in PANK2.

The use of alternative translational start sites as well as dual localization has precedence. For example, NifS, a human iron-sulfur cluster assembly enzyme, is targeted to mitochondria or cytosol/nucleus based on pH-dependent initiation of protein synthesis at different AUGs.[13] Other examples of alternative AUG usage include rat liver fumarase gene,[14] human ubiquitin-activating enzyme E1 gene,[15] mouse LAP gene,[16] mouse GATA-1 transcription factor gene,[17] and yeast tRNA processing enzyme genes *CCA1* and *MOD5*.[18] More to the point, *int*-2, bFGF, and *hck* are similar to PANK2 in that subcellular localizations for all use alternative translation initiation at CUG vs. AUG codons.[19–21] As alternative translational starts may provide flexible response to metabolic changes, CoA intermediates may regulate subcellular distribution of PANK2 through a feedback mechanism.

Interestingly, the well-studied *c-myc* gene provides an excellent example for discussion of PANK2. The two major protein forms of *c-myc*, c-Myc 1 and c-Myc 2, are derived from alternative translational starts at either a CUG or an AUG that is 42 nucleotides downstream in a configuration analogous to that of *PANK2*.[22] Moreover, we previously noted the presence of a stem-loop structure 14 nucleotides downstream from the CUG in *PANK2*,[2] and C-*myc* has a similar structure in the same position. Such stem-loops have been described in other CUG-initiated proteins and shown to enhance translation initiation at non-AUG codons by slowing the scanning complex.[23] A methionine is incorporated as the first residue of c-Myc 1,[22,24] and Met-tRNA$_i^{Met}$ has been shown to recognize and initiate translation at both AUG and non-AUG codons.[25,26] Thus, we would predict that the first amino acid in PANK2 is also a methionine.

We note that another start site for PANK2 translation has been proposed 330 nucleotides upstream of the CUG leucine codon[27] (GenBank accession no. AF494409). However, several lines of evidence argue against this conceptual translation. First, sequence conservation (both nucleotide and predicted amino acid) between the murine, rat, and human genomic sequences is maintained only as far upstream as the CUG start codon, and sequences 5′ of that diverge considerably. This observation is of significance as cross-species sequence conservation is a hallmark of protein-coding capacity. Additionally, "in-frame" stop codons are present in both the murine and rat sequences upstream of the CUG, again signaling lack of protein-coding capacity. Finally, no ESTs can be found to document that these 330 nucleotides are part of a *PANK2* transcript; indeed, we were not able to verify use of this upstream sequence by RACE-PCR.

Mitochondrial localization of PANK2 is predicted to be unique among the PANKs and suggests an organelle-specific function. As CoA is required for mitochondrial function and since the final enzymes in the CoA biosynthetic pathway have been reported in mitochondria,[7,8] our results shed new light on this biochemical pathway. We suggest that the mitochondria likely serve as a second site for complete *de novo* CoA synthesis and predict the presence of the remaining B_5 metabolizing enzymes in the mitochondria. As PANK2 is expressed in all tissues examined,[2] this protein probably fulfills a key function in mitochondria.

If PKAN is a mitochondrial disorder of a widely expressed protein, how do we explain the limited phenotype? One prediction of the effects of defective PANK2 is mitochondrial CoA deficiency leading to increased oxidative stress. In PKAN, the disease primarily impacts the globus pallidus and retina, which share the features of extremely high metabolic demands coupled with cellular environments that are especially prone to oxidative damage. The normally iron-rich globus pallidus demonstrates the earliest and most severe pathological changes in PKAN. The high resting metabolic activity of the globus pallidus makes this tissue especially vulnerable to oxidative stress, explaining its frequent involvement in genetic mitochondrial encephalopathies and exposure to mitochondrial toxins. Similarly, the retina is particularly susceptible to oxidative stress because of its high consumption of oxygen, its high proportion of polyunsaturated fatty acids, and its exposure to visible light. Perhaps the combination of increased metabolic activity in a tissue environment at risk for oxidative damage makes these tissues especially susceptible to PANK2 deficiency.

Historically, the prominent feature of brain iron accumulation formed the basis for theories of disease pathogenesis in PKAN. Now, several lines of evidence suggest that iron deposition is a secondary phenomenon in PKAN, although still one that is likely to contribute to tissue destruction. Based on predictions of the metabolic impact of defective PANK2 and on the observation of elevated brain levels of cysteine in patients,[28] we have proposed an etiology for iron accumulation in PKAN as follows: In normal brain, nonheme iron accumulates regionally and is highest in the medial globus pallidus and the substantia nigra pars reticulata, the two regions most severely affected in PKAN. Phosphopantothenate, the product of pantothenate kinase, normally condenses with cysteine in the next step in CoA synthesis. In PKAN, phosphopantothenate is deficient, theoretically leading to mitochondrial cysteine accumulation. N-Pantothenoyl-cysteine and pantetheine, also substrates for phosphorylation by pantothenate kinase and both containing cysteine moieties, are predicted to accumulate in mitochondria as well. A high cysteine concentration has been found in the globus pallidus in PKAN,[28] and cysteine effectively binds iron. Cysteine, itself cytotoxic, undergoes rapid auto-oxidation in the presence of iron, which results in free radical production. Further, iron-induced lipid peroxidation, a likely mechanism of secondary pathogenesis in PKAN, is enhanced by free cysteine. Since most pathologic studies of brain in patients with PKAN were done following end-stage disease, it has not been possible to discern whether mitochondrial accumulation of iron is an early feature of this disease.

Common allelic variants can confer susceptibility to certain diseases, as shown by the recent association of cardiac arrhythmia with an allele of the cardiac sodium channel gene.[29] Given that defective mitochondrial PANK2 is associated with PKAN, we might reasonably predict that more subtle functional variants of PANK2 might increase the risk for neurodegeneration as well. Five percent of control chromosomes carry the *PANK2* CAG variant, which results in predominantly cytosolic distribution of PANK2. Individuals carrying this allele may have a relative deficiency of mitochondrial PANK2, the effects of which may be exacerbated in tissues with high CoA demand. As a result, this polymorphism may confer susceptibility to neurodegeneration and brain iron accumulation, increasing the risk for Parkinson's disease and related disorders of oxidative stress.

ACKNOWLEDGMENTS

We thank Caroline Enns' lab (OHSU) for kind mentoring regarding immunofluorescence techniques and Dave Koeller's lab (OHSU) for kind mentoring regarding mitochondrial isolation. This work was supported by a grant from the National Eye Institute and by the Sandler Neurogenetics Center at the University of California, San Francisco. J. Gitschier is an Investigator with the Howard Hughes Medical Institute.

REFERENCES

1. DOOLING, E.C., W.C. SCHOENE & E.O. RICHARDSON, JR. 1974. Hallervorden-Spatz syndrome. Arch. Neurol. **30:** 70–83.
2. ZHOU, B., S.K. WESTAWAY, B. LEVINSON et al. 2001. A novel pantothenate kinase gene (*PANK2*) is defective in Hallervorden-Spatz syndrome. Nat. Genet. **28:** 345–349.
3. ABIKO, Y. 1975. Metabolism of coenzyme A. *In* Metabolic Pathway. Volume 7, pp. 1–25. Academic Press. New York.
4. FISHER, M.N. & J.R. NEELY. 1985. Regulation of pantothenate kinase from various tissues of the rat. FEBS Lett. **190:** 293–296.
5. HALVORSEN, O. & S. SKREDE. 1982. Regulation of the biosynthesis of CoA at the level of pantothenate kinase. Eur. J. Biochem. **124:** 211–215.
6. TAHILIANI, A.G. & C.J. BEINLICH. 1991. Pantothenic acid in health and disease. Vitam. Horm. **46:** 165–227.
7. SKREDE, S. & O. HALVORSEN. 1983. Mitochondrial pantotheinephosphate adenyltransferase and dephospho-CoA kinase. Eur. J. Biochem. **131:** 57–63.
8. TAHILIANI, A.G. & J.R. NEELY. 1987. A transport system for coenzyme A into isolated rat heart mitochondria. J. Biol. Chem. **262:** 11607–11610.
9. KARIM, M.A., V.A. VALENTINE & S. JACKOWSKI. 2000. Human pantothenate kinase 1 (*PANK1*) gene: characterization of the cDNAs, structural organization, and mapping of the locus to chromosome 10q23.2–23.31. Am. J. Hum. Genet. **67:** A984.
10. CLAROS, M.B. & P. VINCENS. 1996. Computational method to predict mitochondrially imported proteins and their targeting sequences. Eur. J. Biochem. **241:** 770–786.
11. EMANUELSSON, O., H. NIELSEN, S. BRUNAK & G. VON HEIJNE. 2000. Predicting subcellular localization of proteins based on their N-terminal amino acid sequence. J. Mol. Biol. **300:** 1005–1016.
12. SAMBROOK, J. & D.W. RUSSELL. 2001. Molecular Cloning: A Laboratory Manual. Third Edition. Cold Spring Harbor Laboratory Press. Cold Spring Harbor, NY.
13. LAND, T. & T.A. ROUAULT. 1998. Targeting of a human iron-sulfur cluster assembly enzyme, *nifs*, to different subcellular compartments is regulated through alternative utilization. Mol. Cell **2:** 807–815.
14. SUZUKI, T., T. YOSHIDA & S. TUBOI. 1992. Evidence that rat liver mitochondrial and cytosolic fumarases are synthesized from one species of mRNA by alternative translational initiation at two in-phase AUG codons. Eur. J. Biochem. **207:** 767–772.
15. HANDLEY-GEARHART, P.M., A.G. STEPHEN, J.S. TRAUSCH-AZAR et al. 1994. Human ubiquitin-activation enzyme, E1. J. Biol. Chem. **269:** 33171–33178.
16. DESCOMBES, P. & U. SCHIBLER. 1991. A liver-enriched transcriptional activator protein, LAP, and a transcriptional inhibitory protein, LIP, are translated from the same mRNA. Cell **67:** 569–579.
17. CALLIGARIS, R., S. BOTTARDI, S. COGOI et al. 1995. Alternative translation initiation site usage results in two functionally distinct forms of the GATA-1 transcription factor. Proc. Natl. Acad. Sci. USA **92:** 11598–11602.
18. MARTIN, N. & A. HOPPER. 1994. How single genes provide tRNA processing enzymes to mitochondria, nuclei, and the cytosol. Biochimie **76:** 1161–1167.
19. ACLAND, P., M. DIXON, G. PETERS & C. DICKSON. 1990. Subcellular fate of the *int-2* oncoprotein is determined by choice of initiation codon. Nature **343:** 662–665.
20. BUGLER, B., F. AMALRIC & H. PRATS. 1991. Alternative initiation of translation determines cytoplasmic or nuclear localization of basic fibroblast growth factor. Mol. Cell. Biol. **11:** 573–577.
21. LOCK, P., S. RALPH, E. STANLEY et al. 1991. Two isoforms of murine *hck*, generated by utilization of alternative translational initiation codons, exhibit different patterns of subcellular localizations. Mol. Cell. Biol. **11:** 4363–4370.
22. BLACKWOOD, E.M., T.G. LUGO, L. KRETZNER et al. 1994. Functional analysis of the AUG- and CUG-initiated forms of the *c-myc* proteins. Mol. Biol. Cell **5:** 597–609.
23. KOZAK, M. 1990. Downstream secondary structure facilitates recognition of initiator codons by eukaryotic ribosomes. Proc. Natl. Acad. Sci. USA **87:** 8301–8305.
24. HANN, S.R., K. SLOAN-BROWN & G.D. SPOTTS. 1992. Translational activation of the non-AUG-initiated *c-myc* 1 protein at high cell densities due to methionine deprivation. Genes Dev. **6:** 1229–1240.

25. PEABODY, D.S. 1989. Translation initiation at non-triplets in mammalian cell. J. Biol. Chem. **264:** 5031–5035.
26. DONAHUE, T.F., A.M. CIGAN, E.K. PABICH & B.C. VALAVICIUS. 1988. Mutations at a Zn(II) finger motif in the yeast eIF-2β gene alter ribosomal start-site selection during the scanning process. Cell **54:** 621–632.
27. HORTNAGEL, K., H. PROKISCH & T. MEITINGER. 2003. An isoform of hPANK2, deficient in pantothenate kinase–associated neurodegeneration, localizes to the mitochondria. Hum. Mol. Genet. **12:** 321–327.
28. PERRY, T.L., M.G. NORMAN, V.W. YONG et al. 1985. Hallervorden-Spatz disease: cysteine accumulation and cysteine dioxygenase deficiency in the globus pallidus. Ann. Neurol. **18:** 482–489.
29. SPLAWSKI, I., K.W. TIMOTHY, M. TATEYAMA et al. 2002. Variant of SCN5A sodium channel implicated in risk of cardiac arrhythmia. Science **297:** 1333–1336.

Aceruloplasminemia

An Inherited Neurodegenerative Disease with Impairment of Iron Homeostasis

XUEYING XU, SOKHON PIN, MURAYA GATHINJI, RALPH FUCHS, AND Z. LEAH HARRIS

Department of Anesthesiology and Critical Care Medicine, Johns Hopkins Hospital and School of Medicine, Baltimore, Maryland, USA

ABSTRACT: In 1987, Miyajima *et al.* first characterized an autosomal recessive, adult-onset neurodegenerative disorder resembling Parkinson's disease associated with near-absent circulating serum ceruloplasmin levels. Coined "familial apoceruloplasmin deficiency", they described a patient with a presenting triad of diabetes mellitus, retinal degeneration, and neurodegeneration with blepharospasm. Neuropathological evaluation revealed abundant iron deposition in selected neurons of the basal ganglia and substantia nigra with associated neuronal dropout and spongioform degeneration without evidence of reactive gliosis. Subsequently, mutations in the ceruloplasmin gene have been determined to result in the excessive iron accumulation seen in the pancreas, retina, and brain. Elevated serum ferritin suggests a systemic iron overload syndrome, yet affected patients had low transferrin saturation and a mild anemia. This new disease, "aceruloplasminemia", reveals a role for ceruloplasmin as an essential ferroxidase critical for iron homeostasis. This multicopper oxidase promotes efficient iron efflux such that individuals lacking ceruloplasmin develop a presumed oxidative injury secondary to iron accumulation and significant neuronal damage. Aceruloplasminemic mice provide a valuable model to further study the mechanisms by which ceruloplasmin regulates iron trafficking and the role of iron in oxidative injury. Despite the dependence of ceruloplasmin on copper for its function, aceruloplasminemia represents an iron storage disease and not a defect in copper metabolism. However, recent evidence in *Saccharomyces cerevisiae* indicates that Fet3, the yeast homologue of ceruloplasmin, functions as an essential cuprous oxidase. Further investigation into the mechanisms by which ceruloplasmin regulates iron and copper homeostasis will provide valuable insight into the pathogenesis of metallo-mediated diseases and elucidate mechanisms for transition metal (copper, iron) neuropathology.

KEYWORDS: aceruloplasminemia; iron; homeostasis; ceruloplasmin; Fet3

In 1987, Miyajima *et al.* described a previously healthy 52-year-old woman suffering from blepharospasm, retinal degeneration, diabetes mellitus, and neurodegeneration.[1] Her symptoms included dysarthria, dystonia, ataxia, and mild dementia. Preliminary

Address for correspondence: Z. Leah Harris, Department of Anesthesiology and Critical Care Medicine, Johns Hopkins Hospital and School of Medicine, 600 North Wolfe Street, Blalock 904B, Baltimore, MD 21287. Voice: 410-502-4199; fax: 410-502-5312.
zharris1@jhmi.edu

laboratory analysis revealed an absence of circulating serum ceruloplasmin (Cp). Originally diagnosed as suffering from "atypical Wilson's disease", a liver biopsy was performed that revealed profound hepatic and Kupffer cell iron overload, not copper. Additional laboratory data identified elevated serum ferritin, mild anemia, and low serum iron, suggesting a defect in loading iron onto transferrin. T2-weighted magnetic resonance images of her brain revealed iron overload specific to the basal ganglia and substantia nigra. A mutation was identified in her ceruloplasmin gene.[2] The constellation of her findings suggested a novel disorder of iron metabolism associated with a lack of the copper-containing protein, ceruloplasmin. Termed "aceruloplasminemia", this adult-onset, autosomal recessive disorder has now been associated with multiple distinct mutations in the ceruloplasmin gene.[3–6] The majority of mutations are insertion/deletion events in the ceruloplasmin gene that introduce an early termination sequence and hence have a complete absence of serum ceruloplasmin. In fact, Northern blot analysis reveals a lack of ceruloplasmin message. A patient with a missense mutation in a novel repeat motif critical for appropriate endoplasmic reticulum sorting and trafficking confirmed both that apoceruloplasmin biosynthesis was not dependent on copper availability and that copper incorporation into the nascent ceruloplasmin protein occurs in the trans-Golgi network.[7] The critical physiologic defect in aceruloplasminemia is the absence of enzymatically active holoceruloplasmin. Ceruloplasmin copper incorporation occurs in a precise, cooperative mechanism. Mutations that interfere with a single copper binding site abrogate *in vitro* copper insertion at all copper sites and correlate with disease *in vivo*.[8] Patients with Wilson's disease have a defect in putting copper into ceruloplasmin, but are still able to synthesize and secrete detectable levels of holoceruloplasmin, albeit low (Wilson's disease serum Cp values: 5 mg/dL; normal serum Cp values: 25–40 mg/dL). Of note, 3 patients heterozygous for the nonsense mutation Trp858ter have been described with cerebellar ataxia and hyperreflexia. Brain MRI and autopsy reveal cerebellar atrophy, marked Purkinje cell loss, and cerebellar iron deposition, despite adequate serum ferroxidase activity levels, normal serum iron, and transferring saturation.[9]

Ceruloplasmin, a 132-kDa α_2-glycoprotein, is synthesized predominantly by hepatocytes and secreted into the circulation. This member of the multicopper oxidase family is both synthesized and secreted by astrocytes and exists predominantly in the brain as a glycosylphosphatidylinositol (GPI)–linked astrocyte protein.[10,11] Evidence suggests that RAN-2, a glial lineage marker, is GPI-linked ceruloplasmin.[12] A testicular GPI-linked form of ceruloplasmin has also been identified.[13] The GPI-anchored isoforms are generated by alternative splicing of exons 19 and 20 in both astrocytes and Sertoli cells. Cells of monocyte/macrophage lineage are capable of synthesizing and secreting ceruloplasmin. The holoprotein contains six atoms of copper incorporated during protein synthesis. The facility for electron shuttling by these copper atoms confers ferroxidase activity: the ability to couple substrate oxidation with the 4-electron reduction of oxygen to water. Three members of the multicopper oxidase family possess ferroxidase activity: ceruloplasmin, hephaestin, and Fet3 (the yeast ceruloplasmin homologue). All three are capable of catalyzing the oxidation of multiple substrates, with Fe^{2+} and organic amines being the most favored. It is believed that the iron and organic amines bind different sites. Engineered mutations in the ferroxidase site resulting in a 50% reduction of ferroxidase activity have no effect on amine oxidase activity.[14] Ceruloplasmin has high sequence homology with coagulation factors Va and VIIIa and a similar acute-phase response profile. Ceruloplasmin may also play a role in *S*-nitrosothiol formation and catalyzes the *S*-nitrosolation glutathione.[15] While the ferroxidase activity appears to be the primary physiologic function of ceruloplasmin, it is unclear if this occurs in the context as an essential antioxidant, a component of host defense/bactericidal activity, or a key regulator of iron homeostasis.

An association between the multicopper oxidase, ceruloplasmin, and iron homeostasis had been characterized years earlier by Cartwright[16,17] and Frieden.[18,19] Isolated perfused liver experiments demonstrated a brisk release of hepatic iron following injection of purified holoceruloplasmin, confirming a role for ceruloplasmin as an essential ferroxidase. Holoceruloplasmin oxidizes ferrous iron, Fe^{2+}, to ferric iron, Fe^{3+}, for subsequent binding to transferrin. Only bound to transferrin can iron be allowed to traffic throughout the body (Fe^{3+}-transferrin). Given that 95% of the iron cycle is related to the recycling of iron from senescent red blood cells back to the bone marrow, mobilizing Fe^{2+} to Fe^{3+}-transferrin for delivery to the bone marrow is critical. It is apparent that individuals with aceruloplasminemia suffer from low serum iron and mild anemia because of this ferroxidase defect. However, their anemia is mild (10 mg/dL) and suggests that other mechanisms of ferroxidation occur. The associated elevated serum ferritin represents the degree of tissue iron overload. The potential for oxyradical-mediated injury as described by the Haber-Weiss reaction and predicted by Fenton chemistry is too great to allow for free Fe^{2+} to be found in the circulation. Iron bound to transferrin is rapidly taken up via a classic model of endocytosis via the transferrin receptor. Nontransferrin-dependent iron uptake mechanisms exist predominantly to rapidly clear Fe^{2+} from circulation. This process is believed to be an energy-dependent one controlled in part through a ferric ammonium citrate (FAC)–mediated production of hydroxyl radicals.[20] The details of this process are at present unclear.

It was the generation of a ceruloplasmin null mouse ($Cp^{-/-}$) that provided a critical tool in which to study the role of ceruloplasmin in iron homeostasis.[21] Mice lacking ceruloplasmin develop hepatic reticuloendothelial iron overload, mild anemia, low serum iron, and increased ferritin. Mouse strain difference plays a significant role in phenotype: mice on a pure C57Bl6 background develop a more rapidly abnormal hematologic profile by 2 months of age with increased hepatic iron accumulation, while mice on an SvJ129 background have a similar phenotype by 1 year of age. Using two separate experimental paradigms in which ceruloplasmin is rate-limiting, (1) $Cp^{-/-}$ mice injected with damaged red blood cells (representing an increased reticuloendothelial iron burden) or (2) $Cp^{-/-}$ mice made anemic via serial phlebotomy (causing a state that would encourage reticuloendothelial iron transport to the bone marrow), $Cp^{-/-}$ mice failed to increase serum iron levels in response to either increased iron burden (1) or profound anemia (2). Only upon injection of holoceruloplasmin were $Cp^{-/-}$ mice able to release iron from a storage compartment for delivery to a synthetic compartment, that is, the bone marrow, thus confirming an essential role for ceruloplasmin in regulating efficient iron efflux. Injection with apoceruloplasmin or apotransferrin had no effect on iron mobilization.

The $Cp^{-/-}$ mice develop a mild neurodegenerative phenotype late in life, but never develop a clinical pathology that is equivalent to humans lacking ceruloplasmin.[22] Aceruloplasminemic patients appear to suffer increased plasma lipid peroxidation,[23] increased brain lipid peroxidation,[24] and mitochondrial enzyme abnormalities,[25] and develop oxidatively modified glial fibrillary acidic protein.[26] The human disease is clearly related to abnormal iron deposition in the brain. It is interesting to note that the $Cp^{-/-}$ mouse has very little abnormal central nervous system iron deposition, suggesting that the lack of ceruloplasmin does not cause neurodegeneration, but that the presence of iron within the brain induces damage. Indeed, in aceruloplasminemic patients, the amount of iron detected in tissue correlates with the level of oxysterols measured.[27] Further, positron emission tomography delineates decreased central nervous system glucose and oxygen metabolism.[28] In these same aceruloplasminemic patients, basal gangliar mitochondrial respiratory chain complexes I and IV were measured and determined to be reduced by at least 50%.[28]

With the characterization of hephaestin, a second mammalian multicopper oxidase with ferroxidase activity and high homology to ceruloplasmin, a mouse lacking

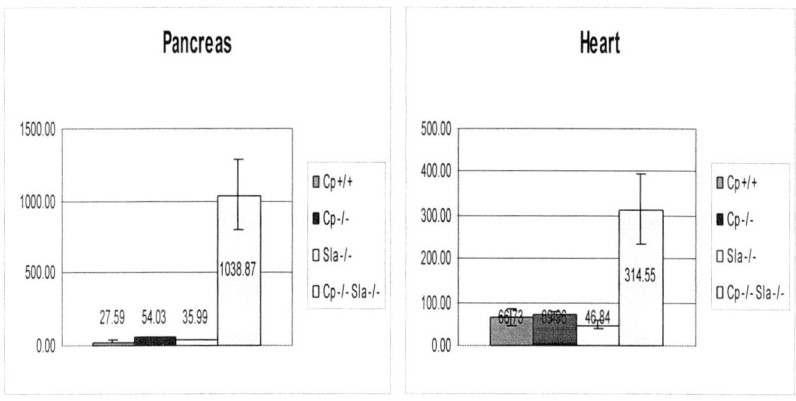

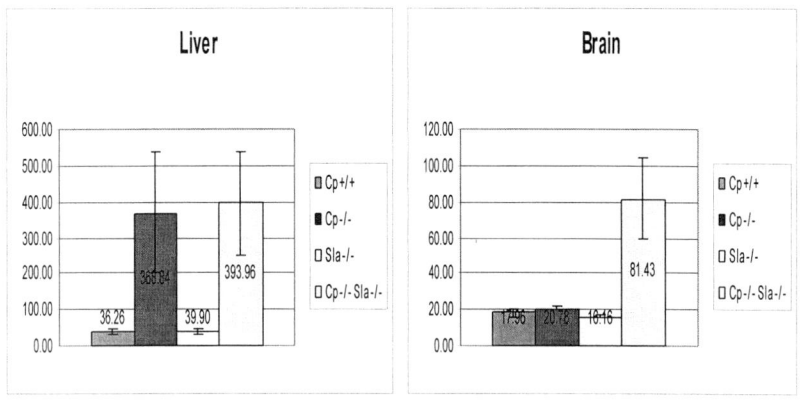

FIGURE 1. Tissue iron content determination following overnight acid digestion (μg Fe/dry weight tissue) in 6-month-old mice. Results are expressed as means ± SD, $n = 6$ minimum. $Cp^{+/+}$ = wild-type mice; $Cp^{-/-}$ = ceruloplasmin-deficient mice; $Sla^{-/-}$ = hephaestin-deficient mice; $Cp^{-/-}Sla^{-/-}$ = multicopper-deficient mice.

ceruloplasmin and hephaestin was generated. Multicopper oxidase (MCO)–deficient mice manifest a neurodegenerative phenotype at approximately 6 months of age (5–9 months). Presenting symptoms included gait abnormalities, weakness, and poor grooming. Tissue iron content studies (μg Fe/dry weight tissue) reveal a significant contribution by hephaestin in regulating iron homeostasis in all tissues except the liver (FIG. 1), presumably reflecting the contribution of ceruloplasmin in regulating hepatic iron metabolism. The neurodegenerative phenotype coupled with the increased whole brain tissue iron content confirmed an essential role for MCOs in central nervous system iron homeostasis. This murine model of aceruloplasminemia suggests either a less significant role for hephaestin in human brain iron homeostasis or an increased role in rodent central nervous system iron metabolism. In either case, the MCOs appear to regulate the redox state of the central nervous system and protect tissues from iron-mediated oxidative damage. We hypothesize that the source of the iron is nontransferrin-bound iron, that is, citrate-Fe^{2+}, ascorbate-Fe^{2+}, or free Fe^{2+}. The lack of ferroxidase activity drives this free iron into neurons and other central nervous system tissue. The autoreducing environment of rodents may predispose them to increased free iron deposition as compared to humans.

Of interest, copper metabolism is normal in aceruloplasminemic mice.[29] Brain copper, as measured by atomic absorption spectroscopy, and the distribution of intravenous ^{64}Cu in brain at 24 h following injection are no different. We initially concluded that none of the clinical pathology of aceruloplasminemia was secondary to copper deficiency or toxicity.[29] A recent publication by Zhu et al. questions this statement.[30] The authors have discovered that, in *Saccharomyces cerevisiae*, Fet3 (the yeast ceruloplasmin homologue) functions as a cuprous oxidase. Ferroxidase-negative Fet3 displays a copper sensitivity and, upon further analysis, the cytotoxic copper species is Cu^{1+}. If ceruloplasmin were to behave similarly, perhaps then cuprous ion toxicity would contribute to the neurodegeneration observed in aceruloplasminemia and the MCO-deficient mouse.

Many questions remain about the mechanisms of neurodegeneration in aceruloplasminemia. Characterizing and delineating the mechanisms of central nervous system iron influx and efflux in the new MCO-deficient mice will be fundamental to answering these questions. Only with this "map" of central nervous system iron homeostasis will the true nature of the associated pathophysiology be addressed.

ACKNOWLEDGMENTS

This work was supported by funds from Grant Nos. DK02464 and DK58086 from the National Institutes of Health.

REFERENCES

1. MIYAJIMA, H., Y. NISHIMURA, K. MIZOGUCHI et al. 1987. Familial apoceruloplasmin deficiency associated with blepharospasm and retinal degeneration. Neurology **37:** 761–767.
2. HARRIS, Z.L., Y. TAKAHASHI, H. MIYAJIMA et al. 1995. Aceruloplasminemia: molecular characterization of this disorder of iron metabolism. Proc. Natl. Acad. Sci. USA **92:** 2539–2543.

3. DAIMON, M., T. KATO, T. KAWANAMI et al. 1995. A nonsense mutation of the ceruloplasmin gene in hereditary ceruloplasmin deficiency with diabetes mellitus. Biochem. Biophys. Res. Commun. **217:** 89–95.
4. HARRIS, Z.L., M.D. MIGAS, A.E. HUGHES et al. 1996. Familial dementia due to a frameshift mutation in the caeruloplasmin gene. Q. J. Med. **89:** 355–359.
5. OKAMOTO, N., S. WADA, T. OGA et al. 1996. Hereditary ceruloplasmin deficiency with hemosiderosis. Hum. Genet. **97:** 755–778.
6. YOSHIDA, K., K. FURIHATA, S. TAKEDA et al. 1995. A mutation in the ceruloplasmin gene is associated with systemic hemosiderosis in humans. Nat. Genet. **9:** 267–272.
7. HELLMAN, N.E., S. KONO, H. MIYAJIMA & J.D. GITLIN. 2002. Biochemical analysis of a missense mutation in aceruloplasminemia. J. Biol. Chem. **277:** 1375–1380.
8. HELLMAN, N.E., S. KONO, G.M. MANCINI et al. 2002. Mechanisms of copper incorporation into human ceruloplasmin. J. Biol. Chem. **277:** 46632–46638.
9. MIYAJIMA, H., S. KONO, Y. TAKAHASHI et al. 2001. Cerebellar ataxia associated with heteroallelic ceruloplasmin gene mutation. Neurology **57:** 2205–2210.
10. KLOMP, L.W.J. & J.D. GITLIN. 1996. Expression of the ceruloplasmin gene in the human retina and brain: implications for the pathogenic model in aceruloplasminemia. Hum. Mol. Genet. **5:** 1989–1996.
11. PATEL, B.N. & S. DAVID. 1997. A novel glycophosphatidylinositol-anchored form of ceruloplasmin is expressed by mammalian astrocytes. J. Biol. Chem. **272:** 20185–20190.
12. SALZER, J.L., L. LOVEJOY, M.C. LINDER & C. ROSEN. 1998. Ran-2, a glial lineage marker, is a GPI-anchored form of ceruloplasmin. J. Neurosci. Res. **54:** 147–157.
13. FORTNA, R.R., H.A. WATSON & S.E. NYQUIST. 1999. Glycosyl phosphatidylinositol–anchored ceruloplasmin is expressed in Sertoli cells and is concentrated in detergent-insoluble membrane fractions. Biol. Reprod. **61:** 1042–1049.
14. BROWN, M.A., L.M. STENBERG & A.G. MAUK. 2002. Identification of catalytically important amino acids in human ceruloplasmin by site-directed mutagenesis. FEBS Lett. **520:** 8–12.
15. INOUE, K., T. AKAIKE, Y. MIYAMOTO et al. 1999. Nitrosothiol formation catalyzed by ceruloplasmin: implication for cytoprotective mechanism in vivo. J. Biol. Chem. **274:** 27069–27075.
16. LEE, G.R., S. NACHT, J.N. LUKENS & G.E. CARTWRIGHT. 1968. Iron metabolism in copper-deficient swine. J. Clin. Invest. **47:** 2058–2069.
17. ROESER, H.P., G.R. LEE, S. NACHT & G.E. CARTWRIGHT. 1970. The role of ceruloplasmin in iron metabolism. J. Clin. Invest. **49:** 2408–2417.
18. OSAKI, S., D. JOHNSON & E. FRIEDEN. 1966. The possible significance of the ferrous oxidase activity of ceruloplasmin in normal human serum. J. Biol. Chem. **241:** 2746–2757.
19. OSAKI, S., D.A. JOHNSON & E. FRIEDEN. 1971. The mobilization of iron from the perfused mammalian liver by a serum copper enzyme, ferroxidase. J. Biol. Chem. **246:** 3018–3023.
20. WESSLING-RESNICK, M. 2000. Iron transport. Annu. Rev. Nutr. **20:** 129–151.
21. HARRIS, Z.L., A.P. DURLEY, T.K. MAN & J.D. GITLIN. 1999. Targeted gene disruption reveals an essential role for ceruloplasmin in cellular iron efflux. Proc. Natl. Acad. Sci. USA **96:** 10812–10817.
22. PATEL, B.N., R.J. DUNN, S.Y. JEONG et al. 2002. Ceruloplasmin regulates iron levels in the CNS and prevents free radical injury. J. Neurosci. **22:** 6578–6586.
23. MIYAJIMA, H., Y. TAKAHASHI, M. SERIZAWA et al. 1996. Increased plasma lipid peroxidation in patients with aceruloplasminemia. Free Radical Biol. Med. **20:** 757–760.
24. YOSHIDA, K., K. KANEKO, H. MIYAJIMA et al. 2000. Increased lipid peroxidation in the brains of aceruloplasminemic patients. J. Neurol. Sci. **175:** 91–95.
25. KOHNO, S., H. MIYAJIMA, Y. TAKAHASHI et al. 2000. Defective electron transfer in complexes I and IV in patients with aceruloplasminemia. J. Neurol. Sci. **182:** 57–60.
26. KANEKO, K., A. NAKAMURA, K. YOSHIDA et al. 2002. Glial fibrillary acidic protein is greatly modified by oxidative stress in aceruloplasminemia brain. Free Radical Res. **36:** 303–306.
27. MIYAJIMA, H., J. ADACHI, S. KOHNO et al. 2001. Increased oxysterols associated with iron accumulation in the brains and visceral organs of acaeruloplasminemia patients. Q. J. Med. **94:** 417–422.

28. MIYAJIMA, H., S. KONO, Y. TAKAHASHI & M. SUGIMOTO. 2002. Increased lipid peroxidation and mitochondrial dysfunction in aceruloplasminemic brains. Blood Cells Mol. Dis. **29:** 433–438.
29. MEYER, L.A., A.P. DURLEY, J.R. PROHASKA & Z.L. HARRIS. 2001. Copper transport and metabolism are normal in aceruloplasminemic mice. J. Biol. Chem. **276:** 36857–36861.
30. SHI, X., C. STOJ, A. ROMEO et al. 2003. Fre1p Cu^{2+} reduction and Fet3p Cu^{1+} oxidation modulate copper toxicity in *Saccharomyces cerevisiae*. J. Biol. Chem. **278:** 50309–50315.

Ironing Iron Out in Parkinson's Disease and Other Neurodegenerative Diseases with Iron Chelators

A Lesson from 6-Hydroxydopamine and Iron Chelators, Desferal and VK-28

MOUSSA B. H. YOUDIM, GALIA STEPHENSON, AND DORIT BEN SHACHAR

Eve Topf and US National Parkinson Foundation Centers of Excellence for Neurodegenerative Diseases Research, and Department of Pharmacology, Technion-Rappaport Faculty of Medicine, Haifa, Israel

ABSTRACT: In Parkinson's disease (PD) and its neurotoxin-induced models, 6-hydroxydopamine (6-OHDA) and *N*-methyl-4-phenyl-1,2,3,6-tetrahydropyridine (MPTP), significant accumulation of iron occurs in the substantia nigra pars compacta. The iron is thought to be in a labile pool, unbound to ferritin, and is thought to have a pivotal role to induce oxidative stress–dependent neurodegeneration of dopamine neurons via Fenton chemistry. The consequence of this is its interaction with H_2O_2 to generate the most reactive radical oxygen species, the hydroxyl radical. This scenario is supported by studies in both human and neurotoxin-induced parkinsonism showing that disposition of H_2O_2 is compromised via depletion of glutathione (GSH), the rate-limiting cofactor of glutathione peroxidase, the major enzyme source to dispose H_2O_2 as water in the brain. Further, radical scavengers have been shown to prevent the neurotoxic action of the above neurotoxins and depletion of GSH. However, our group was the first to demonstrate that the prototype iron chelator, desferal, is a potent neuroprotective agent in the 6-OHDA model. We have extended these studies and examined the neuroprotective effect of intracerebraventricular (ICV) pretreatment with the prototype iron chelator, desferal (1.3, 13, 134 µg), on ICV induced 6-OHDA (250 µg) lesion of striatal dopamine neurons. Desferal alone at the doses studied did not affect striatal tyrosine hydroxylase (TH) activity or dopamine (DA) metabolism. All three pretreatment (30 min) doses of desferal prevented the fall in striatal and frontal cortex DA, dihydroxyphenylacetic acid, and homovalinic acid, as well as the left and right striatum TH activity and DA turnover resulting from 6-OHDA lesion of dopaminergic neurons. A concentration bell-shaped neuroprotective effect of desferal was observed in the striatum, with 13 µg being the most effective. Neither desferal nor 6-OHDA affected striatal serotonin, 5-hydroxyindole acetic acid, or noradrenaline. Desferal also protected against 6-OHDA-induced deficit in locomotor activity, rearing, and exploratory behavior (sniffing) in a novel environment. Since the lowest neuroprotective dose (1.3 µg) of desferal was 200 times less than 6-OHDA, its neuroprotective activity may not be attributed to

Address for correspondence: Professor Moussa B. H. Youdim, Eve Topf and NPF Centers, Technion-Faculty of Medicine, Efron Street, P. O. Box 9697, Haifa 31096, Israel. Voice: +972-4-8295-290; fax: +972-4-8513-145.
youdim@tx.technion.ac.il

interference with the neurotoxin activity, but rather iron chelation. These studies led us to develop novel brain-permeable iron chelators, the VK-28 series, with iron chelating and neuroprotective activity similar to desferal for ironing iron out from PD and other neurodegenerative diseases, such as Alzheimer's disease, Friedreich's ataxia, and Huntington's disease.

KEYWORDS: 6-hydroxydopamine; desferal; nigrostriatal dopamine neurons; iron; iron chelation; neuroprotection; oxygen free radicals; Parkinson's disease

INTRODUCTION

In a number of neurodegenerative diseases, including Parkinson's disease (PD), Alzheimer's disease (AD), Huntington's disease (HD), amyotrophic lateral sclerosis (ALS), and Friedreich's ataxia (FA), significant iron accumulation occurs at the site of neurodegeneration (cf. ref. 1). Indeed, in PD, the increase in iron is specific for substantia nigra pars compacta (SNPC), and not substantia nigra pars reticularis (SNPR),[1–15] and within melanin-containing neurons that specifically degenerate bound to neuromelanin.[5,9,13] Iron in SNPC also accumulates in the proliferated reactive microglia, astrocytes, and oligodendrocytes.[10] This pathology has brought with it the hypothesis of iron-induced oxidative stress[4,16–19] and membrane lipid peroxidation[7] in the SNPC of PD brains, with sparing of SNPR, even though the latter region has a higher content of iron.[1,3] As a consequence of these findings, we[4] suggested that PD could be a progressive oxidative stress siderosis of substantia nigra (SN) resulting from iron-melanin interaction induced neurodegeneration. This is substantiated by the recent studies of Faucheux et al.,[5] where the level of iron redox activity detected in neuromelanin aggregates was significantly increased in parkinsonian patients and was highest in patients with the most severe neuronal loss. Their results support and substantiate our findings that the ratio of $Fe^{+2}:Fe^{+3}$ in parkinsonian SNPC is 1:3 as compared to 1:1 normally observed in control brains. Thus, iron redox is in favor of Fe^{+3}, which can result in generation of reactive oxygen species (ROS).[3] The participation of iron in dopaminergic neurodegeneration is strengthened by the observation that the studies with the neurotoxins, 6-hydroxydopamine (6-OHDA)[20,21] and N-methyl-4-phenyl-1,2,3,6-tetrahydropyridine (MPTP),[22–24] have been shown in several animal species, including rats and monkeys, to induce an increase of iron in SNPC similar to what is observed in PD.[25] The oxidative stress hypothesis of PD has it that free reactive iron (labile iron pool) might participate as a primary event or may have a pivotal role in inducing formation of ROS, the hydroxyl radical. The reactive hydroxyl radical is formed from Fenton chemistry as a consequence of iron interaction with excess hydrogen peroxide, generated by various reactions in the proliferated reactive microglia, and initiates an oxidative stress–dependent cell death resulting in dopaminergic neurodegeneration. Indeed, in SN of PD, a progressive loss of reduced glutathione (GSH) occurs with staging of the disease.[3,18,19,26] This rate-limiting cofactor of glutathione peroxidase is the major hydrogen peroxide–disposing system in the brain. Similarly, both 6-OHDA and MPTP neurotoxicities have been also shown to deplete striatal GSH.[27–30] The consequence of this is thought to be accumulation of hydrogen peroxide and its interaction of the latter with free iron to initiate Fenton chemistry.[31] Support for this hypothesis has come from studies demonstrating that both neurotoxins induce release of iron from iron-bound ferritin,[32–34] and dopaminergic neurotoxicities of 6-OHDA and MPTP are associated with induction of iron-induced oxidative stress[35,36] and GSH depletion.[37]

Recent studies with cDNA microarray gene expression analysis in SN of MPTP-treated mice have confirmed and provided further evidence for the involvement of brain iron dysregulation and oxidative stress in a cascade of events eventually leading to death of nigrostriatal dopamine (DA) neurons.[30] The studies have shown upregulation of gene expression for iron regulatory proteins such as IRP2, huntingtin (an iron regulatory protein), and ferritin, as well as reduction in gene expression for transferrin and transferrin receptor, which are all compatible with iron accumulation in the cells.

Radical scavengers such as vitamin E, lipoic acid, R-apomorphine, EGCG, ebselen, and melatonin pretreatment can protect against the neurotoxins, 6-OHDA and MPTP.[37–46] In a preliminary study, our group was the first to demonstrate that intracerebraventricular (ICV) pretreatment with a prototype iron chelator, desferal, protects against 6-OHDA-induced lesions of DA neurons in the rat.[47] The role of iron in DA neurodegeneration is further supported by the studies of Lan and Jiang,[42] who confirmed the neuroprotective effect of desferal pretreatment against MPTP neurotoxicity in mice, as well as the recent pharmacological chelation and genetic manipulation of iron in MPTP models of PD in mice.[48]

In the present investigation, we have extended these previous studies with desferal and have shown a concentration-dependent neuroprotective action of desferal, both for the nigrostriatal and frontal cortex DA neurons in rats, and prevention of behavioral deficits seen after 6-OHDA. We further showed that complete neuroprotection of striatal DA neurons by desferal is not necessary for complete restoration of behavioral deficits seen with 6-OHDA-induced DA neuron lesion.

MATERIALS AND METHODS

Materials

DA, serotonin (5-HT), noradrenaline (NA), dihydroxyphenylacetic acid (DOPAC), homovalinic acid (HVA), L-dihydroxyphenylalanine (L-dopa), 6-OHDA, pargyline, desmethylimipramine, and desferrioxamine (desferal) were of the highest purity and were purchased from Sigma Chemical (St. Louis, MO).

Animals

All procedures were carried out in accordance with the National Institutes of Health's "Guide for Care and Use of Laboratory Animals" and were approved by the animal ethics committee of Technion (Haifa, Israel). Adult male Sprague-Dawley rats (250 ± 20 g) were housed in groups of four in air-conditioned quarters at room temperature of 22°C and maintained on a 12-h light/dark cycle. They had free access to food and water at all times.

6-OHDA Lesion

Animals were injected ip with desmethylimipramine (25 mg/kg) and pargyline (50 mg/kg) at 30 min prior to surgical incision. The anesthetic agent, Aquasine, was administered ip 20 min prior to operation. The animals were placed into the stereotaxic apparatus, and a point directly dorsal to the left ventricle was marked according to coordinates located in the rat brain atlas[49] and taken to be 1.3 mm lateral to the midline and 0.8 mm posterior to bregma. After making a hole of 1.5-mm diameter, a 26-gauge steel cannula with polypropylene microtubing attached to a 10-µL Hamilton glass microsyringe was slowly lowered into the brain to a depth of 3.6 mm ventral to the surface of the brain, and vehicle and drugs were delivered at a rate of 3 µL/min with a digital pump. Surgical accuracy was determined by the injection of dye, and gross examination showed that the coordinates selected were correct.

Drug Treatment

Rats received various ICV doses of desferal (desferrioxamine) from 1.3 to 131 µg or vehicle alone and followed with similar groups receiving (30 min later) administration of 6-OHDA (250 µg) according to the method of Ungerstedt[50] as adapted by Uretsky and Iversen[51] (described above). Animals had access to food and water, and 6-OHDA groups were given daily doses of glucose (4 mL isotonic) for 4 weeks until they attained the preoperative weight. One group was killed at this time and another was kept for 6 months before neurochemical analysis of brain dopaminergic neurons and animal behavior.

Brains

Animals were killed by rapid brain decapitation after 4 weeks or 6 months and brains were rapidly removed. They were dissected over ice-cold platform; striatum, frontal cortex, and pituitary were isolated; and half were homogenized in 2× 300 µL ice-cold 0.1 M perchloric acid. They were centrifuged at 4°C at 10,000g for 15 min and the supernatant collected and stored at −70°C until HPLC analysis. The other half of the tissues were kept in 0.3 M sucrose at −70°C for enzymatic analysis.

HPLC Analysis of Neurotransmitters

The analysis of NA, DA, 5-HT, and their metabolites, HVA, DOPAC, 5-HIAA, and L-DOPA, was according to the HPLC procedure as adapted by Ben Shachar and Youdim[49] using a hypersil column H30DS-125A, 12.5 cm × 4.6 mm i.d. (Hichrom, U.S.A.). Standard amines and metabolites were carried through in order to evaluate recovery. The HPLC system consisted of an ESA Coulochem Model 5200A electrochemical detector (ESA analytical with a model 5011 dual analytical cell and a model 5021 conditioning cell).

Tyrosine Hydroxylase

For *in vivo* determination of striatal tyrosine hydroxylase (TH) activity, rats of 4 weeks postoperative treatment with 6-OHDA, desferal, or their combination were injected with 100 mg/kg of the dopa decarboxylase inhibitor, NSD1015. The animals were decapitated 30 min later, and L-dopa accumulated in the striatum was analyzed by HPLC as described above. This was taken as the measure of TH *in vivo* activity.

Behavioral Tests

Automated Measurement of Locomotor Activity

The degree of activity exhibited by the control and drug-treated rats was accessed by a Varimax activity meter. Individual rats were placed in unfamiliar clear Perpex cages. Then, their spontaneous activity across the electronic coil was measured and recorded by means of the Varimex meter every 10 s for 10 min, and the readout of each activity for 30 s was recorded. This was done with two animals at a time.

Observation Studies

During the automated activity measurements, the reaction of each animal to the new environment was observed in an experimenter-blind test battery in the following fashion. "Rearing" was defined as the lifting of one or more paws off the floor of the cage and counted in the form of "rear per minute". "Exploratory" behavior was defined as obvious sniffing by the rat in an unfamiliar environment and counted. Working with pairs of rats, observations were made for 60 s per rat, with a 30-s break between each rat study, for a duration of 10 min. Each rat was thus observed for 30 min.

Statistical Analysis

Results are expressed as the mean ± SEM done in duplicate with the number of animals indicated. The significance of the difference between groups was measured by Student's two-tailed *t* test, and analysis of variance (ANOVA) was performed.

RESULTS

Rats treated with 6-OHDA, but not with desferal, lost significant body weight of 56 ± 9 g. They gained their weight with glucose treatment after 18 days and were

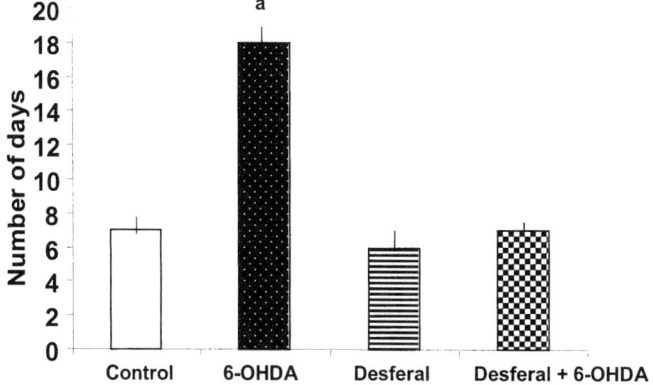

FIGURE 1. The time course of preoperative weight gain after ICV 6-OHDA and response to desferal: control, isotonic saline; 6-OHDA (250 µg ICV); desferal (13 µg ICV). Results are the mean ± SEM of 10–12 rats in each group; a, $p < .001$ vs. control, desferal, and combination.

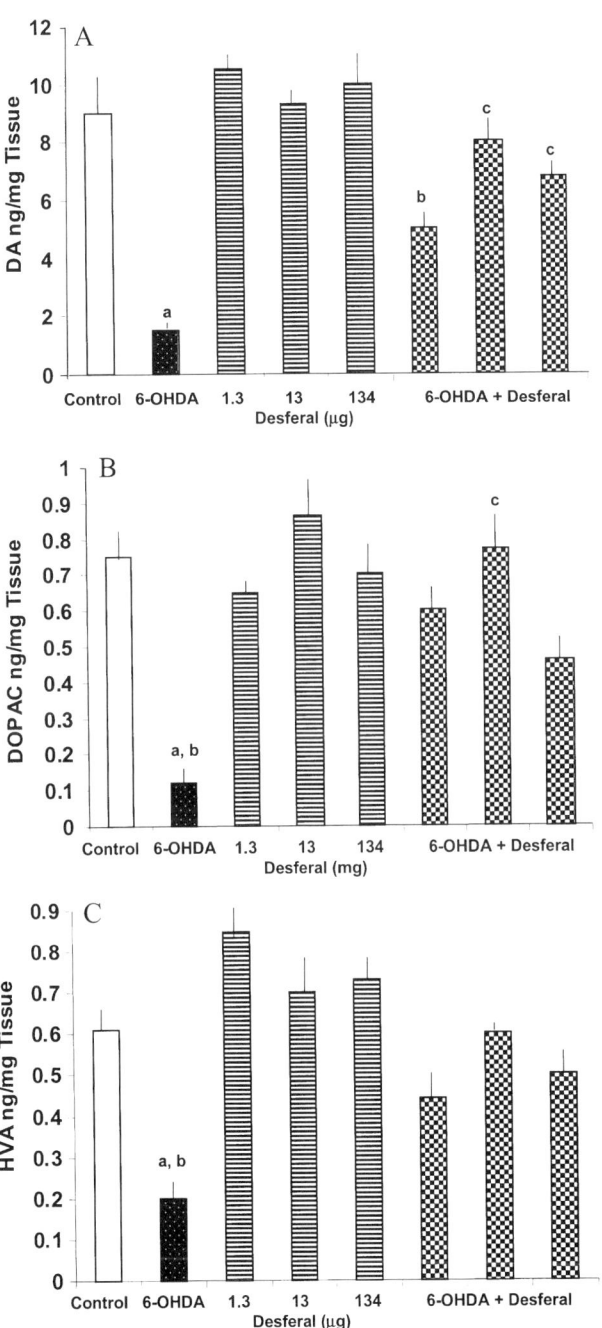

FIGURE 2. *See following page for caption.*

similar to the control and desferal groups. Pretreatment with desferal prevented the 6-OHDA-induced body weight loss (FIG. 1).

Effects of Desferal on 6-OHDA-Induced Depletion of Striatal and Frontal Cortex DA

One month after ICV 6-OHDA, highly significant loss of striatal DA, DOPAC, and HVA occurred (FIGS. 2A, 2B, and 2C), with some 80% depletion of DA. Desferal treatment alone at the three doses had no appreciable effect on striatal DA metabolism. However, ICV pretreatment of rats with the same doses of desferal at 30 min prior to 6-OHDA significantly prevented depletion of DA and its metabolites. The three doses of desferal showed a bell-shaped curve in its neuroprotective action, with 13 µg desferal having a maximal effect in these studies, and the DA value with this concentration of desferal was not significantly different from control. Even at 1.3 µg, desferal was neuroprotective (FIG. 2). These results are supported by the reduced turnover of striatal DA (DOPAC plus HVA:DA ratio), which is increased after 6-OHDA, but it is prevented by desferal pretreatment. Desferal alone did not alter DA turnover (FIG. 3).

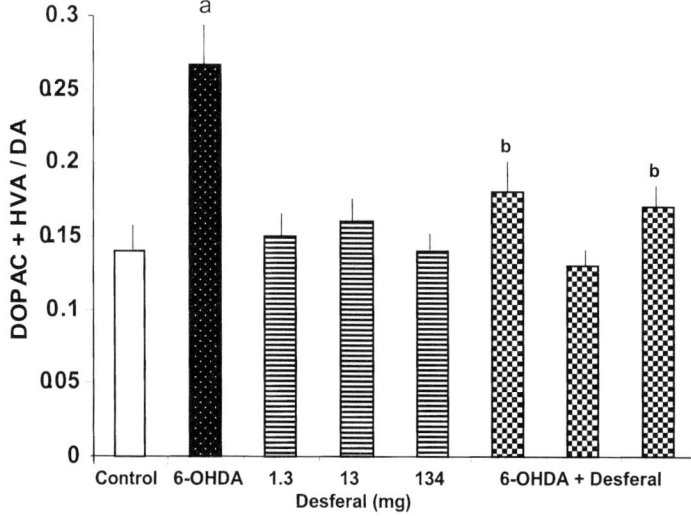

FIGURE 3. Striatal DA turnover at 4 weeks after ICV pretreatment with desferal and 6-OHDA. The turnover (DOPAC plus HVA/DA) values are calculated from FIGURE 2: a, $p < .01$ vs. control, desferal, or desferal (13 µg) plus 6-OHDA; b, $p < .05$ vs. 6-OHDA.

FIGURE 2. Rat striatal **(A)** DA, **(B)** DOPAC, and **(C)** HVA at 4 weeks after ICV pretreatment with desferal and 6-OHDA. The desferal plus 6-OHDA combination consisted of the same dosages of the iron chelator when given alone. Results are the mean ± SEM of 6–11 rats done in duplicate: (A) a, $p < .001$ vs. control and desferal; b, $p < .05$ vs. 6-OHDA or desferal; c, $p < .01$ vs. 6-OHDA; (B) a, $p < .001$ vs. control or desferal; b, $p < .05$ vs. control or 6-OHDA; c, $p < .05$ vs. control or 6-OHDA; (C) a, $p < .001$ vs. control or desferal; b, $p < .01$ vs. desferal plus 6-OHDA.

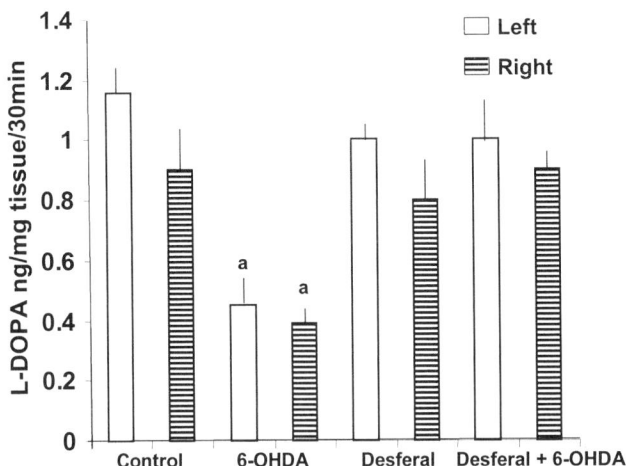

FIGURE 4. Striatal TH activity in ICV desferal and 6-OHDA treated rats. Accumulation of striatal DOPAC after DOPA decarboxylase inhibition was measured as an index of TH activity in the left and right striatum at 4 weeks after ICV desferal (13 μg) and 6-OHDA. Results are the mean ± SEM of 6–8 rats in each group done in duplicate: a, $p < .1$ vs. control, desferal, or desferal plus 6-OHDA.

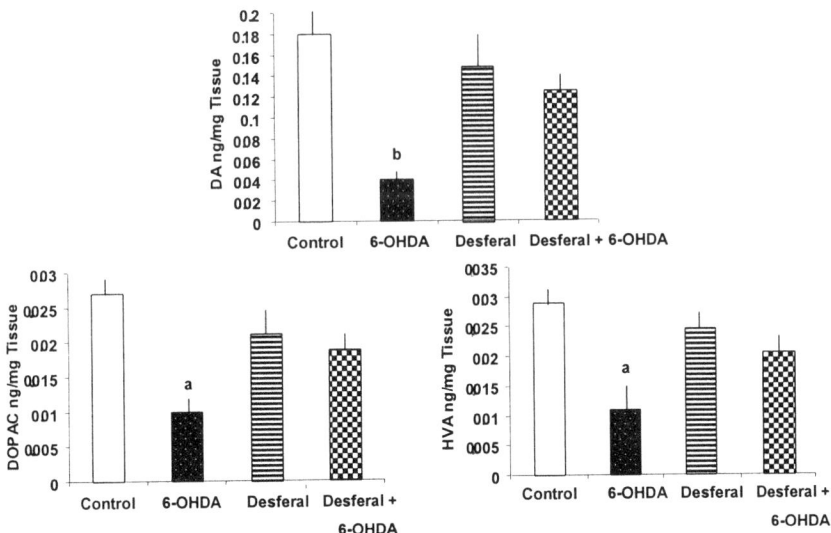

FIGURE 5. Frontal cortex DA (*top*), DOPAC (*lower left*), and HVA (*lower right*) at 4 weeks after ICV pretreatment with desferal and 6-OHDA. The desferal plus 6-OHDA consisted of the same dosage of iron chelator (13 μg) when given alone. Results are the mean ± SEM of 5–6 rats in each group done in duplicate: a, $p < .01$ vs. control, desferal, or desferal plus 6-OHDA; b, $p < .05$ vs. control, desferal, or desferal plus 6-OHDA.

Similar results for DA metabolism were also observed in the frontal cortex after ICV 6-OHDA and desferal treatments. ICV desferal (13 µg) pretreatment was sufficient to protect against 6-OHDA neurotoxicity (FIG. 4) and prevented the 6-OHDA-initiated increased DA turnover (FIG. 5). The results of these studies indicate that ICV 6-OHDA treatment may have a greater effect on striatal DA neurons than on the cortical neurons since DA depletion was greater in the former. Furthermore, the neuroprotection afforded with 13 µg desferal pretreatment was less than that seen with the same desferal dosage in the striatum (FIG. 4).

Striatal TH Activity

Striatal TH activity measured *in vivo* by accumulation of L-dopa as a consequence of dopa decarboxylase inhibition with NSA1015 is shown in FIGURE 6. TH activity on both the left and right striatum was reduced by about 50% after 6-OHDA. ICV desferal (13 µg) alone had no significant effect on L-dopa accumulation in either the left or right striatum. However, pretreatment with deferal (13 µg) prevented the loss of L-dopa accumulation as a consequence of 6-OHDA treatment, and striatal L-dopa concentrations on both sides were similar to those in controls.

Striatal and Frontal Cortex DA Metabolism at 6 Months after 6-OHDA

Six months after ICV 6-OHDA, striatal DA, DOPAC, and HVA are still significantly lower as compared to controls (FIG. 7). It is apparent that there is some recovery from 6-OHDA treatments since DA and its metabolites were roughly 50% of the control values and twice the values observed in 4-week postoperative 6-OHDA animals (FIG. 2A). Further, desferal at the concentration employed almost completely protected against 6-OHDA lesion since values of striatal DA and its metabolites were not significantly different from those of controls. Similar to results obtained with 4-week postoperative 6-OHDA, the increased DA turnover (DOPAC plus

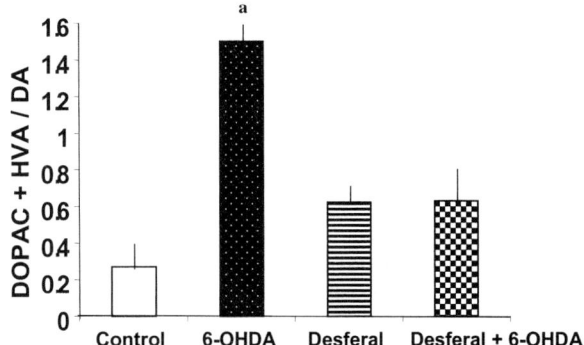

FIGURE 6. Effect of ICV pretreatment with desferal and 6-OHDA on frontal cortex DA turnover. The turnover (DOPAC plus HVA/DA) values are calculated from FIGURE 4: a, $p < .001$ vs. control.

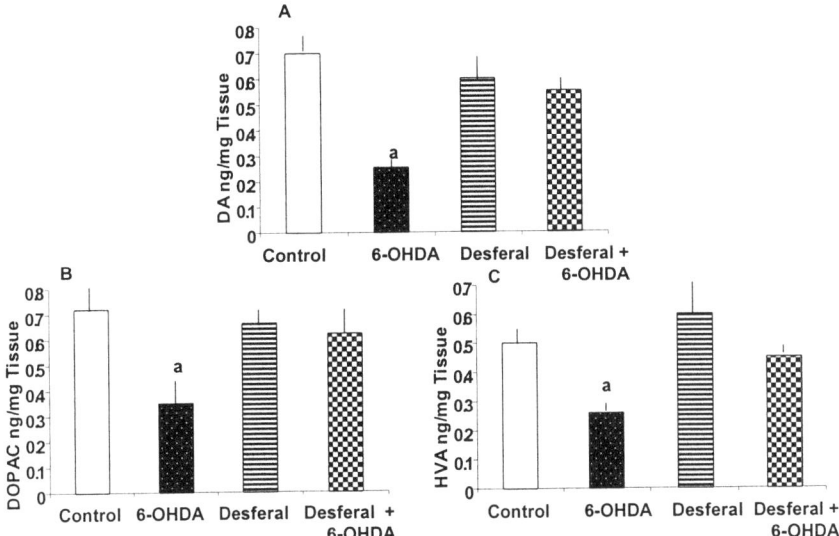

FIGURE 7. Striatal **(A)** DA, **(B)** DOPAC, and **(C)** HVA at 6 months after ICV pretreatment with desferal and 6-OHDA. The desferal plus 6-OHDA consisted of the same dosage of iron chelator (13 µg) when given alone. Results are the mean ± SEM of 3–6 animals in each group done in duplicate: a, $p < .05$.

HVA:DA ratio) was prevented by desferal (13 µg). Similar results were obtained in the frontal cortex, with total protection from 6-OHDA-induced lesions by the same dose of desferal (data not presented).

Striatal 5-HT and NA in Desferal and 6-OHDA Treatments

The specificity of 6-OHDA lesions of striatal DA neurons is confirmed by the absence of its effect on 5-HT and norepinephrine in the striatum of 4- and 6-month post-6-OHDA infusion. No treatments had any appreciable effects, similar to what we had observed previously[47] (data not presented).

Behavioral Responses to Desferal

One month after ICV 6-OHDA, the behavioral responses, namely, locomotor activity time spent exploring, rearing, and sniffing in a novel environment, were examined as described in MATERIALS AND METHODS. While ICV desferal alone at the doses employed in studies to determine its effects on DA metabolism (FIG. 2) had no significant effects on the behavioral parameters in a novel environment, 6-OHDA produced significant reductions in all three parameters (FIGS. 8A, 8B, and 8C). By contrast, desferal pretreatments, even at its lowest dosage (1.3 µg), which partially protected against DA depletion by 6-OHDA (FIG. 2), was able to completely prevent the behavioral deficits induced with 6-OHDA.

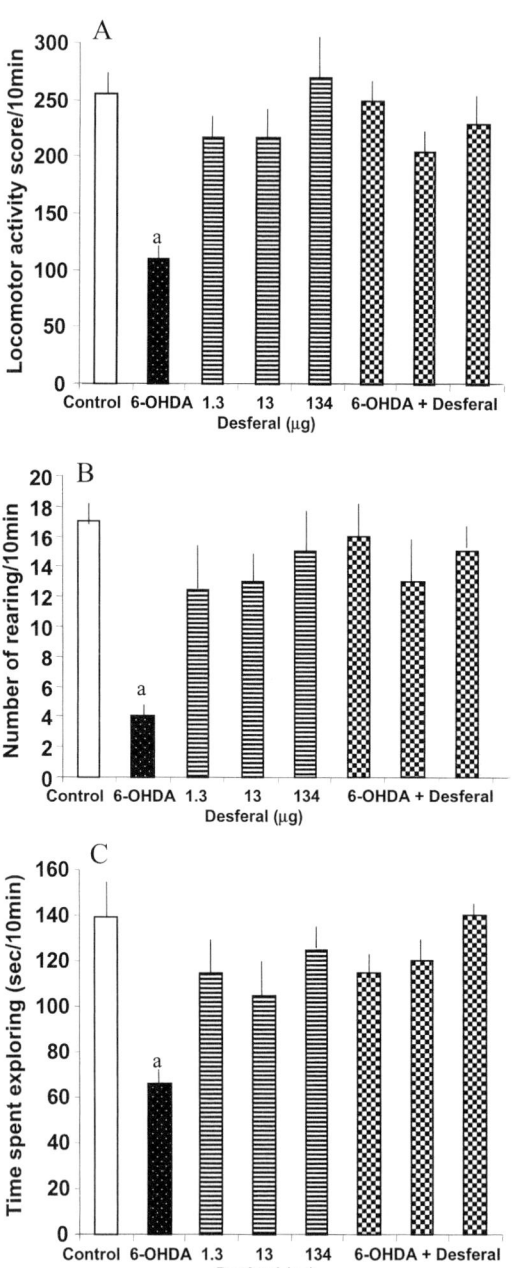

FIGURE 8. Effect of ICV desferal and 6-OHDA treatment on **(A)** locomotor activity, **(B)** rearing, and **(C)** sniffing.

DISCUSSION

Iron and Neurodegeneration

The etiology of PD and the mechanism by which nigrostriatal DA neurons originating from raphe nucleus degenerate are not known. However, substantial evidence from PD brain autopsy and 6-OHDA and MPTP studies has indicated that oxidative stress may have a crucial primary role.[3,17–19,52] One major feature of the pathology of parkinsonian SNPC is the dysregulation of iron homeostasis, resulting in a significant increase of the metal specifically in this region.[1,2,11,53,54] A similar picture is also observed in MPTP- and 6-OHDA-treated monkeys and rats treated with the latter neurotoxins.[20–24] Previously[4,11,55] and more recently,[5,54–56] we have suggested that, as a consequence of its redox state, free (labile pool) iron may have a crucial pivotal, even primary, role in the initiation of oxidative stress–dependent neurodegeneration of nigrostriatal DA neurons. The latter may result from the ability of iron to generate the ROS, hydroxyl radical, from hydrogen peroxide by Fenton chemistry. Such radicals are known to cause membrane lipid peroxidation, resulting in cell membrane fluidity and eventual death of the cells. The support for oxidative stress in PD and its neurotoxin-induced models is further substantiated by the observation that the main pathway of hydrogen peroxide disposition to water, namely, the glutathione peroxidase pathway, is defective.[3,26] Thus, in PD SN and striata of MPTP- or 6-OHDA-treated animals, there is a highly significant diminution of reduced GSH, the cofactor of glutathione peroxidase.[26] Hence, animal and cellular pretreatment studies with antioxidant radical scavengers (e.g., vitamin E, melatonin, and EGCG) have demonstrated neuroprotection against MPP+, MPTP, 6-OHDA, and other neurotoxins. However, few neuroprotective studies in animal models have been done with iron chelators since many such compounds do not cross the blood-brain barrier (BBB).

The misregulation of iron metabolism and its association with a number of neurodegenerative diseases such PD, AD, HD, FA, iron accumulation type 1 (previously Hallervorden-Spatz disease), Wilson's disease, ALS, mad cow disease, aceruloplasminemia, and juvenile parkinsonism have taken center stage with the number of mutated iron metabolism genes described for these diseases.[25,57,58]

We have advocated on several occasions that iron chelation, with brain-permeable iron chelators, may be a valid process for neuroprotection.[40,41,47,55,59] Thus, phenolic compounds, such as *R*- and *S*-apomorphine and green tea polyphenols, EGCG, which have a mixed iron chelating and radical scavenging activity, are relatively potent neuroprotective agents against MPTP and 6-OHDA neurotoxicity in cell culture and *in vivo*.[30,40,41,60] More recently, other investigators have demonstrated neuroprotection with desferal[42] and clioquinol, an antibiotic that crosses the BBB and has iron chelating property[48] in MPTP mice models of parkinsonism. In the present study, we have extended our preliminary studies with the prototype iron chelator, desferal, which is considered the most potent and selective chelator for iron, to determine its neuroprotective activity in the 6-OHDA model of PD. ICV pretreatment with different dosages of desferal was performed since the latter has a very poor penetration into the brain. The results demonstrated that desferal pretreatment alone, at the doses injected, does not affect DA metabolism by inhibiting the iron-dependent rate-limiting enzyme TH. This is validated by comparable striatal DA and

NA levels and accumulation of L-dopa as the index of TH activity as those seen in controls. It is also apparent that it does not affect tryptophan hydroxylase activity, another iron-dependent enzyme, since striatal 5-HT was unaltered. However, desferal pretreatment at the doses employed protected against ICV 6-OHDA lesions of striatal DA neurons as indicated by prevention of the fall in the levels of DA and its metabolites in both striatum and frontal cortex and the fall in TH activity in the striatum.

One consequence of ICV administration of 6-OHDA was the reduction in frontal cortex DA and its metabolites. This may have been a direct result of 6-OHDA neurotoxicity in the soma of frontal cortex neurons or as a result of toxicity in the nucleus accumbens, which has dopaminergic projections to cortical regions. Salamone et al.[61] have shown that behavioral deficits arising from decortication of rats are very similar to those occurring as a consequence of DA depletion by 6-OHDA. This is reinforced in our study by the following observation: although desferal at lower dosages (1.3 µg) did not fully protect (40% protection) against 6-OHDA-induced striatal and cortical DA depletion, it nevertheless completely protected against the behavioral deficit produced by the neurotoxin. We consider this to be extremely important in light of the elegant studies of Zigmond et al.,[61–63] who showed that behavioral deficits with 6-OHDA occur when more than 70% of neurons are lesioned and the dopaminergic compensatory system is not operative. Indeed, the results of our present study are compatible with compensatory mechanisms described by those of Zigmond et al.[61–63] since the reduction in DA produced by ICV 6-OHDA in saline-treated animals was more than 80%.

An important aspect of neuroprotection achieved with desferal is whether it interferes with 6-OHDA uptake at presynaptic DA neurons. The lowest desferal dosage (1.3 µg) that protects against 6-OHDA is 200 times less than the neurotoxin dosage employed. If indeed it would have done so, it should be expected that striatal DA metabolism and turnover in control animals would have been affected, especially at the highest desferal dosage. Of interest is the dose-dependent neuroprotective effect of desferal, which even at its highest concentration did not fully protect against 6-OHDA lesion. Rather, desferal showed a biphasic bell-shaped action, where at the highest concentration used it was less effective. This is not unusual since both desferal and many other antioxidants have been shown to express this type of bell-shaped activity in other tissues, cell lines, and mitochondrial preparations.[64] Their mechanism of action has been attributed to such compounds possessing antioxidant (cytoprotective) and pro-oxidant (cytotoxic) activities at low and high concentrations, respectively. Our recent cDNA microarray gene expression studies with several antioxidant iron chelators have shed light on this behavior. It has been demonstrated in neuronal cell cultures that antioxidants/iron chelators at low concentrations induce antiapoptotic cell survival genes, while at higher concentrations they induce expression of pro-apoptotic cell death–inducing genes.[64] Nevertheless, pretreatment with desferal at the doses employed normalized the turnover of striatal DA and prevented the behavioral deficits induced by 6-OHDA.

Mechanisms of Neuroprotection by Iron Chelators: Mitochondrial Complex I

The most obvious mechanism by which desferal prevents 6-OHDA lesion and protects the DA neurons is via chelation of the free labile iron pool released by the neurotoxin, which is increased in the SNPC. Indeed, several studies have clearly shown that 6-OHDA and a number of other neurotoxins, such as kainite, quisquanate, and paraquat, have the ability to release iron from its inert storage site in ferritin into the labile toxic pool, which can be chelated by desferal.[32–34] Iron is also released from ferritin by nitric oxide and superoxide, and the consequence of this is generation of cytotoxic reactive hydroxyl radical and membrane lipid peroxidation, as seen with 6-OHDA and MPTP and which can be prevented by iron chelators and radical scavengers. Indeed, we[65] have suggested a neurotoxic role for nitric oxide as a releaser of iron and an inhibitor of mitochondrial complex I in PD.

The other mechanism of nigrostriatal DA neurodegeneration in PD has been attributed to the lower mitochondrial respiratory complex I (NADH dehydrogenase) activity observed in SN and more specifically in SNPC.[66,67] Inhibitors of mitochondrial respiratory complex I include neurotoxins such as rotenone, MPTP, and 6-OHDA. They cause a fall in mitochondrial membrane potential and opening of the mitochondrial transition pore that eventually leads to demise of the mitochondria and cell death. Previously, we have shown that 6-OHDA is a potent inhibitor of mitochondrial complex I,[68–70] but significantly more potent than that reported with MPTP.[71] The inhibition of complex I by 6-OHDA has been attributed to 6-OHDA itself rather than its oxidative products, where calcium significantly potentiates this effect. Desferal protects complex I activity against the inhibition by 6-OHDA in *in vitro* studies, but this was not due to its iron chelating or radical scavenging activities. However, desferal was shown to activate NADH dehydrogenase in the absence of 6-OHDA. Thus, it is possible that desferal neuroprotective activity may be a combination of iron chelation and NADH dehydrogenase activation that enhances neuronal survival in the presence of 6-OHDA.[69,70] The inhibition of mitochondrial complex I by MPTP[71] and 6-OHDA[68–70] has been shown to result in dysregulation of mitochondrial iron and glutathione.[37] Thus, drugs that prevent the inhibition of mitochondrial complex I induced by neurotoxins are thought to be neuroprotective.[72] These include iron chelating drugs such as desferal, EGCG, and *R*-apomorphine, which prevent the 6-OHDA inhibition of mitochondrial complex I activity.[68–70]

Iron and Inflammatory Processes

The other pathological feature of SNPC in parkinsonian brain and other neurodegenerative diseases is the profound inflammatory response that is characterized by proliferation of reactive microglia, surrounding the dying neuron and on top of the dead neurons.[73–75] Further, the significant increase in proinflammatory cytotoxic cytokines, IL1, IL6, and TNFα, is observed[11,54,56] and this is thought to contribute to the progression of the disease.[73] This has been attributed to activation and translocation of proinflammatory transcription factor NFkB in SNPC, which is thought to result from the presence of iron and ROS generated by the metal.[74–76] Lin *et al.*[77] in some elegant studies demonstrated that iron can activate and translocate NFkB in rat hepatocytes loaded with iron, which could be inhibited by iron chelators and radical scavengers, including desferal, LM1, and *N*-acetylcysteine. Similar results were reported in the striatum of 6-OHDA lesioned rats and in PC-12 cells treated with this toxin. Similarly, radical scavengers and iron chelators such as *R*-apomorphine and EGCG were neuroprotective and prevented the NFkB activation and translocation.[56,76] Thus, the neuroprotective activity of iron chelators such as desferal and our newly developed brain-permeable iron chelator, VK-28[78] (FIG. 9), against 6-OHDA

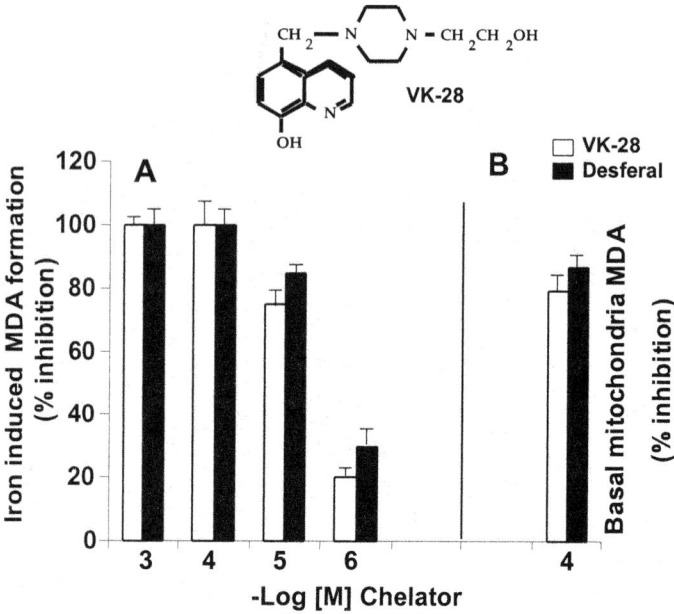

FIGURE 9. (*Top*) Structure of the novel brain-permeable iron chelator, VK-28, and (*bottom*) comparison of its inhibition of iron-induced (100 μM) **(A)** and basal mitochondrial lipid peroxidation **(B)**.[78]

may depend on several mechanisms. These include iron chelation and consequential prevention of ROS generation, inhibition of mitochondrial complex I, and preventions of NFkB activation and translocation, and cytotoxic cytokine formation.[56]

Iron and Iron Responsive Proteins

The pivotal role for iron in neurodegeneration, PD, and other neurodegenerative diseases has been strengthened recently by the identification of iron regulatory proteins 1 and 2 (IRP1, IRP2) in various regions of rat and mice brain, including the striatum and SN,[79] and several mutated iron and ferritin genes. The regulation of iron metabolism in mammalian cells is controlled by the interaction of IRPs with iron responsive elements (IREs) and nitric oxide. When cells are depleted or overloaded with iron, IRPs, which sense cytosolic iron levels, modify the proteins involved in iron metabolism and transport by interacting with IREs. When cells become loaded with iron as a consequence of nitric oxide generation,[65] the labile iron pool within it increases, transferrin receptors are downregulated, and IRP2 is ubiquinated and degraded. The opposite sequence of events occurs when cells are depleted of iron.[25,80,81] IRP1 and IRP2 knockout mice in studies of LaVaute *et al.*[82] have shown that IRP2 knockout mice have misregulated iron metabolism associated with accumulation of iron in the cytosol of neurons in various regions of the brain, including the striatum, which preceded neurodegeneration, ataxia, tremor, and bradykinesia. The striatum has one of the highest concentrations of IRP2 in human and mice brains.[79] We recently have shown that MPTP treatment, which increases iron in SNPC,[22,23] causes the loss of IRP2 and an increase of presynaptic protein α-synuclein in SNPC of MPTP-treated mice.[25,83,84] This process is prevented by pretreatment of mice with iron chelators, *R*- and *S*-apomorphine,[84] which have been shown to protect against MPTP and 6-OHDA neurotoxicity in cell cultures and *in vivo*.[30,31,40–42,44,76] Interaction of free labile iron with α-synuclein is thought to form a highly toxic component that induces liberation of ROS and induction of oxidative stress.[85,86]

Furthermore, support for a role of iron in neurodegeneration as induced by the neurotoxins, 6-OHDA, kainite, and quinolinate, which increase iron as a consequence of their neurotoxicity,[75,87,88] has come from rats made nutritionally iron-deficient.[88] Nutritionally iron-deficient animals show a 30% decrease in brain regional (striatum, hippocampus, cerebellum, and cortex) iron and have recently been shown to be resistant to neurodegeneration induced by kainite or 6-OHDA.[83,88] Histological studies of the brains from animals treated with kainite[78] or 6-OHDA[89] show little neurodegeneration, proliferation of reactive microglia, and accumulation of iron, which are normally seen following treatment with these neurotoxins.[75,88,90] The results from these studies have prompted us to suggest that indeed iron accumulation within the microglia and neurons, as reported in neurodegenerative diseases and their animal models, may be closely linked to the proliferation of the reactive microglia and the inflammatory responses observed in neurotoxin-induced neurodegeneration and neurodegenerative diseases.[75]

In conclusion, the ability of desferal and other iron chelators to protect against 6-OHDA and MPTP neurotoxicity, which may be mediated by iron chelation, is reinforced by our recent studies with the novel brain-permeable iron chelator, VK-28,[78] which has similar iron chelating and iron-induced lipid peroxidation inhibitory activity to that of desferal (FIG. 9). This chelator when injected either ICV or intraperitoneally protects against 6-OHDA nigrostriatal DA neuron lesion. Thus, these studies support the notion that iron chelators can be neuroprotective in PD and other neurodegenerative disease models, where there is iron dysregulation similar to those seen in SNPC of idiopathic and juvenile PD. It reinforces our previous advocation[59] for the use of a brain-permeable iron chelator for ironing iron out of the brain and neuroprotection, not only in PD, but also in other neurodegenerative diseases such as AD, HD, and FA, where similar alteration in iron metabolism and oxidative stress have been reported.[25,91] Since the metabolism of iron within the cell, especially brain neurons, is so tightly controlled,[83] misregulation of iron metabolism may be a primary cause of neurodegeneration.[56,89]

ACKNOWLEDGMENTS

We gratefully acknowledge the support of the National Parkinson Foundation (Miami, FL), the Michael J. Fox Foundation (New York, NY), the Stein Foundation, and the Friedman Foundation (Philadelphia, PA).

REFERENCES

1. YOUDIM, M.B.H. & P. RIEDERER. 2004. Iron in normal and pathological brain. *In* Encyclopedia of Neuroscience. Elsevier. Amsterdam/New York. In press.

2. SOFIC, E. et al. 1991. Selective increase of iron in substantia nigra zona compacta of parkinsonian brains. J. Neurochem. **56:** 978–982.
3. RIEDERER, P. et al. 1989. Transition metals, ferritin, glutathione, and ascorbic acid in parkinsonian brains. J. Neurochem. **52:** 515–520.
4. YOUDIM, M.B.H., D. BEN-SHACHAR & P. RIEDERER. 1989. Is Parkinson's disease a progressive siderosis of substantia nigra resulting in iron and melanin induced neurodegeneration? Acta Neurol. Scand. Suppl. **126:** 47–54.
5. FAUCHEUX, B.A. et al. 2002. Lack of up-regulation of ferritin is associated with sustained iron regulatory protein-1 binding activity in the substantia nigra of patients with Parkinson's disease. J. Neurochem. **83:** 320–330.
6. DEXTER, D.T. et al. 1989. Increased nigral iron content and alterations in other metal ions occurring in brain in Parkinson's disease. J. Neurochem. **52:** 1830–1836.
7. DEXTER, D.T. et al. 1989. Basal lipid peroxidation in substantia nigra is increased in Parkinson's disease. J. Neurochem. **52:** 381–389.
8. JELLINGER, K. et al. 1990. Brain iron and ferritin in Parkinson's and Alzheimer's diseases. J. Neural Transm. Parkinson's Dis. Dementia Sect. **2:** 327–340.
9. JELLINGER, K. et al. 1992. Iron-melanin complex in substantia nigra of parkinsonian brains: an X-ray microanalysis. J. Neurochem. **59:** 1168–1171.
10. JELLINGER, K.A. 1999. The role of iron in neurodegeneration: prospects for pharmacotherapy of Parkinson's disease. Drugs Aging **14:** 115–140.
11. GERLACH, M. et al. 1994. Altered brain metabolism of iron as a cause of neurodegenerative diseases? J. Neurochem. **793:** 793–807.
12. HALLIWELL, B. 2001. Role of free radicals in the neurodegenerative diseases: therapeutic implications for antioxidant treatment. Drugs Aging **18:** 685–716.
13. HIRSCH, E.C. et al. 1991. Iron and aluminum increase in the substantia nigra of patients with Parkinson's disease: an X-ray microanalysis. J. Neurochem. **56:** 446–451.
14. KASTNER, A. et al. 1992. Is the vulnerability of neurons in the substantia nigra of patients with Parkinson's disease related to their neuromelanin content? J. Neurochem. **59:** 1080–1089.
15. TAKANASHI, M. et al. 2001. Iron accumulation in the substantia nigra of autosomal recessive juvenile parkinsonism (ARJP). Parkinson's Relat. Disord. **7:** 311–314.
16. YOUDIM, M.B.H. 1985. Brain iron metabolism: biochemical and behavioral aspects in relation to dopaminergic neurotransmission. In Handbook of Neurochemistry. Plenum. New York.
17. JENNER, P. 1991. Oxidative stress as a cause of Parkinson's disease. Acta Neurol. Scand. Suppl. **136:** 6–15.
18. JENNER, P. & C.W. OLANOW. 1996. Oxidative stress and the pathogenesis of Parkinson's disease. Neurology **47:** S161–S170.
19. COHEN, G. 2000. Oxidative stress, mitochondrial respiration, and Parkinson's disease. Ann. N.Y. Acad. Sci. **899:** 112–120.
20. HALL, S., J.N. RUTLEDGE & T. SCHALLERT. 1992. MRI, brain iron, and experimental Parkinson's disease. J. Neurol. Sci. **113:** 198–208.
21. OESTREICHER, E. et al. 1994. Degeneration of nigrostriatal dopaminergic neurons increases iron within the substantia nigra: a histochemical and neurochemical study. Brain Res. **660:** 8–18.
22. TEMLETT, J.A. et al. 1994. Increased iron in the substantia nigra compacta of the MPTP-lesioned hemiparkinsonian African green monkey: evidence from proton microprobe elemental microanalysis. J. Neurochem. **62:** 134–146.
23. MOCHIZUKI, H. et al. 1994. Iron accumulation in the substantia nigra of 1-methyl-4-phenyl-1,2,3,6-tetrahydropyridine (MPTP)–induced hemiparkinsonian monkeys. Neurosci. Lett. **168:** 251–253.
24. GOTO, K. et al. 1996. An immuno-histochemical study of ferritin in 1-methyl-4-phenyl-1,2,3,6-tetrahydropyridine (MPTP)–induced hemiparkinsonian monkeys. Brain Res. **724:** 125–128.
25. YOUDIM, M.B.H. & P. RIEDERER. 2002. Iron in the brain, normal and pathological. In Encyclopedia of Neuroscience. Elsevier. Amsterdam/New York.
26. BHARATH, S. et al. 2002. Glutathione, iron, and Parkinson's disease. Biochem. Pharmacol. **64:** 1037–1048.

27. PERUMAL, A.S. *et al.* 1989. Regional effects of 6-hydroxydopamine (6-OHDA) on free radical scavengers in rat brain. Brain Res. **504:** 139–141.
28. KUMAR, R., A.K. AGARWAL & P.K. SETH. 1995. Free radical–generated neurotoxicity of 6-hydroxydopamine. J. Neurochem. **64:** 1703–1707.
29. YONG, V.W., T.L. PERRY & A.A. KRISMAN. 1986. Depletion of glutathione in brainstem of mice caused by *N*-methyl-4-phenyl-1,2,3,6-tetrahydropyridine is prevented by antioxidant pretreatment. Neurosci. Lett. **63:** 56–60.
30. GRUNBLATT, E. *et al.* 2001. Effects of *R*-apomorphine and *S*-apomorphine on MPTP-induced nigro-striatal dopamine neuronal loss. J. Neurochem. **77:** 146–156.
31. GRUNBLATT, E. *et al.* 2001. Gene expression analysis in MPTP mice model of Parkinson's disease using cDNA microarray. J. Neurochem. **78:** 1–12.
32. MONTEIRO, H.P. & C.C. WINTERBOURN. 1989. 6-Hydroxydopamine releases iron from ferritin and promotes ferritin-dependent lipid peroxidation. Biochem. Pharmacol. **38:** 4177–4182.
33. LODE, H.N. *et al.* 1990. Release of iron from ferritin by 6-hydroxydopamine under aerobic and anaerobic conditions. Free Radical Res. Commun. **11:** 153–158.
34. LINERT, W. *et al.* 1996. Dopamine, 6-hydroxydopamine, iron, and dioxygen—their mutual interactions and possible implication in the development of Parkinson's disease. Biochim. Biophys. Acta **1316:** 160–168.
35. BORISENKO, G.G. *et al.* 2000. Interaction between 6-hydroxydopamine and transferrin: "let my iron go". Biochemistry **39:** 3392–3400.
36. PEZZELLA, A. *et al.* 1997. Iron-mediated generation of the neurotoxin 6-hydroxydopamine quinone by reaction of fatty acid hydroperoxides with dopamine: a possible contributory mechanism for neuronal degeneration in Parkinson's disease. J. Med. Chem. **40:** 2211–2216.
37. HAN, J. *et al.* 1999. Inhibitors of mitochondrial respiration, iron (II), and hydroxyl radical evoke release and extracellular hydrolysis of glutathione in rat striatum and substantia nigra: potential implications to Parkinson's disease. J. Neurochem. **73:** 1683–1695.
38. CADET, J.L. *et al.* 1989. Vitamin E attenuates the toxic effects of intrastriatal injection of 6-hydroxydopamine (6-OHDA) in rats: behavioral and biochemical evidence. Brain Res. **476:** 10–15.
39. ACUNA-CASTROVIEJO, D. *et al.* 1997. Melatonin is protective against MPTP-induced striatal and hippocampal lesions. Life Sci. **60:** L23–L29.
40. GASSEN, M. *et al.* 1996. Apomorphine is a highly potent free radical scavenger in rat brain mitochondrial fraction. Eur. J. Pharmacol. **308:** 219–225.
41. GASSEN, M., A. GROSS & M.B. YOUDIM. 1998. Apomorphine enantiomers protect cultured pheochromocytoma (PC12) cells from oxidative stress induced by H_2O_2 and 6-hydroxydopamine. Mov. Disord. **13:** 242–248.
42. LAN, J. & D.H. JIANG. 1997. Desferrioxamine and vitamin E protect against iron and MPTP-induced neurodegeneration in mice. J. Neural Transm. (Budapest) **104:** 469–481.
43. LAN, J. & D.H. JIANG. 1997. Excessive iron accumulation in the brain: a possible potential risk of neurodegeneration in Parkinson's disease. J. Neural Transm. **104:** 649–660.
44. LEVITES, Y. *et al.* 2001. Green tea polyphenol (−)-epigallocatechin-3-gallate prevents *N*-methyl-4-phenyl-1,2,3,6-tetrahydropyridine-induced dopaminergic neurodegeneration. J. Neurochem. **78:** 1073–1082.
45. MOUSSAOUI, S. *et al.* 2000. The antioxidant ebselen prevents neurotoxicity and clinical symptoms in a primate model of Parkinson's disease. Exp. Neurol. **166:** 235–245.
46. ROGHANI, M. & G. BEHZADI. 2001. Neuroprotective effect of vitamin E on the early model of Parkinson's disease in rat: behavioral and histochemical evidence. Brain Res. **892:** 211–217.
47. BEN-SHACHAR, D. *et al.* 1991. The iron chelator desferrioxamine (desferal) retards 6-hydroxydopamine-induced degeneration of nigrostriatal dopamine neurons. J. Neurochem. **56:** 1441–1444.
48. KAUR, D. *et al.* 2003. Genetic or pharmacological iron chelation prevents MPTP-induced neurotoxicity *in vivo*: a novel therapy for Parkinson's disease. Neuron **37:** 899–909.
49. BEN-SHACHAR, D. & M.B. YOUDIM. 1991. Intranigral iron injection induces behavioral and biochemical "parkinsonism" in rats. J. Neurochem. **57:** 2133–2135.

50. UNGERSTEDT, U. 1968. 6-Hydroxy-dopamine induced degeneration of central monoamine neurons. Eur. J. Pharmacol. **5:** 107–110.
51. URETSKY, N.J. & L.L. IVERSEN. 1970. Effects of 6-hydroxydopamine on catecholamine containing neurones in the rat brain. J. Neurochem. **17:** 269–278.
52. YOUDIM, M.B.H., D. BEN-SHACHAR & P. RIEDERER. 1993. Parkinson's disease and increased iron in substantia nigra zona compacta. Mov. Disord. **8:** 1–12.
53. BERG, D. et al. 1999. Iron accumulation in the substantia nigra in rats visualized by ultrasound. Ultrasound Med. Biol. **25:** 901–904.
54. BERG, D. et al. 2001. Brain iron pathways and their relevance to Parkinson's disease. J. Neurochem. **79:** 225–236.
55. YOUDIM, M.B., D. BEN-SHACHAR & P. RIEDERER. 1993. The possible role of iron in the etiopathology of Parkinson's disease. Mov. Disord. **8:** 1–12.
56. YOUDIM, M.B.H., E. GRUNBLATT & S. MANDEL. 1999. The pivotal role of iron in NF-kappa B activation and nigrostriatal dopaminergic neurodegeneration: prospects for neuroprotection in Parkinson's disease with iron chelators. Ann. N.Y. Acad. Sci. **890:** 7–25.
57. ROY, C.N. & N.C. ANDREWS. 2001. Recent advances in disorders of iron metabolism: mutations, mechanisms, and modifiers. Hum. Mol. Genet. **10:** 2181–2186.
58. QIAN, Z.M. & X. SHEN. 2001. Brain iron transport and neurodegeneration. Trends Mol. Med. **7:** 103–108.
59. GASSEN, M. & M.B. YOUDIM. 1997. The potential role of iron chelators in the treatment of Parkinson's disease and related neurological disorders. Pharmacol. Toxicol. **80:** 159–166.
60. GRUNBLATT, E. et al. 1999. Apomorphine protects against MPTP-induced neurotoxicity in mice. Mov. Disord. **14:** 612–618.
61. SALAMONE, J.D., M.J. ZIGMOND & E.M. STRICKER. 1990. Characterization of the impaired feeding behavior in rats given haloperidol or dopamine-depleting brain lesions. Neuroscience **39:** 17–24.
62. ZIGMOND, M.J. & E.M. STRICKER. 1980. Supersensitivity after intraventricular 6-hydroxydopamine: relation to dopamine depletion. Experientia **36:** 436–438.
63. ACHESON, A.L., M.J. ZIGMOND & E.M. STRICKER. 1980. Compensatory increase in tyrosine hydroxylase activity in rat brain after intraventricular injections of 6-hydroxydopamine. Science **207:** 537–540.
64. WEINREB, O., S. MANDEL & M.B.H. YOUDIM. 2003. cDNA gene expression profile homology of antioxidants and their antiapoptotic and proapoptotic activities in human neuroblastoma cells. FASEB J. **17:** 935–937.
65. YOUDIM, M.B.H. et al. 1993. The neurotoxicity of iron and nitric oxide: relevance to the etiology of Parkinson's disease. Adv. Neurol. **60:** 259–266.
66. SCHAPIRA, A.H. et al. 2003. Anatomic and disease specificity of NADH CoQ1 reductase (complex I) deficiency in Parkinson's disease. J. Neurochem. **55:** 2142–2145.
67. SCHAPIRA, A.H. 2001. Causes of neuronal death in Parkinson's disease. Adv. Neurol. **86:** 155–162.
68. GLINKA, Y., K.F. TIPTON & M.B. YOUDIM. 1996. Nature of inhibition of mitochondrial respiratory complex I by 6-hydroxydopamine. J. Neurochem. **66:** 2004–2010.
69. GLINKA, Y., K.F. TIPTON & M.B.H. YOUDIM. 1998. Mechanism of inhibition of mitochondrial respiratory complex I by 6-hydroxydopamine and its prevention by desferrioxamine. Eur. J. Pharmacol. **351:** 121–129.
70. GLINKA, Y.Y. & M.B. YOUDIM. 1995. Inhibition of mitochondrial complexes I and IV by 6-hydroxydopamine. Eur. J. Pharmacol. **292:** 329–332.
71. RAMSAY, R.R. et al. 1987. Inhibition of NADH oxidation by pyridine derivatives. Biochem. Biophys. Res. Commun. **1:** 53–60.
72. SEATON, T.A., J.M. COOPER & A.H. SCHAPIRA. 1997. Free radical scavengers protect dopaminergic cell lines from apoptosis induced by complex I inhibitors. Brain Res. **777:** 110–118.
73. HUNOT, S. & E.C. HIRSCH. 2003. Neuroinflammatory processes in Parkinson's disease. Ann. Neurol. Suppl. **3:** S49–S58.
74. HUNOT, S. et al. 1997. Nuclear translocation of NF-kappaB is increased in dopaminergic neurons of patients with Parkinson disease. Proc. Natl. Acad. Sci. USA **94:** 7531–7536.

75. SHOHAM, S. & M.B.H. YOUDIM. 2000. Iron involvement in neural damage and microgliosis in models of neurodegenerative diseases. Cell. Mol. Biol. **46:** 743–760.
76. LEVITES, Y. *et al.* 2002. Attenuation of 6-hydroxydopamine (6-OHDA)–induced nuclear factor-kappaB (NF-kappaB) activation and cell death by tea extracts in neuronal cultures. Biochem. Pharmacol. **63:** 21–29.
77. LIN, M. *et al.* 1997. Role of iron in NF-kappa B activation and cytokine gene expression by rat hepatic macrophages. Am. J. Physiol. **272:** G1355–G1364.
78. BEN-SHACHAR, D. *et al.* 2003. Neuroprotection by a novel brain permeable iron chelator, VK-28, against 6-hydroxydopamine lesion in rats. J. Neurochem. In press.
79. SIDDAPPA, A.J. *et al.* 2002. Developmental changes in the expression of iron regulatory proteins and iron transport proteins in the perinatal rat brain. J. Neurosci. Res. **68:** 761–775.
80. EISENSTEIN, R.S. 2000. Iron regulatory proteins and the molecular control of mammalian iron metabolism. Annu. Rev. Nutr. **20:** 627–662.
81. KIM, S. & P. PONKA. 1999. Control of transferrin receptor expression via nitric oxide–mediated modulation of iron-regulatory protein 2. J. Biol. Chem. **274:** 33035–33042.
82. LAVAUTE, T. *et al.* 2001. Targeted deletion of the gene encoding iron regulatory protein-2 causes misregulation of iron metabolism and neurodegenerative disease in mice. Nat. Genet. **27:** 209–214.
83. YOUDIM, M.B.H., D. BEN-SHACHAR & S. YEHUDA. 1989. The putative biological mechanisms of the effect of iron deficiency on brain biochemistry and behavior. Am. J. Clin. Nutr. **50:** 607–615 (discussion, pp. 615–617).
84. (a) YOUDIM, M.B.H. 2003. What have we learnt from cDNA microarray gene expression studies about the role of iron in MPTP induced neurodegeneration and Parkinson's disease. J. Neural Transm. Suppl. **65:** 73–88; (b) MANDEL, S., G. MAOR & M.B.H. YOUDIM. 2004. Iron and alpha-synuclein in the substantia nigra of MPTP treated mice: effects of neuroprotective drugs *R*-apomorphine and green tea polyphenol, epigallocatechin-3-gallate. J. Mol. Neurosci. In press.
85. GOLTS, N. *et al.* 2002. Magnesium inhibits spontaneous and iron-induced aggregation of alpha-synuclein. J. Biol. Chem. **277:** 16116–16123.
86. OSTREROVA-GOLTS, N. *et al.* 2000. The A53T alpha-synuclein mutation increases iron-dependent aggregation and toxicity. J. Neurosci. **20:** 6048–6054.
87. SHOHAM, S., E. WERTMAN & R.P. EBSTEIN. 1992. Iron accumulation in the rat basal ganglia after excitatory amino acid injections—dissociation from neuronal loss. Exp. Neurol. **118:** 227–241.
88. SHOHAM, S. *et al.* 1996. Brain iron: function and dysfunction in relation to cognitive processes. *In* Iron Nutrition in Health and Disease. Libbey. London.
89. KE, Y. & Z. MING QIAN. 2003. Iron misregulation in the brain: a primary cause of neurodegenerative disorders. Lancet Neurol. **2:** 246–253.
90. LEVENSON, C.W. *et al.* 2003. Effect of dietary iron on motor behaviour and neuronal death in an experimental model of parkinsonism. J. Am. Aging Assoc. In press.
91. PERRY, G. *et al.* 2002. The role of iron and copper in the aetiology of neurodegenerative disorders: therapeutic implications. CNS Drugs **16:** 339–352.

Novel Chelators for Central Nervous System Disorders That Involve Alterations in the Metabolism of Iron and Other Metal Ions

DES R. RICHARDSON

Iron Metabolism and Chelation Program, Children's Cancer Institute Australia for Medical Research, Randwick, Sydney, New South Wales, Australia

ABSTRACT: Recent evidence suggests that iron (Fe) and other metals play a role in a number of neurodegenerative diseases including Friedreich's ataxia, Alzheimer's disease, Huntington's disease, and Parkinson's disease. In this review, the role of Fe and other metals in the pathology of these conditions is assessed and the potential of Fe chelators for treatment is discussed. Lipophilic chelators have been designed that may be capable of crossing the blood-brain barrier, a property lacking in desferrioxamine (DFO), a chelator in widespread clinical use. A far less commonly used chelator, clioquinol, has already shown activity *in vivo* in animal models and also in Alzheimer's disease patients. Considering that there is no effective treatment for many neurological diseases, the therapeutic use of lipophilic Fe chelators remains a potential strategy that requires investigation. In particular, we discuss the development of several series of aroylhydrazone chelators that could have high potential in the treatment of these diseases.

KEYWORDS: Alzheimer's disease; clioquinol; Friedreich's ataxia; Huntington's disease; iron; iron chelators; iron metabolism; Parkinson's disease; pyridoxal isonicotinoyl hydrazone; 2-pyridylcarboxaldehyde isonicotinoyl hydrazone

THE ROLE OF IRON AND OTHER METAL IONS IN NEURODEGENERATIVE DISEASE

General Introduction

In recent years, the role of iron (Fe) and other metal ions [especially Cu(II) and Zn(II)] in a number of neurodegenerative diseases has become an interesting and important area of study.[1-3] While Fe is essential for metabolic functions including DNA synthesis and ATP production, excess Fe results in tissue damage due to the generation of oxygen radicals such as the cytotoxic hydroxyl radical. The toxicity of Fe is apparent in a number of diseases including β-thalassemia and hereditary hemochromatosis, and these remain the most well-characterized conditions of Fe overload.[4,5]

Address for correspondence: Dr. D. R. Richardson, Iron Metabolism and Chelation Program, Children's Cancer Institute Australia for Medical Research, High Street (P. O. Box 81), Randwick, Sydney, New South Wales, 2031 Australia. Voice: +61-2-9314-7924; fax: +61-2-9382-1850.
d.richardson@ccia.org.au

There are a number of neurological problems associated with alterations in the metabolism of Fe and other metal ions, including tardive dyskinesia, hereditary hemochromatosis, aceruloplasminemia, restless legs syndrome, Huntington's disease (HD), Friedreich's ataxia (FA), multiple sclerosis, Parkinson's disease (PD), Alzheimer's disease (AD), and Hallervorden-Spatz syndrome.[6] For some of these diseases, the accumulations or changes in Fe distribution observed could be potentially treated using metal-binding drugs known as chelators.

Iron Metabolism

Before discussing the alterations in Fe metabolism that are found in neurological diseases, I will briefly describe the general model of cellular Fe uptake and storage (for review, see ref. 7). Iron is transported in the serum bound to the glycoprotein, transferrin (Tf), which binds two atoms of high-spin Fe(III).[8] The source of Fe for loading Tf comes from macrophages that recycle hemoglobin Fe.[7]

Diferric Tf donates its Fe to cells by binding to the specific transferrin receptor 1 (TfR1) on the cell surface. The Tf-TfR1 complex is internalized via receptor-mediated endocytosis and the Fe is released from the protein by a decrease in endosomal pH mediated by a proton pump.[8] The Fe(III) is reduced to Fe(II) by an unknown membrane reductase and then transported through the membrane by the natural resistance-associated macrophage protein-2 (Nramp2), which has also been called the divalent metal ion transporter 1 (DMT1).[9,10]

The brain is a sanctuary area that has its own Tf that plays a role in Fe delivery.[6,7] Serum Tf delivers its Fe to the brain via the TfR1 on endothelial cells that comprise the blood-brain barrier.[11] Diferric Tf binds to the TfR1 and is taken up by receptor-mediated endocytosis at the luminal membrane of blood capillaries.[12,13] The Fe then dissociates from Tf in the endosomal compartment and apoTf is rapidly recycled to the blood.[14] The liberated Fe is released from the abluminal membrane by an unknown mechanism and subsequently binds to Tf present in the interstitial space in the brain. The Tf-bound Fe is then delivered to neurons and glia.[13]

Apart from the TfR1, more recent studies have identified a second Tf-binding protein known as the TfR2. The TfR2 has a lower affinity for Tf than the TfR1 and its role in Fe uptake and cellular metabolism remains unclear.[15,16] Apart from the specific high-affinity TfR1, a number of cell types also possess non-receptor-mediated mechanisms of Fe uptake that are consistent with pinocytosis of Tf.[17,18]

Once Fe enters the cell it is incorporated into a poorly characterized compartment known as the intracellular labile Fe pool, the molecular identity of which remains unclear. It has been speculated to be composed of low M_r ligands such as amino acids and citrate.[19] However, low M_r intermediates have never been identified during the Fe uptake process.[7] Alternatively, Fe could be bound to high M_r chaperone molecules, or intracellular Fe transport may require the interaction of organelles.[7,20,21] The Fe in the intracellular pool is critical since it is used for the synthesis of heme and [Fe-S] cluster–containing proteins or can be used for Fe storage in ferritin.[22] Ferritin is a large polymeric protein composed of 24 subunits of two types, H (heavy) and L (light). Ferritin can store up to 4500 atoms of Fe within its protein shell and may be capable of releasing this metal ion when required.[22] However, the mechanism of ferritin Fe mobilization remains unclear.

The intracellular Fe pool regulates two mRNA-binding molecules known as the iron-regulatory proteins 1 and 2 (IRPs; for reviews, see refs. 7 and 23). Both IRP1 and IRP2 are *trans*-regulators that post-transcriptionally control the expression of a variety of molecules that play essential functions in Fe homeostasis.[7,23] The IRPs bind to hairpin-loop structures known as the iron-responsive elements (IREs), which are found in the 5′- or 3′-untranslated regions (UTRs) of mRNAs of a variety of molecules involved in Fe metabolism, including ferritin and TfR1.[7,23]

The IRPs can bind to IREs and either stabilize the mRNA against degradation or inhibit translation.[7,23] In *ferritin* mRNA, the IRE is located in the 5′-UTR, and IRP binding inhibits translation, thereby decreasing Fe storage. However, in *TfR1* mRNA, the IREs (5 stem loops) are located in the 3′-UTR, and IRP binding confers stability to the mRNA, enhancing translation and Fe uptake via the TfR1.[7,23] The IRP1-RNA binding is regulated by the presence of a [4Fe-4S] cluster within the protein.[7,23] When cells are Fe-deplete, the [4Fe-4S] cluster is absent (apo-IRP1), allowing for increased IRP-IRE binding. Conversely, when Fe levels are high, a [4Fe-4S] cluster becomes incorporated into the protein (holo-IRP1), preventing IRP1-IRE binding.[7,23] Unlike IRP1, IRP2 does not contain a [4Fe-4S] cluster. In Fe-replete cells, IRP2 is rapidly degraded by a proteasome-dependent mechanism.[7]

In the following sections, I will describe the alterations in Fe and metal metabolism in HD, PD, AD, and FA. Of these conditions, the role of Fe in the pathogenesis of FA is the most well-characterized and, thus, this condition will be a major focus of this article. The possible role of chelators as a therapeutic strategy for the treatment of neurodegenerative disease will also be discussed.

HUNTINGTON'S DISEASE

This condition is due to selective neuron loss that results from an elongated glutamine tract within the huntingtin protein. HD is an autosomal dominant inherited neurodegenerative condition that becomes evident in midlife and causes progressive motor, psychiatric, and cognitive dysfunction.[24]

The defect in HD is due to a trinucleotide repeat, CAG, in exon 1 of the gene. In unaffected individuals, there are 10–34 CAG repeats; in those affected, there are more than 40 repeats.[24] The function of huntingtin remains unclear, although the protein does have WW domains and caspase cleavage sites.[24] The molecule could interact with other molecules and it may be proteolytically cleaved into other biologically active proteins. Interestingly, huntingtin protein appears to be regulated by intracellular Fe levels as the Fe chelator, desferrioxamine (DFO), resulted in upregulation of huntingtin expression.[25] Moreover, examination of organelles of embryonic stem cells with the Hdh(ex4/5)/Hdh(ex4/5) knockout mutation demonstrated marked changes in six classes of organelles including the nucleus (nucleoli and transcription factor speckles) and perinuclear membranes (mitochondria, endoplasmic reticulum, Golgi, and recycling endosomes).[25] Thus, it has been suggested that the huntingtin protein may play a role in RNA biogenesis, organelle trafficking, and Fe homeostasis.[25] However, it must be noted that the effect of DFO on huntingtin expression is relatively weak evidence to support a role of the protein in Fe metabolism. Indeed, the alteration in huntingtin expression after incubation of cells with DFO may be a secondary effect. Hence, further work is critical to determine a precise molecular role of huntingtin.

PARKINSON'S DISEASE

The etiology of PD remains unclear. However, the aggregation of the protein, α-synuclein, into fibrils probably plays an important role in the pathogenesis of the disease.[26–28] This protein is found in an unfolded α-helical form that can be transformed into β-folds by mutation, environmental factors, or stress.[26] Indeed, α-synuclein along with hyperphosphorylated neurofilaments, lipids, Fe, and ubiquitin are the major constituents of intracellular protein inclusions (Lewy bodies and Lewy neurites) in dopaminergic neurons of the substantia nigra pars compacta.[26] Iron may contribute to the pathology of PD by inducing aggregation of α-synuclein.[28] Toxicity induced by α-synuclein results in an apoptotic mechanism and may be initiated by oxidant stress since hydroxyl radicals are liberated by this molecule in the presence of Fe.[29]

A number of studies have suggested that alterations in cellular Fe metabolism and Fe-induced oxidative stress are important factors in the pathogenesis of PD.[2] For instance, it has been suggested that, in both PD and AD, there are elevated brain Fe accumulations relative to the amounts of ferritin,[30,31] and mice with targeted disruptions in IRP2 suffer Fe accumulation in the brain and develop neurodegeneration resembling PD.[32] However, studies examining polymorphisms in IRP2 have failed to establish any association with sporadic PD,[33] while a polymorphism in the G allele of Tf correlated with the disease.[34] Studies examining IRP1-RNA binding activity failed to show any change in the substantia nigra of PD patients that have increased substantia nigra Fe levels.[30] These authors suggest that the lack of change in IRP1-RNA binding activity could explain the lack of upregulation of ferritin in the presence of increased Fe levels and may suggest a model of compartmentalized Fe accumulation in neurons from PD patients.[30] Considering that disruption of the IRP2 leads to a PD-like illness in mice,[32] it may also be worthwhile to look at changes in IRP2-RNA binding activity in addition to that found for IRP1.

Interestingly, occupational exposure to various metals appears to be a risk factor for PD, and recent studies have suggested that a number of metal ions, including aluminum(III), copper(II), iron(III), cobalt(III), and manganese(II), accelerate the rate of α-synuclein fibril formation.[35] Furthermore, it has been suggested that hydrogen peroxide (H_2O_2) and hydroxyl radicals are generated upon incubation of α-synuclein with small amounts of Fe(II).[27] The authors of this latter publication suggest that the production of H_2O_2 during the formation of abnormal aggregates of α-synuclein could be a major pathological factor in PD.[27]

In PD, there is accumulation of Fe in the brain that selectively involves neuromelanin in substantia nigra neurons,[2,36] decreased levels of glutathione and other antioxidants, increased lipid peroxidation products, and impaired mitochondrial electron transport.[1,2] There is also evidence to indicate that these changes are closely related to interactions between Fe and neuromelanin.[2,37] Further, the major cell types lost in PD, namely, dopaminergic nigrostriatal neurons, are thought to be highly prone to oxidative stress due to the ability of dopamine to auto-oxidize to produce H_2O_2.[37] The H_2O_2 generated can react with Fe to produce the highly toxic hydroxyl radical.[37] It has been suggested that in addition to current therapy for PD, such as levodopa and neuroprotective strategies (e.g., dopamine antagonists), these agents could be supplemented by treatment with chelators and lazaroids that prevent Fe-induced oxidative stress.[2]

ALZHEIMER'S DISEASE

β-Amyloid Is a Metalloprotein That Plays a Role in the Pathogenesis of Alzheimer's Disease

Like many neurological diseases, the precise etiology and pathogenesis of AD have not been fully determined. However, AD is the most common age-dependent neurodegenerative disorder that is characterized by multiple cognitive deficits.[38] The pathological hallmarks of AD are senile plaques, neurofibrillary tangles, neuronal atrophy, and loss of cortical neurons. A major problem in the pathogenesis of AD is the abnormal deposition of β-amyloid.[38] This protein exists as a soluble form in biological fluids that can precipitate extracellularly as a β-pleated sheet-fibrillar form in AD plaque.[39,40]

It has been demonstrated that Fe, Cu, and Zn are enriched within β-amyloid deposits.[38,41–43] Further, there is some evidence of multiple cooperative Cu(II) binding sites on β-amyloid protein that can result in aggregation and precipitation *in vitro*.[44,45] Oligomerization of the β-amyloid into a higher-order complex binds Cu(II) at a Cu/Zn superoxide dismutase–like binding site.[46] When binding up to two equivalents of Cu(II), the protein-metal complex generates H_2O_2 catalytically by utilizing biological reducing agents, such as cholesterol, vitamin C, L-DOPA, and dopamine.[47] In fact, microregional production of H_2O_2 and the depletion of reductants may play a role in the neurotoxicity of β-amyloid in AD.[47]

The coordination of Zn(II) to histidine-13 of β-amyloid induces aggregation of soluble β-amyloid and this effect is totally reversible by chelation or alkalinization.[48,49] More recently, it has been shown that there is a marked decrease in β-amyloid deposition in the brains of Tg2576 mice lacking the synaptic ZnT3 zinc transporter.[50] Moreover, it has been suggested that the amyloid neuropathology of AD is principally caused by Zn(II) released during neurotransmission.[38,50]

Linkage between Iron Metabolism and Alzheimer's Disease: The Expression of Alzheimer's Amyloid Precursor Protein Is Regulated by Iron

The evidence described above has led to the suggestion that chelation therapy may be useful in the treatment of AD and that metal ions are involved in the aggregation of β-amyloid and its subsequent cytotoxicity (see FIG. 1).[51] Furthermore, it has been shown that redox-active Fe can mediate β-amyloid toxicity by resulting in the generation of H_2O_2 and hydroxyl radicals, leading to lipid peroxidation and oxidative stress.[27,52] Pretreatment of β-amyloid *in vitro* with the Fe chelator, DFO, results in diminished neurotoxicity,[52] while reconstitution of the holoprotein with Fe restores its cytopathic effects (FIG. 1).[52]

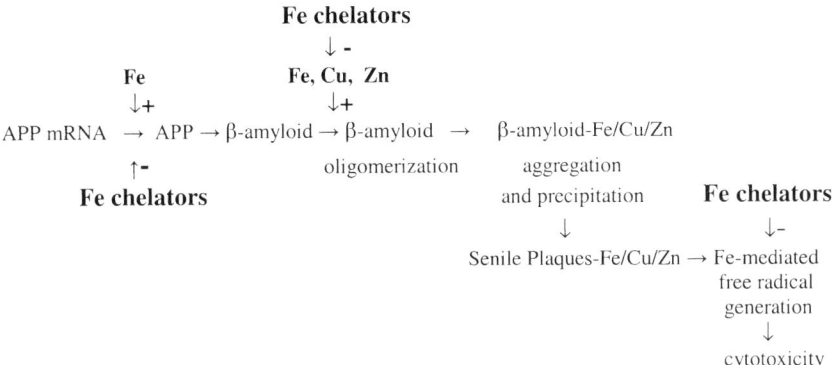

FIGURE 1. Schematic illustration of the possible target sites of chelators in preventing the generation of senile plaques in AD. It is known that APP contains an atypical IRE within the 5'-UTR of its mRNA and thus its translation is inhibited by low intracellular Fe levels. Amyloid deposits are enriched with Fe, Cu, and Zn ions and there is evidence of Cu(II) binding sites on β-amyloid protein that can result in aggregation and precipitation. Chelators can solubilize β-amyloid precipitates via their ability to bind metal ions. Amyloid-Cu/Fe ion complexes generate radicals and may play a role in the neurotoxicity in AD. Chelators may prevent this by binding redox-active Cu and Fe ions. See text for a detailed description.

Additional evidence for a link between the pathogenesis of AD and Fe metabolism is the finding of an atypical "type II" IRE in the 5'-UTR of the Alzheimer's amyloid precursor protein (APP) transcript.[53] The APP is a large membrane protein that is proteolytically cleaved to generate β-amyloid that aggregates to form senile plaques.[54] Importantly, the APP 5'-UTR that contains the IRE becomes specifically bound to IRPs, and the 5'-UTR of APP upon incubation with DFO confers downregulation of APP translation.[53] These authors have suggested that the IRE could be a target for the selection of chelators, such as clioquinol and other drugs, which reduce APP and subsequently decrease the burden of β-amyloid (FIG. 1).[53]

Treatment of Alzheimer's Disease Using the Metal Chelator, Clioquinol

Chelators that are selective for Cu(II) and Zn(II) have been shown to solubilize β-amyloid deposits *in vitro*.[42,43] Studies using APP2576 transgenic mice, a model of AD, demonstrate that oral administration of the lipophilic chelator, 5-chloro-7-iodo-8-hydroxyquinoline (clioquinol; FIG. 2) significantly decreased the deposition of β-amyloid.[43] Clioquinol is a retired USP antibiotic and bioavailable chelator that can cross the blood-brain barrier.[38] Furthermore, health and body weight were more stable in animals treated with clioquinol.[43] A brief 3-week trial of clioquinol in 20 AD patients resulted in some clinical improvement, with a significant increase in comprehension and the GBS rating scale being reported.[55] A preliminary conference report indicated that clioquinol was also capable of arresting cognitive deterioration and significantly lowering plasma β-amyloid in a phase II clinical trial of AD patients.[56]

The potential of clioquinol in the treatment of AD deserves further detailed studies in order to determine its efficacy. Clioquinol has been used in the past as an antiparasitic agent and was associated with an epidemic of neurotoxicity in Japan called subacute myelo-opticoneuropathy (SMON).[57] This condition was related to taking high doses of the drug over long periods[57] and may be due to the formation of a cytotoxic zinc chelate.[58] Therefore, it is important to examine the use of other lipophilic chelators that can permeate the blood-brain barrier, but have less cytotoxic effects than clioquinol. Such compounds are discussed below.

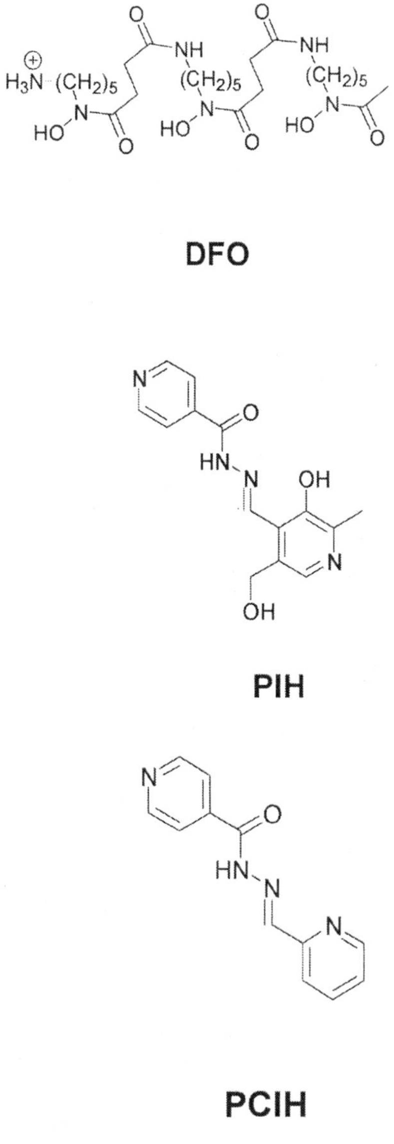

FIGURE 2. Structures of iron chelators described in this review: desferrioxamine (DFO), pyridoxal isonicotinoyl hydrazone (PIH), and 2-pyridylcarboxaldehyde isonicotinoyl hydrazone (PCIH).

FRIEDREICH'S ATAXIA

Friedreich's Ataxia and the Development of Mitochondrial Iron Loading

FA is a severe autosomal recessive neurodegenerative disorder that primarily affects the nervous system and heart.[59,60] The onset of symptoms occurs around the time of puberty and typically before the age of 25 years.[59,60] The disease leads to an early "sentence" in a wheelchair and the development of a dilated cardiomyopathy that results in premature death. The gene responsible for FA has been cloned and is termed *FRDA*, and its product is known as frataxin.[59,60] Patients with FA have, in approximately 97% of cases, a large expansion of an intronic GAA repeat within intron 1 (chromosome 9q13), resulting in a decrease in frataxin expression.[59,60]

Over the last five years, evidence has accumulated to suggest that frataxin plays a role in mitochondrial Fe metabolism, and a loss of frataxin leads to mitochondrial Fe overload.[60] The specific role of frataxin in mammalian mitochondrial Fe metabolism remains unclear at present, with various hypotheses being put forward.[61–64] However, it is likely that the pathology observed in FA may at least be partly due to mitochondrial Fe overload that leads to the production of toxic free radicals, resulting in damage to essential biomolecules.[60] Mitochondrial DNA may be particularly affected by free radicals since it lacks the protection of histone proteins found in nuclear DNA.[65] Thus, in Fe-loaded mitochondria, the opportunity for Fenton-mediated oxidative damage is considerable.

The Use of Iron Chelators as Therapeutic Agents for the Treatment of Friedreich's Ataxia

Apart from the evidence that frataxin plays a role in Fe metabolism,[61–64,66,67] several other facts also suggest that Fe plays a role in the pathogenesis of FA. First, intramitochondrial oxidative stress due to Fe accumulation leads to the loss of activity of critical molecules with [Fe-S] clusters.[68] Second, the sensitivity of cells to the FA mutation is rescued by Fe chelators.[69] Third, antioxidants such as the coenzyme Q analogue, idebenone, have some beneficial effect on the pathogenesis of FA.[70] Indeed, idebenone has been suggested to be effective at controlling the cardiac hypertrophy in FA.[71,72] Collectively, this evidence suggests that an appropriate Fe chelator that targets the mitochondrion could be therapeutically beneficial.

The first criticism leveled at the use of a chelator for FA treatment is obvious. Chelators will diffuse into most cells and non-specifically deplete the patient of Fe, resulting in anemia. This is a valid point, particularly if Fe chelation therapy would be implemented using DFO and standard chelation regimens. This latter approach is not appropriate for FA patients. It is well known that DFO is a hydrophilic drug that does not efficiently permeate plasma membranes to bind cytosolic Fe.[73] Moreover, DFO does not permeate mitochondrial membranes to bind intramitochondrial Fe deposits.[74,75] This reason alone could result in the failure of DFO to successfully treat FA. At present, only idebenone has shown limited success in the treatment of FA. Considering this, further studies are essential to assess the potential of chelation therapy.

Furthermore, FA is not like β-thalassemia, where there is cytoplasmic accumulation of Fe in ferritin and hemosiderin, but not the mitochondrion.[4] In addition, the Fe-loading in FA is not marked, suggesting that a more refined and targeted method is needed. Collectively, all these observations indicate a markedly different approach to Fe chelation therapy for FA patients and other neurological diseases.

A Lipophilic Iron Chelator That Permeates the Mitochondrion: Pyridoxal Isonicotinoyl Hydrazone (PIH)

For the treatment of FA, lipophilic chelators are required that permeate the plasma membrane, the outer and inner mitochondrial membranes, as well as the blood-brain barrier. Over 20 years ago, it was shown that the hydrophobic chelator, pyridoxal isonicotinoyl hydrazone (PIH; FIG. 2), could permeate the mitochondrion to bind accumulated Fe in reticulocytes.[74] In this model, mitochondrial Fe overload was induced using selective inhibitors of the heme synthesis pathway (i.e., isonicotinic acid hydrazide or succinylacetone).[21,74,75] These inhibitors prevent PIX synthesis, and the subsequent production of heme, resulting in a marked accumulation of non-heme mitochondrial Fe (for review, see ref. 76) that can be removed by PIH.[74,75] For these studies, rabbit reticulocytes were used, and this model is subsequently referred to as the reticulocyte Fe-loaded mitochondrial model.

As a possible model chelator for FA treatment, PIH possesses many advantageous characteristics: (1) the ability to permeate mitochondrial membranes and bind intra-mitochondrial Fe accumulations *in vitro* in a cell culture model;[74,75] (2) oral effectiveness in terms of Fe chelation efficacy in animals and humans;[77,78] (3) high affinity and selectivity for Fe(III) over other important biologically relevant metal ions;[79,80] (4) the chelator is predominantly neutral at pH 7.4, allowing easy permeation of biological membranes;[81] (5) near optimal lipophilic/hydrophilic balance of the apochelator and Fe complex, allowing passage into and out of cells;[82] (6) economical and simple to synthesize through a 1-step Schiff base condensation;[74] (7) low cytotoxicity and poor ability to inhibit proliferation and DNA synthesis that is less than that of DFO;[83–85] and (8) high Fe chelation efficacy *in vitro* in cultured cells, *in vivo* in animal models, and in a clinical trial (for review, see ref. 86). The greatest disadvantage with PIH is that most of the results reported with this chelator were published prior to patent protection being established.[86] This prevented interest from pharmaceutical companies, which stifled its development.

In an attempt to improve upon the Fe chelation efficacy of PIH, a range of analogues of this ligand were prepared and some have demonstrated high activity *in vitro* and *in vivo*, being far more effective than DFO.[86] Unfortunately, again, these findings were published prior to patent protection being established.

Analogues of 2-Pyridylcarboxaldehyde Isonicotinoyl Hydrazone (PCIH)

Synthesis and Chemical Properties

Due to the problems associated with the lack of patent protection with PIH and its analogues, a new series of tridentate ligands have been synthesized, characterized, and patented. These new chelators are known as the 2-pyridylcarboxaldehyde isonicotinoyl hydrazone (PCIH) analogues (FIG. 2).[75,85,87] These compounds were simply and economically prepared by a 1-step Schiff base condensation involving the addition of 2-pyridylcarboxaldehyde and a range of acid hydrazides, resulting in a relatively planar tridentate ligand (FIG. 3). Some of these chelators showed high Fe

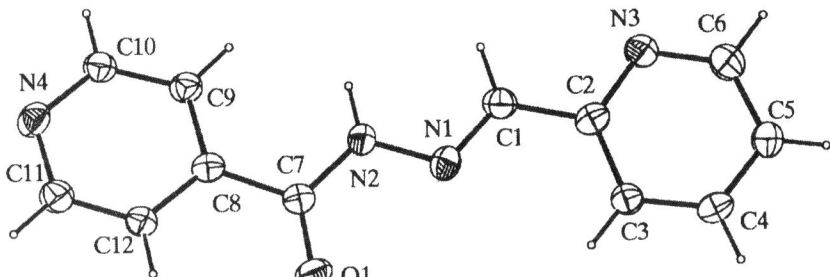

FIGURE 3. X-ray crystal structure of the parent compound of the PCIH class of chelators, namely, 2-pyridylcarboxaldehyde isonicotinoyl hydrazone (PCIH). The 30% probability ellipsoid is shown. [From ref. 73.]

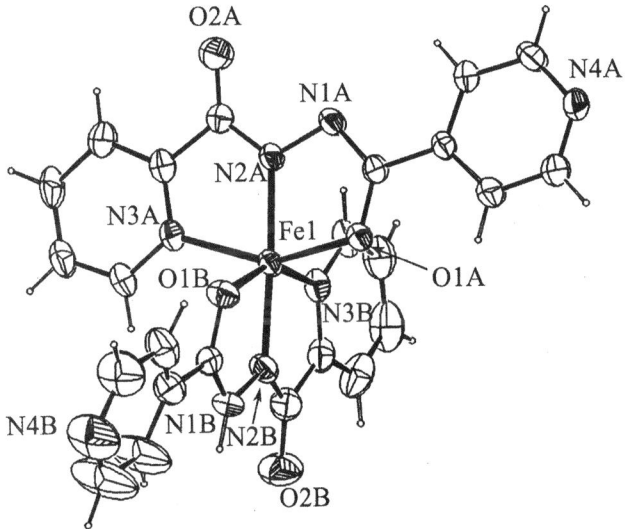

FIGURE 4. X-ray crystal structure of isonicotinoyl picolinoyl hydrazine (IPH)–Fe(III) complex. View of the [Fe(IPH)(HIPH)] molecule. The 30% probability ellipsoid is shown. [From ref. 75.]

chelation activity and low antiproliferative effects and have potential for the treatment of Fe overload disease.[85] The formation constants of PCIH with Fe are not particularly high and there is nothing to indicate that ligands are selective for divalent Fe(II) over other biologically significant metals, including Cu(II) and Zn(II).[87] In fact, the ligands appear to form more stable complexes with Cu(II) and, to a lesser extent, Zn(II)[87]—hence their utility for the possible treatment of not only FA, but also AD and other conditions.

Interestingly and unexpectantly, PCIH has been shown to be oxidized to a diacylhydrazine upon binding Fe, and this chelator is called isonicotinoyl picolinoyl hydrazine (IPH; FIG. 4).[88] Other diacylhydrazine ligands have been synthesized and

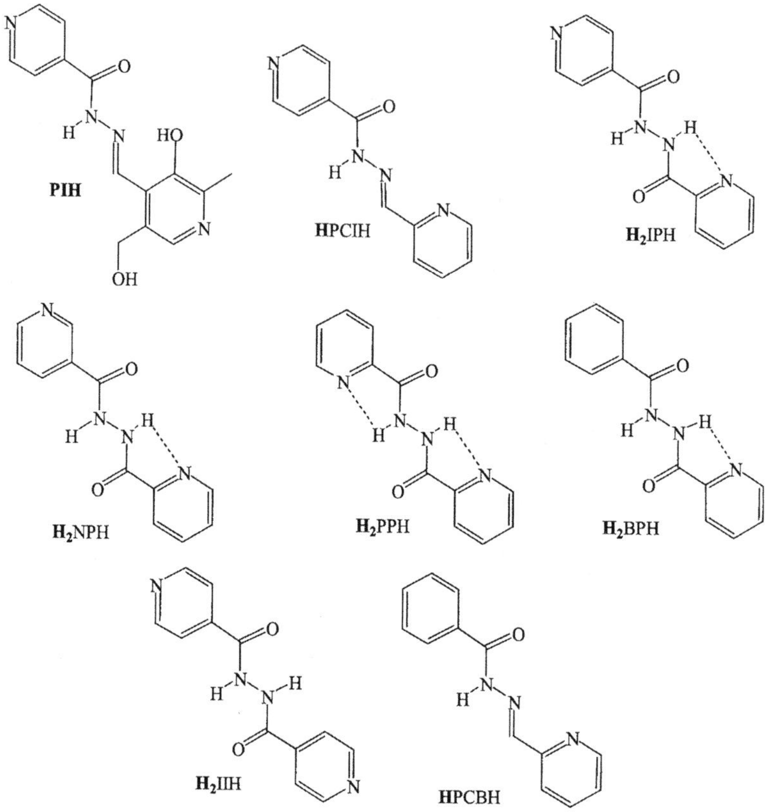

FIGURE 5. Structures of the series of diacylhydrazine analogues showing the parent compound isonicotinoyl picolinoyl hydrazine (H_2IPH) in comparison to 2-pyridylcarboxaldehyde isonicotinoyl hydrazone (PCIH) and pyridoxal isonicotinoyl hydrazone (PIH). The other IPH analogues include H_2NPH, H_2PPH, H_2BPH, H_2IIH, and HPCBH. [From ref. 75.]

show much higher activity than IPH, probably because they had a neutral charge.[88] This has led to the development of the diacylhydrazine series of chelators (FIG. 5) that could have potential in the treatment of Fe-overload disease, FA, AD, and possibly other conditions.[88]

Screening of the PCIH Analogues in the Iron-Loaded Mitochondrial Model of Reticulocytes

The high activity of several PCIH analogues encouraged us to assess their ability to remove Fe from the Fe-loaded mitochondrial system using rabbit reticulocytes.[75] At present, this is the only well-characterized mammalian model of mitochondrial Fe-loading available (for review, see ref. 76).

In studies examining the effects of the PCIH analogues on Fe release from ^{59}Fe-loaded reticulocyte mitochondria, PIH was used as a positive control due to its ability to remove Fe from this organelle. As a function of chelator concentration

(10–200 μM), the three most effective compounds were PCIH > PIH = PCTH.[75] The high activity of the parent chelator, PCIH, was somewhat surprising as this ligand showed relatively less Fe chelation activity than PCTH when examining the neural cell line, SK-N-MC.[85] Interestingly, DFO showed little activity at mobilizing Fe, being no more effective than control medium at concentrations of 10–200 μM.[75]

The identification of ligands that can enter the mitochondrion and bind accumulated Fe is a significant advantage when compared to DFO, which cannot effectively permeate into the mitochondrion. The obvious next step is to use these chelators in an appropriate animal model such as the conditional knockouts of the *Frda* gene in mice.[67] Such work is currently under way in the author's laboratory.

CONCLUSIONS

The role of metal ions in the pathogenesis of AD, PD, HD, and FA is intriguing and, over the last several years, has become increasingly well characterized, particularly for AD and FA. Already, there are preliminary data that the lipophilic chelator, clioquinol, may be useful in the treatment of AD. In this latter condition, there is strong evidence for the role of metal ions, particularly Zn(II), Cu(II), and Fe(II)/(III) in the pathobiology of the disease at a variety of biochemical levels. Further work examining the possible use of chelators in the treatment of FA requires investigation, but the accumulation of Fe in the redox-sensitive environment of the mitochondrion suggests that chelators with mitochondrial access have high potential. These latter compounds include the PIH analogues, PCIH analogues, and the diacylhydrazine series that show high lipophilicity and marked Fe chelation efficacy.

ACKNOWLEDGMENTS

The author is grateful to Juliana Kwok of the Iron Metabolism and Chelation Program for her excellent comments on the manuscript prior to submission. Children's Cancer Institute Australia for Medical Research is affiliated with the University of New South Wales and Sydney Children's Hospital. This project was supported by a fellowship and project grant from the National Health and Medical Research Council of Australia and grants from the Muscular Dystrophy Association (U.S.A.) and National Ataxia Foundation (U.S.A.).

REFERENCES

1. HARLEY, A. *et al.* 1993. Iron induced oxidative stress and mitochondrial dysfunction: relevance to Parkinson's disease. Brain Res. **627:** 349–353.
2. JELLINGER, K.A. 1999. The role of iron in neurodegeneration: prospects for pharmacotherapy of Parkinson's disease. Drugs Aging **14:** 115–140.
3. SHOHAM, S. & M.B. YOUDIM. 2000. Iron involvement in neural damage and microgliosis in models of neurodegenerative diseases. Cell. Mol. Biol. **46:** 743–760.
4. OLIVIERI, N.F. & G. BRITTENHAM. 1997. Iron chelating therapy and the treatment of thalassemia. Blood **89:** 739–761.
5. WONG, C. & D.R. RICHARDSON. 2003. β-Thalassaemia: emergence of new and improved chelators for treatment. Int. J. Biochem. Cell Biol. **35:** 1144–1149.

6. BURDO, J.R. & J.R. CONNOR. 2002. Iron transport in the central nervous system. *In* Molecular and Cellular Iron Transport, pp. 487–508. Dekker. New York.
7. RICHARDSON, D.R. & P. PONKA. 1997. The molecular mechanisms of the metabolism and transport of iron in normal and neoplastic cells. Biochim. Biophys. Acta **1331**: 1–40.
8. MORGAN, E. 1981. Transferrin biochemistry, physiology, and clinical significance. Mol. Aspects Med. **4**: 1–123.
9. GUNSHIN, H. *et al.* 1997. Cloning and characterization of a mammalian proton-coupled metal-ion transporter. Nature **388**: 482–488.
10. FLEMING, M.D. *et al.* 1997. Microcytic anaemia mice have a mutation in Nramp2, a candidate iron transporter gene. Nat. Genet. **16**: 383–386.
11. JEFFERIES, W.A. *et al.* 1984. Transferrin receptor on endothelium of brain capillaries. Nature **312**: 162–163.
12. FISHMAN, J.B. *et al.* 1987. Receptor-mediated transcytosis of transferrin across the blood brain barrier. J. Neurosci. Res. **18**: 299–304.
13. ROBERTS, R.L. *et al.* 1993. Receptor-mediated endocytosis of transferrin at the blood-brain barrier. J. Cell. Sci. **104**: 521–532.
14. TAYLOR, E.M. & E.H. MORGAN. 1990. Developmental changes in transferrin and iron uptake by the brain in the rat. Dev. Brain Res. **55**: 35–42.
15. KAWABATA, H. *et al.* 1999. Molecular cloning of transferrin receptor 2: a new member of the transferrin receptor–like family. J. Biol. Chem. **274**: 20826–20832.
16. KAWABATA, H. *et al.* 2000. Transferrin receptor 2-α supports cell growth both in iron-chelated cultured cells and *in vivo*. J. Biol. Chem. **275**: 16618–16625.
17. RICHARDSON, D.R. & E. BAKER. 1994. Two saturable mechanisms of iron uptake from transferrin in human melanoma cells: the effect of transferrin concentration, chelators, and metabolic probes on transferrin and iron uptake. J. Cell. Physiol. **161**: 160–168.
18. TRINDER, D. *et al.* 1996. Transferrin receptor–independent uptake of diferric transferrin by human hepatoma cells with antisense inhibition of receptor expression. Hepatology **23**: 1512–1520.
19. JACOBS, A. 1977. Low molecular weight intracellular iron transport compounds. Blood **50**: 433–439.
20. RADISKY, D.C. & J. KAPLAN. 1998. Iron in cytosolic ferritin can be recycled through lysosomal degradation in human fibroblasts. Biochem. J. **336**: 201–205.
21. RICHARDSON, D.R. *et al.* 1996. Distribution of iron in reticulocytes after inhibition of heme synthesis with succinylacetone: examination of the intermediates involved in iron metabolism. Blood **87**: 3477–3488.
22. HARRISON, P.M. & P. AROSIO. 1996. The ferritins: molecular properties, iron storage function, and cellular regulation. Biochim. Biophys. Acta **1275**: 161–203.
23. HENTZE, M.W. & L.C. KÜHN. 1996. Molecular control of vertebrate iron metabolism: mRNA-based regulatory circuits operated by iron, nitric oxide, and oxidative stress. Proc. Natl. Acad. Sci. USA **93**: 8175–8182.
24. YOUNG, A.B. 2003. Huntingtin in health and disease. J. Clin. Invest. **111**: 299–302.
25. HILDITCH-MAGUIRE, P. *et al.* 2000. Huntingtin: an iron regulated protein essential for normal nuclear and perinuclear organelles. Hum. Mol. Genet. **9**: 2789–2797.
26. JELLINGER, K.A. 2002. Recent developments in the pathology of Parkinson's disease. J. Neural Transm. Suppl. **62**: 347–376.
27. TABNER, B.J. *et al.* 2002. Formation of hydrogen peroxide and hydroxyl radicals from A(beta) and alpha-synuclein as a possible mechanism of cell death in Alzheimer's disease and Parkinson's disease. Free Radical Biol. Med. **32**: 1076–1083.
28. WOLOZIN, B. & N. GOLTS. 2002. Iron and Parkinson's disease. Neuroscientist **8**: 22–32.
29. EL-AGNAF, O.M. & G.B. IRVINE. 2002. Aggregation and neurotoxicity of alpha-synuclein and related peptides. Biochem. Soc. Trans. **30**: 559–565.
30. FAUCHEUX, B.A. *et al.* 2002. Lack of up-regulation of ferritin is associated with sustained iron regulatory protein-1 binding activity in the substantia nigra of patients with Parkinson's disease. J. Neurochem. **83**: 320–330.
31. THOMPSON, K. *et al.* 2003. Mouse brains deficient in H-ferritin have normal iron concentration, but a protein profile of iron deficiency and increased evidence of oxidative stress. J. Neurosci. Res. **71**: 46–63.

32. LAVAUTE, T. *et al.* 2001. Targeted deletion of the gene encoding iron regulatory protein-2 causes misregulation of iron metabolism and neurodegenerative disease in mice. Nat. Genet. **27:** 209–214.
33. LEE, P.L. *et al.* 2002. Polymorphisms in iron-responsive binding protein 2 and lack of association with sporadic Parkinson's disease. Mov. Disord. **17:** 1302–1304.
34. BORIE, C. *et al.* 2002. Association study between iron-related genes polymorphisms and Parkinson's disease. J. Neurol. **249:** 801–804.
35. UVERSKY, V.N. *et al.* 2001. Metal triggered structural transformations, aggregation, and fibrillation of human alpha-synuclein: a possible molecular NK between Parkinson's disease and heavy metal exposure. J. Biol. Chem. **276:** 44284–44296.
36. BECKER, G. & D. BERG. 2001. Neuroimaging in basal ganglia disorders: perspectives for transcranial ultrasound. Mov. Disord. **16:** 23–32.
37. ANDERSEN, J.K. 2001. Do alterations in glutathione and iron levels contribute to pathology associated with Parkinson's disease ? Novartis Found. Symp. **235:** 11–20.
38. BUSH, A.I. & R.E. TANZI. 2002. The galvanization of β-amyloid in Alzheimer's disease. Proc. Natl. Acad. Sci. USA **99:** 7317–7319.
39. GLENNER, G.G. & C.W. WONG. 1984. Alzheimer's disease: initial report of purification and characterization of a novel cerebrovascular protein. Biochem. Biophys. Res. Commun. **120:** 885–890.
40. GOLABECK, A. *et al.* 1995. Amyloid beta–binding protein *in vitro* and in normal human cerebrospinal fluid. Neurosci. Lett. **191:** 79–82.
41. SU, S.W. *et al.* 2000. Histochemically-reactive zinc in amyloid plaques, angiopathy, and degenerating neurons of Alzheimer's diseased brains. Brain Res. **852:** 274–278.
42. CHERNY, R.A. *et al.* 2000. Chelation and intercalation: complementary properties in a compound for the treatment of Alzheimer's disease. J. Struct. Biol. **130:** 209–216.
43. CHERNY, R.A. *et al.* 2002. Treatment with a copper-zinc chelator markedly and rapidly inhibits beta-amyloid accumulation in Alzheimer's disease transgenic mice. Neuron **30:** 665–676.
44. ATWOOD, C.S. *et al.* 1998. Dramatic aggregation of Alzheimer Aβ by Cu(II) is induced by conditions representing physiological acidosis. J. Biol. Chem. **273:** 12817–12826.
45. ATWOOD, C.S. *et al.* 2000. Characterization of copper interactions with Alzheimer amyloid beta peptides: identification of an attomolar-affinity copper binding site on amyloid beta 1–42. J. Neurochem. **75:** 1219–1233.
46. CURTAIN, C.C. *et al.* 2001. Alzheimer's disease amyloid-beta binds copper and zinc to generate an allosterically ordered membrane-penetrating structure containing superoxide dismutase–like subunits. J. Biol. Chem. **276:** 20466–20473.
47. OPAZO, C. *et al.* 2002. Metalloenzyme-like activity of Alzheimer's disease beta-amyloid: Cu-dependent catalytic conversion of dopamine, cholesterol, and biological reducing agents to neurotoxic H(2)O(2). J. Biol. Chem. **277:** 40302–40308.
48. BUSH, A.I. *et al.* 1994. Rapid induction of Alzheimer A beta amyloid formation by zinc. Science **265:** 1464–1467.
49. LIU, S.T. *et al.* 1999. Histidine-13 is a critical residue in the zinc-ion induced aggregation of the A beta peptide in Alzheimer's disease. Biochemistry **38:** 9373–9378.
50. LEE, J-Y. *et al.* 2002. Contribution by synaptic zinc to the gender-disparate plaque formation in human Swedish mutant APP transgenic mice. Proc. Natl. Acad. Sci. USA **99:** 7705–7710.
51. BUSH, A.I. 2002. Metal complexing agents as therapies for Alzheimer's disease. Neurobiol. Aging **23:** 1031–1038.
52. ROTTKAMP, C.A. *et al.* 2001. Redox-active iron mediated beta-amyloid toxicity. Free Radical Biol. Med. **30:** 447–450.
53. ROGERS, J.T. *et al.* 2002. An iron-responsive element type II in the 5′-untranslated region of the Alzheimer's amyloid precursor protein transcript. J. Biol. Chem. **277:** 45518–45528.
54. DE FELICE, F.G. & S.T. FERREIRA. 2002. Beta-amyloid production, aggregation, and clearance as targets for therapy in Alzheimer's disease. Cell. Mol. Neurobiol. **22:** 545–563.
55. REGLAND, B. *et al.* 2001 Treatment of Alzheimer's disease with clioquinol. Dement. Geriatr. Cogn. Disord. **12:** 408–414.

56. MASTERS, C. 2002 (April 4). Seventh International Geneva/Springfield Alzheimer's Symposium, Geneva, Switzerland.
57. SHIRAKI, H. 1975. The neuropathology of subacute myelo-optico-neuropathy (SMONS) in the humans: with special reference to the quinoform intoxication. Jpn. J. Med. Sci. Biol. **28:** 101–164.
58. ARBISER, J.L. *et al.* 1998. Clioquinol-zinc chelate: a candidate causative agent of subacute myelo-optic neuropathy. Mol. Med. **4:** 665–670.
59. PUCCIO, H. & M. KOENIG. 2000. Recent advances in the molecular pathogenesis of Friedreich ataxia. Hum. Mol. Genet. **9:** 887–892.
60. BECKER, E. & D.R. RICHARDSON. 2001. Frataxin: its role in iron metabolism and the pathogenesis of Friedreich's ataxia. Int. J. Biochem. Cell Biol. **33:** 1–10.
61. RADISKY, D.C. *et al.* 1999. The yeast frataxin homologue mediates mitochondrial iron efflux: evidence for a mitochondrial iron cycle. J. Biol. Chem. **274:** 4497–4499.
62. BECKER, E. *et al.* 2002. Erythroid differentiation and protoporphyrin IX down-regulate frataxin expression in Friend cells: characterisation of frataxin expression compared to molecules involved in iron metabolism and hemoglobinisation. Blood **99:** 3813–3822.
63. MÜHLENHOFF, U. *et al.* 2002. The yeast frataxin homolog Yfh1p plays a specific role in the maturation of cellular Fe/S proteins. Hum. Mol. Genet. **11:** 2025–2036.
64. LESUISSE, E. *et al.* 2003. Iron use for haem synthesis is under the control of the yeast frataxin homologue (Yfh1). Hum. Mol. Genet. **12:** 879–889.
65. EATON, J.W. & M. QIAN. 2002. Molecular basis of cellular iron toxicity. Free Radical Biol. Med. **32:** 833–840.
66. BABCOCK, M. *et al.* 1997. Regulation of mitochondrial iron accumulation by Yfh1p, a putative homologue of frataxin. Science **276:** 1709–1712.
67. PUCCIO, H. *et al.* 2001. Mouse models for Friedreich ataxia exhibit cardiomyopathy, sensory nerve defect, and Fe-S enzyme deficiency followed by intramitochondrial iron deposits. Nat. Genet. **27:** 181–186.
68. ROTIG, A. *et al.* 1997. Aconitase and mitochondrial iron-sulphur protein deficiency in Friedreich ataxia. Nat. Genet. **17:** 215–217.
69. WONG, A. *et al.* 1999. The Friedreich's ataxia mutation confers cellular sensitivity to oxidant stress which is rescued by chelators of iron and calcium and inhibitors of apoptosis. Hum. Mol. Genet. **8:** 425–430.
70. RUSTIN, P. *et al.* 1999. Effect of idebenone on cardiomyopathy in Friedreich's ataxia: a preliminary study. Lancet **354:** 477–479.
71. LODI, R. *et al.* 2001. Antioxidant treatment improves *in vivo* cardiac and skeletal muscle bioenergetics in patients with Friedreich's ataxia. Ann. Neurol. **49:** 590–596.
72. HAUSSE, A.O. *et al.* 2002. Idebenone and reduced cardiac hypertrophy in Friedreich's ataxia. Heart **87:** 346–349.
73. RICHARDSON, D.R. *et al.* 1994. The effect of the iron(III) chelator, desferrioxamine, on iron and transferrin uptake by the human malignant melanoma cell. Cancer Res. **54:** 685–689.
74. PONKA, P. *et al.* 1979. Mobilization of iron from reticulocytes: identification of pyridoxal isonicotinoyl hydrazone as a new iron chelating agent. FEBS Lett. **97:** 317–321.
75. RICHARDSON, D.R. *et al.* 2001. Development of potential iron chelators for the treatment of Friedreich's ataxia: ligands that mobilize mitochondrial iron. Biochim. Biophys. Acta **1536:** 133–140.
76. RICHARDSON, D.R. 2003. Friedreich's ataxia: iron chelators that target the mitochondrion as a therapeutic strategy. Exp. Opin. Invest. Drugs **12:** 235–245.
77. HOY, T. *et al.* 1979. Effective iron chelation following oral administration of an isoniazid pyridoxal hydrazone. Br. J. Haematol. **43:** 443–449.
78. BRITTENHAM, G.M. 1990. Pyridoxal isonicotinoyl hydrazone: an effective chelator after oral administration. Semin. Hematol. **27:** 112–116.
79. RICHARDSON, D.R. *et al.* 1989. Iron chelators of the pyridoxal isonicotinoyl hydrazone class. III. Formation constants with calcium(II), magnesium(II), and zinc(II). Biol. Metals **2:** 161–167.
80. VITOLO, L.M.W. *et al.* 1990. Iron chelators of the pyridoxal isonicotinoyl hydrazone class. Part 2. Formation constants with iron(III) and iron(II). Inorg. Chim. Acta **170:** 171–176.

81. RICHARDSON, D.R. et al. 1990. Iron chelators of the pyridoxal isonicotinoyl hydrazone class. Part 1. Ionisation characteristics of the ligands and their relevance to biological properties. Inorg. Chim. Acta **170:** 165–170.
82. RICHARDSON, D.R. & E. BAKER. 1991. The release of iron and transferrin by the human malignant melanoma cell. Biochim. Biophys. Acta **1091:** 294–302.
83. RICHARDSON, D.R. & P. PONKA. 1994. The iron metabolism of the human neuroblastoma cell: lack of relationship between the efficacy of iron chelation and the inhibition of DNA synthesis. J. Lab. Clin. Med. **124:** 660–671.
84. RICHARDSON, D.R. et al. 1995. The potential of iron chelators of the pyridoxal isonicotinoyl hydrazone class as effective antiproliferative agents. Blood **86:** 4295–4306.
85. BECKER, E. & D.R. RICHARDSON. 1999. Development of novel aroylhydrazone ligands for iron chelation therapy: the 2-pyridylcarboxaldehyde isonicotinoyl hydrazone (PCIH) analogues. J. Lab. Clin. Med. **134:** 510–521.
86. RICHARDSON, D.R. & P. PONKA. 1998. Pyridoxal isonicotinoyl hydrazone and its analogues: potential orally effective iron-chelating agents for the treatment of iron overload disease. J. Lab. Clin. Med. **131:** 306–315.
87. ARMSTRONG, C.M. et al. 2003. Structural variations and formation constants of first-row transition metal complexes of biologically active aroylhydrazones. Eur. J. Inorg. Chem. **2003**(6): 1145–1156.
88. BERNHARDT, P.V. et al. 2001. Unprecedented oxidation of a biologically active aroylhydrazone chelator catalysed by iron(III): serendipitous identification of diacylhydrazine ligands with high iron chelation efficacy. J. Biol. Inorg. Chem. **6:** 801–809.

Redox Neurology: Visions of an Emerging Subspecialty

HYMAN M. SCHIPPER

Center for Neurotranslational Research, Lady Davis Institute for Medical Research, S.M.B.D. Jewish General Hospital, Department of Neurology and Medicine, McGill University, Montreal, Quebec, Canada

ABSTRACT: Recent years have witnessed a dramatic increase in publications implicating free radicals and oxidative stress in virtually every aspect of biology and medicine. Redox Neurology may be defined as the study of the roles of free radicals, transition metals, oxidative stress, and antioxidant defenses in diseases of the nervous system. In this position paper, an argument is presented for recognition of this field as an emerging subspecialty within medical neurology. A program for postresidency fellowship training in Redox Neurology that integrates laboratory experience with specialized clinical practice is proposed. Opportunities for research and teaching careers in the redox neurosciences are outlined. The paper concludes with a forecast of several research themes likely to preoccupy this nascent discipline in the days ahead.

KEYWORDS: antioxidant clinic; free radicals; medicine; metals; neurology; neuroscience; oxidative stress; redox laboratory; therapeutics

FREE RADICALS AND OXIDATIVE STRESS

Free radicals are atoms or molecules that contain unpaired electrons in their outermost orbitals. Their electronic configurations render these chemical species highly reactive with a host of cellular substrates, notably membrane lipids, proteins, and nucleic acids. Free radicals may be generated *de novo* within tissues or may be derived from environmental sources. Examples of common, endogenously produced free radicals and nonradical pro-oxidants include the superoxide anion (O_2^-), hydrogen peroxide (H_2O_2), singlet oxygen, hypochlorous acid (HOCl), peroxynitrite ($ONOO^-$), and the hydroxyl radical (OH$^\bullet$). Transition metals, such as ferrous iron (Fe^{2+}) or cuprous copper (Cu^{1+}), play a vital role in cellular redox chemistry by reducing H_2O_2 to the highly-toxic OH$^\bullet$ radical (Fenton catalysis). Transition metals may also promote the degradation of lipid hydroperoxides within plasma and organeller membranes, or behave as nonenzymatic (pseudo-)peroxidases capable of converting innocuous catechol-containing compounds (such as dopamine) to toxic ortho-semiquinone radicals.

Address for correspondence: Hyman M. Schipper, M.D., Ph.D., FRCPC, Professor of Neurology and Medicine, McGill University, Center for Neurotranslational Research, Lady Davis Institute for Medical Research, S.M.B.D. Jewish General Hospital, 3755 Côte Ste-Catherine Road, Montreal, Quebec H3T 1E2, Canada. Voice: 514-340-8260; fax: 514-340-7502.
hyman.schipper@mcgill.ca

Organisms representing all taxa surveyed to date are endowed with a panoply of defense mechanisms designed to prevent free radical production *in situ*, scavenge reactive oxygen species (ROS) generated within various aqueous and hydrophobic cellular domains, or repair oxidative chemical damage once incurred. In mammalian tissues, key antioxidant enzymes include the superoxide dismutases (disproportionate O_2^- to H_2O_2 and O_2), catalase, the glutathione peroxidases and the peroxiredoxin complex (neutralize H_2O_2 to water), and various reductases (glutathione reductase, thioredoxin reductase) that help replenish stores of key intracellular electron donors [reduced glutathione (GSH), reduced thioredoxin] depleted in the course of oxidoreductase and peroxidase reactions. The antioxidant enzymes operate in concert with a host of nonenzymatic, low-molecular-weight antioxidant compounds (e.g. GSH, ascorbate, the tocopherols, uric acid, melatonin, and bilirubin) to maintain tissue redox homeostasis. By maintaining transition metals in a relatively low-redox state, metal-binding proteins, including ferritin, transferrin, the metallothioneins, and ceruloplasmin, contribute substantially to the antioxidant protection of tissues and body fluids. Oxidative stress (OS) has been defined as "a disturbance in the pro-oxidant/antioxidant balance in favor of the former, leading to possible [*tissue*] damage."[1] Although usually implying the participation of ROS per se, OS may be considered more broadly to encompass the deleterious interactions of nitrogen-based radicals (nitrosative stress) and sulfur-derived thiyl radicals (sulfhydryl stress) with biochemical targets.

Recent years have witnessed an explosion of knowledge implicating free radicals and oxidative stress in virtually every aspect of biology and medicine (FIG. 1). At "physiological" concentrations, ROS may play adaptive roles as signaling molecules involved in cell growth, differentiation, gene regulation, replicative senescence, and apoptosis, and as a primary defense invoked by leukocytes and tissue macrophages against invading pathogens. Moreover, in conditions as disparate as ischemic heart disease, inflammatory bowel disease, cataractogenesis, glomerulonephritis, asthma, cancer, diabetes mellitus, and infertility, OS is being increasingly recognized as a pivotal common pathway for cellular death and dysfunction, and a potentially important target for therapeutic intervention.

OS AND NEUROLOGICAL DISEASE

OS has been implicated in a profusion of disorders affecting the central and peripheral nervous system. The dramatic increase in numbers of published experimental and clinical reports related to "redox neuroscience" and "redox neurology" over the last decade or so (FIG. 1) attests to the burgeoning interest in this field.

Why the brain? There are numerous biochemical and physiological factors that render the CNS particularly prone to OS and associated injury. Although the human brain constitutes only 2% of total body mass, it is responsible for ~20% of total O_2 consumption under basal conditions. The brain is replete with unsaturated fat (e.g., $C_{20:5}$, $C_{22:6}$) and is thus highly susceptible to lipid peroxidation. As amply documented in this compendium, high CNS concentrations of total and free transition metals (mainly Fe and Cu) predispose to Fenton-mediated hydroxyl radical formation and cytotoxicity. Neural tissues are also enriched for low-molecular-weight substrates, such as dopamine, norepinephrine, and 3-hydroxykynurenine, which readily

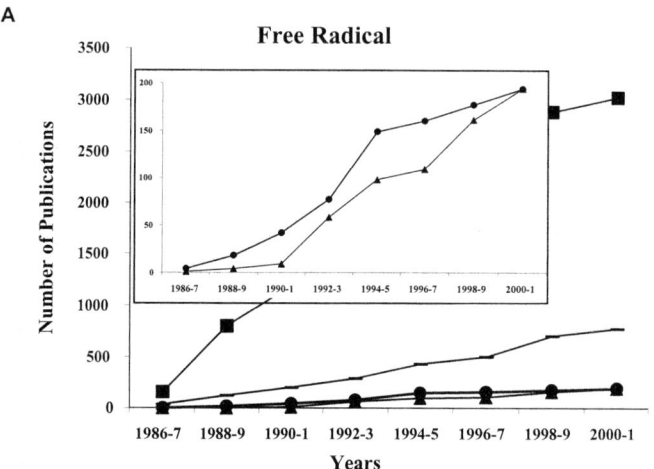

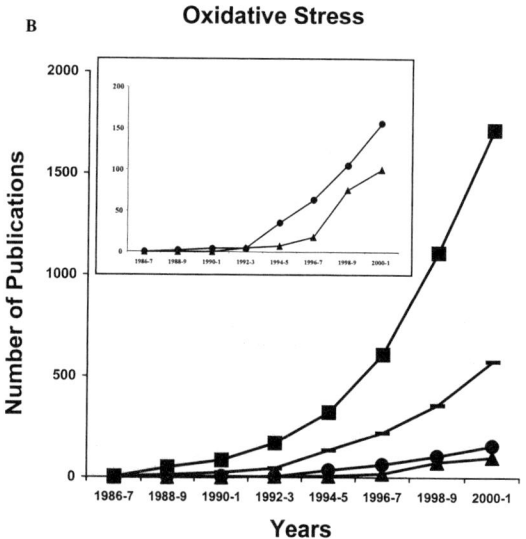

FIGURE 1. Publications in oxidative biomedicine and neuroscience between 1986 and 2001 based on MEDLINE searches of the biological and medical literature. **(A)** Search terms: [FREE RADICAL] and [NEUROLOGY] or [NEUROSCIENCE] or [MEDICINE] or [BIOLOGY]. **(B)** Search terms: [OXIDATIVE STRESS] and [NEUROLOGY] or [NEUROSCIENCE] or [MEDICINE] or [BIOLOGY]. Symbols: (■) Medicine; (−) Biology; (●) Neurology; (▲) Neuroscience.

give rise to ROS via spontaneous and metal-catalyzed autoxidation reactions. Under conditions of ischemia, trauma, or epilepsy, various signaling pathways and enzymes activated by the massive release of the excitatory amino acid, glutamate, promote free radical generation and injury in target neurons as part of the excitotoxicity cascade. A relative dearth of antioxidant enzymes and other defenses further enhances the vulnerability of the nervous system to oxidative damage.[2]

REDOX NEUROLOGY AS A CLINICAL SUBSPECIALTY

Residents completing a standard 3- to 4-year clinical (adult or pediatric) neurology training program in North America may choose to enter into general neurology practice or go on to fellowship training (usually 1–3 years) in one of several classical neurology subspecialties. Recognized neurology subspecialties include Stroke, Movement Disorders, Neuromuscular Diseases, Epilepsy, Neurophysiology, Neuro-ophthalmology, Neuro-oncology, Neuroimmunology, and Behavioral Neurology/Dementia. Some centers offer, in addition, fellowships in Neurointensive Care, Autonomic Neurology, Neuroendocrinology, Neuro-otology, Pain, and Sleep Medicine. Subspecialty certification currently exists in the United States for Child Neurology and Neurophysiology, with ongoing debate as to whether formal certification should be extended to several of the other aforementioned disciplines.[3] In this position paper, the emerging discipline of "Redox Neurology" will be outlined and an argument presented for assigning this field subspecialty status within medical neurology. Redox Neurology may be defined as the study of the roles of free radicals, transition metals, oxidative stress, and antioxidant defenses in diseases of the nervous system.

Clinical Activities

Along the lines of the established subspecialties, the Redox Neurologist would engage in aspects of diagnosis and treatment of adult and/or pediatric patients with neurological conditions implicating antioxidant deficiencies, derangements of transition metal homeostasis, and oxidative tissue injury. It is anticipated that Redox Neurology practice would differ substantially from subspecialties like Movement Disorders and Epilepsy, where referrals to the appropriate practitioner or clinic is primarily driven by the presenting symptoms and complaints (e.g., tremor and convulsions, respectively). Redox Neurology consultations would be contingent on perceived mechanisms of disease pathogenesis (OS) and commensurate treatment strategies (antioxidants, etc.). A common scenario envisioned is the tertiary referral of subjects with established neurological diagnoses to the Redox Neurologist for specialized investigations, antioxidant or metal chelation therapy, and recruitment into clinical trials. For example, family practitioners, geriatricians, and neurologists may refer patients with known Parkinson's disease, Friedreich's ataxia, or Alzheimer's dementia for trials of neuroprotective pharmacotherapy with coenzyme Q10, idebenone, or high-dose vitamin E.[4,5] Similarly, pediatric epileptologists may elect to have children with certain seizure disorders treated adjunctively with N-acetylcysteine or vitamin E.[4,6] In each case, decisions to launch further investigations and implement specific therapeutic interventions would be guided by a knowledge of the pertinent literature and adherence to the principles of evidence-based medicine. In this sense, Redox Neurology would provide a service akin to Neuroimmunology, where evidence of a primary dysimmune state (e.g., as occurs in multiple sclerosis, myasthenia gravis, and the inflammatory myopathies) and expertise in the implementation of immunomodulatory and immunosuppressive therapies (e.g., β-interferons, methotrexate) dictate professional engagement irrespective of the presenting symptoms or region of the neuraxis affected.

Research

The academic potential of this nascent discipline is enormous and practitioners may opt to invest variable amounts of time in research and teaching activities. As in most areas of medicine, research opportunities in Redox Neurology range across the full spectrum of basic, translational, clinical, and epidemiological studies and could be tailored to the investigator's interests and experience. By way of illustration, fundamental redox neuroscience would include such activities as the application of contemporary biochemistry and molecular biology techniques to investigate mechanisms of pathological iron deposition and oxidative mitochondrial damage in aging and degenerating neural tissues.[7] As an example of translational research in this field, we[8] and others[9] have employed proteomic tools to delineate profiles of oxidized plasma proteins as potential biological (diagnostic) markers of sporadic Alzheimer's disease. As an example of a clinical "redox" project, our institution is currently participating in a multicenter phase III trial sponsored by the Alzheimer's Disease Cooperative Study (UCSD) evaluating the potential benefits of vitamin E in forestalling the conversion from mild cognitive impairment to early Alzheimer's dementia. The research venue of the Redox Neurologist may also find utility outside the academic sector. In space exploration and defense applications, for example, she/he could assist in the development of antioxidant prophylaxis to limit CNS and other tissue damage accruing from exposure to cosmic[10] and "dirty bomb" radiation, respectively. The many fine submissions to this volume of the *Annals* underscore the numerous opportunities for fundamental and clinical research in the redox neurosciences.

Teaching

The prospects for a fulfilling teaching career in Redox Neurology would likely be on par with the other neurology subspecialties. The Redox Neurologist would have ample opportunity to impart knowledge of free radical chemistry, transition metal metabolism, redox signaling pathways, antioxidant gene regulation, oxidative tissue injury, and antioxidant pharmacology (see BOUNDARIES OF THE KNOWLEDGE, below) to medical students, residents, clinical fellows, and graduate students interested in oxidative biomedicine. She/he should be well poised to bring to clinical neurology fresh and exciting new perspectives on the nosology, pathogenesis, investigation, and management of a wide range of common and rare nervous system disorders (TABLE 1). The educational mission of the Redox Neurologist may also include provision of timely information concerning the status of free radicals and antioxidants in human health and disease to other health care professionals, nutritionists, government agencies, industrial partners, and the lay public. [During the course of this writing, for example, studies supporting the use of antioxidant (ebselen) therapy in alcoholic brain injury were prominently reported in the lay press.[11]]

TABLE 1. Human nervous system disorders implicating oxidative tissue damage in their etiopathogeneses

I Transition Metal Dysregulation
  Iron
 • CNS senescence
 • Cerebral hemorrhage
 • Alzheimer disease
 • Parkinson disease
 • Huntington disease
 • Progressive supranuclear palsy
 • Multiple sclerosis
 • Friedreich ataxia
 • X-linked sideroblastic
   anemia with ataxia
 • PANK-2 deficiency
 • Aceruloplasminemia
 • Neuroferritinopathy
 • Superficial siderosis
 • Restless legs syndrome?
 • Hemochromatosis?
  Copper
 • Wilson disease
 • Menkes disease
 • Creutzfeldt-Jakob disease?
 	
II Ischemia-Reperfusion Injury
 • Cerebral Infarction
 • Spinal Cord Infarction
 • Peripheral Nerve Ischemia
 • Muscle ischemia
 	
III Trauma
 • Cerebral Contusion
 • Spinal Cord Contusion
 • Peripheral Nerve Injury
 • Muscle Injury
 	
IV Epilepsy
 • Mixed childhood seizures
 • Unverricht-Lundborg disease
 	
V Antioxidant Enzyme Disorders
 • Familial amyotrophic lateral sclerosis
 • Down syndrome
	VI Metabolic Disorders
	 • Vitamin E deficiency
	 • Folate / B12 deficiency
	   (hyperhomocysteinemia)
	 • Diabetic neuropathy
	 • Porphyria
	
	VII Infection and Inflammation
	 • Bacterial meningitis
	 • HIV encephalitis
	 • Cerebral malaria
	
	VIII Mitochondrial Disorders
	 • PEO
	 • MERRF
	 • MELAS
	 • NARP
	 • Secondary mitochondriopathies
	
	IX Disorders of DNA Repair
	 • Ataxia-telangiectasia
	 • Xeroderma pigmentosum
	 • Cockayne syndrome
	 • Trichothiodystrophy
	
	X Neurotoxicants
	 • Manganese parkinsonism
	 • MPTP parkinsonism
	 • Mitochondrial toxins (CO, cyanide)
	 • Excitotoxins (lathyrism,
	   domoic acid)
	 • Inorganic mercury poisoning
	 • Minamata disease
	 • Arsenic neuropathy
	 • Cadmium neuropathy
	 • Cisplatin neuropathy
	 • Radiation damage

NOTE: CO, carbon monoxide; MELAS, mitochondrial encephalomyopathy–lactic acidosis–strokelike episodes; MERRF, myoclonic epilepsy with ragged-red fibers; NARP, neuropathy–ataxia–retinitis pigmentosa; PANK-2, pantothenate kinase–type 2; PEO, progressive external ophthalmoplegia.

THE REDOX NEUROLOGY FELLOWSHIP

Interested candidates completing their neurology residencies would enter a 2- to 3-year fellowship program in Redox Neurology under the tutelage of an established practitioner or clinician-scientist engaged in this area of study. Many of the contributors to this volume of the *Annals* could serve effectively in this capacity. The

TABLE 2. Selected bibliography (1997–2004) for Redox Neurology fellows and practitioners

(A) General texts

Oxidative Stress Biomarkers and Antioxidant Protocols. D. Armstrong, Ed. Humana Press, 2002.

Handbook of Antioxidants, Volume 8. E. Cadenas & L. Packer, Eds. Dekker, 2001.

Reactive Oxygen Metabolites: Chemistry and Medical Consequences. M. K. Eberhardt, Ed. CRC Press, 2000.

Free Radicals in Biology and Medicine, Third Edition. B. Halliwell & J. M. C. Gutteridge, Eds. Oxford University Press, 1999.

Antioxidants in Human Health and Disease. T. K. Basu, M. L. Garg & N. J. Temple, Eds. CAB International, 1999.

Antioxidant and Redox Regulation of Genes. C. K. Sen, P. A. Baeuerle, L. Packer & H. Sies, Eds. Elsevier, 1999.

Interrelations between Free Radicals and Metal Ions in Life Processes, Volume 36. A. Sigel & H. Sigel, Eds. Dekker, 1999.

Free Radicals and Iron: Chemistry, Biology, and Medicine. M. C. Symons & J. Gutteridge, Eds. Oxford University Press, 1999.

Antioxidant Status, Diet, Nutrition, and Health. A. M. Papas & J. H. Quillen, Eds. CRC Press, 1998

(B) Neuroscience texts

Redox-Active Metals in Neurological Disorders, Volume 1012. S. M. LeVine, J. R. Connor & H. M. Schipper, Eds. New York Academy of Sciences, 2004.

Transferrin, Ferritin, and Iron in the Central and Peripheral Nervous System. J. M. Pasquini, Ed. Karger, 2002.

Free Radicals in Brain Pathophysiology. G. Poli, L. Packer & E. Cadenas, Eds. Dekker, 2000.

Nitric Oxide and Free Radicals in Peripheral Neurotransmission, Volume 2. S. Kalsner, Ed. Springer-Verlag, 2000.

Oxidative/Energy Metabolism in Neurodegenerative Disorders, Volume 893. J. P. Blass & F. H. McDowell, Eds. New York Academy of Sciences, 1999.

Oxidative Stress in Skeletal Muscle. A. Z. Reznick, L. Packer & C. K. Sen, Eds. Birkhauser, 1998.

Metals and Oxidative Damage in Neurological Disorders. J. R. Connor, Ed. Kluwer, 1997.

Mitochondria and Free Radicals in Neurodegenerative Diseases. F. M. Beal, I. Bodis-Wollner & N. Howell, Eds. Wiley, 1997.

TABLE 3. Lectures presented in 2003 in an interdisciplinary course on Free Radical Biomedicine offered at McGill University (Montreal)

Introduction/Free Radical Chemistry
Sources of ROS/RNS and Oxidative Stress
Transition Metals I
Transition Metals II
Antioxidant Defenses: Enzymes
Antioxidant Defenses: Nonenzymatic
Antioxidant Gene Regulation
Cell Death
Oxidative Stress (OS) Assays I
OS Assays II
Infection and Inflammation
Atherosclerosis and Ischemia
Aging
Redox Neurology I
Redox Neurology II
OS and Pulmonary Diseases
OS and Diseases of the Skin
OS and Digestive System Disorders
Diabetes Mellitus
OS and Cancer
OS and Diseases of the Eye
OS and Renal Disorders
Free Radicals and Reproduction
Environmental Toxicology

fellowship experience could be weighted towards the clinic or laboratory according to individual preferences. A "generic" 2-year fellowship may be offered along the following lines: The first year would be spent in the laboratory assisting in the design of experimental protocols and mastering techniques for the measurement of oxidative molecular damage and antioxidant enzyme expression profiles in human and animal samples (see THE BIOMEDICAL REDOX LAB, below). The fellow would attain a firm grasp of fundamental concepts germane to free radical chemistry and oxidative biomedicine from assigned and independent readings of the pertinent literature (TABLE 2), by auditing relevant university courses if available (TABLE 3), and by attending lectures and conferences devoted to this subject. In the second year, the candidate would be introduced to translational and clinical trial design and would participate in the workup and management of patients with free radical–related neurological disorders (see THE ANTIOXIDANT CLINIC, below). As per routine for the established neurology fellowships, the candidate would be expected to publish scientific and clinical reports, present findings at international meetings, and assist in the teaching of Redox Neurology to house staff, students, and other health care professionals.

BOUNDARIES OF THE KNOWLEDGE

In broadest terms, the Redox Neurologist would acquire and utilize knowledge in a diverse range of clinical and scientific disciplines at the interface of contemporary neurology and oxidative biomedicine. She/he would gain familiarity with, and build upon, the following core information:

- Oxidative chemistry and free radical generation in biological substrates.
- The special role of mitochondria in oxidative neuropathology.
- Mechanisms of oxidative molecular damage/repair and their registration in neural tissues.
- OS pathways in neural gene regulation, growth, differentiation, apoptosis, and necrosis.
- The molecular biology and enzymology of mammalian antioxidant proteins.
- The sources, neural distribution, and pharmacology of nonenzymatic, low-molecular-weight antioxidants and nutraceuticals.
- The metabolism and neurobiology of transition metals, metal-binding proteins, and metal chelators.
- The roles of OS in environmental and iatrogenic neurotoxicology.
- Neuroimaging of transition metal deposition and oxidative metabolism.
- The gamut of adult and pediatric neurological disorders implicating OS, aberrant metal homeostasis, and antioxidant/chelation therapy in their etiopathogeneses and management (TABLES 1 and 4).

THE ANTIOXIDANT CLINIC

The establishment of autonomous or hospital-based Antioxidant Clinics could facilitate the evaluation and management of patients with OS-related neurological disorders and their recruitment for clinical research. Core clinic personnel may include a director, physician(s), nurse(s), nutritionist, receptionist, and research coordinator. A multidisciplinary staff, united by common interest in oxidative medicine, could represent a mix of specialties including (but not limited to) neurology, geriatrics, internal medicine, ophthalmology, and dermatology. Fellows, residents, and students registered in these diverse programs would assist in patient management and research at the clinic on a rotating basis. The director's office could disseminate a brochure to local hospital department heads, medical specialists, and family practitioners indicating specific conditions that fall within the purview of the Antioxidant Clinic and which may benefit from antioxidant intervention. Upon referral to the clinic, patients may be managed individually and concurrently with the referring doctor or be enrolled in ongoing clinical trials. In either case, the impact

TABLE 4. Neurological conditions reported to benefit from antioxidant or metal chelation therapy

Alzheimer's disease	**Aceruloplasminemia**
Vitamin E[4]	Fresh frozen plasma plus deferoxamine[20]
Gingko biloba[4]	**Vitamin E deficiency**
N-Acetylcysteine[15]	Vitamin E[21]
Idebenone[4]	d-α-Tocopherol-PEG-succinate[22]
Parkinson's disease	**HIV dementia**
Deprenyl[4]	Deprenyl[4]
CoQ10[16]	**Wilson's disease**
Tardive dyskinesia	D-Penicillamine[23]
Vitamin E[17]	Ammonium tetrathiomolybdate[23]
Epilepsy	Triethylene tetramine[23]
N-Acetylcysteine[6]	Zinc acetate[23]
Vitamin E[4]	**Aneurysmal subarachnoid hemorrhage**
Friedreich's ataxia	Ebselen[4]
CoQ10[18]	**Diabetic neuropathy**
Idebenone[18]	α-Lipoic acid[24]
Vitamin E[19]	

NOTE: With the exception of the aceruloplasminemia case report, the studies cited represent published clinical trials.

of specific pharmacotherapies on the natural history of the illness and on biochemical redox indices assayed in tissue samples (e.g. blood, urine, CSF) would be determined. Although the latter specimens could be shipped to one of several commercial entities offering OS-related assays, ideally the clinic would have access to, and collaborate closely with, a local laboratory adept at redox measurements (see THE BIOMEDICAL REDOX LAB, below). In addition to providing medical care, the clinic would serve as a hub for interdisciplinary exchange of ideas and technology related to oxidative biomedicine, an engine for translational and clinical research, and a milieu conducive to the training of students, residents, and fellows with career interests in redox biomedicine.

THE BIOMEDICAL REDOX LAB

As described above, the Redox Neurology fellow may elect to spend 1–2 years in the laboratory learning how to design experiments and conduct assays of relevance to OS research. These objectives may be accomplished in individual hospital or university laboratories engaged in specific projects of immediate interest to the candidate, or in hospital-based diagnostic facilities that offer a wide range of OS-related testing. As an example of the latter, a Biomedical Redox Laboratory was established at the Jewish General Hospital (Montreal) in 2003, which enables monitoring of key

TABLE 5. Assays available or under development at the Biomedical Redox Laboratory of the Jewish General Hospital (Montreal)

Oxidative Tissue Damage	Antioxidant Enzymes
• 8-epiPGF2α assay for lipid peroxidation (ELISA) • TBARS assay for lipid peroxidation (colorimetric) • Protein carbonyl assay for protein oxidation (ELISA) • 8-Hydroxy-deoxyguanosine measurement for DNA oxidation (HPLC)	• CuZnSOD (colorimetric) • MnSOD (colorimetric) • Catalase (colorimetric) • Glutathione peroxidase (colorimetric) • Glutathione reductase (colorimetric) • NADPH diaphorase/nitric oxide synthase (histochemical) • Heme oxygenase-1 (colorimetric, ELISA, Western, Northern) • Thioredoxin reductase (colorimetric)
Antioxidant Compounds	**Transition Metal Metabolism**
• α-Tocopherol (HPLC) • Homocysteine (HPLC) • Uric acid (colorimetric) • Bilirubin (colorimetric) • β-Carotene (HPLC) • Vitamin A (HPLC) • Ascorbate (HPLC) • Glutathione (reduced; oxidized) (HPLC) • Thioredoxin (HPLC)	• Ferritin (colorimetric/IHC) • Ceruloplasmin (colorimetric) • Ferrozine assay for tissue iron (colorimetric) • Transferrin (ELISA) • Total iron binding capacity (colorimetric) • Iron-II/III, copper-I/II, manganese-II/III (atomic absorption spectrometry)
Total Antioxidant Capacity	**Genotyping**
• TRAP assay (colorimetric)	• ApoE (PCR) • MnSOD (PCR) • Hfe/hemochromatosis (PCR)

antioxidant enzymes, nonenzymatic antioxidants, redox-active transition metals, free radical–mediated tissue damage, and redox gene polymorphisms in diverse clinical and experimental samples (TABLE 5). Rotating through such a laboratory, the Redox Neurology fellow (and other trainees) would gain considerable experience in performing a wide range of OS-related biochemical and molecular biological assays. Such expertise would enable the trainee to investigate baseline redox parameters and the impact of therapeutic interventions in patient samples submitted to the redox lab from hospital wards, Antioxidant Clinics, and the pharmaceutical industry, as well as conduct original research using tissue culture and whole animal models.

SUMMARY AND CONCLUSIONS

Free radicals have been implicated in a broad spectrum of normal and abnormal neurological functions. Involvement of OS and/or abnormalities of transition metal homeostasis in the etiopathogeneses of many neurological conditions, such as Alzheimer's disease, Parkinson's disease, familial amyotrophic lateral sclerosis, Wilson's disease, ischemic stroke, and diabetic neuropathy, has become widely accepted in recent years. In other neurological conditions, including such diverse entities as alcoholic brain injury, cerebral and spinal cord trauma, hemorrhagic

TABLE 6. A forecast of research issues in Redox Neurology

Neuropathological footprinting: Do different pro-oxidant species engender unique neuropathological profiles within a given region of the neuraxis? Will the same free radical intermediate evoke differential pathological responses in different brain regions?

Oxidative biomarkers: Can biochemical markers in blood or CSF be exploited for the early diagnosis and prognosis of free radical–related neurological disorders? Will oxidatively modified biomolecules provide surrogate markers for monitoring the efficacy of therapeutic interventions?

Redox neuroimaging: Can imaging techniques be adapted for visualization of free radical intermediates/spin traps and oxidative damage in the living brain?

Redox pharmacogenomics: Do specific antioxidant gene polymorphisms modify the natural history of human neurological diseases and individual responses to pharmacotherapy?

Gene and stem cell therapy: Can cell-based delivery of antioxidant genes or their products help restore a favorable redox microenvironment in human neurodegenerative disorders?

"Designer" antioxidants and metal chelators: Will new generations of antioxidants and metal chelators that safely and effectively target the CNS ameliorate OS-related neurological conditions?[5]

Diet and the environment: Does environmental or dietary exposure to heavy metals or free radical–generating neurotoxins (herbicides/insecticides) constitute a risk factor for human neurodegenerative disorders?[25]

stroke, various movement disorders, certain epilepsies, HIV encephalitis, and the prion diseases, a contributory role of OS to tissue injury appears likely in the light of accumulating clinical, pathological, and biochemical data.[4,12–14] A more thorough understanding of the mechanisms mediating OS-related neural dysfunction should facilitate the refinement of rational antioxidant and metal chelation therapies for many of the conditions alluded to in this compendium.

Establishment of Redox Neurology as a distinct medical subspecialty would confer an important level of organization to this rapidly developing discipline. Following completion of a neurology residency program, formal training in this field could be attained in the guise of a 2- to 3-year Redox Neurology fellowship under the supervision of a mentor specialized in this area of study. The fellowship could readily be tailored to the candidate's career aspirations and include rotations through a Redox Laboratory for familiarization with OS-related assays, an Antioxidant Clinic or equivalent for patient management/clinical research, or both. Perusal of the relevant literature and attendance at courses, lectures, and conferences pertaining to oxidative biomedicine would help round out the knowledge base.

In the author's view, assignment of subspecialty status to Redox Neurology would help define the boundaries of this discipline and propel its maturation in the spheres of clinical practice, research, and education. It is anticipated that the following specific benefits would accrue: (1) Formal training of clinician-scientists in the area of Redox Neurology would create a new breed of specialist adept at integrating clinical and fundamental aspects of oxidative medicine for state-of-the-art management of an expanding array of OS-related neurological conditions, both adult and pediatric. (2) The Redox Neurologist would be ideally poised to carry out innovative basic, translational, and clinical research in this burgeoning field. TABLE 6 lists several areas in the redox neurosciences where fundamental and applied research may prove particularly fruitful in the years ahead. (3) As a consultant and teacher, the subspecialist will lead in educating other health care professionals, students, and the general public about the role of free radicals, transition metals, and antioxidants in diseases of the nervous system. (4) The establishment of successful fellowships in Redox Neurology could provide an impetus for the development of analogous programs in Internal Medicine, Ophthalmology, and other biomedical departments. Exchanges of fundamental scientific knowledge, clinical experience, and technologies related to redox biomedicine among the various specialties, propelled further by the advent of joint conferences, multidisciplinary clinics, and collaborative research projects, could be mutually enriching for all parties concerned and could ultimately contribute to the public health and welfare.

ACKNOWLEDGMENTS

The author thanks Rolando Del Maestro for critical review of this manuscript and Lucia Badolato for skillful secretarial assistance.

REFERENCES

1. SIES, H. 1991. Oxidative Stress: Oxidants and Antioxidants. Elsevier. Amsterdam/New York.
2. HALLIWELL, B. & J.M.C. GUTTERIDGE. 1999. Free Radicals in Biology and Medicine. Third edition. Oxford University Press. London/New York.
3. TYLER, K. et al. 2003. Part 2: History of 20th century neurology—decade by decade. Ann. Neurol. **53**(suppl. 4): S27–S45.
4. DELANTY, N. & M.A. DICHTER. 2000. Antioxidant therapy in neurologic disease. Arch. Neurol. **57**: 1265–1270.
5. BEHL, C. & B. MOOSMANN. 2002. Oxidative nerve cell death in Alzheimer's disease and stroke: antioxidants as neuroprotective compounds. Biol. Chem. **383**: 521–536.
6. EDWARDS, M.J. et al. 2002. N-Acetylcysteine and Unverricht-Lundborg disease: variable response and possible side effects. Neurology **59**: 1447–1449.
7. SCHIPPER, H. 2001. Mitochondrial iron deposition in aging astroglia: mechanisms and disease implications. In Mitochondrial Ubiquinone (Coenzyme Q): Biochemical, Functional, Medical, and Therapeutic Aspects in Human Health and Disease, pp. 267–280. Prominent Press. Scottsdale, AZ.
8. YU, H-L., H.M. CHERTKOW, H. BERGMAN & H.M. SCHIPPER. 2003. Aberrant profiles of native and oxidized glycoproteins in Alzheimer plasma. Proteomics **3**: 2240–2248.
9. CHOI, J. et al. 2002. Identification of oxidized plasma proteins in Alzheimer's disease. Biochem. Biophys. Res. Commun. **293**: 1566–1570.
10. TURNER, N.D., B.L. FORD & J.R. LUPTON. 2002. Opportunities for nutritional amelioration of radiation-induced cellular damage. Nutrition **18**: 904–912.
11. CNN. 2003 (June 3). Antioxidants may protect brain from alcohol. CNN.com.
12. TURCHAN, J. et al. 2003. Oxidative stress in HIV demented patients and protection ex vivo with novel antioxidants. Neurology **60**: 307–314.
13. WAGNER, K.R. et al. 2003. Heme and iron metabolism: role in cerebral hemorrhage. J. Cereb. Blood Flow Metab. **23**: 629–652.
14. WONG, B.S. et al. 2000. Prion disease: a loss of antioxidant function? Biochem. Biophys. Res. Commun. **275**: 249–252.

15. ADAIR, J.C., J.E. KNOEFEL & N. MORGAN. 2001. Controlled trial of *N*-acetylcysteine for patients with probable Alzheimer's disease. Neurology **57:** 1515–1517.
16. SHULTS, C.W. *et al.* 2002. Effects of coenzyme Q10 in early Parkinson disease: evidence of slowing of the functional decline. Arch. Neurol. **59:** 1541–1550.
17. ADLER, L.A. *et al.* 1993. Vitamin E treatment of tardive dyskinesia. Am. J. Psychiatry **150:** 1405–1407.
18. RUSTIN, P. 2002. The use of antioxidants in Friedreich's ataxia treatment. Expert Opin. Invest. Drugs **12:** 569–575.
19. LODI, R. *et al.* 2002. Mitochondrial dysfunction in Friedreich's ataxia: from pathogenesis to treatment perspectives. Free Radical Res. **36:** 461–466.
20. YONEKAWA, M. *et al.* 1999. A case of hereditary ceruloplasmin deficiency with iron deposition in the brain associated with chorea, dementia, diabetes mellitus, and retinal pigmentation: administration of fresh-frozen human plasma. Eur. Neurol. **42:** 157–162.
21. SOKOL, R.J. 1990. Vitamin E and neurologic deficits. Adv. Pediatr. **37:** 119–148.
22. SOKOL, R.J. *et al.* 1993. Multicenter trial of *d*-alpha-tocopheryl polyethylene glycol 1000 succinate for treatment of vitamin E deficiency in children with chronic cholestasis. Gastroenterology **104:** 1727–1735.
23. MENKES, J.H. 1999. Menkes disease and Wilson disease: two sides of the same copper coin. Part I: Menkes disease. Eur. J. Paediatr. Neurol. **3:** 147–158.
24. VAN DAM, P.S. 2002. Oxidative stress and diabetic neuropathy: pathophysiological mechanisms and treatment perspectives. Diabetes Metab. Res. Rev. **18:** 176–184.
25. POWERS, K.M. *et al.* 2003. Parkinson's disease risks associated with dietary iron, manganese, and other nutrient intakes. Neurology **60:** 1761–1766.

Index of Contributors

Aschner, M., 115–128

Bannon, D., 142–152
Bartzokis, G., 224–236
Bondy, S.C., 129–141
Bressler, J.P., 142–152
Bush, A.I., 153–163

Casadesus, G., 179–182
Chakrabarty, A., 252–266
Cheong, J.H., 142–152
Ching, K.H.L., 282–298
Connor, J.R., ix–x, 171–178
Cooperman, S., 65–83
Cummings, J.L., 224–236

Dobson, A.W., 115–128
Dobson, J., 183–192
Double, K., 193–208
Dwyer, B.E., 237–251

Erikson, K.M., 115–128

Fuchs, R., 299–305

Gathinji, M., 299–305
Gerlach, M., 193–208
Ghosh, M., 65–83
Gitschier, J., 282–298
Götz, M.E., 193–208

Halliwell, B., 51–64
HaMai, D., 129–141
Harris, Z.L., 299–305
Hayflick, S.J., 282–298
Honda, K., 179–182
Huang, X., 153–163

Johnson, M.A., 282–298
Jortner, B., 65–83

Kim, Y., 142–152
Kuo, Y.M., 282–298

Land, W., 65–83
LaVaute, T., 65–83
LeVine, S.M., ix–x, 252–266
Lu, P.H., 224–236

Mattson, M.P., 37–50
Messing, A., 65–83
Meyron-Holtz, E., 65–83
Moir, R.D., 153–163
Moos, T., 14–26
Morgan, E.H., 14–26

Olanow, C.W., 209–223
Olivi, L., 142–152
Ollivierre, H., 65–83
Ong, W.-Y., 51–64

Pantopoulos, K., 1–13
Parker, S.M., 282–298
Perry, G., 179–182
Petersen, R.B., 179–182
Pin, S., 299–305
Ponka, P., 267–281

Reynolds, I.J., 27–36
Richardson, D.R., 326–341
Riederer, P., 193–208
Rogers, J.T., 153–163
Rouault, T.A., 65–83

Schipper, H.M., ix–x, 84–93, 342–355
Schöneich, C., 164–170

Shachar, D.B., 306–325
Shin, I.-S., 224–236
Shoham, S., 94–114
Smith, M.A., 179–182
Smith, S.R., 65–83
Stephenson, G., 306–325
Switzer, R., III, 65–83

Tanzi, R.E., 153–163
Tishler, T.A., 224–236
Todorich, B.M., 171–178

Tresser, N., 65–83

Wagner, K.R., 237–251
Westaway, S.K., 282–298

Xu, X., 299–305

Youdim, M.B.H., 94–114, 193–208, 306–325

OHIO UNIVERSITY LIBRARY

Please return this book as soon as you have finished with it. In order to avoid a fine it must be returned by the latest date stamped below. All books are subject to recall after two weeks or immediately if needed for reserve.

MAR 1 4 2005

MAR 1 0 2005

CF